AF670840

Angular Momentum Evolution of Young Stars

NATO ASI Series

Advanced Science Institutes Series

A Series presenting the results of activities sponsored by the NATO Science Committee, which aims at the dissemination of advanced scientific and technological knowledge, with a view to strengthening links between scientific communities.

The Series is published by an international board of publishers in conjunction with the NATO Scientific Affairs Division

A Life Sciences **B Physics**	Plenum Publishing Corporation London and New York
C Mathematical and Physical Sciences **D Behavioural and Social Sciences** **E Applied Sciences**	Kluwer Academic Publishers Dordrecht, Boston and London
F Computer and Systems Sciences **G Ecological Sciences** **H Cell Biology** **I Global Environmental Change**	Springer-Verlag Berlin, Heidelberg, New York, London, Paris and Tokyo

NATO-PCO-DATA BASE

The electronic index to the NATO ASI Series provides full bibliographical references (with keywords and/or abstracts) to more than 30000 contributions from international scientists published in all sections of the NATO ASI Series.
Access to the NATO-PCO-DATA BASE is possible in two ways:

– via online FILE 128 (NATO-PCO-DATA BASE) hosted by ESRIN, Via Galileo Galilei, I-00044 Frascati, Italy.

– via CD-ROM "NATO-PCO-DATA BASE" with user-friendly retrieval software in English, French and German (© WTV GmbH and DATAWARE Technologies Inc. 1989).

The CD-ROM can be ordered through any member of the Board of Publishers or through NATO-PCO, Overijse, Belgium.

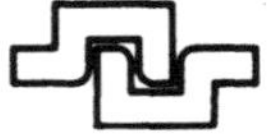

Series C: Mathematical and Physical Sciences - Vol. 340

Angular Momentum Evolution of Young Stars

edited by

S. Catalano

Istituto di Astronomia,
Università di Catania,
Catania, Italy

and

J. R. Stauffer

Center for Astrophysics,
Smithsonian Astrophysical Observatory,
Cambridge, Massachusetts, U.S.A.

Springer Science+Business Media, B.V.

Proceedings of the NATO Advanced Research Workshop on
Angular Momentum Evolution of Young Stars
Noto, Italy
September 17–21, 1990

Library of Congress Cataloging-in-Publication Data

Angular momentum evolution of young stars / edited by S. Catalano and J.R. Stauffer.
p. cm. -- (NATO ASI series. Series C, Mathematical and physical sciences ; vol. 340)
Proceedings of the NATO Advanced Research Workshop on Angular Momentum Evolution of Young Stars, Noto, Italy, Sept. 17-21, 1990.
"Published in cooperation with NATO Scientific Affairs Division."
Includes indexes.
ISBN 978-94-010-5587-1 ISBN 978-94-011-3580-1 (eBook)
DOI 10.1007/978-94-011-3580-1
1. Early stars--Congresses. 2. Stars--Evolution--Congresses. 3. Angular momentum--Congresses. I. Catalano, S. (Santo), 1939- . II. Stauffer, John Richard. III. NATO Advanced Research Workshop on Angular Momentum Evolution of Young Stars (1990 : Noto, Italy) IV. Series: NATO ASI series. Series C, Mathematical and physical sciences ; no. 340.
QB843.E2A54 1991
523.8--dc20 91-18387

ISBN 978-94-010-5587-1

Printed on acid-free paper

TABLE OF CONTENTS

Session V: Observational Perspectives

1 - S. Vauclair, 2 - D. Spadaro, 3 - A. Collier- Cameron, 4 - G. Severino, 5 - V. Reglero, 6 - C. Bertout, 7 - J. Stauffer, 8 - D. Soderblom, 9 - C. Trigilio, 10 - J.C. Mermilliod, 11 - J. Bouvier, 12 - F. Van't Veer, 13 - W. Dziembowsky, 14 - G. Belvedere, 15 - P. Bodenheimer, 16 - R. Kraft, 17 - D. Duncan, 18 - J.R. De Medeiros, 19 - Mrs. Kraft, 20 - M. Mayor, 21 - K. Mc Gregor, 22 - L.W. Hartman, 23 - F. Damiani, 24 - P.F. Chugainov, 25 - M. Barberi, 26 - K.M. Strom, 27 - S.E. Strom, 28 - F. Scaltriti, 29 - Mrs. Scaltriti, 30 - C. Maceroni, 31 - A. Maggio, 32 - M. Rodonò, 33 - G. Micela 34 - E. Schatzman, 35 - Mrs. Catalano, 36 - G.S. Vaiana, 37 - S. Catalano 38 - S. Sciortino, 39 - D. Gough, 40 - S. Sofia, 41 - L. Paternò, 42 - G. Cutispoto, 43 - D. Galli, 44 - V. Geroyannis, 45 - C. Charbonnel, 46 - S. Randich, 47 - M.L. Rapisarda, 48 - F. Grillo, 49 - I.W. Roxburgh, 50 - F. Palla, 51 - E. Marilli, 52 - A. Lanza, 53 - G. Umana, 54 - S. Del Popolo, 55 - D.F. Gray, 56 - R. Ventura, 57 - S. Novello, 58 - A. Frasca, 59 - A. Calì

LIST OF PARTICIPANTS

Anile, A.M.	*Dipartimento di Matematica, Cittá Universitaria, Viale A. Doria 6 , 95125 Catania*, ITALY
Antonuccio, V.	*Osservatorio Astrofisico, Cittá Universitaria, Viale A. Doria 6, 95125 Catania*, ITALY
Belvedere, G.	*Istituto di Astronomia, Cittá Universitaria, Viale A. Doria 6, 95125 Catania*, ITALY
Bertout, C.	*Group d' Astrophysique Observatoire de Grenoble Université J. Fourier - CNRS, CERMO B.P. 53X, 38041 Grenoble Cedex*, FRANCE
Blanco, C.	*Istituto di Astronomia, Citta' Universitaria, Viale A. Doria 6, 95125 Catania*,ITALY
Bodenheimer, P.	*Lick Observatory, University of California, Santa Cruz CA 95064*, USA
Bouvier, J.	*Canada-France-Hawaii Tel.Corp., P.O. Box 1597 Kamuela, HI 96743* USA
Catalano, S.	*Istituto di Astronomia, Cittá Universitaria, Viale A. Doria 6, 95125 Catania*, ITALY
Charbonnel, C.	*Observatoire Midi-Pyrénées, Université Paul Sabatier, 14 Avenue E. Belin, 31400 Toulouse*, FRANCE
Chugainov, P.F.	*Crimean Astrophysical Observatory, P.O. Nauchny, 334413 Crimea*, USSR
Collier-Cameron, A.	*Astronomy Centre, University of Sussex, Falmer, Brighton BN1 9QH*, U.K.
Cutispoto, G.	*Osservatorio Astrofisico, Cittá Universitaria Viale A. Doria 6, 95125 Catania*, ITALY
Damiani, F.	*Osservatorio Astronomico di Palermo, Palazzo dei Normanni, 90134 Palermo*, ITALY
de Medeiros, J.R.	*Observatoire de Geneve, Chemin des Maillettes, 51 - 1290 - Sauverny*, SUISSE
Duncan, D.K.	*Space Telescope Science Institute, 3700 San Martin Drive, Baltimore, MD 21218*, USA
Dziembowski, W.	*Polish Academy of Sciences, Nikolaus Copernicus Astronomy Center, U.L. Bartycka, 18, 00-716 Warsaw*, POLAND
Errico, L.	*Osservatorio Astronomico di Capodimonte, Via Moiarello, 16, 80131 Napoli*, ITALY
Foderà-Serio, G.	*Osservatorio Astronomico di Palermo, Palazzo dei Normanni, 90134 Palermo*, ITALY
Galli, D.	*Osservatorio Astrofisico di Arcetri, Largo E. Fermi 5, 50125 Firenze*, ITALY
Geroyannis, V.	*Department of Astronomy, University of Patras, GR-26110 Patras*, GREECE

Gough, D.O.	*Institute of Astronomy, Madingley Road, Cambridge CB3 OHA*, U.K.
Gray, D.F.	*Department of Astronomy, University of Western Ontario, London, Ontario, N6A 3K7*, CANADA
Grillo, F.	*Osservatorio Astronomico di Palermo, Palazzo dei Normanni, 90134 Palermo*, ITALY
Hartmann, L.W.	*Harvard-Smithsonian Center for Astrophys., 60 Garden Street, Cambridge MA 02138*, USA
Kraft, R.	*Lick Observatory, University of California, Santa Cruz CA 95064*, USA
Lamzin. S.	*Sternberg State Astronomical Institute, University Prospect 13, 119899 Moscow V-234*, USSR
Lanza, A.	*Istituto di Astronomia, Cittá Universitaria, Viale A. Doria 6, 95125 Catania*, ITALY
Lanzafame, G.	*Istituto di Astronomia, Cittá Universitaria, Viale A. Doria 6, 95125 Catania*, ITALY
MacGregor, K.B.	*High Altitude Observatory, N.C.A.R., P.O. Box 3000, Boulder,CO 80307*, USA
Maceroni, C.	*Osservatorio Astronomico di Roma, Viale del Parco Mellini 84, 00136 Roma*, ITALY
Maggio, A.	*Osservatorio Astronomico di Palermo, Palazzo dei Normanni, 90134 Palermo*, ITALY
Marilli, E.	*Osservatorio Astrofisico, Cittá Universitaria, Viale A. Doria 6, 95125 Catania*, ITALY
Mayor, M.	*Observatoire de Geneve, Chemin des Maillettes, 51 CH-1290 Sauverny*, SUISSE
Mermilliod, J.C.	*Institut d'Astronomie de l'Université de Lausanne, CH-1290 Chavannes des Bois*, SUISSE
Micela, G.	*Osservatorio Astronomico di Palermo, Palazzo dei Normanni, 90134 Palermo*, ITALY
Muscato, O.	*Dipartimento di Matematica, Cittá Universitaria, Viale A. Doria 6, 95125 Catania*, ITALY
Palla, F.	*Osservatorio Astrofisico di Arcetri, Largo E. Fermi 5, 50125 Firenze*, ITALY
Paternó, L.	*Istituto di Astronomia, Cittá Universitaria, Viale A. Doria 6, 95125Catania*, ITALY
Peres, G.	*Osservatorio Astrofisico, Cittá Universitaria, Viale A. Doria 6, 95125 Catania*, ITALY
Pinsonneault, M.H.	*Center for Solar and Space Research, Dept.of Astronomy, Yale University, P.O. Box 6666, New Haven, CT 06511*, USA
Randich, S.	*Dipartimento di Astronomia e Scienza dello Spazio, Universitá di Firenze, Largo E. Fermi 5, 50125 Firenze*, ITALY
Reglero, V.	*Universitat de Valencia, Dept. de Matematica y Astronomia, Dr. Moliner 50, 46100 Burjasot - Valencia*, SPAIN

Rodonó, M.	*Istituto di Astronomia, Cittá Universitaria, Viale A. Doria 6, 95125 Catania,* ITALY
Roxburgh, I. W.	*Queen Mary and Westfield College, Mile End Road, London E1 4NS,* U.K.
Scaltriti, F.	*Osservatorio Astronomico di Torino, Via Osservatorio 20, 10025 Pino Torinese,* ITALY
Schatzman, E.	*DASGAL, Observatoire de Meudon, F-92195 Meudon Principal Cedex,* FRANCE
Sciortino, S.	*Osservatorio Astronomico di Palermo, Palazzo dei Normanni, 90134 Palermo,* ITALY
Sserio, S.	*Osservatorio Astronomico di Palermo, Palazzo dei Normanni, 90134 Palermo,* ITALY
Severino, G.	*Osservatorio Astronomico di Capodimonte, Via Moiarello 16, 80131 Napoli,* ITALY
Soderblom, S.	*Space Telescope Science Institute, 3700 San Martin Drive, Baltimore MD 21218,* USA
Sofia, S.	*Center for Solar and Space Research, Dept. of Astronomy, Yale Univiversity, P.O. Box 6666, New Haven, CT 06511,* USA
Spadaro, D.	*Osservatorio Astrofisico, Cittá Universitaria, Viale A. Doria 6, 95125 Catania,* ITALY
Stauffer, J.	*Harvard-Smithsonian Center for Astrophys., 60 Garden Street, Cambridge MA 02138,* USA
Strom, K.M.	*Five College Astron. Dept., University of Massachusetts, Amherst, MA 01003,* USA
Strom, S.E.	*Five College Astron. Dept., University of Massachusetts, Amherst, MA 01003,* USA
Tanzi, E.G.	*Istituto di Fisica Cosmica, CNR, Via E. Bassini 15 , Milano,* ITALY
Trigilio, C.	*Istituto di Radioastronomia, CNR, Stazione di Noto, 96017 Noto,* ITALY
Umana, G.	*Istituto di Radioastronomia, CNR, Stazione di Noto, 96017 Noto,* ITALY
Vaiana, G. S.	*Osservatorio Astronomico, Palazzo dei Normanni, 90134 Palermo,* ITALY
Van't Veer, F.	*Institut d'Astrophysique, CNRS, 98bis Boulevard Arago, 75014 Paris,* FRANCE
Vauclair, S.	*Observatoire Midi - Pyrénées, Université Paul Sabatier, 14 Avenue E. Belin, 31400 Toulouse,* FRANCE
Ventura, R.	*Osservatorio Astrofisico, Cittá Universitaria, Viale A. Doria 6, 95125 Catania,* ITALY
Vittone, A.	*Osservatorio Astronomico di Capodimonte, Via Moiarello 16, 80131 Napoli,* ITALY

FOREWORD

This book reports the Proceedings of the NATO Advanced Research Workshop on "Angular Momentum Evolution of Young Stars" held from 17 to 21 September 1990 at Noto, Italy. The workshop had its immediate origin in a discussion about the availability of stellar rotation data, that took place in 1987 at Viana do Castelo Portugal during the NATO meeting, Formation and Evolution of Low Mass Stars. We recognized that nearly 20 years had passed since the last meeting on stellar rotation and that significant progress in the observation of rotation rates in low mass stars had been made.

During the last 20 years, new efficient instrumentation (CCD and photon counting detectors and echelle spectrographs) and new analysis techniques (profile Fourier analysis) have allowed us to measure rotational velocities as low as 1-2 km/s and to reach low mass stars in young clusters. Even with these advances, rotational velocities of low mass stars would have remained challenging to determine if all single, low mass stars later than G0 had rotational velocities of order or less than 10 km/sec. Evidence that this is not always the case was first provided by the photometric variability data obtained by van Leeuwen and Alphenaar for K dwarfs in the Pleiades and more recently by the *vsini* measurements of low mass stars in several young clusters. Given the availability of a considerable body of new data, we decided that it was appropriate to consider how these data might be used to elucidate basic problems of stellar formation, structure and evolution such as the initial angular momentum distribution, angular momentum evolution and transport, internal rotation, lithium depletion and diffusion, magnetic activity and rotation braking.

This NATO Advanced Research Workshop gathered together about 60 scientists, including both senior "pundits" and young active researchers, whose research interests were directly related to the topic of the meeting. Observers provided new fundamental data on the rotation of T Tauri stars, low mass stars in young clusters, and post-main sequence stars. Additional papers dealt with how new observational data on protostellar disks, lithium abundances in young stars and magnetic field measurements might provide useful constraints on models of rotational velocity evolution. These observational reports provided the basis for a number of theoretical papers which attempted to compare the predictions of angular momentum loss via stellar winds with the observations.

The meeting format left ample space for discussion and included two panel discussion on the key topics:

- Initial Angular Momentum
- Surface Braking and Internal Rotation

The text of the discussions during these panel sessions is provided here particularly to indicate the directions of current research interest.

The book is divided into 5 major sections: **I.** Initial Angular Momentum Distribution, **II.** Angular Momentum Evolution, **III.** Consequences of rotation, **IV.** Internal Rotation and Theoretical Models, V. Observational Perspectives. The volume includes also the complete transcription of discussions following each presentation and the panel discussions as recorded on tapes. There are inevitable gaps due to loss of recording for various reasons. Generally the discussion was transcribed word-by-word, but to improve the language, repetitions, excessive use of colloquialisms, etc. were removed. We are confident that the speakers will recognise themselves and that no unacceptable changes were made.

The Scientific Organizing Committee of the NATO Advanced Research Workshop consisted of S. Catalano (Chair), C. Bertout, D. Gough, M. Rodonó, J. Stauffer.

The local Organizing Committee consisted of S. Catalano(Chair), J. Stauffer(Co-chair), A.M. Anile, E. Marilli, O. Muscato, M. Rodonó, D. Spadaro.

We are very grateful to acknowledge the vital financial support given by the NATO Scientific Affairs Division. We also wish to thank the following Institutions and Companies who supplied additional funds to make the meeting a most rewarding one: The Comune di Noto, the Univeristy of Catania, the Catania Astrophysiscal Observatory, the Ministero della Ricerca Scientifica e Tecnologica, The Consiglio Nazionale delle Ricerche, the Provincia regionale di Siracusa, the Azienda Autonoma Provinciale per l' Incremento Turistico di Siracusa, the Banca Nazionale del Lavoro, and the Convex Computer S.p.A.

The logistics of putting together the meeting, having things run smoothly at Noto, and completing the manuscript for this book were the result of the hard work and dedication of a number of people who we wish to specially thank: A. Cali', S. Del Popolo, P. Massimino, M. Miraglia, S. Novello D. Randazzo, L. Rapisarda, D. Recupero, S. Sardone, C. Spampinato, V. Stancanelli.
We are also indebted to Stepha Genelza, who take care of the vital but exausting job of transcribing the discussions from the tapes.

Most of all, we are grateful to the Workshop participants for making this a lively and profitable meeting. In particular we would like to thank the moderators of the two panel discussions E. Schatzman and P. Bodenheimer who chaired the sessions masterfully by presenting lists of provocative questions which evoked thoughtful responses from the participants.

March 1991 S. Catalano and J. Stauffer

WELCOME ADDRESS

Mr. Mayor, Official representatives of Local Government and Institutions, Collegues, Ladies and Gentlemen:
it is my great privilege and honour, on behalf of the NATO Scientific Affairs Division, of the Scientific and Local Organizing Committees, to welcome you to the Nato Advanced Research Workshop "Angular Momentum Evolution of Young Stars".
First of all let me thank all of you for accepting our invitation to attend this Opening Ceremony, but let me particularly thank all of my colleagues for agreeing to come to this far south part of Sicily, to attend our meeting.

When John Stauffer asked me to organize in Sicily a meeting on stellar rotation I immediately thought about NOTO. Now I am very pleased and proud of the choice. This glorious City, rich in historical and cultural traditions signified by its unique baroque architecture, is experiencing a new vitality in culture and science, thanks to the Centro di Studi Superiori e di Ricerche. It is indeed an appropriate seat for a high quality scientific meeting.

It is now 20 years since the last meeting on stellar rotation, and I think all of us agree that it was time for a new meeting, mainly because of the significant recent efforts in measuring stellar rotation speed and because of the considerable body of other new data now available.
The main purpose of our workshop will be to summarize the new observational results on stellar rotation, magnetic fields and interior rotation of the stars, and to provide a forum to the theoreticians for the confrontation of their models of angular momentum loss and braking mechanisms with the observations. The simplest models of star formation predict that stars should not form because contraction of a molecular cloud of reasonable initial angular momentum would lead to a star rotating much faster than the break-up velocity. A huge amount of angular momentum has to be lost before a star appears on the Hayashi track at the *stellar birthline*. We will discuss how and when forming stars solve this angular momentum problem.
Protoplanetary disks and possibly planetary system formation appears to be one efficient mechanism for the loss of angular momentum of young stars. We will discuss the recent observations of protostellar disks around T Tauri and main sequence stars, of bipolar flows, of magnetic field measurements, and we expect to hear from the theoreticians about progress on stellar formation and evolutionary models incorporating rotation.
We will discuss the time-scale of the instability processes induced by the spin-up and spin-down of low-mass stars as they approach the main sequence. The discussion can take advantage of the recently determined rotational velocities of stars in very young clusters, and should give important hints on the rotation of the radiative core of young stars.
The interface of the radiative core and convective envelope is thought to be the seat of the dynamo process which drives the generation of surface magnetic fields. The theoretical constraints on differential rotation within stellar interiors and on the angular momentum transport mechanisms will be discussed in order to try to understand the observed time evolution of chromospheric and coronal activity, surface Lithium abundance, and magnetic braking.

New observational techniques are being developed and in the near future, powerful facilities will be available for stellar rotation and activity studies, such as the HST, the X-ray observatory ROSAT, the Space platform, and 8-10 meter sized ground telescopes. However important observational data, such as oscillation modes, necessary to study internal rotation and active region mappings, useful to derive the topology of the emergent magnetic fields, require specially dedicated ground or space-born telescopes. Such expensive facilities can be justified only if a large community of well prepared scientists are ready to use them. It is our hope that this workshop will help direct research efforts in this field and stimulate new collaborations among the participants.

Let me gratefully acknowledge the very generous support given by the NATO Scientific Affair Division, which made the workshop possible. Our warmest thanks are due to the Rettore of Catania University, Prof G. Rodolico, to the President of the Science Faculty, prof. G. Santagati and to the Director of the Catania Astrophysical Observatory, Prof M. Rodonó, for their interest in the meeting and daily encouragements.

It is our great pleasure to acknowledge the friendly hospitality offred by the City of NOTO and its Mayor Dr. G. Falconeri, by the Centro di Study Superiori e di Ricerche, by the Provincia Regionale of Siracusa by the Circolo Val di NOTO and its President Dr S. Dejan, our host today, and to acknowledge also the active support in the organization kindly provided by prof. V. Miceli and O. Muscato.

Finally, let me express my personal thanks to the members of the Scientific Organizing Committee, the Local Organizing Committee, and the Local technical staff for the precious, invaluable help to make this workshop a pleasant, interesting and I hope successful conference.
Thank you.

Santo Catalano

Noto, September 17th 1990

ANGULAR MOMENTUM EFFECTS IN STAR FORMATION

PETER BODENHEIMER
University of California Observatories
Lick Observatory
Board of Studies in Astronomy and Astrophysics
University of California, Santa Cruz, CA, USA

ABSTRACT. The star formation era is a period during which rotational effects are of crucial importance in determining the evolution of the star. Also, during this period, significant redistribution of angular momentum must occur in the star, as indicated by the substantial discrepancy between deduced angular momenta in star formation sites in the cores of molecular clouds and in the youngest optically visible stars. A major problem is the short time scale, comparable to a few initial free-fall times of the cloud core, in which this redistribution must take place. The following questions will be considered in this paper: (1) What is the origin of the angular momentum of stars? (2) What is the observational evidence for the existence of angular momentum at the earliest stages of star formation? (3) How serious is the so-called angular momentum problem? (4) For the solution of the angular momentum problem, what kinds of angular momentum transport processes are available and at what stages of star formation do they probably operate? (5) What recent theoretical information is available regarding the evolution of rotating protostars? (6) To what extent can the angular momentum problem be solved by the formation of binary or multiple systems and disks? (7) Given that the T Tauri stars, even near the birthline, are rotating slowly, what is a possible sequence of events in the evolution of the angular momentum distribution that could bring them to this state?

1. The Origin of Angular Momentum in Stars

Interstellar turbulence and galactic differential rotation have been considered as the main mechanisms causing cloud rotation (Tassoul 1978). The absence of a preferred orientation for the rotational axes of single stars or the orbital planes of binaries (Kraft 1970) or, in fact, the rotational axes of molecular clouds themselves (Goldsmith and Arquilla 1985), means that the rotation of the cores in molecular clouds is probably caused by interstellar turbulence. Observations reveal that large molecular clouds are quite inhomogeneous and clumpy, down to the smallest resolvable scales, which are smaller than the likely Jeans length in the clouds. The chaotic internal motions suggest that it is reasonable to consider the clouds as turbulent, taking into account that turbulent flows actually consist of a hierarchy of small-scale irregularities superimposed on larger-scale more systematic motions

S. Catalano and J. R. Stauffer (eds.), Angular Momentum Evolution of Young Stars, 1–20.

such as rotation or expansion.

Larson (1981) pointed out that in the scale range $0.05 \leq L \leq 60$ pc the turbulent velocity, derived from CO linewidths, is v_t (km s^{-1}) $\approx 1.1\ L^{0.38}$, where L is given in parsecs, which is close to the Kolmogoroff spectrum which has a power-law index of 1/3. Subsequent studies have confirmed this relation, but there is appreciable scatter in the data and other power laws have been derived, such that of Solomon and Sanders (1985) where v_t (km s^{-1}) $\approx 0.88\ L^{0.62}$.

The turbulence in the idealized incompressible fluid of classical laboratory hydrodynamics is characterized by a single "cascade," that is, a hierarchy of vortices on a range of length scales down to the scale where dissipation becomes important. However the interstellar turbulence may be quite different and more complicated. The clouds are compressible, the linewidths in general are supersonic, and magnetic fields undoubtedly have some role in supporting the cloud cores against gravitational collapse and in generating the observed line widths (Shu *et al.* 1987). The observed internal velocities could in principle be vortex-free and could refer to the independent motions of clumps, filaments and other wispy condensations under the action of their mutual gravitation. In fact rotation contributes very little to the linewidth and is relatively unimportant in the support of the cloud core against collapse.

Ruzmaikina (1986, 1988) has suggested that the turbulence in molecular clouds could be considered as a superposition of cascades with randomly distributed energy sources on several length scales. One could expect that under such circumstances the interstellar turbulence would have an intermittent character, with, for example, small regions of intense rotational motions, whose locations change with time, separated by extended quiet regions. The importance of intermittency is that the deviation from the average of the angular momenta of the clumps at a given scale would be much larger than for normal turbulence. This variation could account to some extent for the wide range of orbital periods in binary stars and the existence of single stars. Of course this explanation is probably coupled to another effect: the differences in conditions under which the magnetic field becomes unimportant, because of ambipolar diffusion, in providing rotational braking in molecular clouds could result in a spread in the initial rotation rates of protostars (Mouschovias 1977). The existence of slowly-rotating cloud cores has yet to be confirmed by observations; however the picture is undoubtedly more complicated than a simple one-to-one mapping between initial cloud rotation rate and final binary period, since both fragmentation and redistribution of angular momentum must be taken into account.

2. Observed Rotation Rates in Molecular Clouds and Cloud Cores

The rotation of interstellar clouds or their subregions is usually determined by observation of a systematic shift across a cloud of the velocity of the centroid of a spectral line (Goldsmith and Arquilla 1985). The observational limit on the rotational rate depends on the line width, which is determined by turbulence and which increases with the dimension of the cloud. If a cloud possessed a highly ordered velocity field, the observation of a gradient of the radial velocity would be strong evidence for rotation. However, other types of mass motion can produce gradients similar to those resulting from rotation. Therefore the detailed velocity field in a cloud must be determined before a rotation model can assumed to be valid. Arquilla and Goldsmith (1986) have undertaken such a study in eight clouds, six of which were previously referred to in the literature as rotators. They found that rotation is clearly present in only three of them, suggesting that other previous work should also be re-evaluated. The mean specific angular momentum clearly decreases at smaller scales (Goldsmith and Arquilla 1985). However the correlation is not completely clear because there is a range of deduced values of the rotational velocity at each size scale and because many clouds fall below the observational cutoff, which depends on the scale. Typical results indicate that clouds with radius 0.5-1.0 pc and a few hundred solar masses rotate with $\omega \approx 3 - 11 \times 10^{-14}$ rad s^{-1}. The typical specific angular momentum of the clouds that rotate is therefore $j \sim 10^{23}$ cm^2 s^{-1}. Most of the clouds seem to rotate rigidly, although the outer envelope of B361 shows evidence for differential rotation.

The measurement of rotation of cloud cores is particularly difficult because of the small scale (≈ 0.1 pc). Bipolar outflows from embedded low-mass young stars could in principle produce the observed shift across some cores; recent observations reveal a high fraction of outflows in cores with stars. Therefore the measured velocity gradient can be considered only as an upper limit on the rotational velocity projected onto the line of sight (Fuller and Myers 1987, Mathieu *et al.* 1988). In small cores the velocity gradients are less than the observational limit in approximately half of studied cases (Myers and Benson 1983, Ungerechts *et al.* 1982, Heyer 1988). In Heyer's (1988) work on rotational velocities in Taurus the mass range is 0.3 - 38 $M_\odot$, the range in radius is 0.1 - 0.5 pc, the mean density range is 0.1 - 2.6 $\times 10^4$ cm^{-3}, and the observed velocity gradients are 0.2 - 1.5 km s^{-1} pc^{-1}. The angular velocities are 0.6 - 5 $\times 10^{-14}$ rad s^{-1}, corresponding to values of j (at the outer edge) in the range 1 - 45 $\times 10^{21}$ cm^2 s^{-1}, increasing on the larger scales. The small ($0.05 - 0.1$ pc) cores, in which presumably star formation takes place, have $j \approx 10^{21}$ cm^2 s^{-1}, for those cases with measurable rotation. The velocity shift here is comparable to the line width of about 0.45 km s^{-1}. Similar results were found in earlier work (Myers and Benson 1983, Ungerechts *et al.* 1982, Harris *et al.* 1983, Wadiak *et al.* 1985).

3. The Angular Momentum Problem

The typical angular momenta observed in cloud cores may be compared with those in binary systems and single stars (Table 1). Several conclusions may be reached. First, the orbital angular momenta in long-period binaries have the same order of magnitude as those inferred from observations of cloud cores. Second, it is not known but certainly plausible that some cloud cores have angular momenta consistent with those of the systems with the shorter periods. These close systems do have far too much angular momentum to be explained by fission of rotating T Tauri stars. Third, the value of j in the solar nebula, as deduced from Jupiter's orbit, is comparable to that of binary orbits of intermediate period. The question arises whether planetary systems preferentially form from clouds with this value of j, or rather from a very slowly rotating cloud ($j \leq 10^{19}$ cm^2 s^{-1}) with outward transport of angular momentum occurring into a small amount of mass at a later stage, as suggested for example by Safronov and Ruzmaikina (1985). Fourth, the angular momentum problem is apparently associated with the difference of four orders of magnitude in j between the cloud cores and the T Tauri stars. Note, however, that a cloud core with $j \approx 10^{21}$ and 1 $M_\odot$ has a total angular momentum $J \approx 10^{54}$ g cm^2 s^{-1}; an extended disk such as that of HL Tau has a comparable value of J, while a T Tauri star with a moderate size (100 AU) disk has J only one order of magnitude less. Thus a cloud with a rotational velocity near or only slightly below the lowest observed value could form a star-disk system with conservation of total angular momentum. The real problem lies in the redistribution of angular momentum. As Figure 1 shows, a cloud core and a star-disk system with the same total mass and angular momentum differ by 1-2 orders of magnitude in j in the stellar interior and surface layers, if the disk is taken into account. Even so, the angular momentum problem seems to be less dramatic than sometimes stated.

Table 1. Characteristic Values of Specific Angular Momentum

Object	J/M (cm^2 s^{-1})
Molecular cloud (scale 1 pc)	10^{23}
Molecular cloud core (scale 0.1 pc)	10^{21}
Binary (10^4 yr period)	4×10^{20} - 10^{21}
Binary (10 yr period)	4×10^{19} - 10^{20}
Binary (3 day period)	4×10^{18} - 10^{19}
T Tauri star (beginning of contraction)	5×10^{17}
Jupiter (orbit)	10^{20}
Present Sun	10^{15}

4. Mechanisms for Redistribution of Angular Momentum

It is clear that if a cloud core contracts with local conservation of angular momentum, then even negligible initial rotation could be dynamically important during contraction and could stop the contraction in the direction perpendicular to the rotation axis. The extent of redistribution of angular momentum could be a crucial factor in determining what kind of system is produced. The efficiency of redistribution depends strongly on the stage of protostellar evolution. It definitely could be high at the early stages at molecular cloud densities ($\sim 10^3$ cm^{-3}). If the typical magnetic field in a cloud is $\sim 10^{-5}$ gauss and if the magnetic field is frozen in to the material, then the field is effective in braking the rotation of the cloud and in redistributing the angular momentum inside the cloud (Mouschovias 1977, Mestel 1985, Ruzmaikina 1985, Shu *et al.* 1987.) The transport occurs through Alfv́en waves generated by the difference in angular velocity between the cloud and the external outer regions. The time scale is roughly given by the time for an Alfvén wave to traverse a region external to the cloud which has a moment of inertia comparable to that of the cloud. The crossing times have values of 10^6 - 10^7 yr as long as the coupling between the field and gas is maintained. However this coupling decreases because of ambipolar diffusion as the density approaches the values for the dense cores of molecular clouds (Mestel and Spitzer 1956, Lizano and Shu 1989, Ruzmaikina 1985). Magnetic braking becomes less effective, but ambipolar diffusion itself can still be effective in redistributing the angular momentum because of the relatively large effective viscosity associated with the drag of the ionized component moving through the neutral component. Magnetic transport probably accounts for

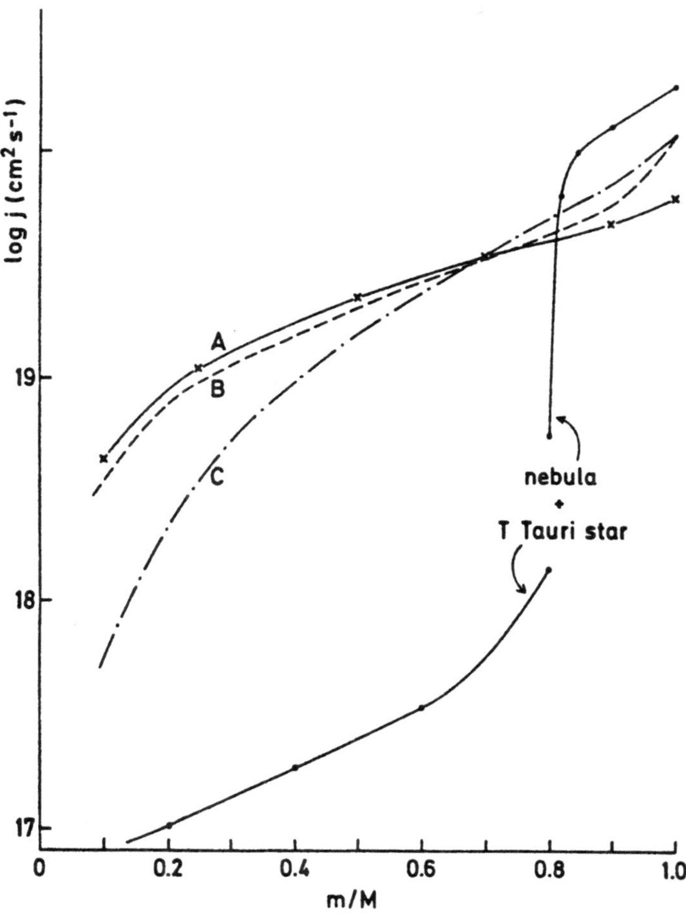

Figure 1: Plausible angular momentum distributions in the rotating core of an interstellar cloud (A, B, and C) are compared with that of a rotating T Tauri star with a surrounding nebular disk. All models have the same total mass and angular momentum. The angular momentum per unit mass (j) is plotted as a function of the mass fraction interior to a given cylinder concentric with the rotation axis. The uniformly rotating star has a mass of 0.8 $M_{\odot}$, a radius of 4 $R_{\odot}$, and a surface rotational velocity of 50 km s^{-1}. The disk is Keplerian with an outer radius of 20 AU and a mass of 0.2 $M_{\odot}$. Curves A, B, and C correspond, respectively, to uniformly rotating clouds with uniform density, with a density distribution of the polytrope of index 1.5, and with a density inversely proportional to the square of the radius (Bodenheimer *et al.* 1988).

the difference of two orders of magnitude in the specific angular momentum of molecular clouds themselves and their cold, dense cores. The similarity of the angular velocities on the two different scales in fact suggests magnetic coupling between the two regions (Shu *et al.* 1987).

At the point where the protostar collapse starts, the magnetic field becomes dynamically unimportant because the degree of ionization of the gas at the appropriate density is extremely low. The evolution time is comparable to the dynamic time, and there will not be appreciable angular momentum transport by any process, unless initial non-axisymmetric perturbations are especially large. The magnetic field remains dynamically unimportant because of the effects of ambipolar diffusion, which restricts the enhancement of the magnetic field. The effective viscosity ,has a time scale longer than that of collapse. Similarly, turbulent velocities are not likely to exceed the sound speed in a given cloud because of greatly increased dissipation at supersonic speeds. Numerical simulations show, however, that the collapse is supersonic at most stages; thus the time scale for angular momentum transport by turbulent viscosity is larger than the collapse time. Also, although small initial non-axisymmetric perturbations can amplify during collapse, the time scale for transport of angular momentum by gravitational torques remains longer than the evolution time until the collapse has been slowed considerably by rotational and pressure effects. After about one free fall time, when the cloud has become rather disklike and is approaching equilibrium, the perturbations can grow appreciably and fragmentation can occur; the process involves conversion of spin angular momentum of the cloud into orbital motion of the fragments.

If the cloud becomes optically thick before rotational effects become important, then the effects of increasing pressure tend to damp out perturbations and to suppress fragmentation. Then the collapse is likely to result in a central object plus a disk. A number of transport processes can be effective in the disk– turbulence, magnetic torques, sound waves with shock dissipation, and gravitational torques produced by spiral density waves (see review by Larson 1989).

Turbulence is likely to be induced in the disk because the low molecular viscosity ν results in a high Reynolds number $R_e \approx (GMR)^{1/2}/\nu \geq 10^{10}$. Mechanisms which could, in principle, result in turbulence include shear instability as a result of differential rotation (Zeldovich 1981), the difference in angular momentum of the accreting material from that of the disk (Cameron 1962), and thermal convection (Lin and Papaloizou 1980). At present the best developed theory is that of turbulent viscosity arising from convective instability in the vertical direction. Calculations based on approximate convection theory yield an equivalent viscosity parameter α (Shakura and Sunyaev 1973) which is $\sim 10^{-2}$ (Lin and Papaloizou 1980), 10^{-2}–10^{-3} (Ruden *et al.* 1988), or 10^{-3}– 10^{-4} (Cabot *et al.* 1987a,b). The corresponding time scale t_r for redistribution of the angular momentum over a length scale of the order of the present-day size of the solar system ranges from 10^5 -10^7 yr. The disk is unstable to convection in general in regions where the opacity increases sufficiently rapidly with temperature (Lin and Papaloizou 1980), such as the grain region at temperatures below 1500 K and the H^- region at temperatures above 3000 K.

Similar time scales can be obtained through the mechanism of wave propagation with excitation, for example, by protogiant planets (Larson 1989). A problem exists in that waves tend to be refracted toward the vertical direction in a thermally stratified disk, so that they can be dissipated near their source; most modes will not propagate over a long range (Lin *et al.* 1990).

Estimates of the turbulent viscosity in the disk arising from shear between accreting material and the disk give $\alpha \sim 10^{-2}$ and $t_r \sim 1 - 5 \times 10^5$ yr (Ruzmaikina 1982). The shear flow will definitely induce turbulence near the surface of the disk, but it is also possible that turbulence can be generated throughout the entire thickness of the disk. This mechanism, of course, operates only during the earlier phases of disk evolution, when it is still accreting material from the infalling cloud. Finally, turbulence associated with differential rotation can proceed through development and interaction of global non-axisymmetric unstable modes (Sekiya and Miyama 1988). The growth time is $\sim 10^3$ - 10^4 yr, but the evolution time t_r is uncertain.

Gravitational torques can result in angular momentum transport if the disk is gravitationally unstable. Conditions for the existence of axisymmetric unstable modes have been investigated by Safronov (1960) and Toomre (1964). Numerical simulations of the evolution of a disk with a central point mass (Cassen *et al.* 1981) indicate that non-axisymmetric modes develop and spiral waves result in transfer of angular momentum if $M_{disk} > M_{central}$. A disk of moderate mass $M_{disk} \sim M_{central}$ can be unstable to non-axisymmetric disturbances having growth rates comparable to the orbital frequency at the outer edge (Adams *et al.* 1989). Lin and Pringle (1987) derived a time scale $t_r \sim 10\ \Omega^{-1}$. However, it is likely that refraction of waves occurs for azimuthal wave number $m \geq 2$ (Lin *et al.* 1990), so that these waves do not travel far in the radial direction. Adams *et al.* (1989) suggest that spiral waves with $m = 1$ may be most effective at transporting angular momentum over large radial distances. These are eccentric modes in which the star and the disturbance orbit around their common center of mass. The unstable modes can encompass the entire disk and be driven non-linearly by these eccentric motions. In general the gravitational instabilities result in transport of angular momentum through spiral waves on a near-dynamical time scale. However, it is possible that under special conditions the $m = 1$ instability could result in the formation of a binary.

Magnetic fields amplified during the collapse of a cloud or generated by a hydromagnetic dynamo can result in angular momentum transport in the inner part of the disk (Lüst and Schlüter 1955, Hoyle 1960, Alfvén and Arrhenius 1976, Levy and Sonett 1978, Ruzmaikina 1981, 1985, Hayashi 1981, Stepinski and Levy 1988). The central region is hot enough for the evaporation of dust particles and thermal ionization of the alkaline metals ($T \geq 1600$ K). The radius of the region where the field is coupled sufficiently to the gas is about 1 AU (Makalkin 1987, Ruzmaikina and Maeva 1986, Hayashi 1981). The approximate time scale for transport is ~ 10 orbital periods (Hayashi 1981).

At larger distances it is generally thought that the degree of ionization near the central plane would be much lower because the thermal ionization is negligible and the disk is opaque to external sources of ionization such as cosmic rays. However, in the region relatively far from the star, where the disk surface density is low so that the layers are transparent to cosmic rays, there can be some effect. Levy *et al.* (1990) have shown that an azimuthal magnetic field can be generated by differential rotation, with or without turbulence; the buildup of field is limited by Ohmic dissipation or ambipolar diffusion. The time scale of redistribution of angular momentum is uncertain and can range upward from 1 yr if the dynamo is effective, or $\sim 10^5$ yr just from the compressed remnant of the interstellar field. Thus the magnetic field could be an effective mechanism for redistribution of angular momentum in a disk except in the dense, cold, essentially unionized region at intermediate distances from the central object (*e.g.* 1–10 AU for central mass $M_c = 1\ M_\odot$ and disk mass $M_d \sim 10^{-2} - 10^{-1}\ M_\odot$).

5. Formation of Disks and Binaries

As Table 1 shows, the specific angular momentum of a typical molecular cloud core is comparable to that in the orbital motion of a long-period binary, pointing towards fragmentation as a possible formation mechanism. Also, if the core collapsed with conservation of angular momentum without fragmentation until it reached the centrifugal barrier, its outer radius would be ≈ 500 AU, comparable in size to the disk around HL Tau (Sargent and Beckwith 1987). It is evident that the formation of disks and binaries can easily solve at least a first stage of the angular momentum problem.

Recent work on fragmentation during the protostar phase has concentrated on numerical calculations of hydrodynamical collapse with rotation in three space dimensions. The evolution of the protostar may be divided into the earlier, optically thin, isothermal phase and the later adiabatic phase, which sets in when the density ρ has increased to values above 10^{-13} g cm^{-3}. The 3-D calculations generally start out with uniform density, uniform angular velocity, and small non-axisymmetric perturbations in the density. During the first free-fall time the cloud collapses, becomes centrally condensed, and develops rotational flattening; however little growth of the perturbations occurs, relative to the collapsing background. The collapse then slows down because of pressure effects parallel to the rotation axis and rotational effects perpendicular to it. During the process of the formation of a disk-like structure, the system fragments, generally through a non-axisymmetric pattern, into two or more orbiting subcondensations. The properties of the fragments depend on the initial perturbation. An $m = 2$ perturbation will generally result in fragments of roughly equal mass, with a total mass of about 15 % of the original cloud mass over the time scale of the calculations, which are generally run only 1-1.5 initial free-fall times (Boss 1986). If low-amplitude random perturbations are imposed on the initial cloud, the number of fragments increases with the number of Jeans masses in the cloud (Larson 1978). The typical fragment has a residual spin angular momentum per unit mass which is an order of magnitude or more smaller than that of the original cloud.

Extensive calculations have been carried out for the isothermal case. Here the initial angular momenta are chosen to be large enough so that rotational effects become important after only a relatively mild degree of collapse; wide systems are produced. This outcome is likely for a cloud with initial $j \approx 10^{21}$ cm^2 s^{-1}. Exploratory calculations with $m = 2$ perturbations of relatively large amplitude produced binaries (Bodenheimer, Tohline, and Black 1980, Boss 1980, Bodenheimer and Boss 1981). Figure 2 gives an example of a case (Boss 1986) that starts in the isothermal phase and includes radiative transfer, with $\alpha_i = 0.25$ and $\beta_i = 0.04$, where α_i and β_i are defined to be the initial ratios of thermal and rotational energy,

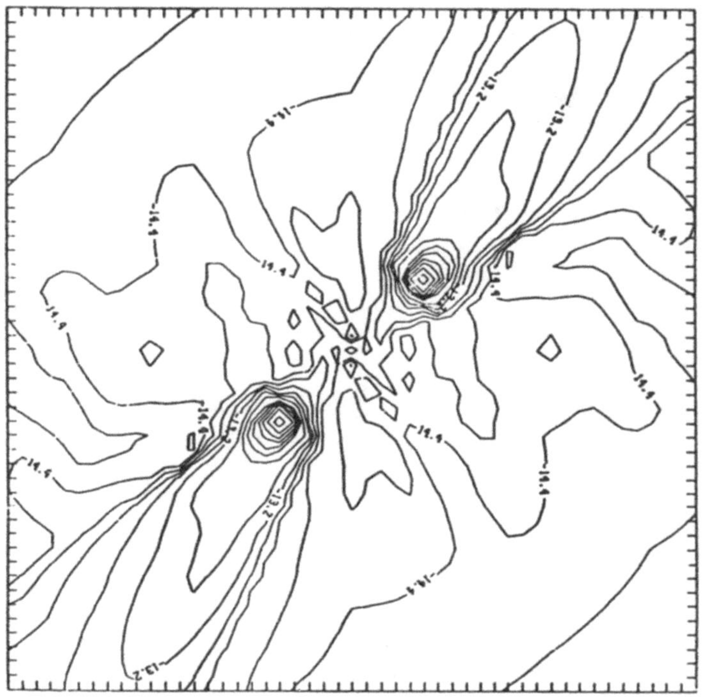

Figure 2. Fragmentation calculation with radiative transfer for $\alpha_i = 0.25, \beta_i = 0.04$ (Boss 1986). Contours of equal density in the equatorial plane are plotted and labelled with values of log ρ. The scale corresponding to the width of the box is 360 AU.

respectively, to the absolute value of the gravitational energy. The initial $j = 1.5 \times 10^{21}$ cm^2 s^{-1} and the initial density is 2.1 $\times 10^{-18}$ g cm^{-3}. In this particular case, fragmentation into a binary occurs just before the adiabatic phase sets in. The distance between the orbiting fragments is 110 AU and the period is $\sim$ 1000 yr, emphasizing the fact that fragmentation in the isothermal phase can explain only the longer-period systems. Calculations carried out with the SPH method with small initial perturbations have produced larger numbers of fragments; we find that the number of fragments increases as α_i decreases. Figure 3 shows the result of Miyama, Hayashi, and Narita (1984) with $\alpha_i = 0.2$ and $\beta_i = 0.3$, calculated with an SPH code starting with density fluctuations $\delta\rho/\rho \sim 0.05$, in which about eight fragments formed. The individual fragments, which have roughly equal mass, also have low enough α_i and β_i so that they can collapse and fragment again, forming hierarchical multiple systems (Larson 1972, Bodenheimer 1978).

For lower values of the initial angular momentum of the cloud ($j \approx 10^{20}$), or for fragments produced in the isothermal phase, the central part of the cloud becomes opaque before rotational effects become significant, and fragmentation will tend to be suppressed (Safronov and Ruzmaikina 1978, Ruzmaikina 1988). Calculations of fragmentation in the adiabatic phase are more difficult, because the cloud develops a large density contrast before the central density reaches 10^{-13} g cm^{-3}, and the inner regions, which contain only a small fraction of the total mass, must be adequately resolved. Calculations for purely adiabatic collapse (Boss 1980) and for the transition between the optically thin and thick regions (Boss 1986) give the general result that if α_i is high at the beginning of adiabatic collapse, fragmentation is suppressed. For $\alpha_i \sim$ 0.05-0.1, fragmentation can occur. The reason is related to the fact that the central part of a protostar becomes optically thick on a scale of 100 AU because of grain opacity. Pressure effects damp out perturbations unless those effects are initially very small.

In a low-angular-momentum cloud rotational effects become important in promoting fragmentation only in the optically thick region, where nearly adiabatic heating and the consequent increase in pressure tend to smooth out perturbations. However, when the central region reaches a temperature of 2000 K, molecular dissociation occurs, the adiabatic Γs of the gas decrease to values close to those appropriate for the isothermal phase, and conditions again become favorable for fragmentation (Larson 1972). The scale of this region is less than 1 AU, and the possibility of forming close binary systems at this stage exists but has not been adequately tested. Problems with this suggestion that must be considered include (1) the amount of mass initially present in the unstable region is very small, (2) later addition of material of higher angular momentum from the outer part of the cloud will tend to separate the system, and (3) generation of spiral waves in the cloud by the binary will tend to bring the components closer together.

General criteria for the fragmentation of rotating collapsing clouds have been developed. For isothermal collapse, Miyama, Hayashi, and Narita (1984) have shown that fragmentation occurs if $\alpha_i\beta_i < 0.12$. Above this limit, non-fragmenting equilibria or disk systems are formed. For an adiabatic gas with γ =1.4 fragmentation occurs if $\alpha_i < 0.09\ \beta_i{}^{0.2}$ (Miyama 1989a, Hachisu *et al.* 1987, Tohline 1981, Boss 1981). This criterion could be applied to fragments whose entire evolution occurs in the adiabatic phase, but not for those which start in the isothermal phase and evolve into the adiabatic phase only at a later time.

Thus wide binaries (those with relatively high angular momentum) can be understood to have been formed by fragmentation in the isothermal phase. But a major problem remains concerning the origin of close binary systems. One possibility is that clouds with relatively low angular momentum collapse through the adiabatic phase without fragmenting, then evolve to a central object plus a relatively massive equilibrium disk. Gravitational instability in such a disk is another possible way of forming close binaries. Investigations of the non-axisymmetric gravitational instability of thin disks have been performed by Adams *et al.* (1989) and Shu *et al.* (1990). They concentrate on the $m = 1$ (one-armed spiral) mode and find that eccentric distortions can grow on a time scale comparable to the orbital frequency at the outer edge of the disk. Because of the eccentric nature of the unstable mode, they speculate that binary formation in the disk is a possible outcome. Other ways of forming close binaries include hierarchical fragmentation starting with low-α

fragments near the beginning of the adiabatic phase (Boss 1988), fragmentation during collapse induced by molecular dissociation (Larson 1972), or orbital decay of a long-period eccentric system by frictional drag induced by the presence of disks around the individual components (Pringle 1989).

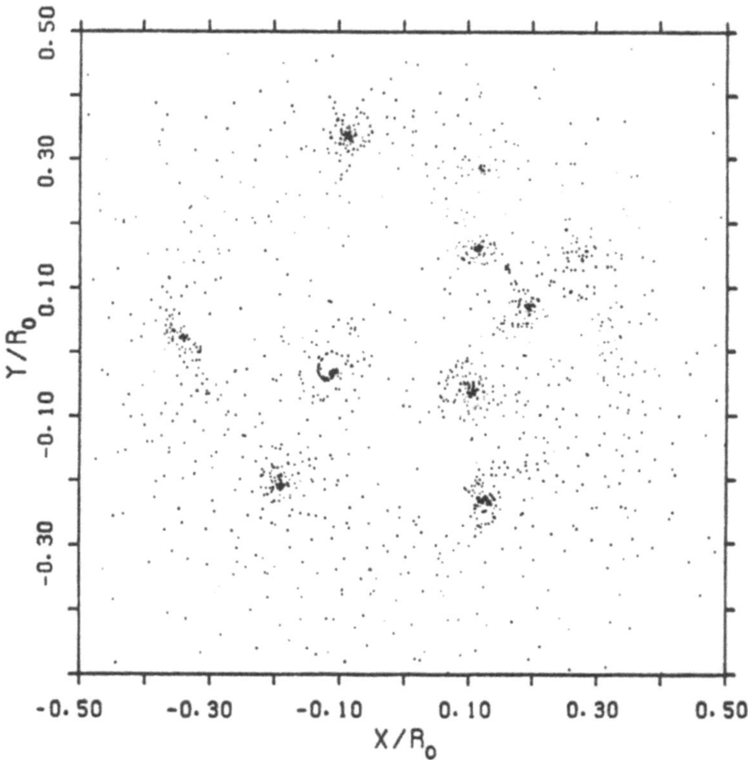

Figure 3. Fragmentation of an isothermal cloud for $\alpha_i = 0.2, \beta_i = 0.3$ (Miyama, Hayashi, and Narita 1984). Particle positions, projected onto the equatorial plane, are plotted after 2.10 initial free-fall times. R_0 is the outer radius.

In any case, the clouds with $j \sim 10^{20}$ will in many cases collapse to a disk that does not fragment. Several recent calculations have considered the details of the process of disk formation for the purpose of comparing theoretical spectra with those of observed embedded IR sources and for investigating the formation of planetary systems. Adams, Lada, and Shu (1987) have constructed semianalytic models of systems consisting of a central object and a disk embedded in an optically thick infalling envelope and have compared the theoretical spectra with those of observed sources. Durisen *et al.* (1989) have constructed full 2-D equilibrium models, under the polytropic approximation, of star-disk systems at various epochs during the process of disk formation. Full 2-D hydrodynamic models of the disk-formation stage with radiation transport have been presented by Morfill, Tscharnuter, and

Völk (1985) and more recently by Bodenheimer *et al.* (1990). For a mass of 1 $M_\odot$, an initial mean density of 4×10^{-15} g cm^{-3}, and $j = 2.5 \times 10^{20}$ cm^2 s^{-1} at the outer edge, the calculation by Bodenheimer *et al.* (1990) results in a central object of 0.6 $M_\odot$ and a warm disk of $\sim$ 0.3 $M_\odot$. Although the disk, of size $\approx$ 50 AU, is gravitationally stable and is probably suitable for the formation of a planetary system, the rapidly rotating central object is unstable to nonaxisymmetric perturbation.

In Figure 4 is shown another calculation under the same physical assumptions, this time with 2 $M_\odot$, an initial mean density of 6×10^{-17} g cm^{-3}, and $j = 7 \times 10^{20}$. Temperatures in the resolved disk region fall in the range 50-100 K and densities in the range 10^{-11} to 10^{-14} g cm^{-3}. The disk in this case contains 65% of the mass and is very likely to be gravitationally unstable, according to Shu *et al.* (1990).

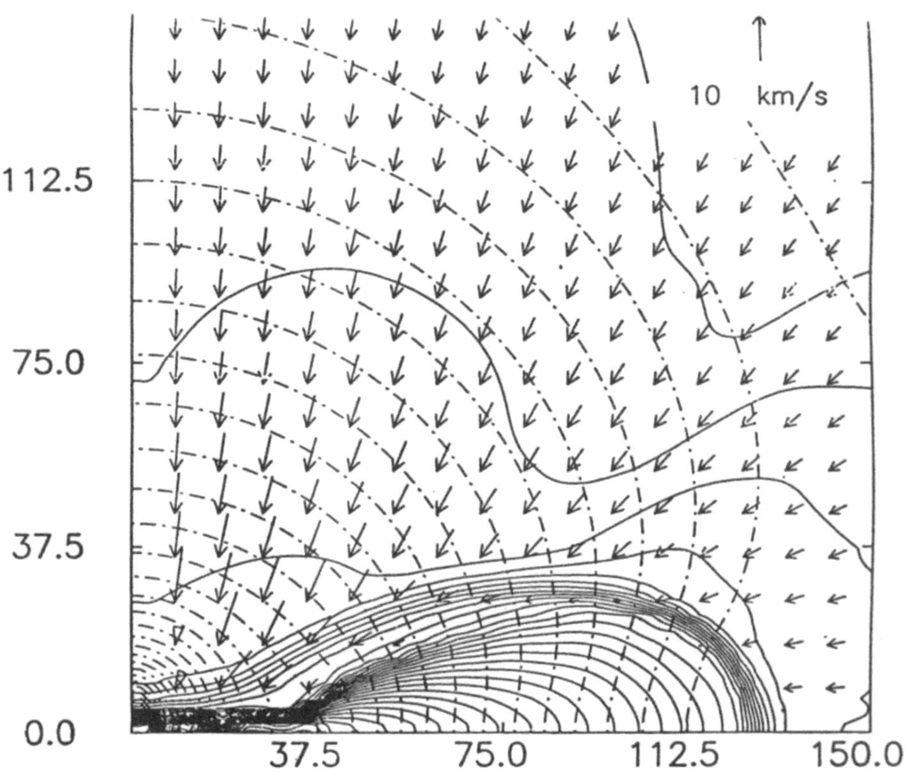

Figure 4. Disk formation in a rotating collapsing cloud 10^4 yr after the beginning of collapse, from a two-dimensional hydrodynamic calculation by Yorke *et al.* (1991). Contours of equal density (solid lines; contour interval $\Delta \log \rho = 0.2$) and equal temperature (dash-dot lines; contour interval $\Delta \log T = 0.03$) are plotted in the (R,Z) plane. The central regions of the disk are thin and unresolved.

Their linear analytic and numerical calculations show that a disk is stable to all gravitational disturbances below a critical value of its mass (which depends also on the temperature in the disk). The critical ratio of disk mass to total mass (disk plus central object) is found to be 0.24 if the Toomre Q value at the outer edge of the disk is 1. The speculation is that disks with lower mass can form planetary systems while those with greater mass may transport angular momentum through spiral waves or may develop a binary companion.

Three-dimensional hydrodynamic calculations of the collapse of a rotating protostar with radiation transport have been reported by Boss (1989), starting at relatively high densities ($10^{-13} - 10^{-14}$ g cm^{-3}) and relatively low specific angular momenta ($j \approx 10^{18} - 3 \times 10^{19}$ cm^2 s^{-1}). Small (1 %) initial non-axisymmetric perturbations are amplified during collapse to produce angular momentum transport times by gravitational torques as short as $10^3 - 10^4$ yr in disks surrounding a central object, although the disk evolution over such a long time scale is not followed. As the mass in the assumed central object is increased relative to the disk mass, a stabilizing effect occurs and the transport times become much longer or shut off altogether. The rapid time scale associated with gravitational instability is important, because the collapse of most clouds is expected to lead to a massive, and probably gravitationally unstable, disk, while observed disks around T Tauri stars have estimated masses of only a few percent of a solar mass.

6. The Spindown of the Central Object

Bouvier's (1990) HR diagram of the currently determined rotation rates of T Tauri stars shows that even near the stellar birthline (Palla and Stahler 1990) the velocities are $\sim$ 20-30 km s^{-1}. However the numerical calculations indicate that the central star should be rotating near breakup when it comes into hydrostatic equilibrium. Several processes may operate to produce the rapid spindown on a time scale of $\sim 10^5$ yr. At first, the star will be rotating rapidly enough so that it will be unstable to nonaxisymmetric perturbations. The 3-D hydrodynamic calculations in this case (Durisen *et al.* 1986, Williams and Tohline 1988) show that the central regions deform into a bar, a spiral pattern develops, and outward transport of angular momentum occurs, ultimately resulting in the ejection of a disk. If a disk is initially present, the waves excited in the disk by the non-axisymmetric object would also transport angular momentum outward (Yuan and Cassen 1985). In either case, the transport would continue until the central object is stabilized; it would still be rotating far too rapidly to be consistent with observations.

By this time the star will have contracted and heated sufficiently so that it will have an extensive convective interior; turbulent viscosity and magnetic fields will transport angular momentum out of the interior, leaving much of the mass in near-uniform rotation. However this process will tend to spin up the surface layers, which were probably already rapidly spinning. Furthermore, the star is accreting matter from a disk, which will also spin it up to breakup (Galli and Shu, this conference). It seems that a powerful wind is required to spin down the star under these circumstances. Magnetic braking as a consequence of a stellar wind which is forced to corotate with the star out to the Alfvén radius has long been known to be an efficient loss mechanism for angular momentum (Schatzman 1962, Weber and Davis 1967, Mestel 1968). The required time scale for angular momentum loss

(t_j) for protostars is 10^5 yr. Theory shows that t_j can be an order of magnitude or so shorter than t_m, the mass loss time scale, which requires mass outflow rates on the order of $10^{-5} - 10^{-6}$ $M_\odot$ yr^{-1}, consistent with observations of bipolar outflow sources. The problem is how to generate the wind. Several mechanisms have been suggested; a few examples follow. The wind energy could be derived from a magnetic dynamo originating in the central star (Natta *et al.* 1988). A magnetic mechanism has also been studied by Draine (1983) in which magnetic field lines from the star are coupled to the external medium, and the star is spun down by the propagation of Alfvén waves in a manner analogous to the spindown of molecular cloud regions. Alternatively, the wind could be centrifugally driven (Hartmann and MacGregor 1982, Shu *et al.* 1988), relying in part on the very low surface gravity near the equator of a star that is rotating near breakup. Magnetic fields on the order of a kilogauss are required to obtain a sufficiently high mass loss rate. Observational work (Cabrit *et al.* 1990) suggests a correlation between accretion luminosity in a disk and diagnostics of wind luminosity in T Tauri stars. They suggest that the gravitational energy released by accretion from a disk onto the star supplies the energy for the wind. Such a correlation may also exist for the embedded infrared sources (Lada 1985, Snell 1987). The determination of the specific mechanism for wind generation and for field generation with the appropriate strength remains an unresolved problem.

7. Summary

Significant angular momentum evolution takes place during the star formation and protostar phases. Before collapse takes place, the dense core of a molecular cloud is held in approximate equilibrium by turbulent and magnetic effects. The time scale is long enough so that angular momentum can be transferred out of the central regions by the propagation of Alfvén waves, leading to a reduction of the specific angular momentum of some material from $\sim 10^{23}$ to $\sim 10^{21}$ cm^2 s^{-1}. The range of densities over which the magnetic field could cease to be dynamically significant and the range of angular velocities of turbulent elements in a gas can account for the wide range in angular momentum in single stars and multiple star systems. Some molecular cloud cores are observed to rotate with $j \approx 10^{21}$ cm^2 s^{-1}; refinement of the observational techniques to allow detection of more slowly rotating clouds is an important priority. During collapse the protostar conserves angular momentum locally, but there is the possibility of converting spin angular momentum into orbital angular momentum through fragmentation. Formation of wide binaries by fragmentation, and formation of disks by the fragments or by the more slowly rotating molecular cloud cores, can solve the angular momentum problem to a considerable extent. However, no matter which path they follow, the stellar cores in most cases will be rotating near breakup at the end of collapse. Furthermore, they tend to accrete matter and angular momentum from a disk. Thus the unsolved part of the angular momentum problem is the redistribution of angular momentum of the stellar core, probably into a wind. Because T Tauri stars are slowly rotating even near the birthline, the problem of the loss of the remaining factor of 10 in angular momentum must be solved during the $\sim 10^5$ years that the star spends as an embedded source in the bipolar outflow phase. The origin of the powerful winds and of the strong magnetic fields needed to produce this rapid loss represents a major theoretical and observational challenge.

ACKNOWLEDGEMENTS. This work was supported in part by a special NASA Theory grant which provides funding for a joint Center for Star Formation Studies at NASA-Ames Research Center, University of California, Berkeley, and University of California, Santa Cruz. Further support was obtained from National Science Foundation grant AST-8914173 and from the San Diego Supercomputer Center in the form of a grant of computing time.

References

Adams, F. C., Lada, C. J., and Shu, F. H. 1987. *Ap. J.*, **312**, 788.

Adams, F. C., Ruden, S. P., and Shu, F. H. 1989. *Ap. J.*, **347**, 959.

Alfvén, H., and Arrhenius, G. 1976. *Evolution of the Solar System* (Washington: NASA SP-345).

Arquilla, R. A., and Goldsmith, P. F. 1986. *Ap. J.*, **303**, 356.

Bodenheimer, P. 1978. *Ap. J.*, **224**, 488.

Bodenheimer, P., and Boss, A. P. 1981. *Mon. Not. R. astr. Soc.*, **197**, 477.

Bodenheimer, P., Różyczka, M., Yorke, H. W., and Tohline, J. E. 1988. In *Formation and Evolution of Low Mass Stars*, eds. A.K. Dupree and M. T. V. T. Lago (Dordrecht: Kluwer Acad. Publ.), p. 139.

Bodenheimer, P., Tohline, J. E., and Black, D. C. 1980. *Ap. J.*, **242**, 209.

Bodenheimer, P., Yorke, H., Różyczka, M., and Tohline, J. 1990. *Ap. J.*, **355**, 651.

Boss, A. P. 1980. *Ap. J.*, **237**, 866.

Boss, A. P. 1986. *Ap. J. Suppl.*, **62**, 519.

Boss, A. P. 1988. *Comments on Astrophys.*, **12**, 169.

Boss, A. P. 1989. *Ap. J.*, **345**, 554.

Bouvier, J. 1990. In *Lecture Notes in Physics: Frontiers of Stellar Structure Theory*, in press.

Cabot, W., Canuto, V., Hubickyj, O., and Pollack, J. B. 1987a. *Icarus*, **69**, 387.

Cabot, W., Canuto, V., Hubickyj, O., and Pollack, J. B. 1987b. *Icarus*, **69**, 423.

Cabrit, S., Edwards, S., Strom, S. E., and Strom, K. M. 1990. *Ap. J.*, **354**, 687.

Cameron, A. G. W. 1962. *Icarus*, **1**, 13.

Cassen, P., Smith, B. F., Miller, R. H., and Reynolds, R. T. 1981. *Icarus*, **48**, 377.

Draine, B. T. 1983. *Ap. J.*, **270**, 519.

Durisen, R. H., Gingold, R. A., Tohline, J. E., and Boss, A. P. 1986. *Ap. J.*, **305**, 281.

Durisen, R. H., Yang, S., Cassen, P., and Stahler, S. 1989. *Ap. J.*, **345**, 959.

Fuller, G. A., and Myers, P. C. 1987. In *Physical Processes in Molecular Clouds*, eds. G. E. Morfill and M. Scholer (Dordrecht: Reidel), p. 137.

Goldsmith, P. F., and Arquilla, R. A. 1985. In *Protostars and Planets II*, eds. D. C. Black and M. S. Matthews (Tucson: Univ. of Arizona Press), p. 137.

Hachisu, I., Tohline, J. E., and Eriguchi, Y. 1987. *Ap. J.*, **323**, 592.

Harris, A., Townes, C. H., Matsakis, D. N., and Palmer, P. 1983. *Ap. J. (Letters)*, **265**, L63.

Hartmann, L., and MacGregor, K. B. 1982. *Ap. J.* , **259**, 180.

Hayashi, C. 1981. *Prog. Theor. Phys. Suppl.*, **70**, 35.

Heyer, M. H. 1988. *Ap. J.*, **324**, 311.

Hoyle, F. 1960. *Q. J. R. astr. Soc.*, **1**, 28.

Kraft, R. P. 1970. In *Spectroscopic Astrophysics*, ed. G. Herbig (Berkeley: University of California Press), p. 385.

Lada, C. J. 1985. *Ann. Rev. Astr. Astrophys.*, **23**, 267.

Larson, R. B. 1972. *Mon. Not. R. astr. Soc.*, **156**, 437.

Larson, R. B. 1978. *Mon. Not. R. astr. Soc.*, **184**, 69.

Larson, R. B. 1981. *Mon. Not. R. astr. Soc.*, **194**, 809.

Larson, R. B. 1989. In *The Formation and Evolution of Planetary Systems*, eds. H. Weaver and L. Danly (Cambridge: Cambridge University Press), p. 31.

Levy, E. H., Ruzmaikin, A., and Ruzmaikina, T. 1990. In *The Sun in Time*, ed. C. P. Sonett (Tucson: Univ. of Arizona Press), in press.

Levy, E. H., and Sonett, C. P. 1978. In *Protostars and Planets*, ed. T. Gehrels (Tucson: Univ. of Arizona Press), p. 516.

Lin, D. N. C., and Papaloizou, J. 1980. *Mon. Not. R. astr. Soc.*, **191**, 37.

Lin, D. N. C., Papaloizou, J., and Savonije, G. 1990. *Ap. J.*, **364**, 326.

Lin, D. N. C., and Pringle, J. E. 1987. *Mon. Not. R. astr. Soc.*, **225**, 607.

Lizano, S., and Shu, F. H. 1989. *Ap. J.*, **342**, 834.

Lüst, R., and Schlüter, A. 1955. *Zeits. f. Astrophys.*, **38**, 190.

Makalkin, A. B. 1987. *Astr. Vestnik (U.S.S.R.)*, **21**, 324.

Mathieu, R. D., Benson, P. J., Fuller, G. A., Myers, P. C., and Schild, R. E. 1988. *Ap. J.*, **330**, 385.

Mestel, L. 1968. *Mon. Not. R. astr. Soc.*, **138**, 359.

Mestel, L. 1985. In *Protostars and Planets II*, eds. D. C. Black and M. S. Matthews (Tucson: Univ. of Arizona Press), p. 320.

Mestel, L., and Spitzer, L. Jr. 1956. *Mon. Not. R. astr. Soc.*, **116**, 505.

Miyama, S. M. 1989. In *Highlights of Astr.*, ed. D. McNally (Dordrecht: Kluwer), vol. **8**, p. 127.

Miyama, S. M., Hayashi, C., and Narita, S. 1984. *Ap. J.*, **279**, 621.

Morfill, G., Tscharnuter, W., and Völk, H. 1985. In *Protostars and Planets II*, eds. D. C. Black and M. S. Matthews (Tucson: Univ. of Arizona Press), p. 493.

Mouschovias, T. Ch. 1977. *Ap. J.*, **211**, 147.

Myers, P. C., and Benson, P. J. 1983. *Ap. J.*, **266**, 309.

Natta, A., Giovanardi, C., and Palla, F. 1988. *Ap. J.*, **332**, 92

Palla, F., and Stahler, S. W. 1990. *Ap. J. (Letters)*, **360**, L47.

Pringle, J. E. 1989. *Mon. Not. R. astr. Soc.*, **239**, 361.

Ruden, S. P., Papaloizou, J., and Lin, D. N. C. 1988. *Ap. J.*, **329**, 739.

Ruzmaikina, T. V. 1981. *Adv. Space Res.*, **1**, No. 7, 49.

Ruzmaikina, T. V. 1982. *Mitt. Astr. Ges.*, **57**, 49.

Ruzmaikina, T. V. 1985. *Astr. Vestnik (U.S.S.R.)*, **19**, 101.

Ruzmaikina, T. V. 1986. *Astr. Zirkular (U.S.S.R.)*, **1439**, 1.

Ruzmaikina, T. V. 1988. In *I.A.U. Colloquium No. 99: Bioastronomy: the next steps*, ed. G. Marx (Dordrecht: Kluwer), p. 41.

Ruzmaikina, T. V., and Maeva, S. V. 1986. *Astr. Vestnik (U.S.S.R.)*, **20**, 212.

Safronov, V. S. 1960. *Ann. Astrophys.*, **23**, 901.

Safronov, V. S., and Ruzmaikina, T. V. 1978. In *Protostars and Planets*, ed. T. Gehrels (Tucson: Univ. of Arizona Press), p. 545.

Safronov, V. S., and Ruzmaikina, T. V. 1985. In *Protostars and Planets II*, eds. D. C. Black and M. S. Matthews (Tucson: Univ. of Arizona Press), p. 959.

Sargent, A., and Beckwith, S. 1987. *Ap. J.*, **323**, 294.

Schatzman, E. 1962. *Ann. Astrophys.*, **25**, 18.

Sekiya, M., and Miyama, S. 1988. *Prog. Theor. Phys. Suppl.*, **96**, 85.

Shakura, N. I., and Sunyaev, R. A. 1973. *Astr. Astrophys.*, **24**, 337.

Shu, F. H., Adams, F. C., and Lizano, S. 1987. *Ann. Rev. Astr. Astrophys.*, **25**, 23.

Shu, F. H., Lizano, S., Adams, F. C., and Ruden, S. P. 1988. in *Formation and Evolution of Low Mass Stars*, eds. A.K. Dupree and M. T. V. T. Lago (Dordrecht: Kluwer Acad. Publ.), p. 123.

Shu, F. H., Tremaine, S., Adams, F. C., and Ruden, S. P. 1990. *Ap. J.*, **358**, 495.

Snell, R. L. 1987. In *IAU Symposium 115, Star Forming Regions*, eds. M. Peimbert and J. Jugaku (Dordrecht: Reidel), p. 213.

Solomon, P., and Sanders, D. 1985. In *Protostars and Planets II*, eds. D. C. Black and M. S. Matthews (Tucson: Univ. of Arizona Press), p. 59.

Stepinski, T. F., and Levy, E. H. 1988. *Ap. J.*, **331**, 416.

Tassoul, J. L. 1978. *Theory of Rotating Stars* (Princeton: Princeton University Press).

Tohline, J. E. 1981. *Ap. J.*, **248**, 717.

Toomre, A. 1964. *Ap. J.*, **139**, 1217.

Ungerechts, H., Walmsley, C. M., and Winnewisser, G. 1982. *Astr. Astrophys.*, **111**, 339.

Wadiak, E. J., Wilson, T. L., Rood, R. T., and Johnston, K. J. 1985. *Ap. J. (Letters)*, **295**, L43.

Weber, E. J., and Davis, L., Jr. 1967. *Ap. J.*, **148**, 217.

Williams, H. A., and Tohline, J. E. 1988. *Ap. J.*, **334**, 449.

Yorke, H., Bodenheimer, P., and Laughlin, G. 1991. In preparation.

Yuan, C., and Cassen, P. 1985. *Icarus*, **64**, 435.

Zel'dovich, Ya. B. 1981. *Proc. R. Soc. London A*, **374**, 299.

DISCUSSION

Strom: All single stars are likely to be built from disks, in which case the observational evidence which you would likely see is first that the mass of the disks should be comparable to that of the stars which they are going to build, and I think that evidence is there, though weakly in that the masses that you derive early on for disks like those estimated for HL Tau or IRS 5 are at least several tenths of a solar mass. The next thing I would guess, and this is where the question comes, is that if you look at some objects like the weak-lined T Tauri stars and the classical T Tauri stars, their J/M's are almost identical, so I guess the implication must be that even though some stars at present seem to lack evidence for optically thick disks the implication is that among single stars, all stars must be built from disks given the picture that you presented. Is that the conclusion I am supposed to draw?

Bodenheimer: I think that is a reasonable picture, yes. First of all, the fraction of cloud cores with angular momentum too low to form reasonable-mass disks I think has to be very small - though that has to be confirmed by more observations. Most cloud cores probably have specific angular momenta of order 10^{20} to 10^{21}, so it seems unavoidable that one has to go through a disk stage. I understand the similarity of the rotational velocities between the weak lined and classical T Tauri stars. It still seems that one has to go through a disk stage early on so the important thing is to look at the earliest time you can look at to see if a massive disk is there - that is, near the birth line if possible. Now HL Tau is a good example, in fact it has a disk of very large radius. If you look at the total angular momentum of HL Tau it is comparable to that of a cloud core, so you don't have to lose any angular momentum at all. Even in that case I think that the mass of the star and the disk are comparable, I think you still have to explain probably some reduction in the disk mass for the very early stage from the time of formation to the time we are observing. I think it seems certainly reasonable that at least those systems which don't fragment into close binaries - and we don't know what happens in that case - probably do have to form out of disks.

Kraft: The emphasis now is on the comparison of pre-main sequence stars with disks with those that do not. Are the stars without disks older than those with disks?

Strom: The interpretation I would make is that the stars with weak disks or no disks had disks early on. Some of the weak-lined stars in fact seem to develop inner holes in their disks, so that you have a ring of material on the outside, but an optically thin, non-accreting inner region, leading to the lack of signatures that you associate with the weak-lined T Tauri stars. What I was trying to suggest was that if we made a census for remnant disks, not only would we find remnant disks around all single T Tauri stars but we would find remnant disks around the weak-lined T Tauri stars as well. And I was trying to imply that the weak-lined T Tauri stars had to be built from disks that just accreted much more rapidly.

Kraft: What about the T Tauri stars that are binary stars - were they built from disks too?

Bodenheimer: Well, there is a lot of work being done on that right now. The wide binaries among T Tauri stars do tend to have disks. I think the close binaries tend to have less massive disks than wide binaries. I think the weak-lined T Tauri stars have a larger fraction of binary stars than classical T Tauri stars; however it is certainly possible for classical T Tauri stars to be in binaries. The wide binaries probably had disks when they formed; in the case of the close binaries we don't really know, but one possibility is that they formed from a disk.

Pinsonneault: There is an intriguing similarity between the angular momentum of the solar system and the angular momentum of the cloud cores. Could you speculate on what potential role planet formation plays in the formation of stars?

Bodenheimer: Consider a system with J/M of 10^{20} say, which can collapse directly to a disk with the same specific angular momentum as the orbits of planets. The numerical calculation I showed was that kind of disk. And the disk mass turns out to relatively large, giving relatively high surface density, which means that one could form planets relatively quickly, which seems to be consistent with some of the observational evidence in our planetary system which suggests that at least Jupiter had to form early. However, one has to consider that since the disk is very massive, it is likely to be unstable and would evolve very rapidly. So the question is whether you can maintain that high density long enough to form planets. But it seems likely that a large number of cloud cores have the right angular momentum and mass distributions to produce disks which at least are favorable for the initial construction of planets.

Roxburgh: In almost all of these scenarios, magnetic fields are important. Could you conceive of a situation where you could form stars with essentially no magnetic field. If not, are all subsequent speculations on the evolution of rotation in forming stars that ignore magnetic fields a waste of time.

Bodenheimer: The final stage of protostar evolution involves the spin-down of the rapidly rotating core to velocities characteristic of T Tauri stars. It is very unlikely that you could do that without magnetic fields. In the early stages, back in the molecular cloud core stage, although it seems that there is plenty of evidence that magnetic fields are present, suppose that they weren't. One can still form binaries and multiple systems just by hierachical fragmentation. Rotation effects alone result in collapse and break-up into fragments, and then repeated breakup of the fragments themselves can form multiple systems, with no magnetic fields at all. I am not saying that this is what usually happens, but it is at least a possibility. During the hydrodynamic collapse and during disk evolution in non-ionized regions the magnetic effects are unimportant.

Geroyannis: If there is a fast mechanism creating differential rotation in protostars, then there will not only be the original magnetic fields but also the magnetic fields produced by differential rotation so that the protostar will be able to remove further angular momentum. Is there any evidence about differential rotation in protostars?

Bodenheimer: Protostars are very hard to observe, and I think there is no evidence for magnetic fields at the present time between the molecular cloud core stage and the T Tauri stage. There are magnetic fields observed in molecular clouds. There is possibly some evidence for differential rotation on the large scales in molecular clouds. I don't think we know anything about differential rotation in molecular cloud cores - observations just aren't good enough. In the protostar stage itself, you expect the star to be heavily embedded. You can't really see into the disks or the objects themselves, so it is really hard to determine anything about that. During the protostar stage, there is undoubtedly differential rotation. However, in that stage there is also very weak coupling to the field, so there will be no further field generated. There is coupling to the field later on when the equilibrium core is formed and becomes ionized, and there additional transport of angular momentum is certainly possible.

EVOLUTIONARY PROPERTIES OF INTERMEDIATE MASS PROTOSTARS

F. PALLA
Osservatorio Astrofisico di Arcetri
L.go E. Fermi, 5
50125–Firenze
Italy

ABSTRACT. The properties of young stars of intermediate–mass are discussed in the light of recent calculations of the protostellar accretion phase. In particular, the problem of the existence of surface convection in these stars will be addressed with reference to the strong activity manifested by the Herbig Ae/Be stars.

Introduction

Recent studies of the evolution of young stars have been focused on the rotational properties of low–mass stars, i.e. stars of (sub-)solar mass, either in isolation or in groups and clusters (cf. the reviews by J. Bouvier and J. Stauffer in this Proceedings). Little attention has been paid to the more massive pre–Main-Sequence (PMS) stars that, as a class, are collectively classified as *Herbig Ae/Be* stars (Herbig 1960) and that, as well as their less massive counterparts, can provide useful information on the evolution of the angular momentum immediately after the star formation process. It is the aim of this contribution to highlight some fundamental issues related to the class of intermediate–mass stars. In particular, results will be discussed that question some well established viewpoints of their structural properties as predicted by the classical theory of PMS stars (for more extended reviews on this topic see Catala 1989 and Palla 1991).

The Herbig Ae/Be stars have masses larger than 2 $M_{\odot}$, as deduced from their location in the H-R diagram, show emission features in the optical spectrum, and are found in the vicinity of reflection nebulosity and regions of obscuration. In addition, they share the signs of activity commonly observed in T Tauri stars, such as strong stellar winds, variability and periodicity in the lines, IR excesses, molecular flows, highly collimated optical jets and Herbig–Haro objects. This activity seems to indicate that star formation and early stellar evolution are marked by the occurrence of the same kind of phenomena over an extended mass interval, say 1 to 10 $M_{\odot}$. However, a significant difference exists between intermediate and low–mass stars, since the former mark the transition from fully convective configurations, that characterize the low-mass stars, to structures in radiative equilibrium that evolve quite rapidly, due to the strong mass dependence of the Kelvin–Helmoltz timescale. Therefore, although not completely unexpected, the strong activity shown by the

S. Catalano and J. R. Stauffer (eds.), Angular Momentum Evolution of Young Stars, 21–25.

Herbig Ae/Be stars is somewhat paradoxical and has remained a puzzle for decades. The basic question of the energy source of the star's activity is still unresolved, but a common assumption is that these stars lack of the fundamental ingredient of the classical theory: ***the presence of deep and extended outer convection zones.*** In fact, an inspection to the H-R diagram of these stars, cf. Figure 1, immediately reveals that they lay closer to the ZAMS than the less massive T Tauri stars, in the radiative portion of the classical evolutionary tracks (Iben 1965). Moreover, recent calculations of the PMS evolution of intermediate–mass stars (Gilliland 1986) have confirmed the old results that convection is limited to a very thin outer region, so that the possilibity of a solar-type dynamo mechanism to explain the chromospheric activity seems hardly tenable (Catala *et al.* 1991).

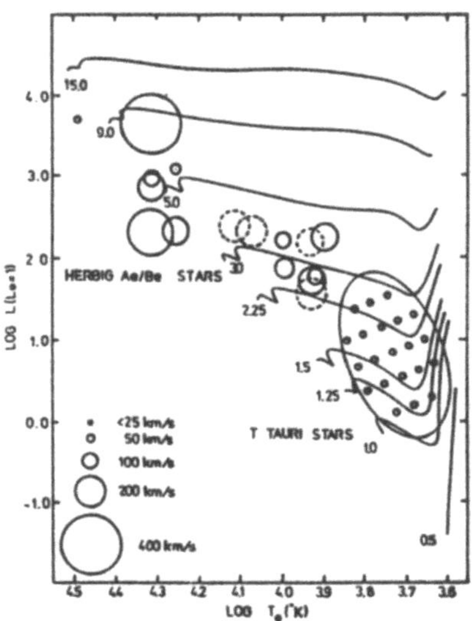

Figure 1. H-R diagram of Herbig Ae/Be (*large circles*) and T Tauri stars (*small circles*). The size of the circles indicates the magnitude of the projected rotational velocities *v sin i* (from Finkenzeller 1985).

Returning to the H-R diagram of Figure 1, another important aspect of Herbig Ae/Be stars is that they are relatively ***fast rotators***, as indicated by the size of the circles. According to Finkenzeller (1985), Herbig stars rotate faster than T Tauri stars, typically with projected rotational velocities between 100 and 200 km s^{-1}, and are depleted of slow rotators. Also, the distribution of the observed *v sin i* is significantly different from that of classical B stars with emission lines, and from A and B Main Sequence stars, so that it is highly unlikely that they could represent evolved objects. On the basis of the angular momentum distribution, it is concluded that Herbig stars should be the predecessors of B-field stars or the "classical" Be stars. It must be noted that, despite the relatively high values of *v sin i*, all the Herbig Ae/Be stars with measured rotational velocities do rotate well below the critical speed, that for stars of spectral type A0–B3 varies between 500 and 650 km s^{-1}. Therefore, the problem of the removal of angular momentum, that has been stressed repeteadly in this Workshop in the context of low–mass star formation, is of significance even in the case of more massive stars.

The fact that Herbig stars are relatively fast rotators has prompted some authors to suggest that rotation can indeed be the energy source for the observed surface activity (e.g. Praderie *et al.* 1986, Catala *et al.* 1986). In the particular case of the best studied case, AB Aur, quantitative analysis of the radiative losses in the chromosphere have shown that they can be accounted for fairly well by the total rotational energy of the stars, spent in the characteristic Kelvin–Helmoltz timescale ($\sim 10^6$yr) of PMS contraction (Catala 1989). Nevertheless, the basic problem of finding an effective mechanism to extract energy from rotation, in the absence of convection, still remains. In an attempt to address this issue, Vigneron *et al.* (1990) have studied the effect of mass loss, through the action of a stellar wind, on the angular momentum distribution in a fully radiative star.

Finally, the observed rapid rotation of the Herbig stars has also been used to explain the rapid short–term variability and periodicity of several lines formed in the stellar wind (e.g. Catala *et al.* 1991). In analogy with the solar wind case, the variation is interpreted as the result of a pattern of slow and fast streams modulated by the star's rotation, implying the presence of a structured magnetic field, whose actual existence has still escaped detection. Alternatively, the same variations have been interpreted by Baade and Stahl (1989) in terms of nonradial pulsations. Once again, the driving mechanism of such oscillations has not been identified, and in order to decide which mechanism is at work more observations are necessary. To conclude this section, it is evident that the Herbig Ae/Be stars represent a highly interesting class of objects, whose fundamental properties need to be rediscussed.

A protostellar approach to intermediate–mass stars

The approach followed by the author in collaboration with S. Stahler at MIT consists of computing the early evolution of intermediate–mass stars starting from the initial protostellar accretion through the entire PMS phase of a selected sample of protostar cores of different mass. The need for such a global study stems from the consideration that, in the case of the protostellar evolution, the only set of detailed calculations available to date is that of Larson (1972): even in the simple case of spherical accretion, the theory of the main accretion phase has been greatly refined since then and needs to be incorporated in more realistic models. As for the PMS phase, the reference is to the classic work of Iben (1965), where the assumptions on the initial conditions of the calculation are rather arbitrary and, more important, totally ignored the previous protostellar evolution.

Our long term goal is to include the new results of protostellar theory in a modern treatment of PMS evolution, in order to improve the comparison with the large body of observations now available. So far, detailed results are available for the initial protostellar phase in a wide variety of assumptions on the geometry of the accretion process and on the rate of mass infall (Palla and Stahler 1990, 1991). One of the key predictions of Larson's models (1972) is that accreting protostars should join the Main Sequence at a mass between 2 and 3 $M_\odot$. The reason being that, in this mass range, gravity becomes so dominant with respect to accretion that the core goes through the entire PMS phase while still gaining matter. The conditions in the interior become soon favorable for the onset of the nuclear burning of hydrogen, and the core begins its evolution as a *bona fide* star. The consequence of such a behavior is that there *should be no optically visible PMS stars above 2–3* $M_\odot$, in obvious disagreement with the obserations of the Herbig Ae/Be stars.

Our calculations show that the main reason of Larson's results is in his neglect of another important nuclear burning process: ***deuterium burning in a shell at the time of the transition from a fully convective configuration to a radiatively stable one.*** The luminosity generated by the D-burning dramatically affects the core evolution. To appreciate this, Figure 2 shows the new core mass–radius relation for accreting protostars at the representative rate of $10^{-5} M_\odot \; yr^{-1}$ with different boundary conditions. The two curves refer to the evolution computed assuming that matter is accreted directly onto the core through a strong shock, and to the case where the core surface is treated as a normal photosphere, as it would be if the accretion were mediated by the presence of a circumstellar disc.

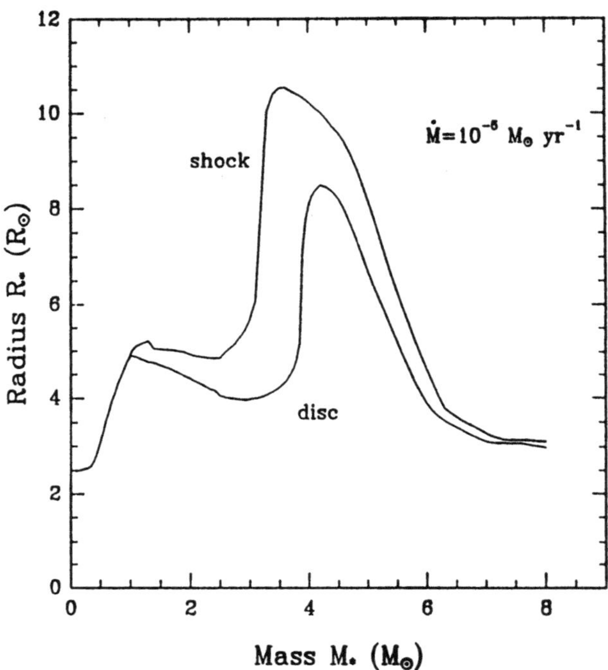

Figure 2. The core mass-radius evolution for accreting protostars in two different geometries (adapted from Palla and Stahler 1991).

It is clear that, apart for minor quantitative differences, the ***qualitative*** behavior is strikingly similar, suggesting that it does not depend on the details of the accretion process. The general evolution can therefore separated into 4 main stages:

$0 \lesssim M_*/M_\odot \lesssim 1$: this is the well known evolution of low–mass protostars. In this range the core is fully convective due to the action of central D–burning that sets the conditions for a steady increase of radius with mass (cf. Stahler 1988);

$1 \lesssim M_*/M_\odot \lesssim 3$: it represents the transitional regime to radiative stability. D–burning in the center can no longer keep the star fully convective; the mass-radius relation of the previous phase breaks down, and the core slowly shrinks under the influence of gravity;

$3 \lesssim M_*/M_\odot \lesssim 4$: it marks the onset of ***D-shell burning*** after the appearance of an internal radiative barrier. In both cases of Fig.2, the core doubles its size in a very narrow mass range;

$4 \lesssim M_*/M_\odot$: for more massive cores, the strength of the gravitational pull prevents a further increase of the radius, and the rapid expansion is replaced by a slow gravitational settling to the conditions of relaxed cores.

The main implication of these results is that, due to the swelling of the core, the interior temperature remains quite low, so that the approach to the Main Sequence is naturally postponed to more massive cores. As shown in Fig.2, central hydrogen burning is delayed until $M_* \sim 6M_\odot$, but the ZAMS is actually reached only at $M_* \sim 8M_\odot$. This finding reconciles the theoretical predictions of the protostellar models with the observations of intermediate-mass PMS stars.

Another important aspect of the present calculations is that, as a consequence of the onset of D–burning in a shell, convection is created and maintained in the outer regions of the core. The extent of the surface convection depends somewhat on the adopted values of the mass accretion rate, but in all cases considered here it encompasses a large fraction of the core mass ($\gtrsim 20\%$). It is tempting to suggest that the role of D–burning in promoting convection in stars that were thought to have none, can also be used to explain the paradoxical activity manifested by the Herbig stars. However, to fully address this important issue, a self–consistent calculation of the PMS evolution of some protostellar cores is in order. Nevertheless, it is easy to anticipate that the evolution will differ markedly from the classical one illustrated by Iben (1965), that assumed completely convective stars with unrealistically large intial radii, and ignored the contribution of D–burning.

Finally, as a last remark, the onset of nuclear reactions near the core surface are also likely to excite instabilities that could possibly develop nonradial oscillations, thus accounting for the rapid spectral line variations discussed before. It is hoped that the results presented in this contribution will foster more research on this field.

References

Baade, D., Stahl, O.: 1989, *Astron. Astrophys.*, **209**, 268.

Catala, C.: 1989, in *Modeling the Stellar Environment: How and Why*, Proc. of the IAP Astrophys. Meeting, Ph. Delache ed., p.207.

Catala, C., Czarny, J., Felenbok, P., Praderie, F.: 1986, *Astron. Astrophys.*, **154**, 103.

Catala, C., Czarny, J., Felenbok, P., Talavera, A., Thé, P.S.: 1991, *Astron. Astrophys.*, in press.

Finkenzeller, U.: 1985, *Astron. Astrophys.*, **151**, 340.

Gilliland, R.L.: 1986, *Astrophys. J.*, **300**, 339.

Herbig, G.: 1960, *Astrophys. J. Suppl.*, **4**, 337.

Iben, I.: 1965, *Astrophys. J.*, **141**, 993.

Larson, R.B.: 1972, *Monthly Not. Roy. Astron. Soc.*, **157**, 121.

Palla, F.: 1991, in *Fragmentation of Molecular Clouds and Star Formation*, Proc. of the IAU Symp. No 147, E. Falgarone ed., Kluwer Acad. Publ., in press.

Palla, F., Stahler, S.W.: 1990, *Astrophys. J. Lett.*, **360**, L47.

Palla, F., Stahler, S.W.: 1991, *Astrophys. J.*, in press.

Praderie, F., Simon, T., Catala, C., Boesgaard, R.M.: 1986, *Astrophys. J.*, **303**, 311.

Stahler, S.W.: 1988, *Astrophys. J.*, **332**, 804.

Vigneron, C., Mangeney, A., Catala, C., Schatzman, E.: 1990, Proc. of the Conf. *Inside the Sun*, Solar Phys., in press.

SMALL–SCALE STRUCTURE AND ANGULAR MOMENTUM TRANSFER IN PROTOSTELLAR ENVIRONMENTS.

V. Antonuccio–Delogu
NORDITA, Copenhagen, DENMARK, and
CNR–GNA, Unita' di Ricerca di Catania, ITALY

F. Atrio–Barandela
NORDITA, Copenhagen, DENMARK

Abstract.

Stochastic fluctuations of the gravitational field in a turbulent, collapsing protostellar clouds are induced by the population of small, high density eddies of the turbulent cascade. We compute the probability distribution of the stochastic force and angular momentum fields induced by these eddies, assuming that the density and velocity correlation functions can be described as power laws. We also show that a point mass moving faster than the velocity of the eddies' system feels a net positive torque, and so its net angular momentum increases: the opposite happens to a particle moving with a velocity slower than the average. This mechanism of angular momentum transfer can be significant in the conditions observed in Molecular Clouds.

1 Introduction

Recent observations support the view that some star–forming environments in our Galaxy can be made of a highly inhomogeneous gas, fragmented into small units probably by turbulent and/or gravitational fragmentation processes. For example, Falgarone and Perault (1988) have observed structures at undercritical density (i.e. below the contraction density) in a nearby Molecular Cloud, down to a scale of 0.02 pc. They conclude that these structures are transients, probably originated from a turbulent state of the gas, and that this situation can be much more frequent in Molecular Clouds than previously suspected (Falgarone and Puget, 1988).

These considerations are of interest to understand the dynamical evolution of a star–forming cloud, which is clearly influenced by the amount of substructure present. The gravitational field inside a collapsing protostar can be decomposed into the sum of two components:

$$\mathbf{F} = \mathbf{F}_{reg}(\mathbf{r}, t) + \mathbf{F}_{stoch}(\mathbf{r}, t) \tag{1}$$

where by $\mathbf{F}_{stoch}$ we have denoted the component of the force induced by randomly distributed, "point–like" objects, while $\mathbf{F}_{reg}(\mathbf{r}, t)$ denotes the contribution from the diffuse

S. Catalano and J. R. Stauffer (eds.), Angular Momentum Evolution of Young Stars, 27–31.

component of the density perturbation. By "point–like" objects we mean those small–scale density irregularities producing a gravitational field which can well be approximated by a Newtonian one at the typical distance of $N(r)^{-1/3}$, where n(r) is the number density. A precise definition requires the introduction of a characteristic length λ over which the density contrast w.r.t. the background varies significantly: $\delta\rho/\rho_{bg} \simeq 4 \div 10$ over the length-scale λ. The regular field $\mathbf{F}_{reg}(\mathbf{r}, t)$ is determined by a proper solution of the set of the fluid equations, constitutive relations and Poisson equation.
Chandrasekhar and Von Neumann (1942) have computed the probability distribution of the force field induced by a random uniform distribution of point–like objects. Later Kandrup (1980) generalized his analysis to the case of a power–law density distribution of the objects. In this work, we will extend their analysis to a **clustered** system of objects characterized by an average density $\rho_0(|\ \mathbf{r}\ |)$ and a correlation function $\xi(|\ \mathbf{r}\ |)$. This means that the probability of finding an object at a given position r is given by:

$$p(r) = \rho_0(r)\,[1 + \xi(r)] \tag{2}$$

This modification is necessary to deal with the structures formed in turbulent, magnetized, self–gravitating flows as those which are thought to be present in Giant Molecular Clouds (see, e.g. Henriksen and Turner (1984)). In what follows we will present the main results of our analysis, leaving to a forthcoming paper the mathematical details of their derivation (Antonuccio–Delogu and Atrio–Barandela, 1990).

2 Force and angular momentum probability distributions.

For brevity, we will use the term "object" for "point–like object". We will suppose that the objects are randomly distributed with an average density $\rho_0(r)$ and a correlation function $\xi(r)$, and generate a stochastic force field with a probability distribution W(**F**). This latter function gives the probability that a test point mass, randomly displaced in space, feels a force **F** from the population of objects. One can then show that this probability is given by the following expression:

$$W(\mathbf{F}) = \frac{2F}{\pi}\int_o^\infty dk \sin(kF) A(k) \tag{3}$$

The functional form of A(k) depends on the particular functional form of $\rho_0(r)$ and $\xi(r)$. Assuming a power law distribution for the density:

$$\rho_0(r) = \frac{c}{r^p} \tag{4}$$

we get:

$$A(k) = A_* \cdot \left\{1 + \frac{1}{2}\frac{\Sigma(k)}{A_{(2)}(k)}\right\} \tag{5}$$

where one has defined:

$$A_* = \left(\frac{4}{15}\right)(2\pi G)^{3/2}\langle m^{3/2}\rangle \rho_0(r), \tag{6}$$

$$\Sigma(k) = \langle m\rangle_p^{(3-p)} \int d\mathbf{r}_1 \int d\mathbf{r}_2 \exp\left[i \cdot \mathbf{k}\mathbf{F}(\mathbf{r}_1)\right] \exp\left[i \cdot \mathbf{k}\mathbf{F}(\mathbf{r}_2)\right] \xi(|\ \mathbf{r}_2 - \mathbf{r}_1\ |)\rho(\mathbf{r}_1)\rho(\mathbf{r}_2) \tag{7}$$

$$A_{(2)}(k) = A^2_{(1)}, A_{(1)} = \frac{3-p}{4\pi R_1^{3-p}} \int_0^{R_1} dr d\Omega r^{2-p} \exp(i\mathbf{k}\cdot\mathbf{F}) \tag{8}$$

(Antonuccio—Delogu and Atrio–Barandela, 1990).The main difference between this formula and those by Chandrasekhar and Von Neumann (1942) and Kandrup (1980) is the second term in the curly brackets in eq.(5) above, which takes into account the fact that the medium is clustered. The integral in eq. (6) can be computed by a stationary phase method, for a power law correlation function, while eq. (7) can be computed exactly (Chandrasekhar and Von Neumann, 1942). As is clear from Fig. 1), the introduction of a clustering changes the behaviour of the probability density, which is now increasing with F at large enough F, while the original Holtsmark's distribution (solid curve) decreases as $F^{-5/2}$ for large F. This behaviour has a simple physical explanation: As Kandrup (1980) and Padmanahban (1990) have shown, the main contribution to the stochastic field at large F comes from random objects at a typical distance of the order of the average interparticle distance $n(r)^{-1/3}$. In a clustered medium, the probability of having a nearby object is enhanced, due to the correlations, so there is a larger probability of having a larger fluctuation in the force.
One can also compute the probability distribution of the torques induced by this stochastic field:

$$W(T) = \frac{1}{2\pi^2} \int_o^{\pi/2} d\theta \sin\theta \int_o^{\infty} dk k^2 A(kR\sin\theta) \cdot \left[1 + \frac{\Sigma(kR\sin\theta)}{2A^2_{(1)}(kR\sin\theta)}\right] J_o(k \mid \mathbf{T} \mid \sin\theta) \tag{9}$$

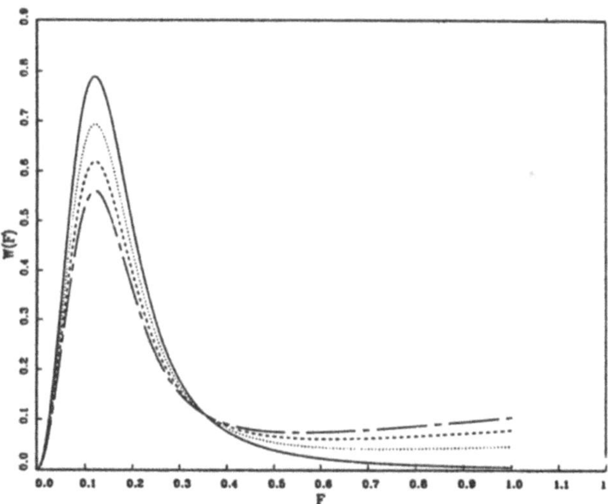

Figure 1. Probability distribution of the force F experienced by a random test particle, for an n= -0.5 power spectrum of the correlation function. The force is in units of $F_{max} \equiv G\langle m\rangle^{-2/3}$. The continuous solid curve is for an uncorrelated system. Dotted, dashed and long–dashed curves are for progressively increasing values of the normalization constant of the correlation functionξ_0. The values are separated by $\Delta\xi_0 = 10^{-1}$.

With these formulas it is possible to compute various average quantities, like the average angular momentum exchanged with the stochastic field by a given mass. One can see that only point masses in motion w.r.t. the system of objects can exchange angular momentum with the population of objects. This becomes evident when one takes into account the characteristic time of a stochastic fluctuation (Chandrasekhar and Von Neumann, 1943), which is given by:

$$t_F \equiv \frac{|\mathbf{F}|}{\langle | \frac{d\mathbf{F}}{dt} |^2 \rangle^{1/2}} = \frac{|\mathbf{F}|}{\frac{2}{3}\pi G \langle m \rangle \rho_0 B' \left(\frac{|\mathbf{F}|}{Q_H} \right) \left[| \mathbf{V}_p^2 | + \frac{3(\mathbf{V}_p \cdot \mathbf{F})^2}{\mathbf{F}^2} \right]} \tag{10}$$

Here $|\mathbf{V}_p|$ is the relative velocity between the test mass and the population of objects and B' is a function whose value is of order unity. Suppose now that the objects have a Maxwellian distribution of velocities. In the reference frame of the test mass the distribution of velocities of the objects will be:

$$f(\mathbf{r}, \mathbf{v}_*) = \rho_0 j^3 e^{-j^2 (\mathrm{f}\, \mathbf{v}_* + \mathbf{V}_p)^2}, \tag{11}$$

so the test star will "see" more objects coming towards it than leaving farther. These objects have a lesser interaction time, so the test mass will finally reach an equilibrium state in which its relative velocity $\mathbf{V}_p = 0$. We refer to another paper for a detailed discussion of this mechanism and for examples (Antonuccio–Delogu and Atrio–Barandela, (1990)).

3 Discussion.

The stochastic field induced by transient features in a region of turbulence and/or fragmentation can be effective in readjusting the original distribution of angular momentum, through a mechanism of purely gravitational interaction. In gaseous environments, like in a Giant Molecular Cloud, viscous transfer of angular momentum can be effective if the density is sufficiently high: the mechanism presented in this contribution, on the other hand, can be effective in low density environments in which a substantial amount of small scale structure is present, a situation which has been recently shown to be possible. This allows us to study the angular momentum transfer during early phases of contraction of a protostellar cloud (Antonuccio–Delogu, in preparation).

References.

Antonuccio–Delogu, V., and Atrio–Barandela, F. (1990) "Stochastic gravitational fields in clustered systems", submitted to *Astrophys. J. Letters.*

Chandrasekhar, S. and Von Neumann, J. (1942),"The statistics of the gravitational field arising from a random distribution of stars. I", *Astrophys. J.* **95**, 489.

Chandrasekhar, S. and Von Neumann, J. (1943), "The statistics of the gravitational field arising from a random distribution of stars. II", *Astrophys. J.* **97**, 1.

Falgarone, E. and Perault (1988), "Structure at the 0.02 pc scale in the molecular gas of low H_2 column density" ,*Astron. Astrophys.* **205**, L1.

Falgarone, E. and Pudget, J.L. (1988) "Long-term support and gravitational stability of molecular clouds", *in* R.E. Pudrits and M. Fichs (eds.), *Galactic and Extra-galactic Star Formation*, Kluwer Academic Publishers, Dordrecht, p. 195–213.

Henriksen, R.N. and Turner, B.E. (1984), "Star cloud turbulence", *Astrophys. J.* **287**, 200.

Kandrup, H. (1980), "Stochastic gravitational field fluctuations in a self–consistent mean field theory", *Phys. Rep.* **63**, 1.

Padmanabahn, T. (1990), "Statistical mechanics of gravitating systems", *Phys. Rep.* **188**, 285.

EKMAN PUMPING IN A ROTATING PROTOSTAR

Daniele GALLI[1,2] and Frank H. SHU[2]
[1] *Osservatorio Astrofisico di Arcetri, Firenze*
[2] *Astronomy Department, University of California at Berkeley*

ABSTRACT. We consider the accretion phase during the formation of a (low mass) protostar. The *spin–up* resulting from accretion through the adjoining protostellar disk generates a meridional circulation (Ekman pumping) inside the protostar directed from the rotational poles to the equatorial plane. The adoption of appropriate analytical techniques allows us to calculate the velocity field in the protostellar interior to the lowest order in a perturbative approach. The consequences for the evolution of an eventual protostellar magnetic field and of a magnetically coupled protostellar wind in the presence of this circulation are briefly discussed.

1. Bipolar outflows from young stellar objects: the dilemma

The two main lines of thought with regard to models of magnetic protostellar winds have been outlined by Pudritz and Norman (1983) and by Shu *et al.* (1988). In the first model, the wind is driven by a large and massive circumstellar disk threadened by an outwards pointing magnetic field, which allows the centrifugal ejection of any material particle attached to it. In the second interpretation, a *stellar* magnetic wind emanates from the cusps of a protostar which is rotating at *break–up*, at which centrifugal and gravitational forces balances exactly.

Without entering into the details of the models, one should note that the possibility of centrifugally or magnetohydrodynamically driven flows from large and massive circumstellar disks has severe mechanistic drawbacks: the requirements of disk accretion rate, disk masses, mass–to–radius ratios, magnetic field strength and geometry are not substantiated by known properties of star forming regions. On the other hand, a strong magnetic field with a component directed perpendicular to the rotation axis is not a problem for a (proto)star. However, T Tauri stars, at least, do not possess the requisite *break–up* rotation speeds.

2. Compromise solution: Ekman pumping?

A compromise solution might be represented by the mechanism described in the following sections. The basic process, Ekman pumping, utilizes the effects of spin-up induced by accretion from a circumstellar disk to drive an internal circulation in the protostar that can react on the stellar magnetic field, establishing the appropriate conditions for the onset of a wind.

Here, we limit ourselves to consider the problem of determining the internal motions in a convective protostar during the accretion phase of low mass star formation (Shu, Adams and Lizano 1987). The calculations of the effects of these large scale motions on the stellar magnetic field is currently under study by the authors.

S. Catalano and J. R. Stauffer (eds.), Angular Momentum Evolution of Young Stars, 33–40.

3. The basic state

During the phase of spherical accretion, the radiative interior of the protostar is likely to rotate *differentially*, and since material of low angular momentum falls in first, the protostar is built up with a mechanically stable stratification (a distribution of angular momentum which increases outward). When deuterium ignites, at a temperature $T \simeq 10^6$ K, convection does occur in a major way and bouyant motions across shearing layers would probably try to bring the material in the convection zone in a state of more nearly uniform rotation.

If this is the correct physical picture, we can model a protostar after the onset of deuterium burning as a $n = 3/2$ polytrope uniformly rotating (with angular velocity Ω_0, say) and to consider the effects of the slow increase of angular momentum of the protostar produced by disk *spin-up* as a perturbation to this basic state.

In fact, if dM_*/dt is the protostar accretion rate through the disk, the increase in angular velocity of the protostar (assuming the disk to be in keplerian rotation) is

$$\dot{\Omega}_0 \simeq \frac{1}{I}\frac{dJ_*}{dt} \simeq 1.0 \times 10^{-16} \left(\frac{\dot{M}_*}{10^{-6} M_\odot \mathrm{yr}^{-1}}\right) \left(\frac{M_*}{M_\odot}\right)^{1/2} \left(\frac{R_*}{R_\odot}\right)^{-3/2} \mathrm{s}^{-2},$$

where I is the momentum of inertia ($\simeq 0.6\ M_* R_*^2$ for a $n = 3/2$ polytrope) and J_* the angular momentum of the protostar. Inserting typical values ($dM_*/dt \simeq 10^{-6}\ \mathrm{M}_\odot\ \mathrm{yr}^{-1}$, $M_* \simeq 1\ M_\odot$, $R_* \simeq 3\ R_\odot$), the resulting angular acceleration is

$$\dot{\Omega}_0 \simeq 10^{-17}\,\mathrm{s}^{-2}.$$

If the protostar's angular velocity is of the order or less than the critical (*break-up*) velocity ($\Omega_c \simeq 10^{-4}\ \mathrm{s}^{-1}$, for the values of the stellar parameters quoted above), the effects of $d\Omega_0/dt$ can be considered a small perturbation to the state of uniform rotation with angular velocity Ω_0 ($d\Omega_0/dt << \Omega_0^2$).

In the following, we shall assume axial symmetry and neglect, in this first approach, the deviation from sphericity of the star introduced by rotation.

4. The linear perturbation

In a convective structure, momentum transport due to irregular fluctuations in the fluid velocity (i.e. "turbulence") is often represented by an *eddy viscosity*, and the Navier–Stokes equations are used to describe the *mean* turbulent flow. Assuming the Reynolds stress tensor $\overleftrightarrow{\Pi}$ of the form

$$\left\{\overleftrightarrow{\Pi}\right\}_{ij} = \nu_t \left[\left(\frac{\partial U_i}{\partial x_j} + \frac{\partial U_j}{\partial x_i} - \frac{2}{3}\delta_{ij}\frac{\partial U_s}{\partial x_s}\right) + \delta_{ij}\frac{\partial U_s}{\partial x_s}\right],$$

where ν_t is the coefficient of eddy viscosity, the basic hydrodynamic equations in an inertial reference frame are:

- the equation of continuity

$$\frac{\partial \rho}{\partial t} + \nabla \cdot (\rho \mathbf{U}) = 0,$$

- the Navier–Stokes equations

$$\frac{\partial \mathbf{U}}{\partial t} + (\mathbf{U} \cdot \nabla)\mathbf{U} = -\frac{1}{\rho}\nabla p - \nabla V + \frac{1}{\rho}\nabla \cdot \overleftrightarrow{\mathbf{\Pi}},$$

- the Poisson equation

$$\nabla^2 V = 4\pi G \rho.$$

- the polytropic equation

$$p = K\rho^{1+\frac{1}{n}}.$$

The coefficient of eddy viscosity ν_t is many orders of magnitude larger than the coefficient of ordinary (atomic or molecular) viscosity, typically

$$\nu_t \simeq 10^{14}\ \mathrm{cm}^2\ \mathrm{s}^{-1}.$$

After non–dimensionalization of the above equations, the problem is seen to contain only one natural (small) non–dimensional parameter, the *Ekman number* ϵ, defined as

$$\epsilon \equiv \frac{\nu_t}{\Omega_0 R_*^2} \simeq 10^{-4}.$$

This suggests the viability of a perturbative approach to the solution of the hydrodynamical equations in the present context.

5. Solution of the perturbed equations

The solution of the perturbed equations is greatly simplified by the use of two analytical techniques:

- Separation of variables by expansion of the angular part of the velocity field in *vector* spherical harmonics $\mathbf{Y}_{l,0}^{(-1)}(\theta)$, $\mathbf{Y}_{l,0}^{(1)}(\theta)$ and $\mathbf{Y}_{l,0}^{(0)}(\theta)$ (the photon wavefunctions in QED, see Landau and Lifshitz [1977] for a definition). Vector spherical harmonics allow to work directly on the hydrodynamical equations in the much more compact vector form instead of expanding their complicated scalar components.

- Integration of the set of ordinary differential equations by the technique of *matched asymptotic expansions*. Matched asymptotic expansions are required by the *boundary layer* character of the velocity field, as suggested by the presence of the small parameter ϵ multiplying the highest derivatives of the velocity. In fact, a "viscous" boundary layer will, in general, be formed at any limiting surface at which the boundary conditions are not satisfied exactly by the velocity field derived from the "inviscid" fluid equations.

Assuming stress–free boundary conditions at the surface, the "inviscid" solution obtained by putting $\epsilon = 0$ provides a good description of meridional streaming in the main bulk of the protostar's interior except near the surface boundary, where "viscous" effects are significant only in a thin layer adjoining the surface, the thickness of which (δ_E, say) is small compared to the protostellar radius. The thickness of this surface boundary layer (Ekman layer) can be found by a suitable scaling of the variables near the surface boundary, and comes out as

$$\delta_E \sim \left(\frac{\nu_t}{\Omega_0}\right)^{1/2} \simeq 10^{-2}\ R_*.$$

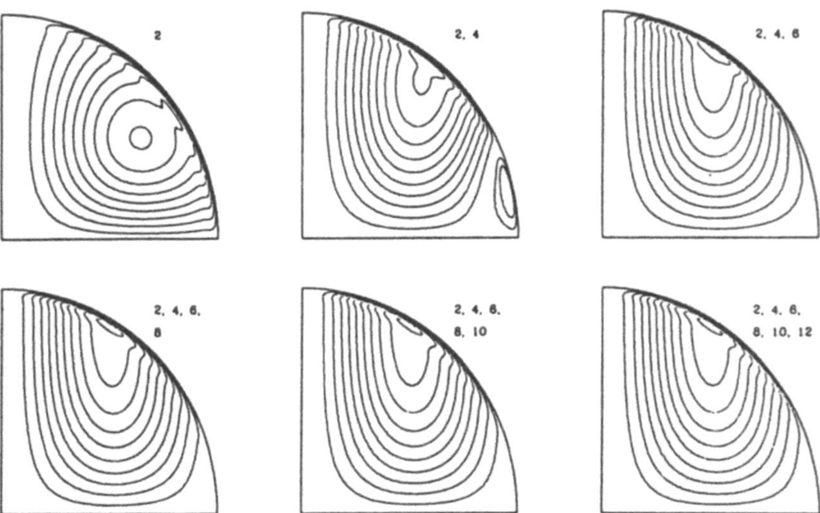

Fig. 1 - Streamlines of meridional circulation assuming $\nu_t = 10^{14}$ cm^2 s^{-1}. The expansion of the angular component of the velocity field in vector spherical harmonics is truncated at $l_{max} = 2$, $l_{max} = 4, \ldots, l_{max} = 12$.

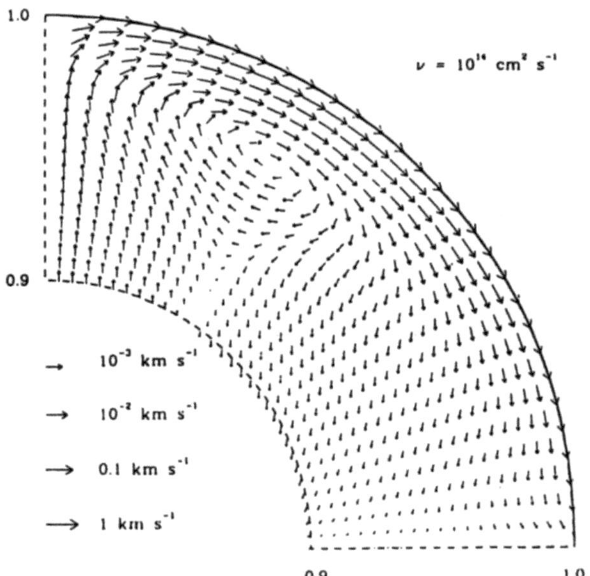

Fig. 2 - Particular of the subsurface region for $\nu_t = 10^{14}$ cm^2 s^{-1} and $l_{max} = 12$. Note that the Figure is not on scale and the arrows represent the velocity field on a *logarithmic* scale.

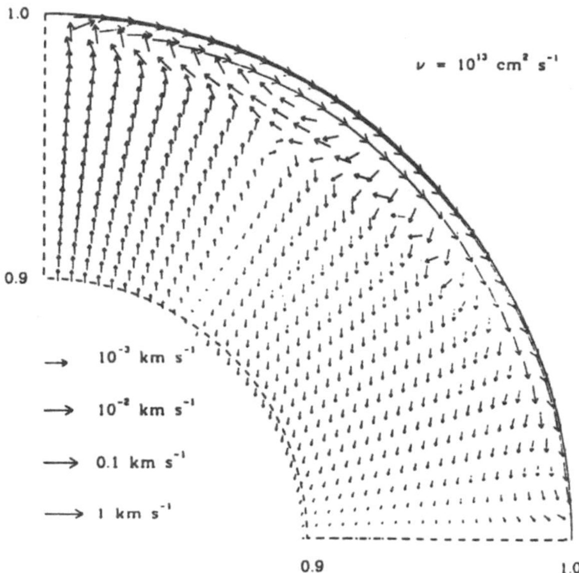

Fig. 3 - **As in Fig. 2 for $\nu_t = 10^{13}$ cm^2 s^{-1}.**

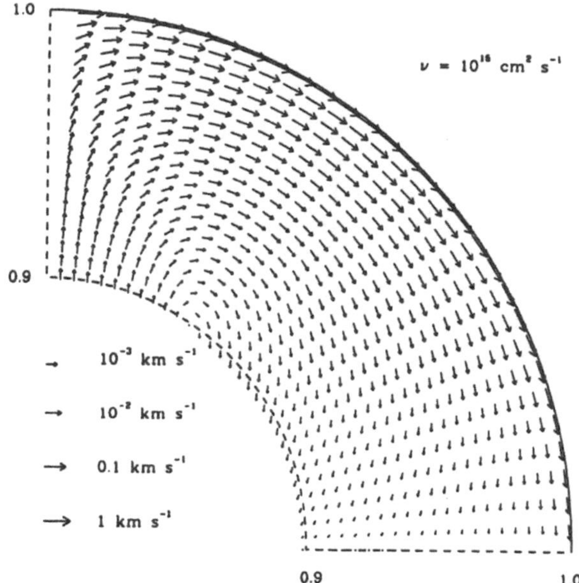

Fig. 4 - **As in Fig. 3 for $\nu_t = 10^{15}$ cm^2 s^{-1}.**

		interior		*boundary layer*	
ν_t $cm^2\ s^{-1}$	δ_E km	V_c $cm\ s^{-1}$	B_e Gauss	V_c $cm\ s^{-1}$	B_e Gauss
10(13)	1(4)	2(−2)	3(−2)	4(2)	1(1)
10(14)	3(4)	2(−2)	3(−1)	4(2)	3(1)
10(15)	9(4)	2(0)	3(0)	3(2)	5(1)

Table 1: Relevant characteristics of the Ekman circulation. See description in the text.

6. Results and discussions

The streamlines of meridional circulation in a $n = 3/2$ polytrope with $\nu_t = 10^{15}$ $cm^2\ s^{-1}$ and $\Omega_0 = 0.1\ \Omega_c$ are displayed in Fig. 1. Each figure is labelled with the index l of the angular expansion in vector spherical harmonics. For $l = 2, \ldots, 12$ the convergence of the series solution is better than 1%.

A blown–out view of the subsurface region (ranging from $\frac{9}{10} R_*$ to R_*) is shown in Figs. 2, 3 and 4 for three different values of the coefficient of eddy viscosity. Note that the figures are not in scale and the velocity field is represented on a logarithmic scale.

The relevant characteristics of the meridional circulation are summarized in Table 1 for different values of the coefficient of eddy viscosity ν_t. The quantities shown are: (i) the depth of the surface boundary layer δ_E, (ii) the velocity of the circulation V_c at half stellar radius and at the bottom of the boundary layer, and (iii) the *equivalent* magnetic field strenght B_e obtained from the formula $B_e = \sqrt{4\pi\rho_0}\, V_c$ in the same regions.

In the case of a perfectly conducting star, the relationship between magnetic field and meridional circulation is expressed by the condition of frozen field $\mathbf{B}_p = \beta\rho_0\mathbf{U}$, where $\mathbf{B}_p$ is the poloidal component of the magnetic field and β is a scalar dependent in general on the position. As we have found in our analysis, the circulation has the quadrupole as the primary component, plus harmonics of even parity. On the basis of the frozen field condition alone, we should therefore expect that the internal magnetic field also possesses a primary quadrupolar component. For example, if the star had an initial magnetic field with dipole symmetry or containing other harmonics of odd parity, the eminently quadrupolar mass flow will not satisfy the frozen field condition and the field will be progressively distorted and eventually submerged by the Ekman circulation, leaving the quadrupole as the dominant component. Such a quadrupolar magnetic field could have field lines emerging from the equatorial region (possibly with enhancement of flux density), a configuration favorable to the onset of a magneto–centrifugally driven protostellar wind, according to the model proposed by Shu *et al.* (1988). Clearly, the infinite conductivity approximation fails very near to the protostellar surface, where the temperature decreases rapidly, allowing the lines of force to cross the streamlines and emerge from the surface. In a more realistic picture, once the fluid does begin to circulate a dynamo mechanism should operate that amplify magnetic fields (Parker 1979), which can eventually pump energy into driving a stellar wind and coincidentally account for the high levels of surface activity that characterize, for example, the atmospheres of T Tauri stars. A complete magneto–hydrodynamical analysis of the problem is currently under study by Galli and Shu (1990).

References

Galli, D. and Shu, F. H. (1990) *in preparation.*

Landau, L. D. and Lifshitz, E. M. (1977) *Quantum Electrodynamics*, Pergamon Press, Oxford.

Parker (1979) *Cosmical Magnetic Fields*, Oxford Univ. Press, Oxford.

Pudritz R. E. and Norman C. A. (1983) *Astrophys. J.*, **274**, 677.

Shu, F. H., Adams, F. C. and Lizano, S. (1987) *Ann. Rev. Astr. Astrophys.*, **25**, 23.

Shu, F. H., Lizano, S., Ruden, S. P. and Najita, J. (1988) *Astrophys. J. Lett.*, **328**, L19.

DISCUSSION

Gough: What was the boundary condition which you applied at the surface where you were accreting from the infalling material. Was it a point?

Galli: Just at this point here. It is a point like condition. But, we have stress-free conditions here.

Gough: So, isn't it surprising that you get apparent convergence with so few terms?

Galli: We get convergence up to twelve harmonics. I can show you the specifics.

Gough: Yes, I know, you showed that. I still do not understand. Is your little gamma 5/3?

Galli: Yes.

Gough: I don't understand why the circulation can go on a timescale of 10 years if you have such a strong buoyancy force. If the gamma for the gas is 5/3, you have a strong buoyancy force unless you have a polytrope of index 1.5. This goes totally against expectation.

Vauclair: What is the temperature of the interior of the stars which you study? I mean the central temperature.

Galli: One million degrees.

Vauclair: I am surprised that you do not have the centrifugal potential in your Poisson equation (at the surface of the star).

Galli: I just did not write it. I included Poisson's equation just to complete the set of equations, but I do not solve it.

ROTATION IN PRE-MAIN SEQUENCE STARS: PROPERTIES AND EVOLUTION

J. BOUVIER
Canada-France-Hawaii Telescope Corporation
P.O. Box 1597, Kamuela, HI 96743

ABSTRACT. The rotational properties of pre-main sequence stars with a mass between 0.5 and 9 $M_{\odot}$ are reviewed. The impact of binarity and disk accretion upon early rotational evolution is discussed and unsolved problems related to the evolution of stellar angular momentum prior to the main sequence are outlined.

1. Introduction

At the beginning of the last decade, very little was known about the rotation of pre-main sequence stars. The reason is that these stars are relatively faint and rotate slowly, so that it was very difficult to obtain good enough high-resolution spectrograms needed to measure the small rotational broadening of the line profiles. As more efficient detectors became available, this difficulty was overcome and, in the last 5 years, the rotational velocities of more than 150 pre-main sequence stars could be measured. This provides a large enough database to start investigating the rotational properties of pre-main sequence stars on a statistical basis.

At the same time, new concepts emerged regarding the physical properties of young stars (see the review by Bertout 1989). The major advance in our understanding of young stars was to recognize that the majority of pre-main sequence stars are surrounded by circumstellar disks which may or may not interact with the star. The evidence for accretion disks is well-documented for "classical" T Tauri stars (CTTS), a class of low-mass pre-main sequence stars ($M < 2\ M_{\odot}$) characterized by a strong emission-line spectrum (e.g., EW(Hα) > 10 Å, and up to 200 Å) and continuum emission at UV and IR wavelengths in excess of photospheric levels. Most of the peculiar properties of CTTS are now thought to result from the interaction between the star and its circumstellar disk. In the same mass range, "weak-line" T Tauri stars (WTTS) only exhibit faint Hα emission (EW(Hα)< 10 Å), show no UV excess and have small or no near-IR excess. Although their properties can usually be interpreted as the result of solar-type magnetic activity alone, a number of WTTS have recently been detected at far-IR and sub-mm wavelengths, wich indicates the presence of cool circumstellar material far from the stellar surface (Strom et al. 1989, Beckwith et al. 1990). Hence, even some WTTS may possess circumstellar disks which, however, do not extend down to the stellar photosphere. It is only very recently that circumstellar disks have been

S. Catalano and J. R. Stauffer (eds.), Angular Momentum Evolution of Young Stars, 41–62.

searched for around more massive pre-main sequence stars, the so-called Herbig Ae-Be stars ($2 < M_* < 9\ M_\odot$). Clues for the existence of massive circumstellar disks around these stars have been obtained from IR and sub-mm observations (see Strom et al., this volume). Hence, the occurence of circumstellar disks with a size ranging from a few hundred to a few thousand A.U. and a mass between 10^{-3} and $1\ M_\odot$ appears to be a widespread phenomenon in pre-main sequence stars.

Recent reviews dealing with rotation in pre-main sequence stars have been published by Bouvier (1990a) and by Stauffer and Soderblom (1990). The present review includes the most recent measurements of rotation in pre-main sequence stars (Bouvier et al. 1991) to provide an updated account of their rotational properties (Section II), and concentrates on new developments related to the impact of binarity (Section III) and disk accretion (Section IV) upon early rotational evolution. Finally, unsolved problems regarding the evolution of stellar angular momentum from the T Tauri phase to the ZAMS are outlined in Section V.

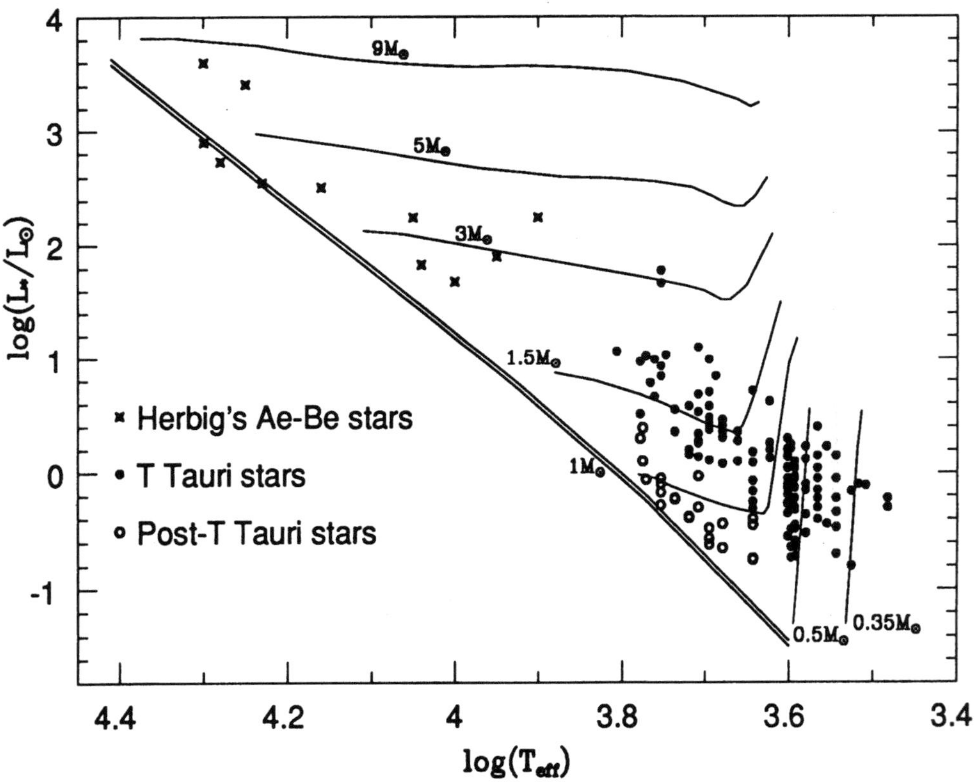

Figure 1. The location of pre-main sequence stars with known rotational velocities in the H-R diagram. Theoretical pms evolutionary tracks for stars with a mass between $0.35 M_\odot$ and $9 M_\odot$ are shown as solid curves. The theoretical zero-age main sequence is shown as a double line.

2. The rotational properties of pre-main sequence stars

Figure 1 shows the location of pre-main sequence stars with known rotational velocities in the H-R diagram. T Tauri stars, with an age less than 10^7 years, are shown as black dots, while the so-called "post-T Tauri" stars, older than 10^7 years, are shown as white dots. Ae-Be Herbig stars have an age comparable to that of T Tauri stars but lie closer to the main sequence since, being more massive, their contraction timescale is shorter. For most of the stars in Figure 1, the stellar rotational velocity projected onto the line of sight, *vsini*, was derived from the Doppler broadening of photospheric line profiles. In addition, rotational periods were directly measured for 34 T Tauri stars from the modulation of their luminosity by starspots (e.g., Bouvier and Bertout 1989, Vrba et al. 1989). Because rotational periods are not affected by random projection effects they provide a much more accurate measurement of rotation than does *vsini*, and are therefore most useful to investigate subtle rotational effects in a small stellar sample. In contrast, projected velocities have been measured for many more stars, which makes them very valuable to study the rotational properties of young stars on statistical grounds. Before the distribution of projected velocities can be confidently used to characterize the rotational properties of young stars, however, one must check whether it correctly describes the distribution of true equatorial velocities, i.e., that the geometric term *sini* does not introduce any systematic bias in the observed *vsini* distribution. In addition, since the stars included in the rotational database belong to different stellar formation regions, one is concerned with the possible variation of rotational properties from one molecular cloud to the other. These two issues are addressed below before the rotational properties of pre-main sequence stars are discussed.

2.1. STELLAR ROTATION IN VARIOUS STAR-FORMING REGIONS

Figure 2 shows the distribution of projected velocities for low- and intermediate-mass TTS belonging to various stellar formation regions. The majority of low-mass TTS with known *vsini* belong to the Taurus-Auriga region, and the small number of such stars in other molecular clouds makes difficult a detailed comparison. Nevertheless, it is seen that low-mass TTS span a range of *vsini* from less than 10 to 30 km s^{-1}, with no indication for substantial differences from one star forming region to the other. More massive TTS with known *vsini* are more evenly distributed in the Taurus-Auriga, Orion, and Chameleon dark clouds. Statistical tests indicate similar *vsini* distributions for stars in the Taurus-Auriga and Orion regions, as well as for stars in the Orion and Chameleon regions, at a significance level of 63%. The rotational properties of young stars thus seem to be fairly universal in the sense that they do not strongly depend upon the star's birthplace. This result already suggests that the stellar rotational axes are most probably randomly orientated. Any preferential orientation in a given stellar formation region would result in a systematic bias in the *vsini* distribution, which is not observed.

2.2. THE ORIENTATION OF ROTATIONAL AXES

When both the projected velocity and the rotational period are known, the inclination of the stellar rotational axis on the line of sight can be calculated as:

$$sini = \frac{vsini}{v} = \frac{vsini \cdot P}{2\pi} \cdot \frac{1}{R_\star}.$$

The main source of uncertainty here comes from the determination of the stellar radius. So far, it has mainly been derived from the star's bolometric luminosity and effective temperature. However, because TTS have strong non-stellar continuum excesses at UV and IR wavelengths, the bolometric luminosity is often much larger than the stellar luminosity and this results in overestimated stellar radii. This is the reason why previous studies of axial inclination in TTS (e.g., Weaver 1987) led to anomalous distributions with an excess of low inclinations. Here, the *stellar* luminosities published in the literature (e.g., Strom et al. 1989) were used in order to derive the stellar radius, thus reducing any systematic bias connected with non-stellar continuum excesses. The resulting distribution of axial inclinations is shown in Figure 3 for 28 TTS with known projected velocity and rotational period, while the dashed histogram shows the expected distribution for randomly orientated rotational axes. A chi-square test indicates that the observed and expected distributions are the same at a 99% confidence level, even though a small deficiency of stars seen at high inclination is clearly apparent in Figure 3. A possible explanation for this deficiency is that a small error on *sini* leads to a large uncertainty on the axial inclination when

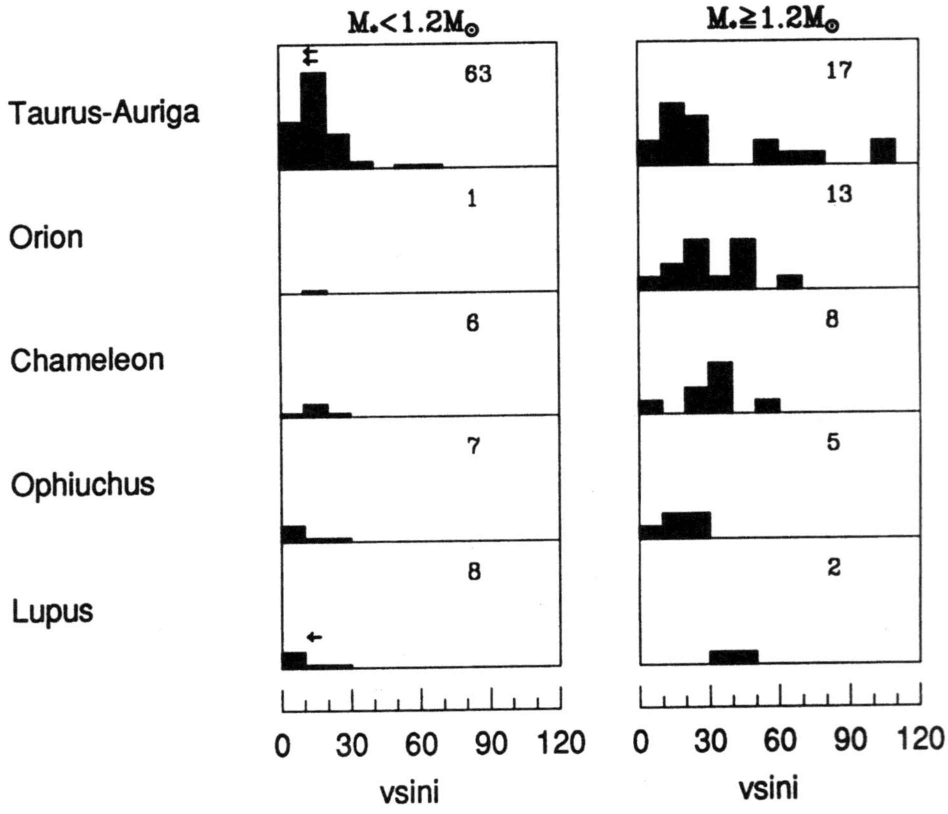

Figure 2. The *vsini* distributions of T Tauri stars in various star-forming regions. The number of stars is indicated in each panel. Horizontal arrows indicate *vsini* upper limits of 15 km s^{-1}.

sini is large (see Soderblom 1985). This is illustrated in Figure 3 where the size of the error bars on the axial inclination is seen to rapidly increase toward high inclinations (this is also the reason why a larger bin size has been chosen for high inclinations in Fig. 3). Also, stellar luminosity estimates for CTTS may still be slightly overestimated if continuum emission from the star/disk boundary layer contributes to the observed flux at optical wavelengths (Hartigan et al. 1990).

That the rotational axes of TTS are randomly orientated may appear somewhat surprising since it is widely believed that magnetic fields play an important dynamical role during the early evolution of molecular clouds. For instance, Heyer et al. (1987) found that the rotational axes of 5 low-density clouds in the Taurus molecular complex were aligned with the direction of the local magnetic field. Besides, polarization maps of dark clouds reveal large-scale patterns of aligned polarization vectors, which suggests that the direction of magnetic field lines remains fairly constant over distance scales of the order of 1pc (e.g., Goodman et al. 1990). Hence, if the magnetic field remains dynamically important during the whole star formation process, one would expect the angular momentum vector of protostars to be aligned along the preferential direction

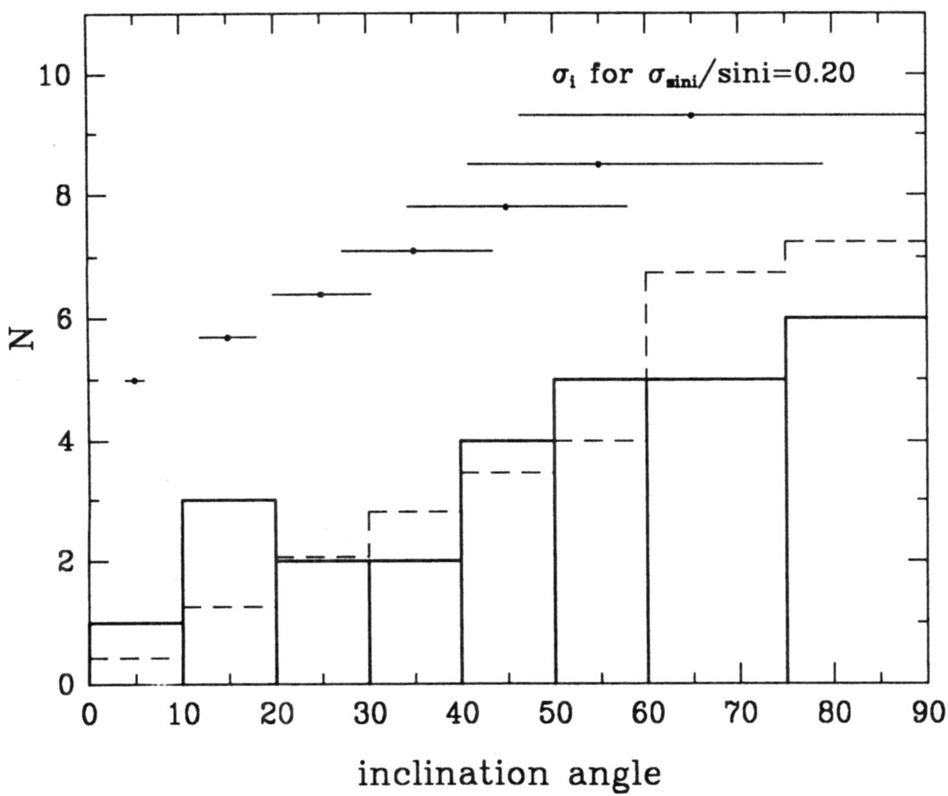

Figure 3. The distribution of axial inclinations for T Tauri stars. The dashed histogram represents the expected distribution for randomly orientated rotational axes. Error bars on the inclination angle are shown assuming a 20% uncertainty on *sini*.

of the cloud's magnetic field. Observations yield somewhat contrasting results on this issue. On one hand, collimated mass outflows associated with 5 young stellar objects in the Lynds 1641 cloud were found to be directed parallel to the orientation of the local magnetic field (Vrba et al. 1988). If mass outflows emerge through the polar regions of protostars, as is usually assumed, this would suggest that magnetic stresses dictated the orientation of the protostar's angular momentum vector. On the other hand, Heyer (1988) found that the rotational axes of dense, pre-stellar cloud cores in the Taurus complex are randomly orientated with respect to the ambient magnetic field direction. He concluded that the magnetic field ceases to be dynamically important before the protostellar collapse takes place and suggested that randomization of the rotational axes occurs at this point via gravitational interaction or physical collisions between cloud cores. An intriguing possibility in order to reconcile these results is to assume that the mass outflows associated with protostars are not systematically directed along the protostar's rotational axis. Alternatively, it may be that the magnetic field orientation on small distance scales is much more random than suggested by large-scale studies. Obviously, many more observations are needed to settle the role played by magnetic field during the protostellar collapse.

2.3. PRE-MAIN SEQUENCE ROTATION AS A FUNCTION OF STELLAR MASS

The distribution of projected velocities as a function of mass is shown in Figure 4 for T Tauri stars and Herbig Ae-Be stars. It is known since the pioneering work of Vogel and Kuhi (1981) that TTS with a mass larger than $1.5M_{\odot}$ tend to rotate faster than lower mass TTS, and Bouvier et al. (1986) suggested that the transition between slow and rapid rotation occurs at $M_{\star} \simeq 1.25M_{\odot}$. Figure 4 indeed suggests a smooth increase of mean rotation rate with mass among pre-main sequence stars: the average *vsini* is approximately 14 km s^{-1} in the mass range from 0.5 to 1 $M_{\odot}$, 30 km s^{-1} between 1 and 2 $M_{\odot}$, 50 km s^{-1} between 2 and 3 $M_{\odot}$, and 170 km s^{-1} for Herbig Ae-Be stars.

Most likely, part of the observed increase of mean rotation rate with mass is an evolutionary effect. Herbig Ae-Be stars are already near or on the main sequence and their rotation rates are similar to those of B0–B9 field dwarfs. Similarly, most TTS with a mass larger than $1M_{\odot}$ have already developed a large radiative core while lower-mass TTS are still mainly convective. If no significant angular momentum loss occurs, more massive TTS are thus expected to spin-up faster than low-mass ones, as a result of their shorter contraction timescale to the main sequence and because their radiative core develops earlier, thus leading to a rapid reduction of the stellar moment of inertia.

The observed dependence of rotation upon mass in pre-main sequence stars may also indicate that, on the average, more massive stars form with higher angular momentum. It is usually assumed that the relationship Kraft (1970) derived for massive field dwarfs, $J \sim M^{5/3}$ for $M_{\star} \geq 2M_{\odot}$, can be extrapolated to lower mass stars in order to estimate their initial angular momenta. As shown by Hartmann et al. (1986), angular momentum estimates for TTS are statistically consistent with the values expected from the extrapolation of Kraft's relationship to solar-mass stars. However, mass and radius estimates for pre-main sequence stars are still too uncertain to rule out slightly different relationships. Kraft's relationship has been recently re-interpreted by

Kawaler (1987) as indicating that the mean rotational velocities of high-mass stars are a constant fraction of break-up velocity, namely, <v> $\simeq$ 0.3 v(break-up). Obviously, this new relationship does not extend to lower mass stars since the mean rotational velocity of low-mass pre-main sequence stars is at least 10 times less than break-up velocity. Finally, it is worth noting that these relationships only deal with the *mean* rotation rates of stars. It is well-known, however, that massive main sequence stars exhibit a very large spread of rotation rates around the mean, whose origin is still poorly understood (see, e.g., Wolff, Edwards and Preston 1982).

Similarly, although the trend of increasing mean rotational velocity with mass is well-defined, TTS exhibit a large spread of *vsini* around the mean at any mass. That the observed spread cannot be entirely accounted for by observational errors or *sini* effects is best realized when considering rotational periods rather than projected velocities. Rotational periods are not affected by projection effects and are directly measured with an accuracy usually better than 10%. TTS with a mass less than $1M_\odot$ have rotational periods ranging from 1.5 to 8.5 days (but a large majority have rotational periods between 5 and 8.5 days), while more massive TTS have rotational periods between 1.2 and 6 days. The total spread in rotational periods at any mass thus amounts roughly

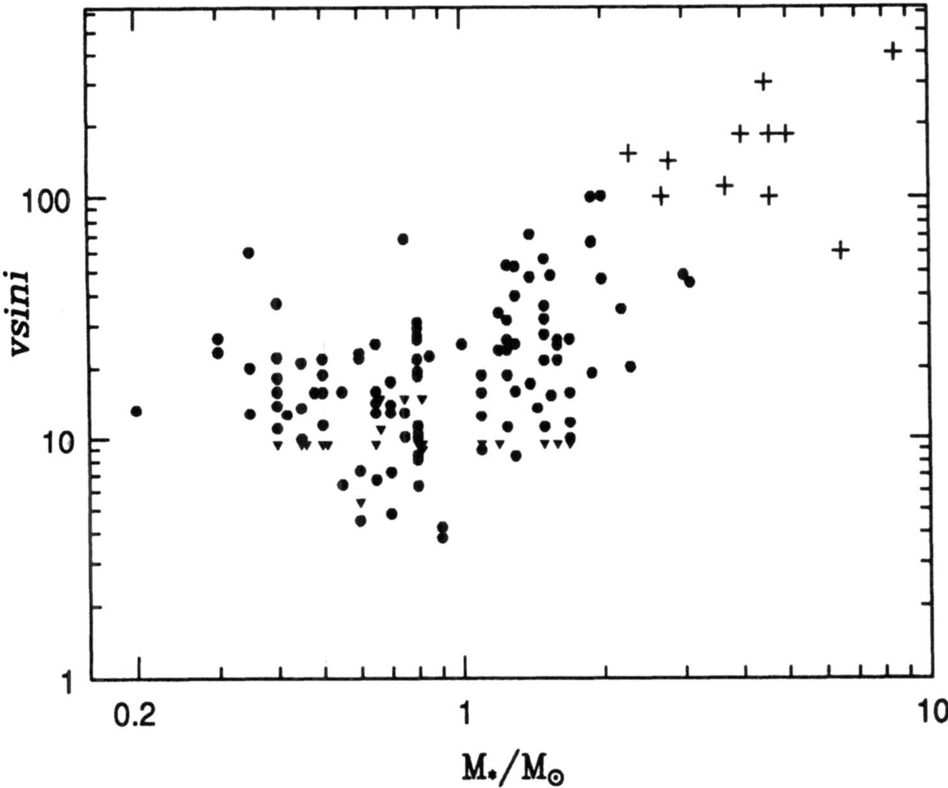

Figure 4. The dependence of *vsini* upon stellar mass for T Tauri stars (•) and Herbig Ae-Be stars (+). Inverted triangles indicate *vsini* upper limits.

a factor of 6. This estimate should actually be regarded as a lower limit to the true rotational dispersion since the lack of detected rotational periods longer than 10 days is most likely due to an observational selection bias. In turn, this result indicates that the intrinsic scatter of specific angular momenta at any mass in young stars amounts at least a factor of 2 and probably more. Such a scatter in initial stellar angular momenta may be expected to result from the complex evolution of angular momentum prior and during the protostellar collapse (see Bouvier 1990a for a more detailed discussion). Other conceivable physical mechanisms which may contribute to the observed spread are i) tidal interactions in close binary systems, and ii) accretion of high-angular momentum material from a circumstellar disk. These two issues are discussed in the next sections.

3. Rotation and binarity

Presumably, the evolution of stellar angular momentum may be quite different in single stars and in close binary systems. Zahn and Bouchet (1989) investigated theoretically the tidal evolution of close pre-main sequence binaries and showed that tidal effects are most important at this stage because pre-main sequence stars have larger radii than their main sequence counterparts and are mostly convective. On the observational side, major efforts have been made in the last few years to determine the frequency of spectroscopic binaries among pre-main sequence stars. So far, 11 such binaries have been found (Mathieu, Walter, and Myers 1989, and references therein), 9 of which are weak-line TTS, the 2 other being CTTS. According to Mathieu et al., the frequency of short-period binaries (P_{orb}< 100 days) is the same in WTTS as in Pop.I solar-mass stars, namely $\simeq$10%. The binary frequency among CTTS is still unknown but is suspected to be less. From the data at hand, it appears that binary systems with an orbital period shorter than 4 days have circular orbits, which suggests that their orbital and rotational motions are synchronized, while longer-period systems have highly eccentric orbits.

In order to investigate the impact of binarity upon rotation, the distributions of rotational periods for low- and intermediate-mass TTS are shown in Figure 5a and 5b, respectively. The corresponding average projected velocities, calculated from the rotational period assuming a $2R_{\odot}$ stellar radius, are given in the upper part of the figure. Known spectroscopic binaries are indicated either by "S", which stands for "synchronized", when the orbital period is within 10% of the rotational period, or by "NS" when the orbital period is much longer than the rotational period, and "?" means that the orbital elements are unknown. Before interpreting these histograms, one must note that there exists a strong observational bias against the detection of rotational periods longer than about 9 days. This is because most rotational modulation studies have been made over a timescale of 2 weeks at most, which is too short to detect modulation periods longer than about 9 days. The abrupt wall seen at that period in the distribution of low-mass TTS is most likely accounted for by this observational bias. In fact, many low-mass TTS have *vsini* less than 10 km s^{-1}, and are thus expected to have rotational periods significantly longer than this observational limit.

The most striking feature in the rotational period distribution of low-mass TTS is its bimodal shape, with an apparent deficiency of stars with a period around 4 days. Although the reality of the gap between stars with a period shorter and longer than 4 days may be questioned in regard to the small sample size, it remains that most low-mass TTS have rotational periods longer than 4 days (probably many more, in fact, than is apparent from Figure 5a owing to the observational bias discussed above). Interestingly enough, the 4-day period corresponds to the orbital period at which circularization of the orbit occurs. In other words, binary systems with an orbital period shorter than 4 days are expected to experience strong tidal effects which will tend to synchronize the orbital and rotational periods, thus enforcing rapid rotation. Indeed, among the 7 low-mass TTS with a rotational period shorter than 4 days, 2 are known to be synchronized binaries while another one may be so. These considerations suggest that most, if not all, low-mass TTS with a rotational period less than 4 days might be members of close, synchronized binary systems while slower rotators are either single stars or members of wider binary systems.

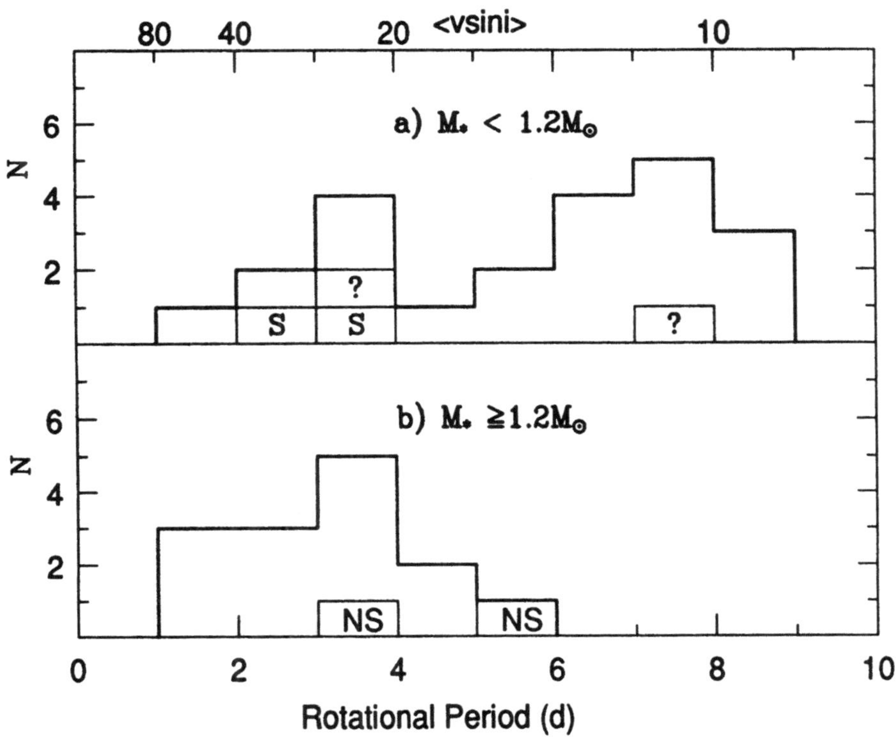

Figure 5. The distribution of rotational periods for low-mass and intermediate-mass T Tauri stars. Known spectroscopic binaries are indicated by "S" (synchronized), "NS" (non synchronized), or "?" (unknown orbital elements). The corresponding *vsini* scale, assuming $R_\star=2R_\odot$, is given in the upper part of the figure.

Such a dichotomy between the rotational periods of synchronized binaries and single stars would be much less apparent among TTS with a mass larger than 1.2 $M_\odot$. The reason is that most TTS in this mass range have *vsini* between 15 and 60 km s^{-1} corresponding to rotational periods between 1.5 and 6 days. Hence, most single stars are expected to have rotational periods similar to those of synchronized binaries, thus strongly decreasing the contrast between the two groups in the rotational period distribution. This is consistent with the observed rotational period distribution shown in Figure 5b for these stars. Note that the lack of TTS with a period longer than 6 days in this mass range is not an observational bias, since, in principle, rotational periods up to 9 days could be detected. Instead, this points to a real deficiency of relatively slow rotators among intermediate-mass TTS, which confirms the observed trend of increasing *vsini* with stellar mass in pre-main sequence stars.

4. Rotation and disk accretion

It is now well-established that CTTS, which accrete material from their Keplerian disk at a rate between 10^{-8} and 10^{-6} $M_\odot$ yr^{-1} (Kenyon and Hartmann 1987, Bertout, Basri and Bouvier 1988, Basri and Bertout 1989), and WTTS, which lack accretion disks, have similar rotation rates (Hartmann, Soderblom and Stauffer 1987, Hartmann and Stauffer 1989, Bouvier 1990a). This result first came as a surprise since Hartmann and Stauffer (1989) showed than, in the absence of angular momentum loss, a fully convective, solar-mass CTTS accreting from its Keplerian disk at a rate of 10^{-7} $M_\odot$ yr^{-1} would spin up to half of its break-up velocity, i.e., $\simeq$120 km s^{-1}, in 10^6 years. At first, it was suggested that CTTS do accrete large amount of high-angular momentum material from their disk but that the angular momentum excess is carried away by a strong, magnetized stellar wind. In this framework, however, there is no intimate connection between the stellar wind and the accretion process, so that it is difficult to understand how the stellar wind adjust in order to exactly balance the angular momentum transferred from the disk to the star.

A new, more promising way out of this dilemma is to assume that the disk material actually loses its angular momentum *before* it reaches the stellar surface. This may be the case, for instance, if the stellar magnetic field is strong enough to thread the inner part of the disk. As was first shown by Ghosh and Lamb (1979), disk accretion onto strongly magnetized objects does not necessarily lead to stellar spin-up. The torque exerted on the star by the disk consists in two parts: i) a torque due to the disk material falling onto the star which always tends to spin it up, and ii) a magnetic torque which results from the interaction of the inner disk material with the star's magnetosphere. If the stellar magnetosphere extends beyond the distance at which the disk material has the same angular velocity than the star, the magnetic torque is negative, thus spinning the star down. For a solar-mass TTS rotating at 20 km s^{-1}, surface magnetic fields of the order of 10^3 G are required to produce such a negative magnetic torque (see Bouvier 1990a). So far, the only direct measurement of surface magnetic field on a TTS was obtained by Basri and Marcy (1990) who found a magnetic field of 1700 G averaged over the surface of a WTTS. Other, indirect evidence for magnetic fields of a few thousand gauss at the surface of TTS comes from the properties of their stellar spots (Bouvier and Bertout 1989) and from the level of magnetic activity they exhibit (Bouvier 1990b). Hence, both direct measurements and indirect evidence suggest that magnetic fields in T Tauri stars are strong enough to play an important dynamical role in the accretion process near the stellar surface. The topology of the magnetic field at the surface of TTS is probably quite complex. Dipolar magnetic fields of a few thousand gauss would certainly not have escaped detection from polarization studies. In fact, in order to explain the existence of a surface spot much hotter than the surrounding photosphere, Bertout, Basri and Bouvier (1988) proposed that the disk material accreted at the surface of the CTTS DF Tauri is channeled along a large magnetic loop. Further evidence for giant magnetic loops in TTS are provided by X-ray and radio observations (see Montmerle et al. 1990).

Recent models of magnetized accretion disks around young stars have been developed by Camenzind (1990) and Königl (1989). In agreement with empirical evidence reported by Cabrit et al. (1990) and Cohen et al. (1989), these models predict the existence of a fundamental connection between mass-loss and mass-accretion processes. In both models, centrifugally-driven MHD winds are powered by the accretion process in the inner disk regions and carry away the excess of angular momentum. An interesting consequence is that the accretion and mass-loss mechanisms are self-regulated, which ensures a constant equilibrium between angular momentum gain and loss, thus leaving the star's rotation essentially unaffected. Such self-regulated mechanisms are indeed required to explain the similar rotation rates of WTTS and CTTS.

5. Angular momentum evolution to the zero-age main sequence

Ideally, one could trace observationally the evolution of stellar angular momentum from the T Tauri phase to the zero-age main sequence (ZAMS) by measuring the rotation rates of stars of increasing age along a given pre-main sequence evolutionary track. Unfortunately, the relative paucity of known post-TTS, with an evolutionary status intermediate between TTS and ZAMS dwarfs, has prevented such an empirical approach to be followed so far. Theoretical models are then required to fill the observational gap between the T Tauri phase and the ZAMS: starting from initial conditions defined by TTS rotation (at t = 0.1-1 10^7 yrs), they aim at reproducing the observed rotation rates of low-mass stars on the ZAMS (at t = 3-7 10^7 yrs) by taking into account physical mechanisms relevant for the evolution of angular momentum such as rotational braking by magnetic winds, angular momentum transport in the stellar interior, and the reduction of the star's moment of inertia during its contraction to the main sequence. Current observational constraints are reviewed below and theoretical problems related to the evolution of angular momentum prior to the main sequence are discussed.

The dispersal time of T Tauri associations is of the order of 10^7 years (e.g., Hartmann et al. 1990). Hence, by the time TTS have evolved into ZAMS dwarfs, they have moved far away from their parental cloud and are extremely difficult to identify among the much more numerous evolved dwarfs of the field. In contrast, the lifetime of stellar clusters is much longer, so that young open clusters contain a large number of low-mass dwarfs on or near the ZAMS. Since these stars are concentrated over a small area on the sky, they are easily identified from proper motion surveys. As a result, the rotational properties of stars on the ZAMS are commonly derived from statistical studies of low-mass dwarfs in young open clusters (see Stauffer, this volume). At an age of 50 millions years, the α Persei cluster is the youngest cluster for which a significant amount of rotational data have been obtained (Stauffer et al. 1985, Stauffer, Hartmann and Jones

1989). Solar-mass stars in this cluster have recently reached the main-sequence while lower mass stars are still slightly above it. Hence, the rotational properties of dwarfs with a mass between 0.5 and 1.0 $M_\odot$ in the α Persei cluster provide clues to the angular momentum distribution of low-mass stars at the time of their arrival onto the main sequence.

Figure 6 shows the rotational velocity distribution of TTS in the mass range from 0.5 to 1$M_\odot$ (panel a), the rotation rates these stars would have on the ZAMS assuming no angular momentum loss and solid-body rotation (panel b), and the observed rotation rates of ZAMS dwarfs in the α Persei cluster (panel c) (a similar analysis has been made by Stauffer and Hartmann 1987 who compared the rotation rates of TTS with those of low-mass dwarfs in the slightly older Pleiades cluster). Under the above assumptions, the predicted (panel b) and observed (panel c) *vsini* distributions on the ZAMS are quite different. Nevertheless, the existence of very fast rotators in the α Persei cluster indicates that TTS must be strongly accelerated as they contract toward the main sequence. Theoretical models do predict that TTS spin up on their radiative tracks as a result of their rapidly decreasing moment of inertia. However, they also show that, starting from rotational velocities typical of low-mass TTS, the spin-up will hardly be sufficient to yield ZAMS rotation rates in excess of 100 km s^{-1} if angular momentum loss occurs (see, e.g., Figure 3 of Pinsonneault et al 1989, and Figure 5 of MacGregor, this volume). A first strong constraint then arises from the observations: *if TTS are the progenitors of rapidly-rotating, low-mass dwarfs in the α Persei cluster, at least some of them must retain most of their initial angular momentum and remain in quasi-solid rotation up to the ZAMS.* Other explanations for the existence of rapid rotators in the α Persei cluster, e.g., tidal interaction in close binary systems or spin-up by disk accretion have been shown to be inadequate (for a more detailed discussion, see Bouvier 1990a).

That at least some TTS do not experience significant angular momentum loss suggests that magnetic braking is relatively inefficient during pre-main sequence evolution. A rough estimate of the pre-main sequence braking timescale can be derived from the rate of angular momentum loss due to a magnetic stellar wind:

$$\frac{dJ}{dt} = \frac{2}{3}\frac{dM}{dt}\Omega R_\star^2 (\frac{r_a}{R_\star})^n \quad (1)$$

where $\frac{dM}{dt}$ is the mass-loss rate, Ω the angular rotation rate, $R_\star$ the stellar radius, r_a the Alfvén radius, and the exponent n depends upon the magnetic field geometry (see, e.g., Mestel 1984). Typical values for TTS are Ω=10^{-5} s^{-1}, which corresponds to *vsini*=15 km s^{-1}, r_a=5$R_\star$=10$R_\odot$ assuming kilogauss surface magnetic fields, and n=2 for radial field lines, which maximizes the angular momentum loss rate. CTTS mass-loss rates are of the order of 10^{-8} $M_\odot$ yr^{-1} while WTTS have much weaker winds with associated mass-loss rates perhaps as low as 10^{-10} $M_\odot$ yr^{-1}. Putting these numbers into equation (1) yields a braking timescale of 1.5 10^6 yrs for CTTS and 1.5 10^8 yrs for WTTS, the difference being due to the much weaker winds of WTTS. According to this estimate, WTTS are not significantly braked until they reach the ZAMS, while CTTS are expected to be braked on a timescale much shorter than their contraction timescale to the main sequence. However, the braking timescale of CTTS may be much longer than this estimate suggests if their winds are accretion-driven rather than originating from the star itself since accretion-driven winds remove angular momentum from the disk but do not directly contribute to the star's braking (see, e.g., Königl 1989). Thus, pending more accurate derivations of the strength of *stellar* winds in TTS, the hypothesis that magnetic braking is not very efficient during pre-main sequence evolution does not seem to pose insuperable difficulties.

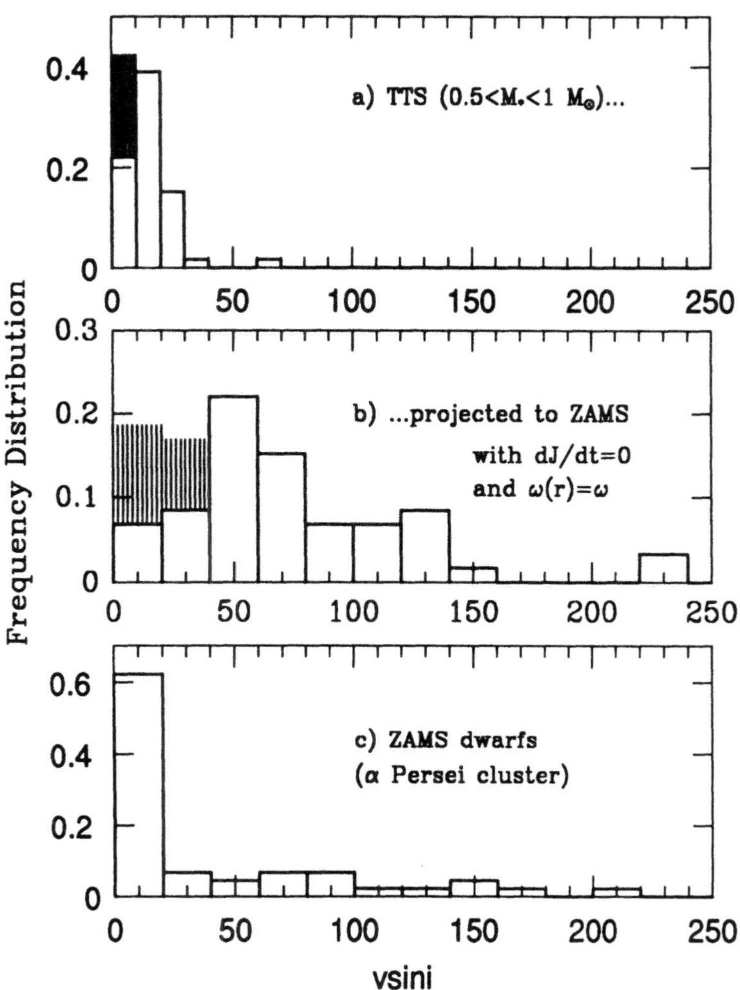

Figure 6. a) The observed *vsini* distribution of T Tauri stars with a mass between 0.5 and 1.0$M_\odot$; b) the predicted *vsini* distribution of TTS at their arrival onto the ZAMS, assuming no angular momentum loss and rigid rotation; c) the observed *vsini* distribution of low-mass, ZAMS dwarfs in the Alpha Persei cluster. The hatched region of the histograms represents stars for which only a *vsini* upper limit of 10 km s^{-1} or less has been measured.

It then remains to understand why approximately 2/3 of the α Persei dwarfs are slow rotators, with *vsini* less than 20 km s^{-1}. If no angular momentum loss occurs, the progenitors of such slow rotators must be TTS with an initial rotation rate of a few km s^{-1} at most. As can be seen from Figure 6, there are clearly too few TTS with such a low *vsini* to account for the high frequency of slow rotators among α Persei dwarfs. Hence, the spread in initial angular momentum as observed in T Tauri stars is too small to account for the very wide distribution of rotation rates observed among ZAMS dwarfs, unless a peculiar form of rotational braking is postulated (Stauffer and Hartmann 1987). Moreover, the recent proper motion survey of the central part of the Taurus-Auriga stellar formation region by Hartmann et al. (1990) indicates that known TTS constitute a representative sample of the entire population of pre-main sequence stars in this molecular cloud. This seems to rule out the existence of very slowly-rotating pre-main sequence stars not belonging to the T Tauri class which would be the progenitors of the slow rotators in the α Persei cluster. Alternatively, a combination of rapid main sequence spin-down and a significant age spread among cluster dwarfs has been proposed by Stauffer et al. (1984) to explain the coexistence of slow and rapid rotators in young open clusters. Rotational studies of low-mass dwarfs in young open clusters of increasing age show that solar-mass stars are braked to rotational velocities less than 20 km s^{-1} within 2 10^7yrs after their arrival upon the main sequence (e.g., Stauffer 1987). According to theoretical models, only the outer convective envelope is spun down at this stage as it dynamically uncouples from the radiative core which remains in rapid rotation. Since the moment of inertia of the convective envelope is small, rapid braking of the surface layers results. Then, assuming an age spread of a few 10^7yrs among stars in the α Persei cluster, one can hypothesize that stars which first reached the main sequence have been already braked thereupon and are now slow rotators while those which last arrived onto the main sequence have not had time yet to experience significant braking and, therefore, are still rapid rotators. This hypothesis, however, was severely questioned by Stauffer et al. (1989) who found both slow and rapid rotators among low-mass stars in a cluster even younger than α Persei, IC 2391 with an age of 3 10^7 years, even though the age spread seems to be less than 2 10^7 years.

The coexistence of slow and rapid rotators within the same mass range in young stellar clusters is a real challenge for theoretical models. So far, models have mainly aimed at reproducing the observed run of *mean* rotational velocity with time (e.g., Pinsonneault et al. 1989, see also MacGregor in this volume) and have been sucessful in predicting pre-main sequence spin-up and rapid main sequence spin down, though complete agreement with observations has not yet been reached. However, because the observed range of rotational velocities is so wide, especially for ZAMS stars in young clusters, the definition of a *mean* rotational velocity is hardly meaningful. Hence, the next step is to simultaneously account for the increasing *spread* of rotational velocities observed from the pre-main sequence to the ZAMS and the subsequent rapid convergence toward low rotation rates on the main sequence. Indeed, theoretical models have to face two major difficulties. One is to explain how a solar-mass pre-main sequence star is able to spin up from typically 20 km s^{-1} to 150 km s^{-1} in 2 10^7 years and then spin down from 150 km s^{-1} to less than 20 km s^{-1} in roughly the same amount of time on the main sequence (see Soderblom and Stauffer 1990). The other is to understand why some pre-main sequence stars are rapid rotators as they reach the main sequence while the majority are slow rotators.

The failure of current models to successfully address these issues may indicate that some important physical mechanism has been overlooked. For instance, it has been suggested that slow rotators in young open clusters might have formed planets and transferred part of their spin angular momentum into planetary orbital motion (Stauffer and Soderblom 1990). Alternatively, slow and rapid rotators in young clusters may have similar amount of total angular momentum, but slow rotators hide it in a rapidly rotating core, while rapid rotators maintained nearly rigid rotation during their pre-main sequence evolution. Why, then, does the coupling between the convective envelope and the radiative core in rapid rotators become suddenly weak upon the arrival onto the main sequence, thus enabling the convective envelope to spin down rapidly, even though the radiative core of a solar-mass star is already nearly fully developed at an age of 10^7 yrs, i.e., well before the star reaches the main sequence? Could other mechanisms of angular momentum transport in the stellar interior account for such a behaviour, such as, e.g., magnetic fields which have not been considered so far?

In view of the above difficulties, one may reasonably question the validity of the whole approach followed so far, namely: can T Tauri stars be considered as the progenitors of dwarfs in young open clusters, or, more precisely, is it relevant to compare the rotation rates of T Tauri stars with those of low-mass stars in young clusters? The basic assumption here is that the initial distribution of angular momentum and its subsequent evolution are the same in stellar clusters, where low-mass ZAMS dwarfs are observed, and in stellar associations, where T Tauri stars are found. Obviously, this issue directly bears upon the origin of stellar angular momentum. Once thought to result from differential galactic rotation, the build-up of stellar angular momentum is now more likely seen as the result of interstellar turbulence in molecular clouds (see, e.g., Fleck and Clark 1981). The random orientation of the rotational axes of single stars and of the orbital plane of binary systems (Kraft 1970, see also Huang and Struve 1954, Huang and Wade 1966, and Guthrie 1985) would be difficult to understand if angular momentum originated in differential galactic rotation but is a natural consequence of turbulence in molecular clouds. Then, the question arises whether the pattern of interstellar turbulence is the same in dense proto-clusters and in molecular clouds which will form stellar associations. The properties of turbulence in molecular clouds have been recently reviewed by Bodenheimer et al. (1990) who concluded that it is highly non-uniform in time and space. Unfortunately, similar studies are lacking for protocluster clouds. Nevertheless, studies of rotation in massive stars have often revealed significant differences in the *vsini* distribution of field and cluster dwarfs. For instance, Guthrie (1982) found the *vsini* distribution of late B-type stars in galactic clusters to be bimodal with a lack of intermediate rotators while field stars in the same mass range have a Maxwellian rotational distribution (see also Abt 1970 and Guthrie 1984). Other intriguing results include claims that, in some clusters, rapid rotators are concentrated toward the cluster's center (Struve 1945, Slettebak 1968, Abt et al. 1969) and that the more rapid rotators tend to have the larger space motions (Huang and Struve 1954). These results are usually interpreted as indicating that stars acquire most of their angular momentum through gravitational encounters with other stars at a very early stage of evolution (e.g., Wolff, Edwards, and Preston 1982). Since the density of protostars may be expected to be much higher in proto-clusters than in proto-associations, gravitational encounters will be much more frequent in the former, thus possibly leading to significant differences in the distribution of initial angular momentum. Much more observational and theoretical work is needed to settle this issue. However, one must keep in mind that current ideas related to the evolution of stellar angular momentum prior to the main sequence rest upon an assumption whose validity has still to be demonstrated.

6. Summary and Conclusions

Important observational clues for the rotational properties of pre-main sequence stars have been obtained in the recent years. The main conclusions are as follow:

- the rotational properties of young stars do not significantly vary from one stellar formation region to the other
- the rotational axes of T Tauri stars appear to be randomly orientated
- the mean rotational velocity of T Tauri stars is of the order of 1/10 the break-up velocity and it increases with stellar mass, from $\simeq$14 km s^{-1} in solar-mass TTS to more than 150 km s^{-1} in massive Herbig Ae-Be stars
- there is a one order of magnitude spread in *vsini* for TTS at any mass, part of which reflect an intrinsic spread in angular momentum
- some (all?) of the most rapidly rotating, low-mass TTS are members of close binary systems in which tidal interactions have enforced synchronization between orbital and rotational motions. The total angular momentum of such systems is at least one order of magnitude larger than that of single stars
- weak-line TTS and classical TTS have similar rotation rates, which suggests that disk accretion does not lead to significant stellar spin-up
- the average angular momentum of solar-mass TTS is consistent with the extrapolation of Kraft's relationship toward low-mass stars. However, the derivation of angular momentum for pre-main sequence stars is still uncertain, so that slightly different relationships cannot be ruled out. Taking into account current uncertainties on the determination of stellar mass and radius for pre-main sequence stars, an intrinsic spread of at least a factor of 2 in the initial distribution of angular momentum at any mass is suggested
- *if low-mass TTS are the progenitors of late-type dwarfs in young open clusters*, the most rapidly rotating TTS must retain most of their initial angular momentum and rotate like solid-bodies up to the main sequence in order to account for the fast rotators observed in young clusters.

Additional measurements of rotational velocities for T Tauri stars are not expected to change these conclusions drastically. Nevertheless, observational studies are still needed to address some important issues. In particular, observational efforts should be directed toward the measurement of rotation rates for TTS that only have upper limits of the order of 10 km s^{-1} set on their *vsini*. This will allow one to completely define the rotational distribution of low-mass pre-main sequence stars. Another important observational contribution will be the determination of the binary frequency among pre-main sequence stars. Various groups are working on this issue, and should soon provide reliable estimates of the fraction of binaries among T Tauri stars. Finally, one should aim at reducing the observational gap which exists between T Tauri stars and ZAMS stars by measuring the rotation rates of low-mass stars in open clusters much younger than the α Persei cluster. This would provide very valuable insight into the evolution of angular momentum on radiative pre-main sequence tracks where the most dramatic variation of surface rotation is expected to occur.

While further observational work is required, the most significant advance in our understanding of the evolution of stellar angular momentum prior to the main sequence is expected to result from theoretical studies. Indeed, observational results are now well-enough established to severely constrain theoretical models of angular momentum evolution from the star formation process up to the main sequence. For instance, although models of star formation are not yet able to handle the complete evolutionary process from diffuse interstellar clouds to visible pre-main sequence stars, it appears that rotational braking by magnetic fields must be instrumental during the protostellar phase in order to account for the low rotation rates of visible pre-main sequence stars compared to break-up velocity. At a later stage of evolution, current parametrized models have reached qualitative agreement with observations in predicting both pre-main sequence spin-up and subsequent rapid main-sequence spin down. However, no convincing explanation has been proposed so far to account for the increasing spread in rotation rates observed between the T Tauri phase and the ZAMS. Undoubtedly, this is the greatest challenge theoretical models will have to face in the forthcoming years.

Acknowledgments: I thank R. D. Mathieu for communicating results on the spectroscopic binary GW Ori prior to publication.

References

Abt, H.A. 1970, "Stellar Rotation in Open Clusters", in *Stellar Rotation*, ed. A. Slettebak (D. Reidel: 1970), p.193-203

Abt, H.A., Clements, A.E., Doose, L.R., and Harris, D.H. 1969, "Rotational Velocities in NGC 2516", *Astron. J.* **74**, 1153-1154

Basri, G., and Bertout, C. 1989, "Accretion Disks around T Tauri Stars. II. Balmer Emission", *Astrophys. J.* **341**, 340-358

Basri, G., and Marcy, G. 1990, in *The Sun and Cool Stars: Activity, Magnetism, Dynamos*, Helsinki 1990, in press

Beckwith, S.V.W., Sargent, A.I., Chini, R.S., and Güsten, R. 1990, "A Survey for Circumstellar Disks around Young Stellar Objects", *Astron. J.* **99**, 924-945

Bertout, C. 1989, "T Tauri Stars: Wild as Dust", *Ann. Rev. Astron. Astrophys.* **27**, 351-395

Bertout, C., Basri, G., and Bouvier, J. 1988, "Accretion Disks around T Tauri Stars", *Astrophys. J.* **330**, 350-373

Bodenheimer, P., Ruzmaikina, T., and Mathieu, R.D. 1990, "Stellar Multiple Systems: Constraints on the Mechanism of Origin", in *Protostar and Planets III*, Tucson, March 1990, in press

Bouvier, J. 1990a, "Stellar Rotation prior to the Main Sequence", in *Rotation and Mixing in Stellar Interiors*, ed. M.-J. Goupil and J.-P. Zahn (Springer-Verlag), *Lecture Notes in Physics* **366**, 47-68

Bouvier, J. 1990b, "Rotation in T Tauri Stars. II. Clues for Magnetic Activity", *Astron. J.* **99**, 946-964

Bouvier, J., and Bertout, C. 1989, "Spots on T Tauri Stars", *Astron. Astrophys.* **211**, 99-114

Bouvier, J., Bertout, C., Benz, W., and Mayor, M. 1986, "Rotation in T Tauri Stars. I. Observations and Immediate Analysis", *Astron. Astrophys.* **165**, 110-119

Bouvier, J., Lindgren, H., Mayor, M., and Reipurth, B. 1991, "Rotation in T Tauri Stars. III. The Slow Rotators", in preparation

Cabrit, S., Edwards, S., Strom, S.E., and Strom, K.E. 1990, "Forbidden-Line Emission and Infrared Excess in T Tauri Stars: Evidence for Accretion-Driven Mass-Loss?", *Astrophys. J.* **354**, 687-700

Camenzind, M. 1990, "Magnetized Disk-Winds and the Origin of Bipolar Outflows", in *Reviews in Modern Astronomy*, Vol.3, in press

Cohen, M., Emerson, J.P., and Beichman, C.A. 1989, "A Reexamination of Luminosity Sources in T Tauri Stars. I. Taurus-Auriga", *Astrophys. J.* **339**, 455-473

Fleck, R.C., and Clark, F.O. 1981, "A Turbulent Origin for the Rotation of Molecular Clouds", *Astrophys. J.* **245**, 898-902

Ghosh, P., and Lamb, F.K. 1979, "Accretion by Rotating Magnetic Neutron Stars. III. Accretion Torques and Period Changes in Pulsating X-Ray Sources", *Astrophys. J.* **234**, 296-316

Goodman, A.A., Bastien, P., Myers, P.C., and Ménard, F. 1990, "Optical Polarization Maps of Star-Forming Regions in Perseus, Taurus, and Ophiuchus", *Astrophys. J.* **359**, 363-377

Guthrie, B.N.G. 1982, "The Bimodal Distribution of Rotational Velocities of Late B-type Stars in Galactic Clusters", *Monthly Notices Roy. Astron. Soc.* **198**, 795-810

Guthrie, B.N.G. 1984, "The Rotation of Early B-type Stars and the Problem of Star Formation", *Monthly Notices Roy. Astron. Soc.* **210**, 159-171

Guthrie, B.N.G. 1985, "The Distribution of the Rotation Axes of A-type Stars", *Monthly Notices Roy. Astron. Soc.* **215**, 545-553

Hartigan, P., Hartmann, L., Kenyon, S.J., Strom, S.E., and Skrutskie, M.F. 1990, "Correlations of Optical and Infrared Excesses in T Tauri Stars", *Astrophys. J. Letters* **354**, L25-L28

Hartmann, L.W., and Stauffer, J.R. 1989, "Additional Measurements of Pre-Main Sequence Stellar Rotation", *Astron. J.* **97**, 873-880

Hartmann, L.W., Soderblom, D.R., and Stauffer, J.R. 1987, "Rotation and Kinematics of Pre-Main Sequence Stars in Taurus-Auriga with CaII Emission", *Astron. J.* **93**, 907-912

Hartmann, L., Hewett, R., Stahler, S., and Mathieu, R.D. 1986, "Rotational and Radial Velocities of T Tauri Stars", *Astrophys. J.* **309**, 275-293

Hartmann, L.W., Jones, B.F., Stauffer, J.R., and Kenyon, S.J. 1990, "A Proper Motion Survey for Pre-Main Sequence Stars in Taurus-Auriga", *Astron. J.* in press

Heyer, M.H. 1988, "The Magnetic Evolution of the Taurus Molecular Clouds. II. A Reduced Role of the Magnetic Field in Dense Core Regions", *Astrophys. J.* **324**, 311-320

Heyer, M.H., Vrba, F.J., Snell, R.L., Schloerb, F.P., Strom, S.E., Goldsmith, P.F., and Strom, K.M. 1987, "The Magnetic Evolution of the Taurus Molecular Clouds. I. Large-Scale Properties", *Astrophys. J.* **321**, 855-876

Huang, S.-S., and Struve, O. 1954, "Stellar Rotation", *Annales d'Astrophys.* **17**, 85-93

Huang, S.-S., and Wade, C. 1966, "Galactic Distribution of Eclipsing Binaries and its Significance", *Astrophys. J.* **143**, 146-149

Kawaler, S.D. 1987, "Angular Momentum in Stars: the Kraft Curve Revisited", *Publ. Astron. Soc. Pac.* **99**, 1322-1328

Kenyon, S.J., and Hartmann, L.W. 1987, "Spectral Energy Distribution of T Tauri Stars: Disk Flaring and Limits on Accretion", *Astrophys. J.* **323**, 714-733

Königl, A. 1989, "Self-Similar Models of Magnetized Accretion Disks", *Astrophys. J.* **342**, 208-223

Kraft, R.P. 1970, "Stellar Rotation", in *Spectroscopic Astrophysics*, ed. G.H. Herbig (Univ. California Press), p.385-422

Mathieu, R.D., Walter, F.M., and Myers, P.C. 1989, "The Discovery of Six Pre-Main Sequence Spectroscopic Binaries", *Astron. J.* **98**, 987-1001

Mestel, L. 1984, "Angular Momentum Loss during Pre-Main Sequence Contraction", in *3rd Cambridge Workshop on Cool Stars, Stellar Systems, and the Sun*, ed. S.L. Baliunas and L. Hartmann (New York: Springer), p.49-59

Montmerle, T., Feigelson, E.D., Bouvier, J., and André, P. 1990, "Magnetic Fields, Activity, and Circumstellar Material around Young Stellar Objects", in *Protostars and Planets III*, Tucson, March 1990, in press

Pinsonneault, M.H., Kawaler, S.D., Sofia, S., and Demarque, P. 1989, "Evolutionary Models of the Rotating Sun", *Astrophys. J.* **338**, 424-452

Slettebak, A. 1968, "Stellar Rotation in the Scorpio-Centaurus Association", *Astrophys. J.* **151**, 1043-1050

Soderblom, D.R. 1985, "Determining the Spatial Orientation of Stellar Rotation Axes", *Publ. Astron. Soc. Pac.* **97**, 57-59

Soderblom, D.R., and Stauffer, J.R. 1990, "The Evolution of the Sun's Angular Momentum", in *Inside the Sun*, ed. G. Berthomieu and M. Cribier (Kluwer Academic Publisher), p.403-413

Stauffer, J.R. 1987, "Rotational Velocity Evolution On and Prior to the Main Sequence", *Lect. Notes in Phys.* **291**, 182-191

Stauffer, J.R., and Hartmann, L.W. 1987, "The Distribution of Rotational Velocities for Low-Mass Stars in the Pleiades", *Astrophys. J.* **318**, 337-355

Stauffer, J.R., and Soderblom, D.R. 1990, "The Evolution of Angular Momentum in Solar Mass Stars", in *The Sun in Time*, ed. C.P. Sonett and M.S. Giampapa, in press

Stauffer, J.R., Hartmann, L.W., and Jones, B.F. 1989, "Rotational Velocities of Newly Discovered, Low-Mass Members of the Alpha Persei Cluster", *Astrophys. J.* **346**, 160-167

Stauffer, J.R., Hartmann, L.W., Burnham, J.N., and Jones, B.F. 1985, "Evolution of Low-Mass Stars in the Alpha Persei Cluster", *Astrophys. J.* **289**, 247-261

Stauffer, J.R., Hartmann, L.W., Jones, B.F., and McNamara, B.R. 1989, "Pre-Main Sequence Stars in the Young Cluster IC 2391", *Astrophys. J.* **342**, 285-294

Stauffer, J.R., Hartmann, L.W., Soderblom, D.R., and Burnham, N. 1984, "Rotational Velocities of Low-Mass Stars in the Pleiades", *Astrophys. J.* **280**, 202-212

Strom, K.M., Strom, S.E., Edwards, S., Cabrit, S., and Strutskie, M.F. 1989, "Circumstellar Material Associated with Solar-Type Pre-Main Sequence Stars: a Possible Constraint on the Timescale for Planet Building", *Astron. J.* **97**, 1451-1470

Struve, O. 1945, *Pop. Astron.* **53**, 259

Vogel, S.N., and Kuhi, L.V. 1981, "Rotational Velocities of Pre-main Sequence Stars", *Astrophys.*

J. **245**, 960-976

Vrba, F.J., Strom, S.E., and Strom, K.M. 1988, "Optical Polarization Measurements in the Vicinity of Nearby Star-Forming Complexes. I. The Lynds 1641 Giant Molecular Cloud", *Astron. J.* **96**, 680-694

Vrba, F.J., Rydgren, A.E., Chugainov, P.F., Shakovskaya, N.I., and Weaver, W.B. 1989, "Additional UBVRI Photometric Searches for Periodic Light Variability from T Tauri Stars", *Astron. J.* **97**, 483-498

Weaver, W.B. 1987, "Anomalous Distribution of Axial Inclinations of T Tauri Stars", *Astrophys. J. Letters* **319**, L89-L92

Wolff, S.C., Edwards, S., and Preston, G.W. 1982, "The Origin of Stellar Angular Momentum", *Astrophys. J.* **252**, 322-336

Zahn, J.-P., and Bouchet, L. 1989, "Tidal Evolution of Close Binary Stars. II. Orbital Circularization of Late-Type Binaries", *Astron. Astrophys.* **223**, 112-118

DISCUSSION

van't Veer: What can you say about duplicity and rotation?

Bouvier: Fortunately, there are very few known spectroscopic binaries among the T Tauri stars. According to Mathieu and collaborators, there are only nine whose orbital elements have been published. The frequency of short period binaries with periods less than 100 days is only about 10% among the weak-lined T Tauri stars, which is similar to the frequency of short period spectroscopic binaries among population I stars. But, the frequency may be considerably less for classical T Tauri stars, because only two spectroscopic binaries have been detected for them.

Duncan: I guess tomorrow I am going to make things even a little more interesting for theoreticians, because I will present data on some stars which are between the ages of the T Tauri stars and the main sequence. By observing as faint as I could in the Orion nebula region, I have a fairly good sample of stars most of which are post-T Tauri stars and most of which are even post weak lined T Tauri stars (mass range 0.8 to 1.0 solar masses). The shortest summary of what I see is that there is no indication of any rapid rotation among any of my stars. In fact there is really very little indication of change in the surface rotation, which is what you measure since the stars were T Tauri stars. My distribution looks very similar to your distribtuion of the T Tauri stars. Now the stars have a range in ages, but the majority of them are well along on the radiative track, and they don't yet show the speed up.

Bouvier: Yes, I agree with that. Although there are a few rapid rotators in the T Tauri phase there is no striking difference between the rotational velocity distribution for the classical T Tauri stars and the weak line T Tauri stars. So, I completely agree with you.

Schatzman: Just one question, what is the size for the accreting disk in the classical T Tauri stars and what is the mass?

Bouvier: The mass estimates come from submilleter and millimeter observations, and range from one hundredth to one tenth of a solar mass. The size is a few hundred astronomical units.

Vaiana: There may be additional problems when you compare the rotational velocities of stars in different associations. You showed a bar chart diagram of various clusters; it seems to be that in a the number of cases you explored there are mostly upper limits except in the Taurus-Auriga and in the higher mass stars in Orion. I wonder what sort of statistical tests were applied, in particular when there are upper limits. I wonder whether tests such as ones we are using for generating luminosity functions in the case of X-ray emission were applied to cases of this kind. So therefore, you could make a quantitative comparison of the similarities between the distributions. So that if we find the similarities with age or other things later on, we could clearly find out whether there is a contradiction or not.

Bouvier: Well, there aren't really many upper limits here. For the diagram you are referring to, there are only three stars for which we only have upper limits to the rotational velocity. You can now do statistical tests for data which includes upper limits - there is a paper by Feigelson and co-workers which discusses this. Here for now I will not make any statistical tests, because the number of points is so small that it really does not make any sense.

Vaiana: Yes I agree with that. What I meant is for future times. The similarities you have indicated are possible, but the statistical significance is at yet limited. We are finding more and more, that as we improve the statistical techniques in the analysis of the X-ray data, some previous conclusions must be reviewed quantitatively.

Bouvier: Yes, I think the best way to do quantitative statistics is with the cumulative distributions.

DISKS ASSOCIATED WITH INTERMEDIATE MASS STARS[1]

STEPHEN E. STROM

Five College Astronomy Department and Physics and Astronomy Department, University of Massachusetts Amherst, MA 01003

JOCELYN KEENE

California Institute of Technology, Pasadena, CA 91125

SUZAN EDWARDS[2], LYNNE HILLENBRAND, KAREN STROM, LISA GAUVIN and GARY CONDON

Five College Astronomy Department

ABSTRACT This review summarizes 1) the observations which provide evidence of disks surrounding young, intermediate mass stars ($1.5 < M/M_\odot < 10$); 2) the sizes and masses of such circumstellar disks; 3) evidence for accretion through these disks and estimates of mass accretion rates; 4) the relation between mass accretion and mass outflows during early evolutionary phases; and 5) the timescales required for such disks to evolve from the massive, optically thick structures which apparently surround the youngest intermediate mass stars, to optically thin disks, perhaps analogous to the β Pictoris disk. It appears as if intermediate mass stars are built via accretion of material from large ($r \sim 300$ AU), massive circumstellar disks ($M \sim 1\ M_\odot$) on timescales $t < 1$ Myr (for stars with $M > 3\ M_\odot$). The initial stellar angular momentum reflects a balance between the angular momentum added from accreting disk material, and that carried away by a wind.

1 Based in part on observations carried out at the Cal Tech Submillimeter Observatory

2 Also at Smith College, Northampton,MA 01065

S. Catalano and J. R. Stauffer (eds.), Angular Momentum Evolution of Young Stars, 63–81.

INTRODUCTION

Observations of young, optically visible pre-main sequence (PMS) stars show that many solar-type stars ($0.2 < M/M_{\odot} < 1.5$) are surrounded by disks of solar-system dimension ($5 < r < 200$ AU), and masses $0.01 < M_{disk}/M_{\odot} < 0.1$ as evidenced by:

- infrared, sub-millimeter, and millimeter continuum radiation in excess of photospheric levels, which finds most straightforward interpretation in terms of emission arising in circumstellar disks (Myers et al., 1987; Adams, Lada and Shu, 1987; Kenyon and Hartmann, 1987; K.M. Strom et al. 1989; Strom, Edwards and Strom, 1989; Skrutskie et al., 1989; Beckwith et al., 1990). Such excess radiation arises both from photospheric radiation absorbed and re-radiated by circumstellar dust grains ("reprocessing"), and from heating of circumstellar gas and dust via accretion of material through the disk (see Lynden-Bell and Pringle, 1974; Kenyon and Hartmann, 1987; Adams, Lada and Shu, 1987; Bertout, Basri and Bouvier, 1988; Bertout,1989). The resulting wide range of disk temperatures (from the warm material located near the stellar surface, to the cold material located in the outer disk region) provides a straightforward explanation for the broad spectral energy distributions characteristic of the excess radiation (see Figure 1; also Myers et al., 1987).
- broad forbidden line profiles, which diagnose outflowing gas driven by PMS stars and show only *blue-shifted* components; no red-shifted emission is observed. The receding gas in the outflow is presumed to be occulted by an optically thick disk of dimension ~10 AU to 100s of AU (e.g. Appenzeller et al., 1984; Edwards et al., 1987, Cabrit et al., 1990).

No solar-type *main sequence* stars exhibit the strong infrared excess emission characteristic of the optically thick disks which surround many stars during earlier evolutionary phases. Hence, the mass of disk material in the form of small dust grains must decrease with time. The observation of infrared and mm-continuum emission arising in massive, optically thick disks surrounding young solar-type PMS stars, combined with the absence of such signatures among their older stellar counterparts provides the basis for establishing *timescales for disk evolution* from studies of intermediate age stars. Such timescales provide an upper limit to the time available for building large grains and possibly planetesimals (K.M. Strom et al., 1989; Skrutskie et al., 1990).

While the masses, sizes and evolutionary timescales are now fairly well established for disks associated with solar type stars, our current knowledge regarding disks associated with *intermediate mass* stars ($1.5 < M/M_{\odot} < 10$) is minimal. Our goals in this review are:

- to establish whether disks form around intermediate mass stars;
- to determine the masses and sizes of disks surrounding the youngest visible intermediate mass stars;

- to determine the accretion rates through circumstellar disks from the observed excess infrared luminosities produced by disk emission;
- to compare the relationship between accretion rate and mass outflow rate found for intermediate mass stars with that established for solar-type PMS stars;
- to estimate disk survival times from the derived disk masses and mass accretion rates;
- to report on the status of searches for the evolutionary descendents of the massive disks which appear to surround intermediate stars at birth;
- to understand the role of disks in the formation and early angular momentum history of intermediate stars.

THE EXISTENCE OF DISKS SURROUNDING INTERMEDIATE MASS STARS

Establishing the existence and characteristics of disks associated with intermediate mass stars requires that we first examine B and A stars at an age young enough so that disk gas and dust which may surround such stars at birth, will not yet have been accreted, dissipated or possibly assembled into larger bodies. The youngest, optically visible intermediate mass stars are the Herbig Ae/Be stars (Herbig, 1960), whose masses and ages lie in the ranges $1.5 < M/M_{\odot} < 10$, and $0.1 < t < 1$ Myr respectively (Strom et al., 1972; Finkenzeller and Mundt, 1984). As is the case for solar-type stars, the best, albeit indirect, evidence for disks comes from observation of 1) infrared, sub-mm and mm-continuum excesses, and 2) blue-shifted forbidden line emission.

Infrared and Millimeter Continuum Excesses in Young B and A Stars

The most persuasive indirect evidence for disks surrounding young intermediate mass stars is provided by the observed infrared spectral energy distributions of Herbig Ae/Be stars. As can be seen in Figure 1, these objects exhibit excess emission with spectral shapes similar to those characterizing strong emission-line T-Tauri stars, for which the presence of disks is well established from both the observed forbidden line profiles, and from the success of detailed disk models in reproducing the excess emission (Beckwith et al., 1990).

The case for disks is further strengthed by the recent mm-continuum observations for 7 Herbig Ae/Be stars obtained at the Cal Tech Submillimeter Observatory (CSO) by Strom, Edwards and Keene (1990; unpublished). The observed excess mm-continuum fluxes require objects in their sample to be surrounded by circumstellar envelopes with masses ranging from ~0.01 to 2 $M_{\odot}$. Were the gas and dust corresponding to a mass M(env) ~ 0.1 $M_{\odot}$, distributed in a spherical envelope of radius 300 AU, the optical depth at 0.55μm would be ~1000, thus rendering the Ae/Be star invisible. However, by confining the material to a disk, one can produce the same mm continuum fluxes while preserving an optically thin line of sight to the star at most viewing angles (see as well the similar argument advanced by Myers et al., 1987).

Forbidden Line Emission in Young B and A Stars

Observations of the Herbig Ae/Be stars R Mon and V645 Cyg, reveal broad, blue-shifted emission lines (Edwards et al., 1987; Hamann and Persson, 1988) analogous to those characterizing many T Tauri stars. The forbidden line fluxes, combined with electron density estimates obtained from [S II] line ratios, provide the basis for computing the emitting volume of the wind region, and thus a minimum radius for the occulting disk. For R Mon and V645 Cyg, the putative disks must have radii $r > 1000$ AU in order to obscure the receding portion of the outflowing material traced by the forbidden lines. These disks are ~10× the typical sizes of the structures thought to surround solar-type PMS stars. The majority of the relatively small number of Ae/Be stars observed to date, however, show forbidden lines centered near the rest velocity of the star (Finkenzeller, 1985; Catala, 1989). Hence, forbidden line profiles cannot alone be invoked to support the hypothesis that disks are ubiquitous among the precursors of main sequence B and A stars.

DISK PROPERTIES: MASSES, SIZES, ACCRETION RATES, SURVIVAL TIMES

Studies of the disk properties characterizing young intermediate mass stars are at a more primitive stage than are similar investigations of disks associated with solar-type PMS stars, largely because adequate spectrophotometric data is lacking. In the discussion below, we present preliminary estimates of fundamental disk properties.

Disk Masses from mm-continuum and Infrared Measurements

Determinations of disk masses and sizes depend on modeling of far-infrared, sub-mm and mm-continuum excesses arising in circumstellar disks (Beckwith et al., 1990; Adams et al., 1990). Sub-mm and mm-continuum measurements are particularly critical because disks with masses $M \geq 0.1\ M_{\odot}$, and sizes $r \leq 300$ AU will be optically opaque at wavelengths $\lambda \leq 100\mu m$; flux measurements below 100μm will thus be insensitive to the disk mass. However, the masses derived from sub-mm and mm-continuum measurements depend upon knowledge of the dust grain emissivity and its variation with wavelength, and the disk temperature structure. Current uncertainties in grain emissivities make absolute mass determinations uncertain at the 0.5 dex level, although estimates of relative disk masses should be considerably more accurate.

At present, obtaining spectral energy distributions over the range 300μm to 3 mm represents a formidable observational challenge, particularly if large samples must be observed. Consequently, published measurements are available for fewer than 10 solar-type PMS stars, and no intermediate mass stars. As noted above, Strom, Edwards and Keene (1990; unpublished) recently obtained 1.3mm fluxes for 7 Herbig Ae/Be stars at the CSO. *Rough* disk mass estimates can be made by using their measured 1.3mm continuum fluxes and

$$M_{disk}/M_{\odot} = 0.8\, F_{1.3mm}\,(Jy)\,[\,\exp(13.1/T\,(disk)\,-1]\times(d\,/\,140\,pc)^2$$

where d is the distance of the star in parsecs (Adams, Emerson and Fuller (1990). Note that this is equation 7 of Adams et al., adjusted to the mass scale of Beckwith et al. (1990) by applying a multiplicative factor of 0.6. We assume a disk temperature, T(disk) = 50 K, which corresponds to the highest disk temperatures derived by Adams et al. from sub-mm and mm-continuum observations of luminous T Tauri stars. By assuming T(disk) values of 100 K and 25 K, derived disk masses would be decreased and increased respectively by ~2×. Disk masses for 7 Herbig Ae/Be derived with the above assumptions are listed in Table 1.

Table 1

Mass Estimates From Millimeter
Continuum Measurements for Herbig Ae/Be Stars

Star	$F_{1.3mm}$ (Jy)	M_{disk} ($M_{\odot}$)	M_* ($M_{\odot}$)	M_{disk}/M_*	$\dot{M}_{acc}$ ($10^{-6}\,M_{\odot}/yr$)	$M_{disk}/\dot{M}_{acc}$ (Myr)
KK Oph	0.04±0.01	0.01	1.5	0.008	0.3	0.04
LkHα 198[1]	0.4±0.1	1.9	4.5	0.4	8	0.2
LkHα 234[2]	0.9±0.1	<12	5.0	<2.3	30	0.4
Par 21	0.06±0.02	0.1	1.4	0.09	0.2	0.7
V 376 Cas[1]	0.23±0.06	1.1	3.3	0.3	6	0.2
V 594 Cas	0.06±0.02	0.3	4.2	0.08	4	0.08
V1686 Cas	0.5±0.1	2.2	2.0:	1.1:	...	...

[1]LkHα 198 and V376 Cas are separated by θ = 18″ (as compared to the 28″ CSO beam); a map of the region enabled rough division of the 1.3mm flux between these sources.

[2]emission is extended; the mass listed must therefore be considered an upper limit to the disk mass.

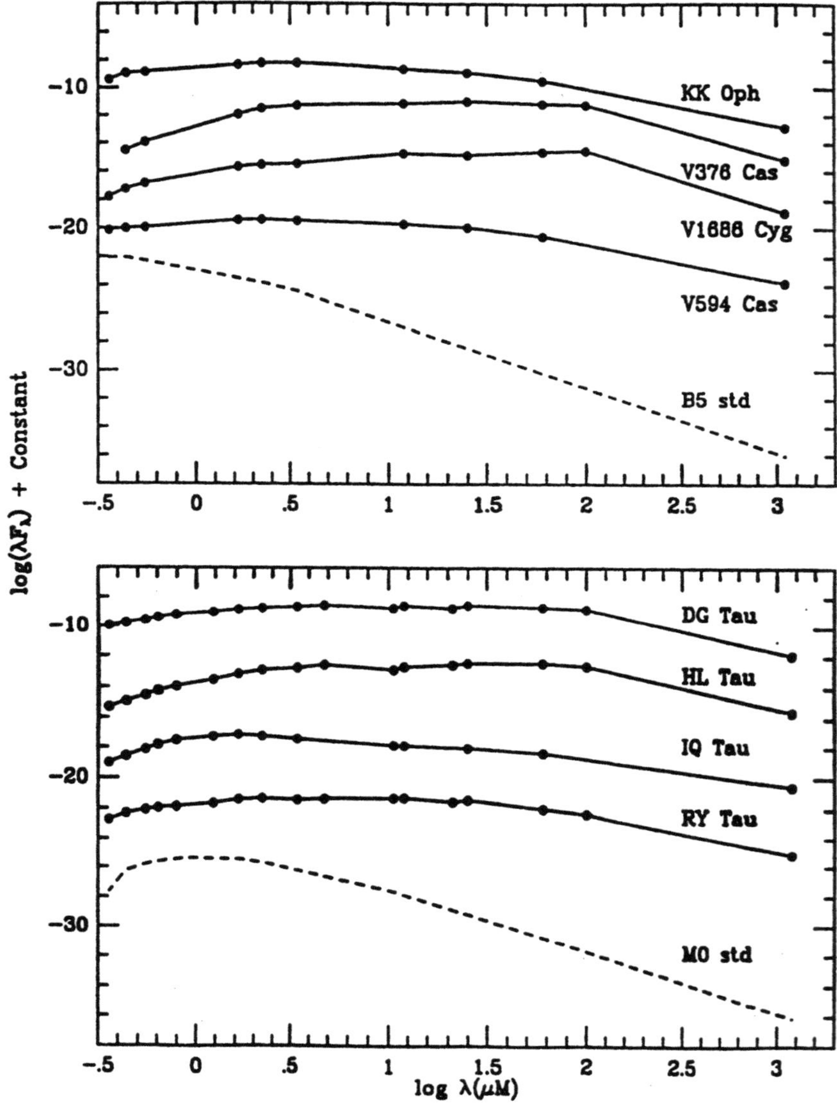

Figure 1: Spectral energy distributions for a selection of Herbig Ae/Be stars (top) and T Tauri Stars (bottom). Superposed on this figure are spectral energy distributions for unreddened B5 (top) and K0 stars (bottom). The mm-continuum points plotted are from Beckwith et al. (1990) for the T Tauri stars, and from unpublished data obtained at the Cal Tech Submillimeter Observatory by Strom, Edwards and Keene. Note that the shapes of the excess emission above photospheric levels at $\lambda > 2.2\mu m$ are similar for both young intermediate mass and solar-type pre-main sequence stars.

We caution that these mass estimates are preliminary; a more detailed and complete discussion will be forthcoming. It should be noted as well that the beam-size of the CSO at 1.3mm is ~28". Thus we cannot exclude the possibility that a significant contribution to the observed 1.3mm flux arises from ambient molecular cloud material. The above masses should therefore be treated as upper limits until higher angular resolution observations become available. If we assume that the disk masses listed in Table 1 are reliable, we can estimate the ratio of disk to stellar mass; stellar masses follow from the observed location of the star in the HR diagram and a comparison with computed evolutionary tracks. From examination of Table 1, it appears that *in some cases, the masses of disks surrounding intermediate mass stars are comparable to the mass of the central star.*

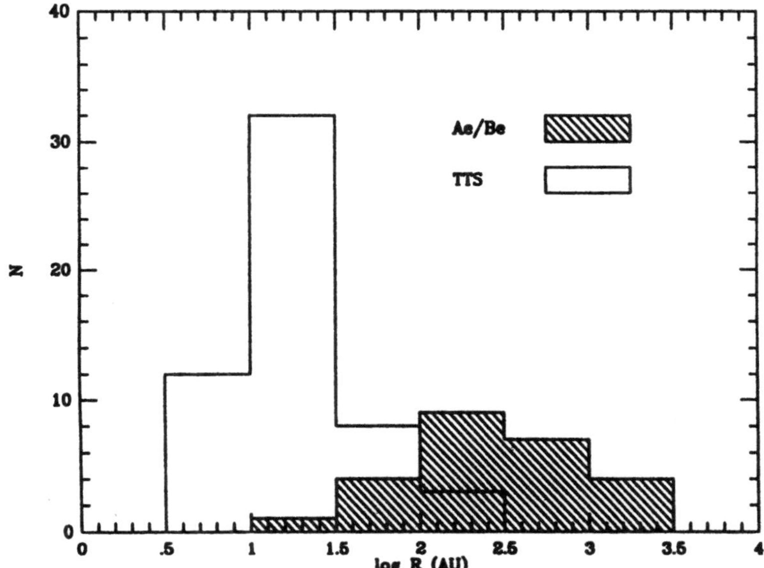

Figure 2: A histogram depicting the frequency distribution of disk radii estimated from F(60μm), the disk flux at λ = 60μm for Herbig Ae/Be stars (shaded region), and for the sample of solar-type pre-main sequence stars in Taurus-Auriga studied recently by K.M. Strom et al. (1989).

Disk Sizes from IRAS Measurements

IRAS measurements at far-infrared wavelengths probe emission arising from the colder, outer regions of circumstellar disks. If the disk is optically thick at λ = 60μm, we can make a rough estimate of the disk radius from the projected radiating area required to account for the observed 60μm flux:

$$2\,\pi\,r\,(\mathrm{disk})^2\,B_\lambda\,(T\,(\mathrm{disk}))\,/\,4\,\pi\,d^2 = F(60\mu m)$$

T (disk) is the characteristic temperature in the region of the disk contributing the largest fraction of the 60μm flux, d is the distance of the star/disk system, and F(60μm) is derived from the IRAS survey.

In Figure 2, we plot the frequency distribution of radii derived from the IRAS 60μm fluxes measured for a sample of 25 of the 57 Herbig Ae/Be stars identified by Finkenzeller and Mundt (1984). The 60μm fluxes are derived from detailed examination of ADDSCANS centered at the position of these objects. In all cases, we assume a disk temperature T(disk) = 50 K. Increasing or decreasing the disk temperature by 2× will result in decreasing or increasing the derived radii by ~1.5× respectively. For comparison, the frequency distribution of disk radii computed under identical assumptions is also presented for the sample of solar-type PMS stars in Taurus-Auriga discussed by K.M. Strom et al. (1989). Note that the radii for the disks surrounding intermediate mass stars range from $30 < r < 3000$ AU, or approximately 10× the characteristic sizes of disks surrounding solar-type PMS stars.

We note that for solar-type PMS stars, the radii derived from our simple assumptions are ~ 4× the radii, r_1, listed by Beckwith et al. (1990). Their radius estimates derive from fitting the fluxes predicted from a disk model (parameterized by power law surface density and temperature distributions) to the observed infrared and mm-continuum fluxes. In the context of their models, r_1 represents a "transition radius", interior to which the disks are optically thick, and outside of which they become optically thin. When sufficient mm- and sub-mm continuum data become available for Ae/Be stars, similar model fitting should be carried out in order to effect a comparison of r_1 values for disks associated with intermediate mass stars, with the Beckwith et al. (1990) sizes for optically thick disk regions in solar-type PMS stars.

Estimates of Disk Accretion Rates from Infrared Measurements

The magnitude of the infrared excess luminosity can provide an estimate of the accretion rate through the disk. Grains embedded within perfectly flat reprocessing disks (no internal disk heat source) will reradiate 0.25× the luminosity of the central star. Flared reprocessing disks (in which the disk scale height increases with radius) will reradiate up to ~0.5× times the stellar luminosity (Kenyon and Hartmann, 1987). Therefore, B and A stars which exhibit total luminosities in excess of $0.5 \times L_*$ are most probably surrounded by disks which are heated via accretion of material (Lynden-Bell and Pringle, 1974). The *minimum* luminosity contributed by accretion is thus given by (L_{IR} - 0.5 L_*), where L_{IR} is the excess luminosity above photospheric levels over the wavelength range 0.64μm to 100μm. The corresponding mass accretion rate is

$$\dot{M}_{acc} = (L_{IR} - 0.5\, L_*)\, R_* / (G\, M_*)$$

Following the procedure outlined for solar-type PMS stars by Cabrit et al. (1990), we use the spectral type and observed optical colors to derive reddening-corrected fluxes from 0.36μm to 100μm for Ae/Be stars. The flux distribution for a standard star of identical spectral type is normalized to the observed flux distribution at $\lambda = 0.55\mu m$. L_{IR} is then computed from the observed excess radiation above photospheric levels, while L_* is derived from the reddening-corrected flux at $\lambda = 0.55\mu m$, and the bolometeric correction appropriate to the spectral type. Finkenzeller and Mundt (1984) list 57 candidate Herbig Ae/Be stars of which 34 have photometry adequate to estimate infrared luminosities; of these 34, 14 exhibit infrared luminosities in excess of $0.5\, L_*$. The accretion rates estimated for these 14 stars range from 10^{-4} to $10^{-7.8}\, M_\odot$/yr; the median value is $10^{-5.5}\, M_\odot$/yr; accretion rates for solar-type PMS stars range from $10^{-5.7}$ to $10^{-8.1}\, M_\odot$/yr.

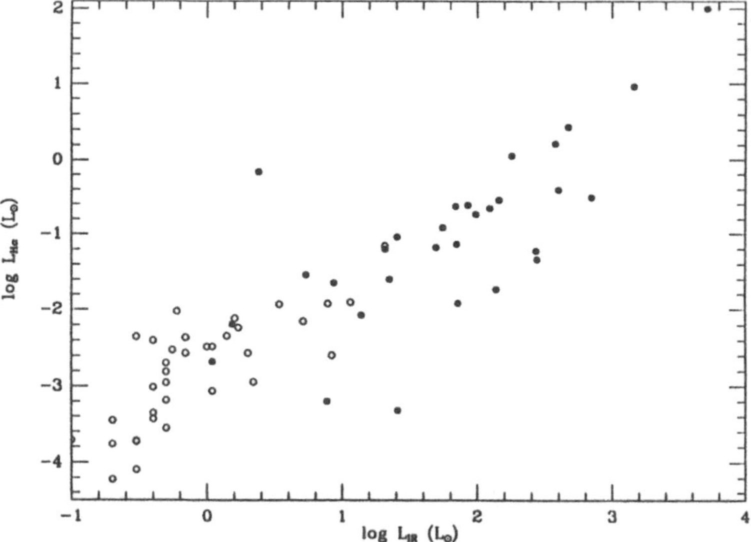

Figure 3: A plot of the Hα luminosity against L_{IR}, the infrared luminosity in excess of photospheric levels integrated over the wavelength range $0.64\mu m \le \lambda \le 100\mu m$ for Herbig Ae/Be stars (filled circles) and for solar-type PMS stars in Taurus-Auriga (open circles; Cabrit et al., 1990).

Correlation between Accretion and Mass Outflow

The accretion luminosity generated by disks surrounding solar-type PMS stars appears well correlated with the wind mechanical luminosity (Cabrit et al., 1990). L_{IR} provides a measure of accretion luminosity, while the wind mechanical luminosity is estimated from the luminosity derived for the broad ($\Delta v > 200$ km/sec) Hα emission line in solar-type PMS stars; this line is formed in the inner regions of the wind, at a distance of ~10 stellar radii from the surface of the star. In Figure 3, we plot the Hα luminosity against L_{IR} for all Herbig Ae/Be stars for which both quantities can be derived. Also included in Figure 3 are the sample of solar-type PMS stars in Taurus-Auriga studied by Cabrit et al. (1990). The correlation between L(Hα) and L_{IR} for the Herbig Ae/Be is strong, and *extends and overlaps the correlation between these quantities derived for solar-type PMS stars*. It is most economical to assume that the fundamental mechanism coupling mass accretion and mass outflows is identical for the two groups of stars. This relationship may be critical to determining the early angular momentum evolution of both intermediate mass and solar-type stars.

Disk Lifetimes

The accretion rate and the disk mass can provide an estimate ($M/\dot{M}_{acc}$) of the lifetime of a disk as a massive, optically thick structure. In Table 1, we list this quantity for the 7 Herbig Ae/Be stars having disk masses deriving from mm-continuum flux measurements. Disk lifetimes estimate in this way lie in the range $0.04 < t < 0.7$ Myr. Comparable values for solar-type PMS stars range from 1 to 10 Myr.

In principle, we should be able to compare these estimates of disk lifetimes, with independent estimates deduced from disk mass determinations for B and A stars of differing ages. For example, disk lifetime could be defined as the average age of the youngest intermediate mass stars for which disk masses are reduced to 0.1× the disk mass characteristic of the Herbig Ae/Be stars. As noted above, however, direct measurements of disk masses are available for only a few young, intermediate mass stars. Hence, we must depend instead on estimates of disk optical depth derived from infrared excesses to provide surrogate measurements of disk mass.

Disks with masses $M \geq 0.1\ M_\odot$ and radii $r < 300$ AU will be optically thick at all wavelengths $\lambda \leq 100\mu m$ provided that the gas/dust ratio and grain size distribution in such disks is identical to that in the interstellar medium; disks with smaller masses will have optical depths $\tau < 1$ at some or all wavelengths $\lambda \leq 100\mu m$. The observed infrared excess relative to the photospheric flux can determine whether the disk is optically thick or thin at a radius corresponding to the wavelength of observation: near-infrared observations probe disk optical depths in the warmer, inner regions of the disk, while far-IR observations probe the cooler, outer disk regions. For example, Figure 4 shows the computed spectral energy distribution for a main sequence A0 star (located at a distance of 150 pc), and the spectral energy distribution for such a star surrounded by a perfectly flat, optically thick reprocessing disk viewed at inclination angles, $i = 0°$ (pole-on) and $i = 80°$. Additional heating due to accretion, or an increase in the emitting solid angle from the disk (e.g. if the disk is slightly flared, with its vertical scale height increasing with distance from the star) can increase the observed infrared spectral energy distribution above the solid line (Kenyon and Hartmann, 1987); inclination effects reduce the observed disk contribution from its maximum for pole-on stars, by a factor ~cos(i).

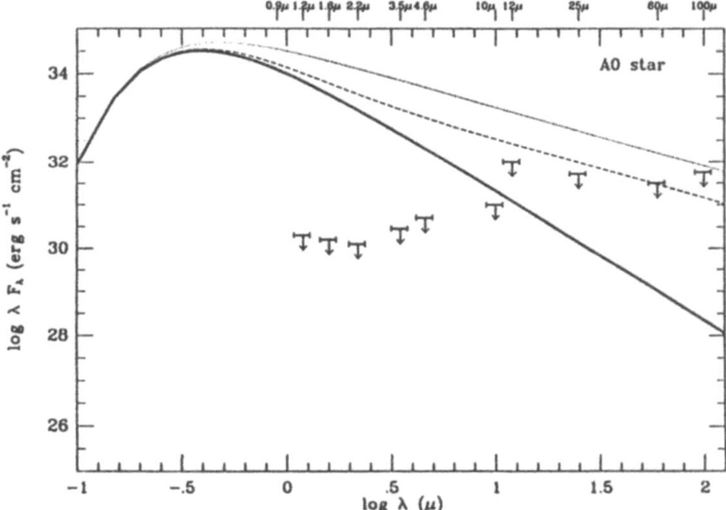

Figure 4: A plot of the spectral energy distribution for a "diskless" A0 star (dark solid line) of radius 3 $R_\odot$, and an A0 star surrounded by a flat reprocessing disk viewed at inclinations $i = 0°$ (pole-on; light dotted line) and $i = 80°$ (light dashed line). Superposed on this plot are the 5σ limiting fluxes which can be achieved in an integration time of 1 hour on the 3-meter diameterr IRTF at wavelengths $\lambda \leq 10\mu m$, along with the limits reached by the IRAS for $\lambda \geq 12\mu m$ for such a star observed at a distance, d = 150 pc.

We will consider stars of a given spectral type showing excess emission falling below the expected emission for a flat, optically thick disk viewed at $i = 80°$ at a given wavelength, λ, to be *candidates* for objects surrounded by disks that are *optically thin* at those radii which dominate contributions to the observed flux at that λ. Among a large sample of such candidates, 1/6 may in fact be optically thick disks viewed at $i > 80°$. We will consider stars showing excess emission falling above this level as *candidates* for objects surrounded by *optically thick* disks, recognizing that among a large sample of such candidates, we could include some optically thin disks with $0.2 < \tau < 1$ (if such disks exist).

With this definition in mind, we can make use of observed infrared excesses to determine the frequency with which optically thick disks surround B and A stars of differing ages. Of the sample of candidate 34 Ae/Be stars listed in the Finkenzeller and Mundt (1984) catalog having infrared and optical photometry adequate to determine excess emission above photospheric levels, all but 9 show infrared excesses ($\lambda \geq 2.2\mu m$) consistent with that expected from optically thick disks. The range of ages represented among this sample is $0.1 < t < 1$ Myr.

The next "snapshot" of disk optical depth is provided by near-infrared ($\lambda \leq 3.5\mu m$) observations of B and A stars in the young, nearby Orion (d = 460 pc) and NGC 2264 (d = 800 pc) clusters (Orion: Penston, 1973; McNamara, 1976; NGC 2264: Warner, Strom and Strom, 1977). In both cases, the youngest cluster stars have ages $t \sim 3$ Myr. For proper motion members (probability ≥ 80%) of these clusters in the mass range $1.5 \leq M/M_{\odot} \leq 3$, and with ages $t \leq 3$ Myr, we find that in Orion, the fraction of stars with optically thick disks (as judged from $2.2\mu m$ flux excesses) is 9/23 (Penston sample), and 10/19 (McNamara sample); in NGC 2264, the corresponding fraction is 7/18 (Warner et al. sample). If 40-60% of the stars in this mass range are still surrounded by optically thick disks, the typical disk survival times must be $t \geq 3$ Myr. Observations of A and late B stars in somewhat older clusters and associations will be required in order to determine an upper limit to the disk survival time for stars in this mass range.

Both Orion and NGC 2264 contain PMS stars with masses $M < 3\ M_{\odot}$ and with ages as small as $t \sim 1$ Myr as judged by their location in the HR diagram. While age determinations are impossible for more massive stars already on the main sequence, it is likely that *at least some* of these stars have ages comparable to those of the youngest PMS stars ($t \sim 1$ Myr). Nevertheless, *none of the 11 stars in NGC 2264 with $M > 3\ M_{\odot}$ have infrared excess emission signifying the presence of optically thick disks* (neither the McNamara nor Penston sample in Orion include a significant number of main sequence B stars). Thus, the massive, optically thick disks which surround the young Herbig Ae/Be stars ($0.1 < t < 1$ Myr) must evolve; by ages $t \sim 3$ Myr, at least the inner regions of such disks (probed by near-infrared flux excesses) are no longer present. Hence, the upper limit on disk survival time provided by observations of B stars in young clusters ($t < 3$ Myr) is consistent with the survival times deduced from disk masses and accretion rates derived for Herbig Ae/Be stars (see Table 1).

These lifetimes should be compared with the survival times estimated from the fraction of solar-type PMS stars surrounded by optically thick disks as a function of age. Skrutskie et al. (1990) find survival times ranging from $t << 1$ Myr to $t \sim 10$ Myr, with a most likely value of $t \sim 3$ Myr.

What is the fate of disk material surrounding intermediate mass stars? Is it all accreted on timescales $t < 1$ Myr, or is some assembled into larger bodies as appears to be the case for solar-type PMS stars?

LATER STAGES OF DISK EVOLUTION

One of the most dramatic revelations of the IRAS survey was the "Vega" phenomenon: the discovery of extended infrared emission associated with α Lyr and several other nearby A stars. Aumann et al. (1984), Gillett, Aumann and Low (1984), and Aumann (1985) argue that this emission very likely arises in optically thin circumstellar disks. Persuasive evidence that this excess emission indeed arises from disks (as opposed to circumstellar shells) is provided by optical observations of light scattered by small dust grains in the nearly edge-on ($i > 76°$) disk surrounding β Pictoris (Smith and Terrile, 1984; Paresce and Burrows, 1987; Artymowicz et al., 1989). It is noteworthy that the observed *size* of the β Pic disk (r ~ 1000 AU) is comparable to the disk sizes estimated from 60μm fluxes for Herbig Ae/Be stars (see above), since an Ae/Be star is presumed to be the evolutionary predecessor of β Pic.

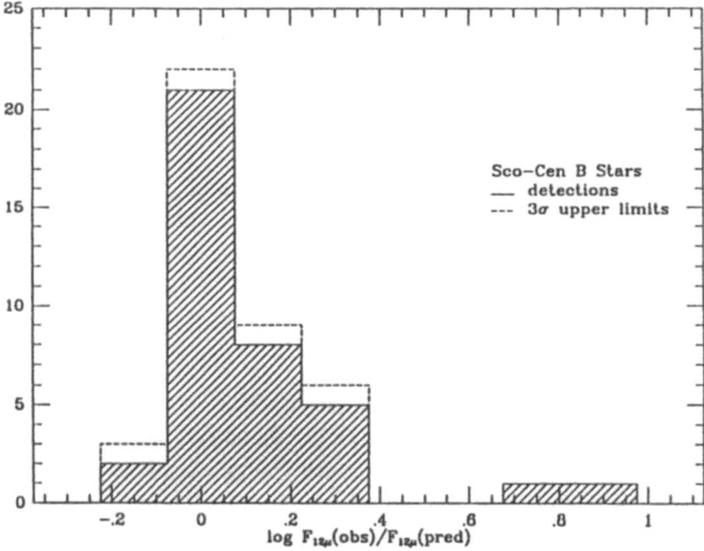

Figure 5: A histogram depicting the frequency distribution of 12μ excesses ($\log_{10}$ [F(observed)/F(photosphere)]) derived for main sequence B stars in the young (t ~7-10 Myr) Scorpio-Centaurus Association; only objects with detections at a level 3σ or greater are included. Where the detection lies below this limit, we estimate the relevant upper limit from the observed background fluctuations in the vicinity of the source; the locations of such objects are indicated by the dotted, open regions of the histogram. We note that candidate optically *thick* disks surrounding a B5 star would have 12μm flux excesses exceeding 1.5 dex; all objects which exhibit significant mid-infrared excesses have excesses *below* this value, and thus could be intermediate mass stars optically *thin* disks.

Examination of the IRAS fluxes for main sequence B stars 1) bright enough to be included in the Catalog of Bright Stars, and 2) for which rotational velocities are available reveals significant infrared excesses (12μm flux excess ≥ 0.2 dex) for 11/94 stars (Waters, 1986). Known Be stars, for which excess infrared emission can be attributed to free-free emission in gaseous envelopes, have been excluded from this sample. Thus, ~12% of *normal* main sequence B stars show 12μm excesses of a magnitude consistent with that expected from optically thin disks. While Water's (1986) result suggests that optically thin disks indeed surround intermediate mass stars, they do not provide the basis for understanding how many such structures form and how long they last: disks are likely to evolve, and ages for individual stars in this visual-brightness-selected sample cannot be determined in general (main sequence lifetimes for stars in the Waters' sample range from 20 Myr at B0 to 500 Myr at A0).

To better understand the later evolution of disks associated with intermediate mass stars, we have embarked on a program aimed at systematically searching for and characterizing optically thin disks associated with B and A stars of known age. Candidate optically thin disks will show infrared emission which lies below the excess emission produced by a flat, optically thick reprocessing disk, but above photospheric emission (see Figure 4). Extant spectral types, optical/near-infrared photometry and IRAS flux measurements allow us to construct reddening-corrected spectral energy distributions, and to search for the relatively small infrared excesses expected in optically thin disks. In Figure 5, we present a histogram depicting the frequency distribution of 12μm flux excesses derived from IRAS co-adds for members of the nearby (d = 160 pc) Sco-Cen association (t ~7-10 Myr) having known rotational velocities and excluding all known Be stars. Nearly 1/3 of this sample shows evidence of excess infrared emission consistent with that expected from optically thin disks; 7/38 (18%) show excess 12μm emission ≥ 0.2 dex, as compared with 12% for the more heterogeneous sample studied by Waters (1986).

If these observed excesses can be confirmed from ground-based 10μm measurements, and if they arise in optically thin disks, the implications could be profound. Excess infrared radiation must be produced by small (a ~1 μm diameter) radiating grains. If these grains are located in a gas-free region, then Poynting-Robertson drag will cause them to spiral from a distance of 20 AU into a central star of spectral type A5 on a timescale, t < 1000 yr (e.g. Burns, Lamy and Soter, 1979). If the grains are suspended within a residual quiescent gas layer, sedimentation and coagulation of small grains into $a > 1$ cm-sized objects can occur in ~1000 $(r/\mathrm{AU})^{3/2}$ yr (Weidenschilling, 1980), where r is the distance from the central star. Aerodynamic drag on these larger bodies forces them to spiral into the protostar in < 10^4 yr (Weidenschilling, 1977). Hence, the detection of near-infrared emission from small grains surrounding stars with t >> 0.1 Myr means that their *inner disks must be re-supplied continuously with fresh grain material* (see Gillett, 1986; Backman and Gillett, 1988).

Fresh grains could be introduced into the inner disk regions via Poynting-Robertson drag operating on micron-size grains located in isolated outer disks; such cold grains would be detectable only from sensitive mm-continuum measurements. The magnitude of the observed near- and mid- infrared excesses produced by small grains introduced into the inner disk will reflect a balance between the rate of injection of fresh grains from the outer disk, and the rate of depletion by Poynting-Robertson or aerodynamic drag in the inner disk. Grains could also be produced *in situ* as a byproduct of collisions which fragment pieces of larger parent bodies (e.g. large grains; planetesimals), or as a result of evaporation of cometesimals gravitationally scattered into the inner disk by one or more massive planets.

It is currently believed that earth-mass planets were built by inelastic collisions of planetesimals over timescales t > 100 Myr; micron-size dust grains are byproducts of such collisions. If the presence of cold, outer disk regions containing a significant store of small grains can be excluded from mm-continuum observations, then measurement of infrared excesses diagnosing optically thin inner disks might establish the rate at which micron-size dust is replenished via planetesimal collisions: a decrease in the infrared excess surrounding older stars would indicate a decrease in the rate of planetesimal collisions. Determining the timescale for infrared excess emission to become undetectable would *thus place an upper limit on the time available to build planets from planetesimals*. It is thus critical 1) to determine disk optical depths around main sequence stars having a wide range in ages; and 2) to establish from mm-continuum observations whether stars that show evidence of infrared emission arising from optically thin inner disks, are surrounded by outer disks comprised in part of cold micron-size dust which might "feed" small grains into the inner disk regions.

It is worth noting again that the optically thick disks which initially surround intermediate mass stars may be quite large (see Figure 2). If so, then their optically thin descendents may also be large, (as is indeed the case for β Pic) and potentially *resolvable from the ground*. For example, a disk of radius 300 AU at the distance of the Sco-Cen association will have an angular radius $\theta \sim 2''$. Coronographic observations in the near-infrared using new generation such as the ESO NTT may thus be able to provide images of radiation scattered by dust grains embedded within such disks.

CONCLUSIONS AND IMPLICATIONS FOR ANGULAR MOMENTUM EVOLUTION

Current observations lead to the following conclusions regarding disks associated with intermediate mass ($1.5 < M/M_{\odot} < 10$) stars:

- disks appear to be a natural outcome of the star formation process for intermediate mass stars;
- rough estimates of disk masses (based on mm-continuum measurements) range from $0.013 < M/M_{\odot} < 2$ for the small sample of Ae/Be stars thus far observed (see Table 1);

- the ratio of disk to stellar masses ranges from 0.008 to 1 (see Table 1). Some young intermediate mass stars appear to be surrounded by disks of mass comparable to the stellar mass;

- rough estimates of disk sizes (based on 60μm flux measurements) range from $30 < r < 3000$ AU; the median value is $r \sim 300$ AU.

- at least 40% of the disks surrounding young intermediate mass stars (Ae/Be stars) must be accreting material, as judged from the magnitude of their infrared excess luminosity compared with the bolometric luminosity of the star. The mass accretion rates characteristic of these stars range from 10^{-4} to $10^{-7.8}$ $M_{\odot}$/yr; the median value is $\sim 10^{5.5}$ $M_{\odot}$/yr.

- accretion luminosity and wind luminosity appear to be well correlated, suggesting that mass accretion and mass outflow are linked;

- massive, optically thick disks which initially surround many, if not all, B and A stars evolve. The survival times for disks as massive, optically thick structures is estimated to lie in the range $0.04 < t < 0.7$ Myr for B stars ($M > 3$ $M_{\odot}$); for stars with masses $1.5 \leq M/M_{\odot} \leq 3$, the survival times are longer, $t >$ 3 Myr. These conclusions are based both on 1) the ratio of disk mass to disk mass accretion rate, and on 2) a comparison of the fraction of young ($t < 1$ Myr) and older ($t > 3$ Myr) B and A stars surrounded by optically thick disks;

- the descendents of massive, optically thick disks must be optically thin structures, perhaps analogous to the structures found associated with the nearby A stars, Vega and β Pic. Candidate optically thin disks are found among field B stars, and among the B and A stars in the nearby Sco-Cen association ($t \sim 7$-10 Myr).

- The presence of optically thin circumstellar disks surrounding a significant number of intermediate mass stars requires continuous replenishment of small grains in the inner disk regions. Candidate sources of replenishment are 1) small grains or cometesimals resident in the outer disk and introduced into inner disk regions; and 2) collisions between large grains or planetesimals.

- if large particles or planetesimals grow in disks surrounding stars with $M > 3$ $M_{\odot}$, they must grow quickly, given the short survival times for disks as massive, optically thick structures ($t < 1$ Myr).

The properties of disks associated with intermediate mass stars provide important clues regarding the role played by disks in the star formation process. It is particularly significant that disk masses for some optically visible Ae/Be stars are comparable to the stellar mass. Given the relatively short survival times ($t < 1$ Myr) for disks surrounding intermediate mass stars implied by high disk accretion rates, it is unlikely that a sample of even the youngest, optically visible B and A stars will show evidence of a significant number of massive disks unless the initial amount of circumstellar material is large. Hence *a significant fraction of the material which ultimately forms an intermediate mass star must pass through a disk.*

It is also noteworthy that the angular momentum/unit mass (J/M) for a "typical" disk surrounding a Herbig Ae/Be star ($M \sim 1\ M_{\odot}$; $r \sim 1000$ AU) is ~0.007 km/sec/pc. The corresponding value of (J/M) for molecular cores with masses in the range $1 < M/M_{\odot} < 10$ is ~0.02 km/sec/pc (Goldsmith and Arquila, 1984). Hence, *disks surrounding intermediate mass stars contain a significant fraction of the initial angular momentum of a typical protostellar core.*

A typical Herbig Ae/Be star has $M \sim 3\ M_{\odot}$, $R \sim 3\ R_{\odot}$ and a rotational speed v ~ 150 km/sec (~0.3× breakup; Finkenzeller, 1985). Hence $(J/M)_* \sim 10^{-5}$ km/sec/pc, nearly 3 orders of magnitude smaller than $(J/M)_{disk}$. If most of the material which comprises intermediate mass stars passes through a disk, a highly efficient mechanism for removing angular momentum from the star is required. The observed correlation between accretion luminosity and wind luminosity suggests that *stellar winds may provide the mechanism by which angular momentum from accreting disk material is carried away from the star.* The detailed physical process(es) which link mass accretion and mass outflow remain to be explored.

Observations of disks associated with intermediate mass stars thus lead to the working hypotheses that (1) stars are built via accretion of material from large ($r \sim 300$ AU), massive ($M \sim 1\ M_{\odot}$) circumstellar disks, and (2) that the initial stellar angular momentum reflects a balance between angular momentum added from the disk, and carried away by a wind.

ACKNOWLEDGMENTS

The authors wish to acknowledge support from the National Science Foundation, the NASA Astrophysics Data Program, and Planetary Program.

REFERENCES

Adams,F., Lada,C.J., and Shu, F. 1987, Ap.J. **312**, 788.

Adams,F., Emerson,J. and Fuller,G. 1990, Ap.J. **357**, 606.

Appenzeller,I., Jankovics,I. and Ostreicher, R. 1984, Astr. Ap. **141**, 108.

Artymowicz,P., Burrows, C. and Paresce,F. 1989, Ap.J. **337**, 494.

Aumann,H.H., Gillett,F.C., Beichmann,C.A., de Jong,T., Houck,J., Low,F.J., Neugebauer,G., Walker,R., Wesselius,P.R. 1984, Ap.J. **278**, L23.

Aumann,H.H. 1985, Pub. Astr. Soc. Pacific, **97**, 885.

Backman,D.E. and Gillett,F.C. 1988, in *Cool Stars, Stellar Systems and the Sun*, ed. J. Linsky and R. Stencel, Springer-Verlag:Berlin.

Beckwith,S., Sargent,A., Chini,R., and Gusten,R. 1990, A.J. **99**, 924.

Bertout,C., Basri,C., and Bouvier,J. 1988, Ap.J. **330,** 350.

Bertout,C., 1989, Ann. Rev. Astr. Ap. **27,** 351.

Burns, J.A., Lamy,P., and Soter,S. 1979, Icarus **40,** 1.

Cabrit,S., Edwards,S., Strom,S.E. and Strom, K.M. 1990, Ap.J. (in press).

Catala,C. 1989 in *ESO Workshop on Low Mass Star Formation and Pre-Main Sequence Objects* ed. B. Reipurth (ESO; Garching bei Munchen)

Edwards,S., Cabrit,S., Strom,S.E., Heyer,I., Strom,K.M., and Anderson,E. 1987, Ap.J. **321,** 473.

Edwards,S., Cabrit,S., Ghandour,L. and Strom,S.E. 1989 *in* ESO Workshop on *Low Mass Star Formation and Pre-Main Sequence Objects,* held in Garching, F.R.G. (July, 1989).

Finkenzeller,U. and Mundt,R. 1984, Astr. Ap. Suppl. **55,** 109.

Finkenzeller,U. 1985, Astr. Ap. **151,** 340.

Gillett,F.C., Aumann,H.H., and Low,F.J. 1984, in Protostars and Planets II, ed. D. Black and M.S. Matthews, University of Arizona Press: Tucson,AZ.

Gillett,F.C. 1986, in *Light on Dark Matter,* ed. F. Israel, Dordrecht: Reidel.

Goldsmith,P. and Arquila,R. 1984, in Protostars and Planets II, ed. D. Black and M.S. Matthews, University of Arizona Press: Tucson,AZ.

Hamann,F. and Persson, S.E. 1989, Ap.J. **339,** 1078.

Herbig,G.H. 1960, Ap.J. Suppl. **4,** 337.

Iben,I. 1967, Ann. Rev. Astr. Ap. **5,** 571.

Kenyon,S.J. and Hartmann,L.W. 1987, Ap.J. **323,** 314.

Kenyon,S.J. and Hartmann,L.W. 1989, Ap.J. **349,** 197.

Lorenzetti,D., Saraceno,P. and Strafella,F. 1983, Ap.J. **264,** 554.

Lynden-Bell,D. and Pringle,J.E. 1974, MNRAS, **168,** 603.

McNamara,B.J. 1976, AJ **81,** 845.

Myers,P., Fuller,G.A., Mathieu,R.D., Beichmann,C.A., Benson,P.J., Schild,R.E., and Emerson,J.P., 1987, Ap.J. **319,** 340.

Paresce,F. and Burrows,C. 1987, Ap.J. **319,** L23.

Penston, M.V. 1973, Ap.J. **183,** 505. Skrutskie,M.F., Dutkevitch, D., Strom,S.E., Edwards,S. and Strom,K.M. 1990, AJ **99,** 1187.

Strom,S.E., Strom,K.M., Yost,J., Carrasco,L. and Grasdalen,G.L. 1972, Ap.J. **173,** 353.

Smith,B.A. and Terrile,R.J. 1984, Science, **226,** 1421.

Strom,S.E., Edwards,S. and Strom,K.M. 1989, in *The Formation and Evolution of Planetary Systems,* ed. H.A.Weaver and L. Danly, Cambridge University Press (Cambridge, UK).

Strom,K.M., Strom,S.E., Edwards,S., Cabrit,S. and Skrutskie,M. 1989, A.J. **97,** 1451.

Strom,S.E., Edwards,S. and Skrutskie,M.F. 1990 in *Protostars and Planets III,* ed. J. Lunine and E. Levy (University of Arizona Press, Tucson).

Waters,L.B.F.M. 1986, Astr. Ap. **159,** L1.

Warner,J.W., Strom,S.E. and Strom,K.M. 1977, Ap.J. **213m** 427.

Weidenschilling,S.J. 1977, MNRAS **180,** 57.

Weidenschilling,S.J. 1980, Icarus **44,** 172.

DISCUSSION

Palla: What are the statistics of the forbidden lines with asymmetries which you mentioned at the beginning of the talk? How many stars show asymmetries? I have a second question. Have you searched for binaries in the these intermediate mass stars?

Strom : I can give you approximate statistics. There are three cases of blue shifted emission of which I am aware out of a sample of 12 stars which have been looked at both by Finkenzeller and Mundt and by us. I should point out that among the T Tauri stars, while it is true that there are a number of dramatic cases of blue shifted emission, a large number - for example in the Cabrit et al sample - have very small blue shifts or forbidden lines near rest velocity as well. So, again, in the cases where you do have blue-shifted emission, it is clear that there is no analogous red-shifted component and you have to invoke a disk. It is also true that the statistics suggest that there are not many objects with blue-shifted emission compared to the sample as a whole, but there are no cases with red shifted emission. With respect to the second question, the best test would be to determine what fraction of the objects in Sco-Cen have companions? This could be answered with an imaging survey, which we have not yet done yet.

Dziembowski: Could you tell me if there are any objects intermediate between Ae/Be stars and T Tauri stars?

Strom: T Tauri is the best. The star, not the class. RY Tau would be another, as would SU Aur. These would all be intermediate in mass between the typical T Tauri stars and Ae/Be stars. Let me tell you what I believe and I believe the observations support. I believe that all stars are surrounded initially by massive optically thick disks. Secondly, that in their early stages, these disks are accretion disks. Thirdly, that the accretion rate and the mass outflow rates are coupled, one to the other - higher mass accretion rate, higher mass outflow rate - and that it is the wind that probably is the regulation mechanism that keeps the angular momentum of the central object from increasing. I don't think that there is any distinction at all between the low mass stars and the high mass stars - that they are totally continuous and that probably both sets of stars are built from disks.

Dziembowski: Are there any pre-main sequence stars which are F stars?

Strom: SU Aur. Well, it is going to be an F star.

Differential Rotation of Fully Convective Pre-Main Sequence Stars

IAN W ROXBURGH
Astronomy Unit,
Queen Mary and Westfield College,
University of London
Mile End Road,
London E1 4NS., UK.

ABSTRACT. The internal rotation of fully convective pre-main sequence stars is investigated using anisotropic and inhomogeneous turbulent viscosity and heat transport coefficients. The models have a large scale circulation and differential rotation. Models with standard isotropic transport also have circulation and differential rotation; the circulation being driven by the latitudinal variation of convective and radiative heat flux in rotating stars in an analogous way to the driving of meridian circulation in radiative layers. Simple model calculations suggest that the differential rotation in pre-main sequence fully convective stars may be large and the stars may be heavily distorted.

1. Introduction

There is no a priori reason to expect fully convective stars to rotate uniformly. The outer layers of the solar convective zone are observed to rotate differentially and the indications from helioseismology are that this differential rotation persists in the interior of the convective region. There have been several attempts to explain the solar differential rotation, ranging from numerical simulations of convection in a rotating shell to turbulent averaged models with anisotropic and/or inhomogeneous turbulent transport coefficients (cf Kippenhahn 1963, Roxburgh 1970). I here apply these turbulent averaged models to rotating fully convective pre-main sequence stars

The phenomenon of differential rotation is readily explained. For equilibrium there must be a balance between the azimuthal component of the frictional force and the rate of advection of angular momentum by any large scale circulation. In the case of anisotropic viscosity, the viscosity tensor is taken to be diagonal with components in spherical polar coordinates of $\eta_{\theta\theta} = \eta_{\phi\phi} = s\eta_{rr}$, with $s \neq 1$ the vanishing of the azimuthal component of the viscous force requires non uniform rotation this in turn requires a meridional circulation v which advects angular momentum, equilibrium is established with $\Omega = \Omega(r,\theta)$ and v non zero.

With inhomogeneous transport coefficients the latitudinal variation of the energy flux drives a large scale circulation which advects both energy and angular momentum; in equilibrium this advection of angular momentum is balanced by viscous forces, again giving $\Omega = \Omega(r,\theta)$ and v non zero. The essential features of these models are therefore governed by the steady state azimuthal component of the equation of motion which can be expressed in the form:

$$\mathrm{div}\,(\rho\,\underline{v}\,\Omega\, r^2 \sin^2\theta + \mathcal{F}) = 0 \tag{1}$$

$$\mathcal{F}_r = r^2 \eta \sin^2\theta \frac{\partial\Omega}{\partial r} + 2\,(1\text{-}s)\,\eta\, r \sin^2\theta\, \Omega, \qquad \mathcal{F}_\theta = s\, r\, \eta \sin^2\theta \frac{\partial\Omega}{\partial\theta} \tag{2}$$

S. Catalano and J. R. Stauffer (eds.), Angular Momentum Evolution of Young Stars, 83–88.

2. Anisotropic Viscosity Models

The models of the solar convective zone by Kippenhahn (1963) and Cocke (1967) have s constant in a spherical shell. For a fully convective star the centre is a point of symmetry where $s = 1$. I therefore take a simple generalisation with $s = 1 - \alpha\,(r/R)^2$. When the Reynolds number is small (Kippenhahn 1963), the vanishing of the frictional force gives $\Omega = \Omega\,(r)$ and hence (Figure 1)

$$r\frac{d\Omega}{dr} + 2(1-s)\,\Omega = 0, \qquad \Omega(r) = \Omega_0 \exp(-\alpha\, r^2/R^2) \tag{3}$$

When the Reynolds number is large, $\Omega = \Omega(\varpi)$ (Cocke 1967, Smith 1970), where ϖ is the distance from the rotation axis. On integrating the azimuthal component of the equation of motion along cylinders ϖ = constant from $z = 0$ to $z = (R^2 - \varpi^2)^{1/2}$ the equation of motion takes the form:

$$\Lambda_1(\varpi)\frac{d^2\Omega}{d\varpi^2} + \Lambda_2(\varpi)\frac{d\Omega}{d\varpi} + \Lambda_3(\varpi)\,\Omega = 0 \tag{4}$$

where Λ_i are integrals over the viscosity distribution. For η = constant this equation has an exact solution (Figure 2):

$$\Omega = \lambda^5 \frac{[\lambda^2 + \varpi^2/4]}{[\lambda^2 - \varpi^2]^{7/2}}, \qquad \lambda^2 = \left(1 + \frac{3}{\alpha}\right)R^2 \tag{5}$$

A third class of such models has $s = s(\theta)$ (Roxburgh 1974); taking $s = 1 + \varepsilon_0\, r^2 + \varepsilon_2\, r^2 \sin^2\theta$, with $\varepsilon_0, \varepsilon_2 << 1$, then $\Omega = \Omega_0\,(1 + \omega_0(r) + \omega_2(r)\sin^2\theta)$ and $\omega_0(r)$ and $\omega_2(r)$ satisfy

$$\frac{1}{r^2}\frac{d}{dr}\left(r^3\eta\frac{d\omega_0}{dr} - 2\varepsilon_0 r^2\right) + 8\eta\omega_2 = 0, \qquad \frac{1}{r^2}\frac{d}{dr}\left(r^3\eta\frac{d\omega_2}{dr} - 2\varepsilon_2 r^2\right) + 10\eta\omega_2 = 0 \tag{6}$$

For η constant this has the solution:

$$\omega_0 = \left(\frac{r}{R}\right)^2\left[\varepsilon_0 - \varepsilon_2\left(\frac{8}{7}\ln\left(\frac{r}{R}\right) + \frac{4}{35}\right)\right], \qquad \omega_2 = \left(\frac{r}{R}\right)^2 \varepsilon_2\left(\frac{10}{7}\ln\left(\frac{r}{R}\right) - \frac{6}{7}\right) \tag{7}$$

Numerical solutions for variable η have also been calculated.

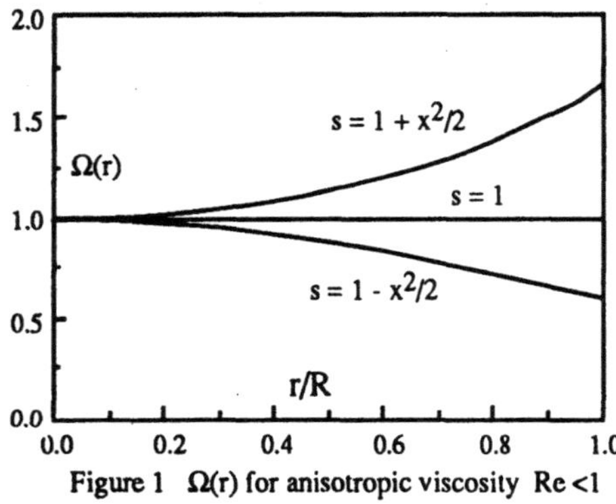

Figure 1 $\Omega(r)$ for anisotropic viscosity Re <1

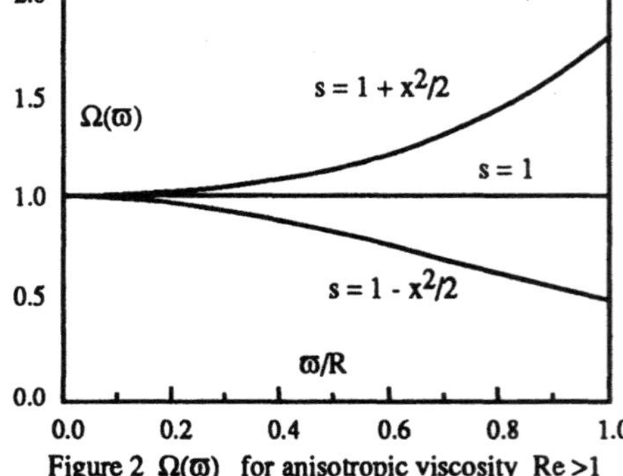

Figure 2 $\Omega(\varpi)$ for anisotropic viscosity Re >1

3. Isotropic turbulent transport coefficients and a von Zeipel problem

The von Zeipel problem, and resulting Eddington-Sweet circulation currents, follows from the result that in a uniformly rotating star, or one where the angular velocity $\Omega(\varpi)$ depends only on distance ϖ from the rotation axis, the surfaces of constant pressure, temperature, and total potential coincide. The radiative energy flux can then be written as $\underline{F_R} = -K(T)\, \nabla T$ and so

$$\mathrm{div}\,(\underline{F_R}) = -K(T)\,\nabla^2 T - \frac{\partial K(T)}{\partial T}(\nabla T \,.\, \nabla T) \tag{8}$$

This is not everywhere zero unless ∇T is constant on surfaces T = constant; this is only the case for spheres and planes (Roxburgh 1965). Equilibrium is established through a circulation that carries the excess energy.

A similar problem arises in convective regions; there the convective energy transport is, to a first approximation $\underline{F_c} = \eta\, \nabla S$ where S is the entropy and η the turbulent (or eddy) transport coefficient; the divergence of the total flux div $(\underline{F_c} + \underline{F_R}) \neq 0$ everywhere and a circulation is required to transport the excess energy. In the radiative layers the entropy gradient is large and the typical circulation speed, V_{ES}, is small, whereas in a convective region the entropy gradient is small and the typical circulation speed, V_{CZ}, is large. To order of magnitude

$$V_{ES} = \left(\frac{\Omega^2 r}{g}\right) V_{Thermal} \approx 10^{-4}\ \mathrm{cm/sec}, \qquad V_{CZ} = \left(\frac{\Omega^2 r}{g}\right) V_{Conv} \approx 10^{+4}\ \mathrm{cm/sec} \tag{9}$$

where g is the acceleration due to gravity. The circulation advects angular momentum which in equilibrium is balanced by the azimuthal component of the viscous force. Details of the solution are contained in section 4 below.

4. Isotropic and anisotropic energy transport models

In this class of models (which includes the isotropic convective von Zeipel models), the circulation velocity v is given by the energy balance equation (cf Durney and Roxburgh 1971). The determination of this velocity requires a detailed modelling of the convective and radiative energy transport in a contacting fully convective star which is beyond the scope of this preliminary investigation. I will here take the velocity to be given by a stream function $\Psi(x,\theta) = \psi(x)\sin^2\theta$, which would follow from the non spherical perturbation to the flux in a uniformly rotating star and which is of the form used by Durney and Roxburgh ($x = r / R$). Expanding the angular velocity as $\Omega = \Omega_o\,[1 + \omega_o(x) + \omega_2(x)\,(3\cos^2\theta - 1)/2]$ where Ω_o is constant, then with $|\omega_o|, |\omega_2| < 1$, the azimuthal component of the equation of motion reduces to

$$\frac{d^2\omega_0}{dx^2} + \frac{d\omega_0}{dx}\left[\frac{4}{x} + \frac{\eta'}{\eta}\right] - 2\frac{\omega_2}{x^2} = -\frac{2}{3}\frac{\psi'}{\eta x^2} \tag{10}$$

$$\frac{d^2\omega_2}{dx^2} + \frac{d\omega_2}{dx}\left[\frac{4}{x} + \frac{\eta'}{\eta}\right] - 10\frac{\omega_2}{x^2} = 4\frac{\psi}{\eta x^3} - \frac{4}{3}\frac{\psi'}{\eta x^2} \tag{11}$$

subject to the boundary conditions $\omega_o' = 0$, $\omega_2' = 0$ at $r = 0, R$. Since the azimuthal component of the equation of motion (Equation (1) above) can be expressed as div $(\underline{\mathbf{A}} + \underline{\mathbf{B}}) = 0$ these equations admit a first integral, obtained by integrating over a sphere of radius r, which is

$$\frac{d\omega_0}{dx} = \frac{1}{5}\frac{d\omega_2}{dx} - \frac{2}{5}\frac{\psi}{\eta\, x^2} \tag{12}$$

A simple model is given by prescribing taking $\psi(r) = \psi_0\, x^4(1 - x)(1 - \beta x)$ where $x = r/R$, where β is determined by the surface boundary condition on ψ, $\psi'' = (2/x + \rho'/\rho)\psi'$ at $x = 1$. If ρ and η are taken to be constant, $\beta = 3/4$ and these equations have the analytic solution:

$$\omega_0(x) = -\frac{1}{270}x^2 - \frac{1}{6}x^3 + \frac{49}{216}x^4 - \frac{2}{25}x^5 \tag{13}$$

$$\omega_2(x) = -\frac{1}{54}x^2 - \frac{1}{6}x^3 + \frac{7}{27}x^4 - \frac{1}{10}x^5 \tag{14}$$

Equations (11) and (12) that determine the differential rotation were solved for variable viscosity η which was taken to be proportional to the density ρ (Figure 3), the density distribution was that of an n=3/2 polytrope (a good approximation for fully convective stars), and the stream function of the circulation $\psi(x)$ was that given in Figure 4. The solutions for ω_0 and ω_2 are given Figure 5.

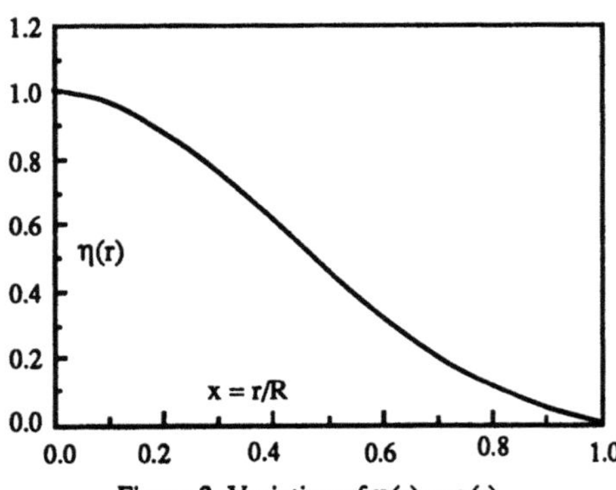

Figure 3 Variation of $\eta(r) \propto \rho(r)$

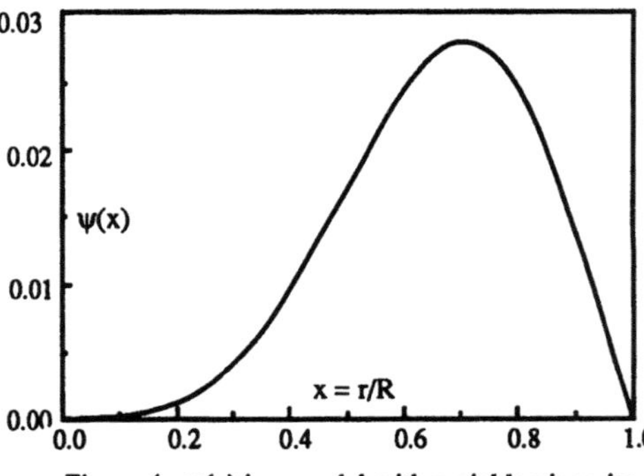

Figure 4 $\psi(x)$ in a model with variable viscosity

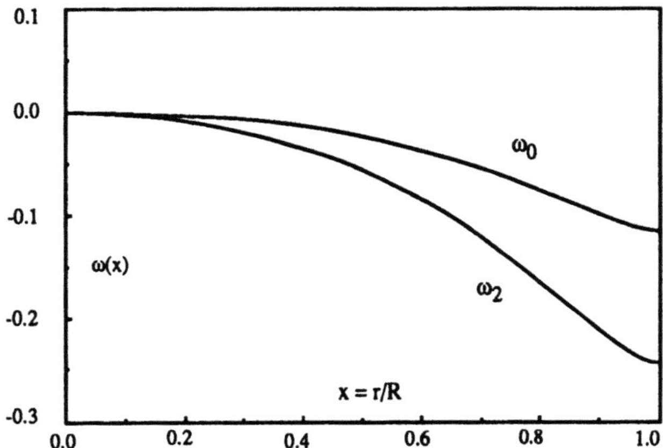

Figure 5 Variation of $\omega_0(x)$ and $\omega_2(x)$ in a model with variable viscosity and stream lines ascending up the rotation axis

5. Conclusion

For Kippenhahn type models where the differential rotation is primarily determined by the anisotropy in the turbulent viscosity, the magnitude of the differential rotation varies with (1 - s). If the horizontal correlations $<v_\theta v_\theta>$, $<v_\phi v_\phi>$ are greater than $<v_r v_r>$, $s > 1$, and with $s = 1$ at the centre s increases outwards; the angular velocity in the surface layers is greater than that in the interior. For Cocke type models the rotation increases with distance from the rotation axis. If $s < 1$, then with $s = 1$ at the centre and $(1 - s) > 0$ the angular velocity decreases outwards by at most a factor of e. For the models with $(s - 1) = \varepsilon_2 r^2 \sin^2\theta$, if $\varepsilon_2 > 0$ then the surface angular velocity is about $0.5\varepsilon_2\, \Omega_0$ smaller than the central value of Ω_0, and conversely greater if $\varepsilon_2 < 0$.

In the new isotropic transport models, and in Durney-Roxburgh anisotropic energy models, the differential rotation is determined by the large scale circulation velocities. Since in the bulk of the star most of the energy is carried by turbulent convection, the circulation velocities are smaller than the convective velocities by the factor which determines the relative angular variation of the energy flux. For the anisotropic Durney-Roxburgh models this is supposed due to the interaction of rotation with convection and may be large, in the new isotropic models this perturbation is essentially the ratio of centrifugal force to gravity, $(\Omega^2 R/g)$, which is small but may approach unity in rapidly rotating stars.

If the dimensionless stream function $\psi > 0$ the circulation velocities are towards the surface along the rotation axis and inwards on the equator, giving an equatorial acceleration and an inward increase of angular velocity. The magnitude of ψ is essentially the ratio of the circulation velocity to the convective velocity, so an estimate of the magnitude of the differential rotation for a given circulation velocity may be obtained by scaling the results in displayed in Figures 5 and 6. The maximum value of ψ in figure 6 is about 0.03, hence for a given value of λ, the ratio of circulation velocity to convective velocity, the differential rotation is obtained by scaling the results in Figure 5 by $\lambda/0.03$. For isotropic transport coefficients λ is essentially $(\Omega^2 R/g)$ so that rapidly rotating stars would have a very large radial and latitudinal differential rotation. If the transport coefficients are isotropic, then by analogy with the von Zeipel problem, $\psi > 0$, and the centre and equator rotate more rapidly than the polar and surface regions. However it is clear that this conclusion is not valid for large rotation since it was obtained from a perturbation analysis in which the differential rotation was assumed to be small and a more careful analysis, including a detailed consideration of the energy transport is needed. However it does suggest that rapidly rotating fully convective stars may have substantial differential rotation and raises the possibility that rapidly rotating young stars may be disc-like. These and other issues will be addressed elsewhere.

References

Cocke W J , 1967, 'On the Soalar Differential Rotation and Meridional Currents', *Astrophys J*, **150**, 1041

Durney B R and Roxburgh I W, 1971,'Inhomogeneous Convection and the Equatorial Acceleration of the Sun', *Solar Physics*, **16**, 3.

Kippenhahn R, 1963, 'Differential Rotation in Stars with Convective Envelopes', *Astrophys J*, **137**, 664.

Roxburgh I W, 1966, 'The Extension of Von Zeipel's Theorem to Rotation Laws with Cylindrical Symmetry', *Appendix to On Stellar Rotation III*, *Mon Not Roy Ast Soc* , **132**, 201.

Roxburgh I W, 1970. 'On the Equatorial Acceleration of the Sun', in *Stellar Rotation* ed A Sletteback, p 318, Reidel

Roxburgh I W, 1974. 'A Note on Anisotropic Convection and the Rotation of Stellar Convective Zones', *Astrophysics and Space Science*, **27**, 419

Smith B 1970, 'Anisotropic Convection and Hydromagnetic Stability in Stars' Ph D Thesis, University of London.

DISCUSSION

MacGregor: Would your results apply equally well to a convective core, if the boundary conditions were the same?

Roxburgh: Yes, in the sense that if you just take the boundary of the convective core as stress-free, etc. etc.

Kraft: An observer would also be interested with the distribution of flux over the surface of a rotating convective star. Of course, Sweet and Roy and what you did and others many years ago did that for a radiative star. But what happens for convective zones?

Roxburgh: I don't know,and I don't think anybody knows because you don't have to ask only what happens in convective zones, you have to ask what happens in the photosphere. You have to model the atmosphere. There have been various attempts including one by myself and a student, but nothing I think would be appropriate to give an observer.

Rodonó: What consequences would your model have for dynamo models?

Roxburgh: The only honest answer is to say that I don't know.

THE X-RAY EMISSION FROM PMS STARS IN TAURUS-AURIGA, AND THE RELATIONSHIP WITH OTHER DIAGNOSTICS OF ACTIVITY

F.DAMIANI, G.MICELA, G.S.VAIANA
Osservatorio Astronomico di Palermo
Palazzo dei Normanni
90134 Palermo
Italy

ABSTRACT. We present a systematic study of the X-ray emission of pre-main-sequence stars in the Taurus-Auriga complex; the sample consists of all catalogued stars in this region observed with the *Einstein* Observatory, resulting in a total of 69 observed stars, 53 of which identified with X-ray sources. We present the X-ray luminosity functions, based on detections and upper limits, for the total sample and for suitably chosen subsamples. We find a definite inverse relation between the X-ray luminosity and the square of the rotational period for the classical T Tauri stars. Moreover an inverse relation exists between the X-ray luminosity and the H-alpha emission, expressed as the ratio between the H-alpha and 25-micron far-infrared luminosity, indicator of disk emission. We argue that this indicates a real inhibition of the wind emission in the most X-ray active T Tauri stars and we suggest a picture in which this phenomenon is likely to take place.

1. X-Ray Observations of Taurus-Auriga Stars

The Taurus-Auriga region is the nearest star formation region, and it has been extensively studied from the point of view of pre-main-sequence stellar activity; here we are mainly concerned with the X-ray emission of the young stars contained in it. The choice of a particular star-formation region allows us to avoid possible inhomogeities resulting from simultaneously considering stars belonging to various regions, which can in principle differ in age, chemical composition, and dynamics. We have selected all stars in this region listed in the Herbig and Bell (1988) catalog of PMS stars (HBC), and among these we have searched all stars which fall into the fields of view of the Imaging Proportional Counter (IPC) onboard the *Einstein* Observatory, with a passband in the range 0.2-3.5 keV; this resulted in a total sample of 69 stars (out of 161 PMS stars in Taurus-Auriga). We have used the latest revision of the reduction algorithm for the IPC data, the so-called REV1B (Harnden *et al.* 1984), in deriving X-ray fluxes and luminosities in an uniform way for the whole sample we study, and independently from previous studies on the same stars. Details about this work can be found in Damiani *et al.* 1990.

T Tauri stars in Taurus-Auriga have been studied in X-rays since the early '80, by various authors; among these are Gahm (1980). Feigelson and DeCampli (1981), Feigelson and Kriss (1981), Walter and Kuhi (1981,1984). The original scope of these studies was an examination of the X-ray properties of optically selected well-known T Tauri stars, without any attempt to do a systematic or complete investigation, and moreover they had been made utilizing earlier and less refined versions of the data reduction algorithm than the one we use here. During these studies were discovered the first so-called Naked T Tauri Stars (NTTS), PMS stars selected just because of their strong X-ray emission, but otherwise lacking the spectral peculiarities of the 'classical' T Tauri stars (CTTS). Systematic studies

S. Catalano and J. R. Stauffer (eds.), Angular Momentum Evolution of Young Stars, 89–95.

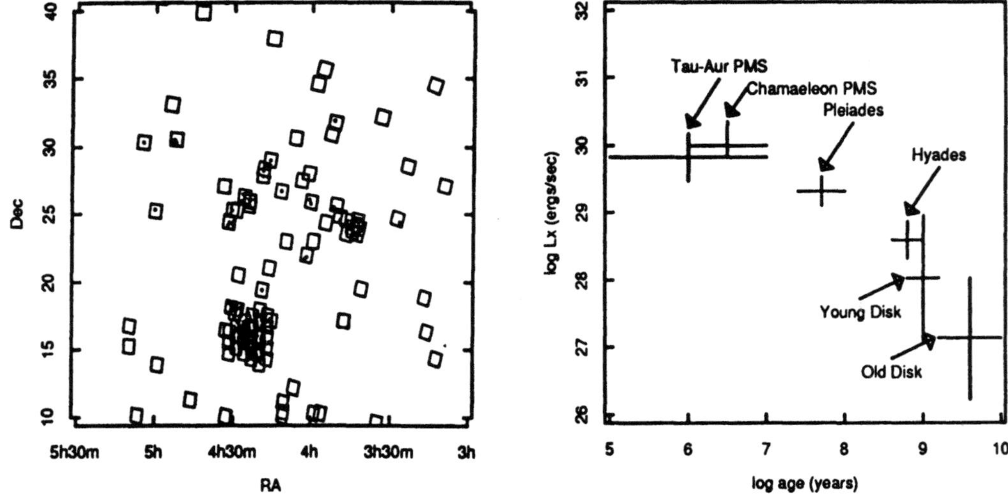

Figure 1: Map of the Taurus-Auriga region: dots are PMS stars, squares are the IPC fields of view.

Figure 2: X-ray luminosity (median and $\pm 1\sigma$ values for each distribution) as a function of age for various groups of stars; the data not pertaining Taurus-Auriga are taken from Feigelson and Kriss (1989)(Chamaeleon region), Micela *et al.* (1990)(Pleiades), Micela *et al.* (1988)(Hyades), Bookbinder (1985)(Young and Old Disk Stars).

on these stars have been more recently carried out, for example by Feigelson *et al.* (1987) and by Walter *et al.* (1988). The latter authors in particular find 28 new stars, bringing the total number of Tau-Aur NTTS to 44 stars, of which 31 in common with our sample. Our work is a systematic and homogeneous study of the whole known PMS population in Tau-Aur, including both CTTS and NTTS, and the first attempt of getting an X-ray luminosity function for these stars.

Figure 1 is a spatial map of the region we are interested in: we can note the inhomogeneous distribution of the PMS stars and the incomplete sky coverage of the IPC observations; this, and the fact that many of the IPC observation in this region are in fact pointed observations on well-known CTTS, could introduce some selection effect on the composition of the sample we study, beyond limiting the completeness of our study. Furthermore a more important kind of bias in the sample we study arises from the composition itself of the HBC catalog we started from; in fact, the majority of the NTTS there contained have been selected just on the basis of their X-ray properties, because their PMS properties are barely recognizable by means of other diagnostics, so that we don't know if there exists a greater population of stars with otherwise similar optical properties, but without such an intense X-ray emission. So that while we can reasonably say that we have studied the X-ray properties of the CTTS in an unbiased way, as these stars are selected independently from their X-ray emission, we cannot be sure to be unbiased with respect to the NTTS. If such class of X-ray-quiet NTTS really exists, one should regard the results we have obtained here as pertaining only to the high-luminosity tail of the whole X-ray luminosity distribution, and the average (or median) X-ray luminosity we find as an upper limit to the true value. In total we have found PMS stars in 35 IPC fields in Tau-Aur, out of 113 fields falling in a 34° × 30° box centered as in figure 1, we have detected 53 of them as X-ray sources, and we have calculated upper limits for the X-ray luminosity for the remaining 16 stars.

Taking into account both the detections and the upper limits, we have built the Maximum Likelihood X-ray luminosity distribution (or briefly luminosity function), and we have compared it with analogous luminosity functions for other groups of stars, namely the Pleiades (Micela *et al.* 1990), Hyades (Micela *et al.* 1988), Young Disk and Old Disk stars (Bookbinder 1985), and the Chamaeleon stars (Feigelson and Kriss 1989). The latter stars have X-ray luminosities and ages very close to those of the Taurus-Auriga stars, but constitute a smaller sample; our result therefore essentially strenghtens what has been found by Feigelson and Kriss (1989), extending its validity also to the Taurus region. This means that the striking difference which is seen between the average X-ray luminosities of the various groups of stars results very likely from an evolutionary effect; this is clear from figure 2, where we plot values of the median X-ray luminosity, with an indication of the dispersion ($\pm 1\sigma$) of the actual values around it, as a function of a representative age range for each group of stars; the trend of the X-ray emission to decrease with the stellar age is a well-established result for stars like the Pleiades and older, now we agree with Feigelson and Kriss (1989) in extending this trend to the PMS stars, so *stars become weaker X-ray emitters as they evolve, starting from the PMS phase* at Taurus-Auriga and Chamaeleon ages.

2. Relations with Other Diagnostics of Activity

In order to examine the possible relations of the X-ray emission with other observed properties of these stars, we have divided our sample of stars in two subsamples, according to different criteria First, the most commonly used criterion to discriminate CTTS from NTTS is the value of the EW of the Hα emission line; we have built X-ray luminosity functions for the stars with EW(Hα)>10 Å and for the other stars; by applying a Wilcoxon test to the two distributions (which takes into account the presence of the upper limits) we find that these are different with a confidence level of 99.81%, so stars which have a different emission-line activity (as measured from the EW(Hα)) have a different behaviour from the X-ray point of view, the most emission-line active stars being the least X-ray luminous ones. This result is in agreement with the statements made by Walter and Kuhi (1981), who consider a smaller sample of stars, and follow a different route to reach this conclusion; however it should be noted that in neither case a tight anticorrelation is found between L_x and EW(α).

We have analogously divided our whole sample according to another diagnostic, namely the far infrared emission, a proxy for the disk luminosity, which is typically observed for the CTTS, and much more rarely present in the NTTS, and we choose for our study the 25μ emission, as observed from the IRAS satellite; by separating stars which have been revealed at this wavelength from those which are not (what amounts to fix a cutoff of 0.1 Janskys for the observed flux), we obtained two subsamples of stars, and applying again a Wilcoxon test to the X-ray luminosity functions of these subsamples we find them to be different at the 93.87% confidence level. In this case we see that stars which have strongly different far-infrared emissions are only marginally different in their X-ray emission levels.

An EW of the Hα line is a quantity much easier to measure than a 25μ flux, and we should ask which of these indicators is the most meaningful from a physical point of view.

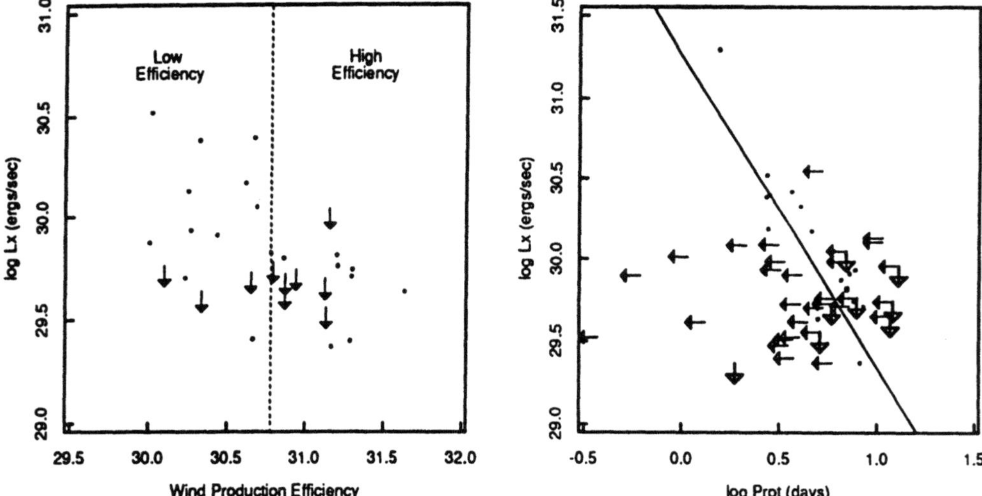

Figure 3: L_x vs. wind production efficiency; the vertical line separates high- and low-efficiency stars.

Figure 4: L_x vs. P_{rot}: dots are photometric periods, arrows are period estimates from $v\sin i$; the straight line is parallel to a best fit to the former data (see text).

It is known that there is a qualitative relation between the presence of an infrared excess and of emission lines (Cohen, Emerson and Beichman 1989), moreover a quantitative correlation exists between the far-IR emission and the Hα *luminosity* (Cabrit *et al.* 1990, who study a sample with a wide range of values of EW(Hα)), and, by using for the IR and the Hα line data existing in the literature, we recover such a correlation for our sample (actually, a direct proportionality, with a scatter of $\sim$ 1.5 orders of magnitude). According to current ideas, the copious IR emission from these stars originates from a circumstellar disk, probably accreting matter on the stellar surface, while the Hα emission (in the CTTS) arises very likely in a stellar wind. Moreover, because these diagnostic appear to be correlated, as we have just seen, has been suggested (Cabrit *et al.* 1990) that there can be some mechanism which relates these two phenomena, i.e. which is capable of using the matter and energy the star gains from the disk to power a wind, detectable in the Hα, [O I], Na I, Ca II, and other emission lines. For this reason we think of the accretion disk and its associated IR emission as a more fundamental property of a T Tauri star than a wind with its Hα emission, as the latter depends on a complex energy transfer at the stellar surface, and may not be a very good indicator of activity, but only one of the possible results of the underlying accretion process. Therefore we prefer to think in terms of strong-IR or weak-IR stars, rather than strong-emission-line or weak-emission-line stars.

In this framework, we can investigate if some meaning can be attached to the slight difference of the X-ray properties of the Hα- or IR-selected classes of stars, as we have previously shown. We have said that L_x is inversely related to the EW(Hα); in the above physical picture, one may look for a quantity which measures how effective the energy transfer process is, and we choose to express this quantity, which can be called 'wind production efficiency', as the ratio between the Hα luminosity and the 25μ flux[1]. The X-ray luminosity does really inversely relate with this efficiency, as it is seen in figure 3;

[1]This ratio is on the average numerically close to the EW(Hα) (apart from constant factors).

indeed if we divide these stars in two classes (low and high efficiency respectively), their X-ray luminosity functions are in fact different at the 99.79% confidence level according to a Wilcoxon test. We conclude that, on the average, the most efficient Hα emitters are the least X-ray active stars, and vice-versa; the scatter of the data points can be attributed to the fact that the measurements of Hα and X-ray luminosity are not simultaneous, and while for many stars variability in Hα has been reported, a given variability in X-rays can be expected to exist if for example these stars have cycles similar to the Solar one, but has not been detected because of the low S/N and the paucity of observational data; we therefore expect that simultaneous measurements in these two diagnostics could produce a much tighter anticorrelation between the L_x and the wind production efficiency.

Another very important relation we have found for our sample stars is the anticorrelation of L_x with the rotational period (figure 4); if we consider only those stars for which we know the period from photometric lightcurves we see a quite definite relationship, of an approximate form $L_x \sim P_{rot}^{-2}$; if we then add all the other stars, for which a period can be estimated from the $v\sin i$ (actually, an upper limit for the true period), the situation becomes more confuse, and no relation is clearly observed. However we like to think of it as an evolutionary effect; in fact, if we consider a best-fit straight line across the datapoints for which we have photometric periods, and we shift it downward (i.e. towards lower L_x) until it divides our sample in two equally-sized subsamples, we find that the stars above and below it have age distributions which are different at the 3σ level, the stars lying above (and so more X-ray luminous for equal P_{rot}) being the youngest ones. This result suggesting a relation between X-ray luminosity and rotational period in these stars is indeed similar to what has been found recently by Bouvier (1990), but with some differences: first, he finds the X-ray surface flux as the best quantity with which P_{rot} is correlated; we have also searched for a correlation between the X-ray surface flux and P_{rot}, and it appears to exist, but is much less tight than the correlation previously seen between L_x and P_{rot} (the correlation coefficients are $r_c \sim 0.6$ and 0.9 respectively), so we think of it as only a byproduct of the relation $L_x - P_{rot}$. Second, Bouvier considers together stars belonging to different regions (Taurus-Auriga, Orion, Chamaeleon, Lupus, Ophiuchus: this can affect the homogeneity of the studied sample), and excludes almost all of the NTTS, retaining only those for which a photometric period is available, without attempting to estimate it from the $v\sin i$ values in the other cases. Third, for the 31 stars in common between his sample and ours the values of L_x appear to be different, and systematically higher in Bouvier's sample; we at present do not know the reason for this, and suspect it may reside in a different correction for the line-of-sight absorption of X-rays.

3. Discussion and Conclusion

There are two main conclusions we can draw from all these observational data, the first regarding the origin of the X-ray emission from T Tauri stars, and the second one about the relations which tie the X-ray emission with diagnostics of circumstellar activity like the Hα or IR emission.

The relation between L_x and P_{rot} can be taken as an indication of a magnetic origin for the X-ray emission (in analogy with other late-type stars, where similar relations between the stellar $v\sin i$ and L_x have been interpreted in the sense of a dynamo-related magnetic activity underlying the X-ray emission processes), i.e. X-rays are emitted by hot plasma confined in closed magnetic structures (loops) above the stellar surface; these loops are well known to exist on the Sun, where of course it is also possible to resolve them spatially. Other circumstantial evidences for such a coronal origin of the X-ray emission from these stars can be the following: first, the emitting plasma is too hot ($T \sim 10^7$ K, as we have inferred from the Hardness Ratios of the X-ray sources) to be confined only by the gravity of these stars; second, the X-ray luminosity is not directly related to any wind or disk indicator; third, there is a high fraction of spotted stars which have been detected in X-rays (18 out of 19, mainly CTTS), as compared with the fraction of detected CTTS (20 out of 34).

If (1) the previous picture of the surfaces of these stars is correct, namely if these surfaces are actually partially covered by closed-field regions, in which the X-ray emission originates, and moreover if (2) the assignments of a strong Hα emission to a stellar wind and of a 25μ emission to an accretion disk are correct, and (3) if the wind is powered by the accretion energy, and the ratio (Hα luminosity)/(25μ flux) is a reliable measure of the wind production efficiency, *then* the anticorrelation we observe between the X-ray luminosity and the wind production efficiency for the CTTS suggests that at the surface of these stars a stellar wind can only be produced in the regions with open field topology, while in the closed field regions the magnetic pressure overcomes the deposited momentum flux, inhibiting the formation of a wind. We note that in this picture a positive feedback effect could also result, because the rotation of the star determines what the level of surface magnetic activity is, in turn the surface closed-field structures act to reduce the intensity of the stellar wind, which in turn is likely to be a braking agent for the star rotation.

REFERENCES

Bookbinder,J. 1985, Ph.D. Thesis, Harvard University.

Bouvier,J. 1990, *Astron. J.*, **99**, 946.

Cabrit,S., Edwards,S., Strom,S.E., Strom,K.M. 1990, *Ap. J.*, **354**, 687.

Cohen,M., Emerson,J.P., Beichman,C.A. 1989, *Ap. J.*, **339**, 455.

Damiani,F., Micela,G., Vaiana,G.S. 1990, in preparation.

Feigelson,E.D., Kriss,G.A. 1981, *Ap. J. Lett.*, **248**, L35.

Feigelson,E.D., DeCampli,W.M. 1981, *Ap. J. Lett.*, **243**, L89.

Feigelson,E.D., Jackson,J.M., Mathieu,R.D., Myers,P.C., Walter,F.M. 1987, *Astron. J.*, **94**, 1251.

Feigelson,E.D., Kriss,G.A. 1989, *Ap. J.*, **338**, 262.

Gahm,G.F. 1980, *Ap. J. Lett.*, **242**, L163.

Harnden,F.R., Fabricant,D.G., Harris,D.E., Schwartz,J. 1984, SAO Special Report, **393**.

Herbig,G.H., Robbin-Bell,K. 1988, *Third Catalog of Emission-Line Stars of the Orion Population*, Lick Obs. Bull. n. 1111.

Micela,G., Sciortino,S., Vaiana,G.S., Schmitt,J H.M.M., Stern,R.A., Harnden, F.R., Rosner,R. 1988, *Ap. J.*, **325**, 798.

Micela,G., Sciortino,S., Vaiana,G.S., Harnden,F.R., Rosner,R., Schmitt,J.H.M.M. 1990, *Ap. J.*, **348**, 557.

Walter,F.M., Kuhi,L.V. 1981, *Ap. J.*, **250**, 254.

Walter,F.M., Kuhi,L.V. 1984, *Ap. J.*, **284**, 194.

Walter,F.M., Brown,A., Mathieu,R.D., Myers,P.C., Vrba,F.J. 1988, *Astron. J.*, **96**, 297.

DISCUSSION

Lamzin: As I remember, only two T Tauri stars which were observed with Einstein show variability in X-rays less than a factor of two. All other T Tauri stars show greater variability than that. For example, for Rho Oph, we have clear evidence for large amounts of variability - sometimes by a factor of 10. How does this affect your relation? If you have these uncertainties due to variability, does not your relation disappear altogether?

Damiani: You are correct that there is variability, and that we generally do not have multiple observations.

Vaiana: These T Tauri stars do not have variability at the level of the Rho Oph stars. The variation we have seen in our sample is less than a factor of two, which would not seriously affect our relation.

Lamzin: As I understand, this is just old data, you have no new data. I know that there are many cases for PMS stars which have very large amounts of variability. Sometimes upper limits, sometimes detections. If you observe the star only once, you cannot say anything.

Vaiana: The fact is that in Taurus we have looked and seen only a factor of two variation, and that would not affect our results.

Hartmann: Have you made any correction for interstellar absorption?

Damiani: Yes, the data reduction takes into account interstellar absorption with a correction factor to the conversion between counts and flux.

Hartmann: But doesn't that correction factor depend very sensitively on the temperature of the spectrum you assume?

Damiani: We have estimated the temperature of the emission spectra for the sources we have observed from the hardness ratio, and from this one can infer the temperature of the emitting plasma.

Hartmann: What is the uncertainty in that temperature?

Damiani: It is better to ask what the uncertainty in the conversion factor is. The uncertainty in the temperature is a little bit higher. The uncertainty in the conversion factor is at most 15%-20%.

MULTIFREQUENCY MONITORING OF RU LUPI: OBSERVATIONAL RESULTS AND A MODEL

F. GIOVANNELLI[1], C. ROSSI[2], L. ERRICO[3], A.A. VITTONE[3], G.S. BISNOVATYI-KOGAN[4], V.G. KURT[4], S.A. LAMZIN[5], E.K. SHEFFER[5]

[1] *Space Astrophysical Institute, CNR, I-00044 Frascati, Italy;*
[2] *Astronomical Institute, Rome University, I-00185 Roma, Italy;*
[3] *Astronomical Observatory of Capodimonte, I-80131 Napoli, Italy;*
[4] *Space Research Institute, Moscow, URSS;*
[5] *Sternberg Astronomical Institute, Moscow, URSS*

ABSTRACT. A large campaign of multifrequency observations of the extreme T Tauri star RU Lupi was performed from 1983 to 1988.

We present some observational results and a first attempt of their interpretation. The various phenomena observed in RU Lupi can be explained in term of a disk accretion onto an young star with a moderately strong global magnetic field. The role of axial rotation is also discussed.

1 Introduction

Angular momenta like magnetic fields, accretion of matter and mass loss processes play a fundamental role in the evolution of pre-main-sequence (PMS) stars. T Tauri stars (TTSs) are young low mass PMS stars. Exhaustive reviews on TTSs have been recently published (Bertout, 1989; Appenzeller and Mundt, 1989), which describe their observed properties and discuss the current theoretical models developed in recent years. Our knowledge on the physical structure of the TTSs is mainly based on observations carried out during the past 10 years at X-ray, UV, optical, IR, submm, and radio wavelengths.

In particular a long term multifrequency campaign on the TTS RU Lupi has been carried out by us, from 1983 to 1988, with the ASTRON and IUE satellites and ESO 1.5 m (low and medium resolution optical spectrophotometry), 1.4 m CAT (high resolution optical spectroscopy), 0.5 m (UBVRI photometry), and 1.0 m (JHKLM photometry and low resolution IR spectrophotometry) telescopes respectively.

RU Lupi is an extreme TTS (with $W_{H_\alpha} \approx 200$ Å) of spectral type late K, $T_{eff} \approx 4400$ K, $L_{bol} \approx 2L_\odot$, $R \approx 2.4R_\odot$, $d \approx 150pc$ (Gahm *et al.*, 1974).

X-ray observations of RU Lupi by HEAO-B satellite did not detected a X-ray flux above $1.2 \times 10^{-13}\, erg\, cm^{-2} s^{-2}$ (Gahm, 1980). Three X-ray observations were obtained, in 1983, 84 and 85 respectively, by ASTRON in the energy range $2 \div 25$ Kev. In one case a positive detection of flux was obtained in the range $2 \div 6$ Kev while in the other two cases an upper limit was only found (Giovannelli *et al.*, 1986).

Like other TTSs, RU Lupi shows strong UV and blue excesses. In IUE band the total luminosity is $L \approx 0.3L_\odot$. Typical IUE spectra are shown in fig.1. The fluxes of the emission lines (as CIII, CIV, SiIII, SiIV and NV) are about 3×10^5 times the solar values (Giampapa, 1984). Strong variations up to 30% $\div$ 40% in the continuum and line intensities occur in

S. Catalano and J. R. Stauffer (eds.), Angular Momentum Evolution of Young Stars, 97–108.

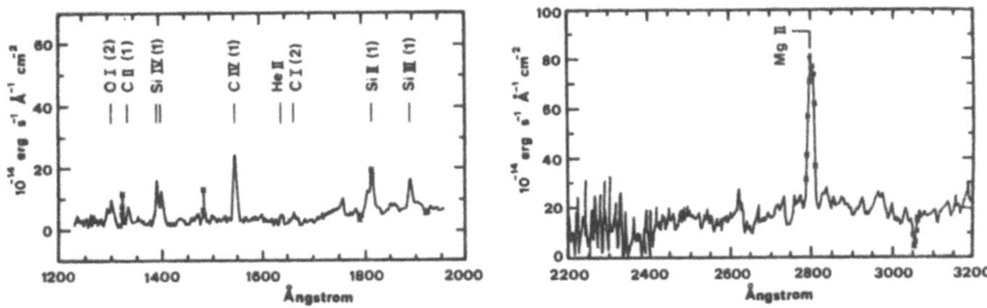

Figure 1: UV spectrum of RU Lupi.

different time scales (days ÷ years) (Giovannelli *et al.*, 1988).

The optical spectrum (see fig.2) of RU Lupi is characterized by strong emission lines of H, CaII, HeI, FeI and FeII. The Balmer lines have sometimes P Cygni profiles indicating the presence of an extended and expanding envelope (Gahm *et al.*, 1979; Schwartz and Heuermann, 1981; Lago, 1982). The blueshifted forbidden [OI] line found in the RU Lupi spectrum (Gahm *et al.*, 1981) can also indicate the existence of a powerful gas outflow. In the optical range the continuum and line intensity variations are smaller than in UV (Giovannelli *et al.*, 1988). Furthermore the detection of a H_2O maser source near RU Lupi (Sealise *et al.*, 1981) and the presence of the Herbig-Haro object HH55 placed 2' South-West of the star, strongly supports a stellar wind model.

In the infrared the energy distribution of RU Lupi shows a strong excess increasing with increasing wavelength (Giovannelli *et al.*, 1987).

In this paper we present some observational results of our campaign of multifrequency observations. The various phenomena observed in RU Lupi can be explained in terms of a disk accretion onto an young star with a moderately strong (about 100 G) global magnetic field.

2 Observational results

2.1 Ultraviolet

We observed RU Lupi with the IUE satellite in 8 shifts between July 1983 and April 1988. Several short wavelength (SW) and long wavelength (LW) spectra have been collected in the low resolution mode. Detailed identifications and an analysis of the emission lines were performed (Giovannelli *et al.*, 1986, 1988, 1990a). We dereddened the spectra using the Seaton's extinction law (Seaton, 1979) with a value of $E_{b-v} = 0.1$, which gave the best correction for the 2200 Å interstellar absorption band.

In this paper we should like to outline the correlation found between the total line fluxes and the total flux of the continuum in the SW range. The correlation (see fig.3) is well fitted by a straight line of slope 0.28. A similar correlation was also found between the continuum level and the intesity of the main lines. It was not possible to find a correlation in the LW range due to problems of continuum definition and accurate identifications of the faint lines.

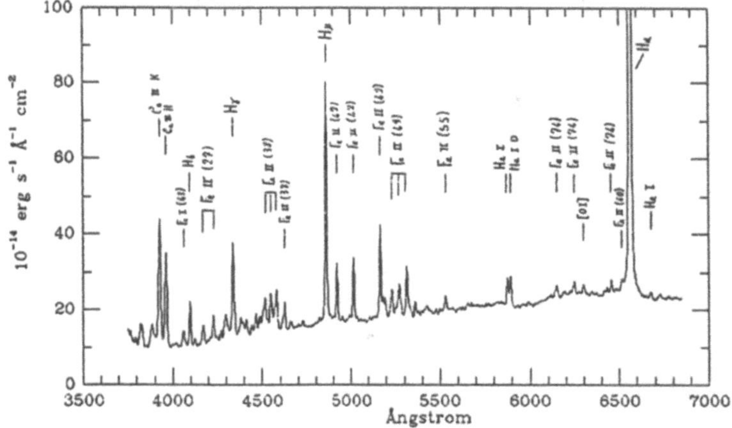

Figure 2: Optical spectrum (dispersion 171 Å/mm) of RU Lupi.

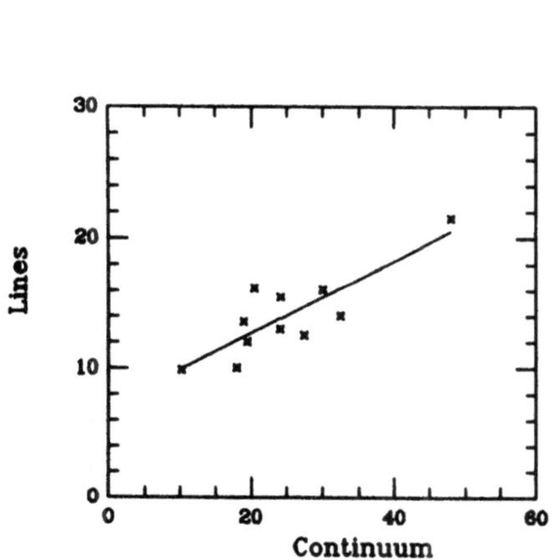

Figure 3: Total line fluxes vs. continuum in the 1200 ÷ 2000 Å range and the best fit with a straight line.

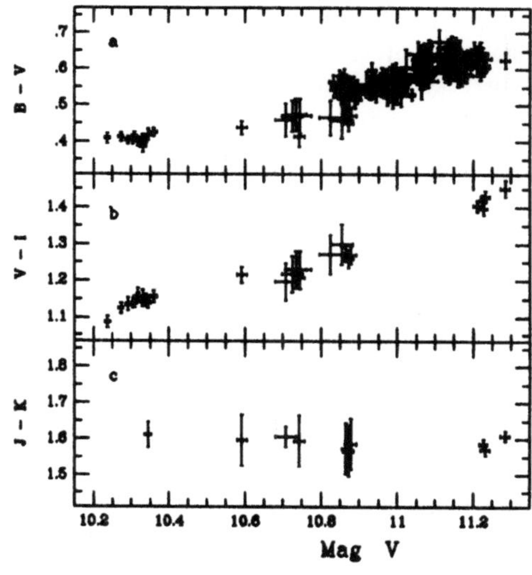

Figure 4: Colour-magnitude diagrams of RU Lupi. a: data of both 1983 and 1986 observation runs; b,c: data of 1986 observation run.

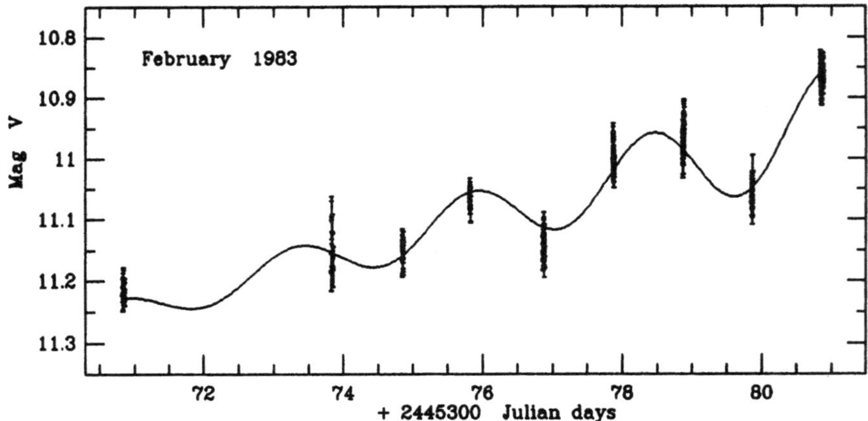

Figure 5: Best fit of RU Lupi light curve assuming the following function: $m_v = Ae^{\lambda t} sin(\omega t + \alpha) + Bt + C$. The best values are: $A = 0.02\,mag$; $B = -0.3\,mag/days$; $C = 11.27\,mag$; $\omega = 2.6\,days$; $\alpha = 3.1\,rad$.

2.2 Optical and infrared

Two simultaneous optical and IR photometric observation runs of RU Lupi were carried out at ESO in February 1983 and June 1986 (Giovannelli et al., 1990b). In June 1986 the star was, on the average, 0.5 *mag* brighter than in February 1983. Furthermore in June we observed a brightness increase up to 1 *mag* on a time scale of 4 nights.

The $(B - V)$ *vs* V diagram for RU Lupi shown in fig.4a clearly indicates that the star is redder when fainter. The $(J - K)$ colour index was costant on all time scales covered by our observations. It is not sensitive to the V variations (see fig.4c).

A correlation between the magnitude and its variations is present in February 1983 light curve (see fig.5). The best fit gives a periodicity of 2.6^d, probably connected with the rotation period. Anyway the period found in our light curve can't be assumed as the rotational one owing to the scantiness of our observations.

The only periodicity of RU Lupi reported in literature is 3.7^d (Plagemann, 1969). In order to better evidentiate a possible periodicity in the light curve of RU Lupi, we analyze two homogeneous data sets with the technique proposed by Horne and Baliunas (1986) for detecting the presence and significance of a period in unequally sampled time series data. Fig.6 shown the light curve from August 1952 to October 1953 (Hoffmeister, 1958), the relative periodogram and magnitude vs phase plot according with $P = 3.7^d$. In our periodogram the 3.7^d peak has a false alarm probability less than 50%. Plagemann found a more outstanding 3.7^d peak analyzing only a subset of the data used in our study. Furthermore there is no evidence of a 3.7^d period in the mag-phase diagram shown in fig.6. Fig.7 shows the RU Lupi light curve from June to September 1959 (Hoffmeister, 1965) and relative periodogram. No 3.7^d periodicity was found in this light curve.

2.3 Energy distribution

During this campaign, we got simultaneous observations from UV to IR in five occasions (Giovannelli *et al.*, 1988). Fig.8 shows the energy distribution of RU Lupi at different epochs. The observed calibrated fluxes were corrected for interstellar reddening using a

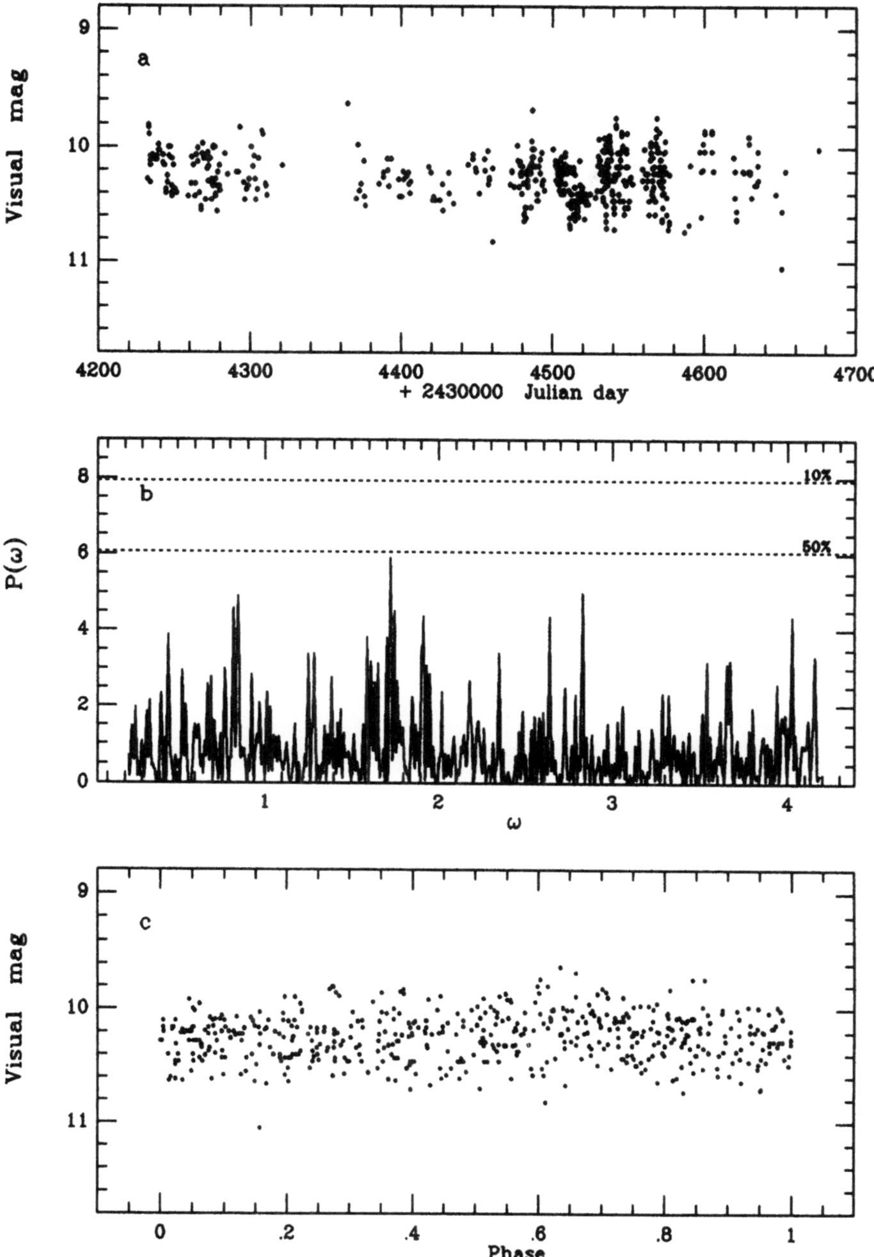

Figure 6: a: RU Lupi light curve from August 1952 to October 1953 (data by Hoffmeister 1958); b: relative Fourier trasform periodogram; c: mag-phase diagram according with the 3.7^d period.

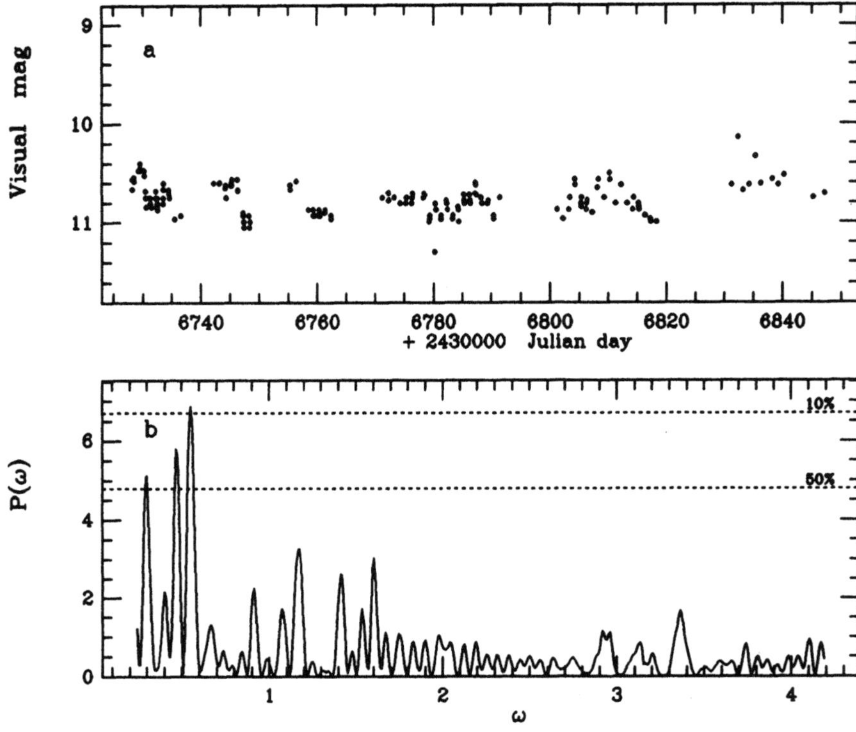

Figure 7: a: RU Lupi light curve from June to September 1959 (data by Hoffmeister 1965); b: relative Fourier trasform periodogram in which two false alarm probability levels are marked.

colour excess $E_{b-v} = 0.1$. A strong variability in UV and optical region is present and, very roughly, RU Lupi may be seen in two different states: a "low" state and an "high" one.

3 The model

The observed energy distribution and other photometric and spectroscopic properties of RU Lupi can be very successful explained by a model in which an young star with a moderately strong global magnetic field is actively accreting from a circumstellar disk. We show in fig.9 the general sketch of the disk model which we propose for RU Lupi.

The problem of angular momentum excess due to the accretion from disk is solved if we assume (at least for the "low" state of RU Lupi) that most of infalling matter does not reach the stellar surface, but is carried off by stellar wind. Only a small part of the accreted plasma freeses into the global stellar magnetic field lines and then slides along them onto the polar regions of the star. This gas is decelerated and heated via shock wave near stellar surface forming, therefore, hot polar ring-like spots. According to us, the relative contributions of the above mentioned regions at different spectral regions are summarized in tab.1.

According to our model the continuum energy distribution of RU Lupi at $\lambda > 1\mu$, in the "low" state, is due to a blackbody radiating star and an optically thick accretion

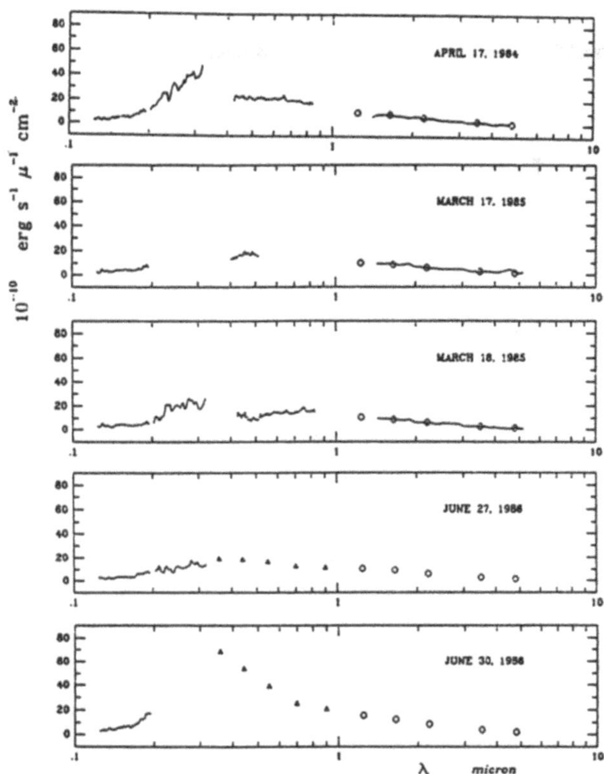

Figure 8: Simultaneous UV, optical and IR energy ditributions of RU Lupi. The lines indicate spectroscopic observations, the triangles and the circles indicate UBVRI and JHKLM photometric observations respectively.

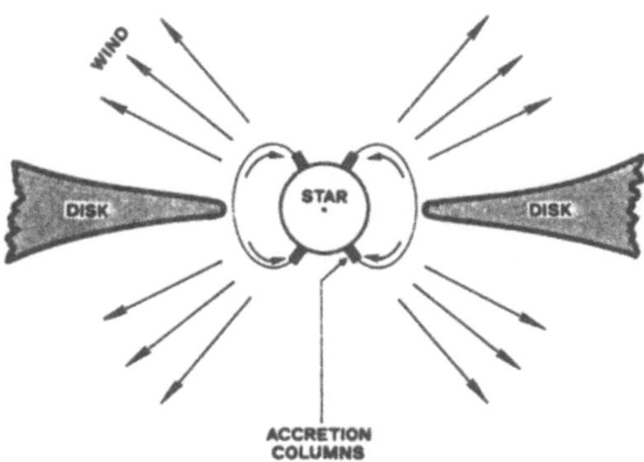

Figure 9: General sketch of our model.

SPECTRAL BAND	MAIN SOURCES OF EMISSION	
	Low state	High state
Far IR	Disk	Disk
Near IR	Disk + Star	Disk + Spots
Optical:		
continuum	Wind + Star	Spots + Wind
lines	Wind + Spots	Spots + Wind
Ultraviolet:		
continuum	Wind + Spots	Spots
lines	Spots	Spots

Table 1: Main sources of emission in our model.

disk. The sources of heating of the disk are the viscosity and absorption of star light. We use the same approximation adopted by Kenyon and Hartmann (1987). The results of our computations are presented on fig.10, where R_* and T_e are the radius and effective temperature of the star rispectively, R_1 the inner boundary of disk in R_* units, $\dot{M}_d$ the accretion rate, h_0 the height of disk at $r=R_1$ in R_* units; z an exponent characterizing the disk shape ($h = h_0 \cdot r^z$). For the outer boundary of disk we obtained only a low limit $R_2 > 300R_*$. Using this value and taking into account the small value of the interstellar absorption ($A_v = 0.3$) found for RU Lupi we can obtain a low limit for the inclination angle: $i < 90^o - \arctan(h_0 \cdot R_2^{z-1}) \approx 60^o$. The value $i = 0^o$ was assumed, for simplicity, in our computations.

If we suppose that this model can describe a middle level of the star activity, adopting the value $P_r=3^d$ for rotational period, it is possible to estimate the characteristic time τ of star spin up: $\tau \sim J_*/\dot{J} < 10^4$ yrs (Bouvier, 1990). This value is less than the lifetime of RU Lupi derived from Hayashi isochrons and lifetime of the disk $t_d \sim M_d/\dot{M}_d > 10^4$ yrs. The disk mass $M_d \simeq 10^{-2}M_\odot$ of RU Lupi was derived from submm observations (Weintraub *et al.*, 1989). This means that the main part of angular momentum must go away via stellar wind. We suppose that this wind must be relatively cold (T $\sim 5 \div 8 \cdot 10^3 K$), dense ($N_e > 10^{11}\, cm^{-3}$), neutral ($N_e/N_H \ll 1$), and with a terminal velocity and a mass loss rate $V_\infty \geq 100\, km/s$ and $\dot{M}_w \approx \dot{M}_d$ respectively.

At the moment we can't directly confirm our supposition, because it is difficult to separate the relative contribution of wind and hot spots in the optical continuum and/or Balmer emission lines. Anyway similar parameters were obtained for the gas outflow of some classical TTSs (Giovanardi *et al.*, 1990). The relatively high value of $\dot{M}_w \sim 6 \cdot 10^{-7}\, M_\odot/year$ is in agreement with the large H_α equivalent width ($\approx$ 200 Å) observed in RU Lupi.

The blueshifted (up to 250 km/s) [OI] 6300 Å line in the RU Lupi spectrum is another independent argument in favour of the existence of a powerful gas outflow with high values of V_∞ and $\dot{M}_w$. The mechanisms of acceleration and heating of the wind as well as its geometry are unknown up to now.

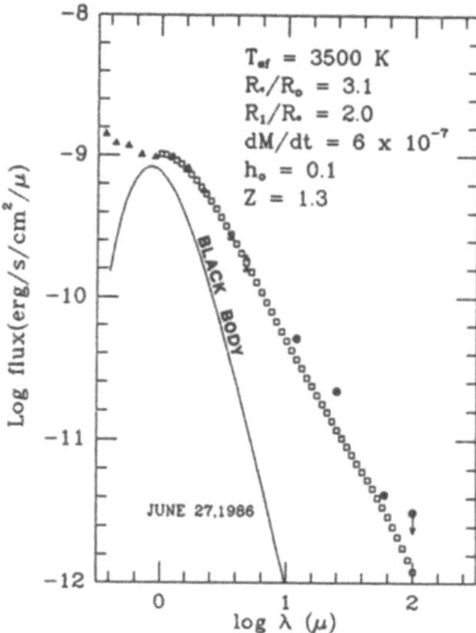

Figure 10: Comparison between theoretical and observational spectra. The square show our model, the triangles indicate UBVRIJHKLM photometric observations and the filled circles indicate IR measurements by IRAS.

The set of parameters obtained for our disk model is not unique. Anyway the crucial point of our model is the existence of a solution with a large cavity enough between the star surface and the inner boundary of accretion disk. We suppose that the cavity is due to the interaction of a global stellar magnetic field with the accretion disk. In fact, the fast rotational speed $V_e = 2\pi R_*/P_{rot} \sim 50\,km/s$ and an huge convective zone are good reasons to assume that RU Lupi has at least a global magnetic field of the order of $100\,G$. It seems very naturally, that only a small part of the gas accreted from the disk freeses into magnetic field lines of the magnetosphere and then slides down on to polar regions of the star. This gas, accelerated by gravity up to velocity $V_{ff} \sim 300km/s$, falls onto stellar surface and then decelerates and heats via shock waves. The maximum value of ion temperature in the shock front is of order of $10^6\,K$, on the contrary the maximum value of electron temperature must be less, due to electron heat conductivity and large radiative losses (Zeldovich and Rayzer, 1966). This is a possible explanation of the relatively small X-ray luminosity found for RU Lupi.

The regions, where emission lines of highly ionized ions such as NV, CIV, SiIV, etc. are formed, must be placed both after shock front and before it.The ionization source is due to a preheating of the infall plasma via electron heat conductivity. This circumstance and the specific orientation of RU Lupi relative to the Earth may be the explanation of the observed redshift of HeII 4686 Å line on the April 16, 1985 spectrum of RU Lupi as shown in fig.11. The HeII 4686 Å line is optically thin and its redshift is a direct indication of a gas infalling on the line formation region (Lamzin, 1989)

As we can see in fig.10 even in the "low" state gas free-free and free-bound radiation is the chief source of continuum emission in the UV and optical spectral regions. According to this the changes of energy distribution shown in fig.8 can be due to variations of relative emission contribution of hot polar spots and cold outflowing gas. More precisely, we suppose that the variability is mainly due to a variation of the accretion mass rate $\dot{M}_{ac}$ onto the

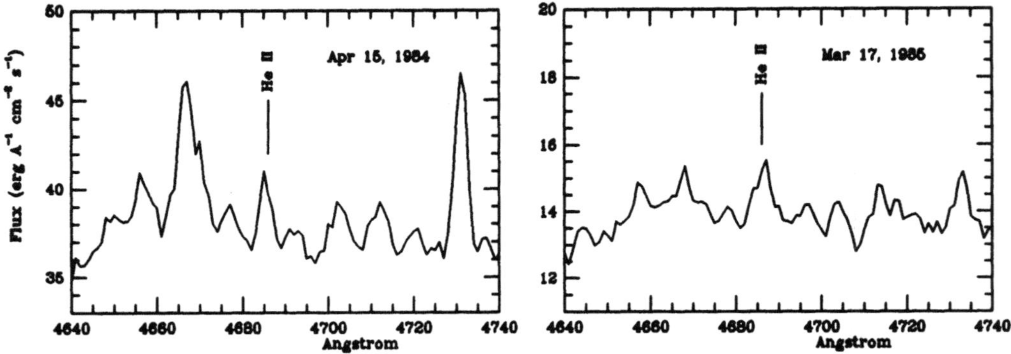

Figure 11: HeII 4686 line profiles of RU Lupi at different epochs.

stellar surface. This variation may be due to some instabilities in the magnetosphere near the inner boundary of accretion disk and/or to the variability of the strength of the stellar magnetic field. At the "low" state $\dot{M}_{ac} \ll \dot{M}_d$ and the chief source of continuum radiation at $\lambda < 1\mu$ is the gas emission in the wind. $\dot{M}_{ac}$ increases both the intensity of UV lines and continuum as it is clearly shown in fig.3 At the same time the star becomes bluer (see fig.4a) owing to an increasing contribution of the hot gas emission. As far as the disk is a main source of IR emission and taking into account that the color index $J - K$ does not changes significantly (see fig.4c), we suppose, that the disk accretion rate $\dot{M}_d$ is more or less stationary. Let us suppose now that the RU Lupi rotational axis is not coincident with the magnetic one. On this assumption the periodic variations of the star brightness are due to the hot spot movements relative to the Earth, as in the case of X-ray pulsars. When the star becomes fainter ($\dot{M}_{ac}$ decreases) the contrast of hot spots must fall down as it is shown in fig.5.

Finally, we suppose that different orientations of (magneto) polar accretion columns relative to the Earth can explain the HeII 4686 Å line profile variations shown in fig.11. We have a redshifted line profile (fig.11a) when we see one magnetic polar region preferentially while a nearly symmetric one (fig.11b) is observed when we see two polar regions.

Acknowledgements: One of us (S. Lamzin) would like to thank the director of Naples Astronomical Observatory of Capodimonte for hospitality and the local organizing committee of the NATO advanced research workshop "*Angular momentum evolution of young stars*" for invitation and grant.

References

Appenzeller I., Mundt R.: 1989, *Astron. Astrophys. Rev.* **1**, 291

Bertout C.: 1989, *Ann. Rev. Astron. Astrophys.* **27**, 351

Bouvier J.: 1991, *this proceeding*

Gahm G.F., Nordh H.L., Olofsson S.G., Carlborg N.C.J.: 1974, *Astron. Astrophys.* **33**, 399

Gahm G., Lago M.T.V.T., Penston M.: 1981, *Monthly Notices Roy. Astron. Soc.* **195**, 59p
Gahm G.F.: 1980, *Astrophys. J.* **242**, L163
Giampapa M.S.: 1984, *Lecture Notes in Physics* **193**, 14
Giovannardi G., Gennari S., Natta A., Stanga R.: 1986, *Astrophys. J.* in press
Giovannelli F., Vittone A.A., Rossi C., Nastari I., Bisnovatyi-Kogan G.S., Golinskaya I.M., Kurt V.G., Misyakina I.A., Shafer E.Yu., Shamolin V.M., Sheffer E.K., Smirnov A.S., Lamsin S.A., Larionov M.G., Sidorenko V.N.: 1986, in *New Insight in Astrophysics - 8 years of UV Astronomy with IUE*, ESA SP-263, 95
Giovannelli F., Rossi C., Covino E., de Martino D., Errico L., Vittone A.A.: 1987, *Mem. Soc. Astron. Ital.* **58**, 205
Giovannelli F., Rossi C., Covino E., Errico L., Vittone A.A., Bisnovatyi-Kogan G.S., Kurt V.G., Sheffer E.K., Lamsin S.A.: 1988, in *A decade of UV Astronomy with IUE*, ESA SP-281, **2**, 101
Giovannelli F., Rossi C., Errico L., Vittone A.A., Bisnovatyi-Kogan G.S., Kurt V.G., Sheffer E.K., Lamsin S.A.: 1990a, *In the Proceeding of the IUE Conference, Tolouse, France*
Giovannelli F., Errico L., Vittone A.A., Rossi C.: 1990b, *Astron. Astrophys. Suppl. Ser.* in press
Hoffmeister C.: 1958, *Veröff. der Sternwarte in Sonneberg* **3**, 333
Hoffmeister C.: 1965, *Veröff. der Sternwarte in Sonneberg* **6**, 93
Horne J.H., Baliunas S.L.: 1986, *Astrophys. J.* **302**, 757
Kenyon S., Hartmann L.: 1987, *Astrophys. J.* **323**, 714
Lago M.T.V.T.: 1982, *Monthly Notices Roy. Astron. Soc.* **198**, 445
Lamsin S.: 1989, *Soviet Astronomy* **66**, 1330 (in Russian)
Plagemann S.: 1969, *Mémoires Soc. Roy. Sciences Liège, 5th series* **19**, 331
Schwartz R.D., Heuermann R.W.: 1981, *Astron. J.* **86**, 1526
Sealise E., Gahm G.F., Sandell G.: 1981, *Astron. Astrophys.* **104**, 166
Seaton H.J.: 1979, *Monthly Notices Roy. Astron. Soc.* **187**, 73p
Weintraub D., Sandell G., Duncan W.: 1989, *Astrophys. J.* **340**, L69
Zeldovich Ya.B., Rayzer Yu.P.: 1966, *Physics of shock waves and high temperature hydrodynamics phenomena*, Academic Press

DISCUSSION

Hartmann: If you had a high velocity accretion shock, you might expect to see high temperature ions, in particular NIV. I thought in earlier spectra that it wasn't clear if it was present. Is it present in your spectra?

Vittone: Yes, it is present.

Kraft: I couldn't see very well in the pictures. Can you confirm whether the HeII emission line 4686 was symmetrical in RU Lupi, as opposed to DR Tauri which was quite unsymmetrical? In DR Tauri, it looks to me that you have a great deal of continuous emission superposed on this emission line, and it is covering up what looks to me like what might be an incipient P Cygni line profile, which means that the underlying emission is not so red-shifted as you think it is. Now, I cannot interpret RU Lupi - it justs looks more or less symmetrical. Are you saying that it is different or is it the same?

Vittone: That is what I am saying. I think it is the same. Of course, we must obtain high resolution spectroscopy to speak about this more. We will try to detect periodicity in the line profile shape. What about P Cygni profile? Maybe, but I don't believe. Because I am sure this line is optically thin. But, maybe.

Dziembowski: I have difficulty to understand the physics behind your assumptions. Like the rate of accretion is exactly the same, or almost exactly the same, as the wind mass loss rate. My physical difficulty is that you require the following situation. You use all the energy from the rotation to heat material and then in a very efficient way to transfer again the mechanical energy of the wind. Do you see the problem? Is there enough energy?

Lamzin: In principal, yes there is a problem. If you have accretion dissipation, radial velocity is very low, so not much energy is lost in stopping the accretion. But, I agree that the calculation must be done. No doubt that we have disk accretion, no doubt that we have wind. But, about the theory, we have only speculation.

Belvedere: What is the origin of your mass loss rate variation with time?

Lamzin: I suppose that this is due to the variation in strength of the magnetic field. If we have alfvenic radius smaller than corotation radius, accretion must stop. This means that we must have some difference in the process when the alfvenic radius changes.

Belvedere: I believe it is quite difficult to explain things with this sort of idea. Also, in the case of cataclysmic variables, sometimes one must give some dark models to try to explain these observations. The magnetic field is quite difficult to understand.

Lamzin: Yes, but please understand that our J-H and H-K color are stable, and our infrared part of spectra depends mainly on mass loss rate. And if it is constant, that means that the mass loss rate is mainly constant.

POLARIMETRY AND CCD IMAGING OF HERBIG Ae/Be STARS AND STAR FORMING REGIONS

F. Scaltriti (1), V. Piirola (2), M. Robberto (1), M. Busso (1)

(1) Osservatorio Astronomico di Torino - Pino Torinese (Torino) - Italia
(2) Observatory and Astrophysics Laboratory - Helsinki - Finland

1. INTRODUCTION

The sample of Ae/Be stars listed by Herbig (1960) are prime candidates for early-type, pre-Main Sequence stars. Their characteristics, as described by Herbig (1960) and the spectroscopic studies by Strom et al. (1972) and by Garrison and Anderson (1977), indicate that the objects are young with features such as emission lines, association with dust clouds and reflection nebulosity, irregular variability, infrared excess and molecular line emission. These objects are recognized as representing a group of higher-mass analogs of the T Tauri stars.

Many of these features suggest that the presence of circumstellar dust, perhaps left-over from contraction, is common along the objects of this group. This could cause a net polarization of starlight as the result of scattering from material which is distributed non uniformly around the star. Also scattering by free electrons in the partially ionized circumstellar envelope contributes. Several polarimetric investigations of these objects exist; even if there is a lack of systematics in this kind of research, the main characteristics that can be drawn are the following:

1) the Herbig Ae/Be stars generally show appreciable intrinsic polarization;
2) the wavelength dependence of polarization seems to vary from star to star;
3) for an appreciable amount of the objects we may infer a substantial time variability in polarization.

In general, we may notice that polarimetry constitutes a technique which probe directly the geometry, optical depth and composition of the circumstellar matter.

Herbig-Haro (H-H) objects are semi-stellar emission nebulae associated with star forming regions whose characteristics have been extensively discussed by several authors (see, for example, Mundt 1987). The existing observations indicate that many H-H nebulae are the product of supersonic bipolar mass outflows from young stellar objects; the exciting stars are probably low-mass stars, perhaps of the T Tauri class (Cohen and Schwartz 1983).

Linear polarization measurements of several H-H objects can be found in Strom et al. (1974), Vrba et al. (1975), Schmidt and Vrba (1975). Generally, quite high values (ranging 5-30%) of linear polarization have been observed as a characteristic of the reflection nebulae. However, for HH1 and HH2 (Strom et al. 1974) quite low detections have been obtained.

S. Catalano and J. R. Stauffer (eds.), Angular Momentum Evolution of Young Stars, 109–115.

2. THE OBSERVING PROGRAM

In 1990 we started an observational program in order to get UBVRI polarization data on Herbig Ae/Be stars and on H-H objects. Our sample includes BD+67°1283, BD+40°4124, BD+41°3731, BD+46°3471, BD+65°1637, HD 200775, PP2-3, PP99, GY2, GY5. The last four objects belong to the compilations by Parsamian and Petrosian (1979) (PP objects) and Gyul'budagyan (1982) (GY objects); they have been selected because we got for them also V,R,I,H-alpha CCD imaging and near-infrared (JHKL) photometry (Persi et al. 1988a,b; Origlia et al. 1990).

The magnitude range in our sample is 7.4-14.5 mag in V light.

Here we present preliminary results concerning HD 200775 and BD+65°1637.

The polarimetric observations have been performed during August, September and October 1990 in two sites:
1) at Asiago Astrophysical Observatory, employing the 1.82-m telescope;
2) at La Palma, using the 2.56-m NORDIC telescope.

Table 1. Polarization data for HD 200775 and NGC 7129-S

P(%)	P.A.	P(%)	P.A.
- HD200775		- NGC 7129-S (Object II)	
U) 0.778 +/- 0.019	97.9 +/- 0.7	U) 0.378 +/- 0.131	114.1 +/- 9.5
B) 0.889 +/- 0.033	94.9 +/- 1.1	B) 0.287 +/- 0.053	44.1 +/- 5.2
V) 0.903 +/- 0.014	96.0 +/- 0.5	V) 0.358 +/- 0.060	70.0 +/- 4.8
R) 0.784 +/- 0.013	97.0 +/- 0.5	R) 0.128 +/- 0.093	47.6 +/- 18.0
I) 0.716 +/- 0.011	96.7 +/- 0.4	I) 0.447 +/- 0.208	61.0 +/- 12.5
- NGC 7129-S (Object I)		- NGC 7129-S (Object III)	
U) 0.240 +/- 0.148	123.7 +/- 15.9	U) 1.067 +/- 0.036	95.3 +/- 1.0
B) 0.202 +/- 0.049	108.7 +/- 6.8	B) 1.067 +/- 0.029	102.1 +/- 0.8
V) 0.174 +/- 0.104	76.1 +/- 15.4	V) 1.140 +/- 0.041	103.6 +/- 1.0
R) 0.151 +/- 0.036	81.1 +/- 6.7	R) 0.949 +/- 0.024	104.8 +/- 0.7
I) 0.030 +/- 0.085	100.6 +/- 35.3	I) 0.815 +/- 0.033	110.6 +/- 1.2

In both cases we used the simultaneous five colour (UBVRI) polarimeter constructed according to the original design by Piirola (1973,1988). By means of observations on high and null polarization standards we have taken into account the contribution of the instrumental polarization and the correction for the zero reference of the position angle.

In June 1990 we obtained deep V,R,I CCD images of the region NGC 7129-S (where BD+65°1637 is located). The observations were made using the RCA CCD at the 1.52-m telescope of Loiano (the observing site of Bologna Astronomical Observatory). The scale of the system is 0.51 arcsec/pixel; the field of view on the 320x512 array turns out to be about 163x261 arcsec.

3. POLARIMETRY OF HD 200775

This object has been recently studied polarimetrically by Pfau et al. (1987, hereafter PPR), who determined the contributions from the intrinsic (circumstellar) and interstellar polarization components. The method was based on the assumption that the position angle of the intrinsic polarization does not change during the time interval of the observations. Accordingly, when the degree of the intrinsic polarization changes, the observed polarization points define a straight line in the (Px, Py) plane, where Px = P cos[2(theta)] and Py = P sin[2(theta)] are the components of the polarization vector. The wavelength dependence of the intrinsic

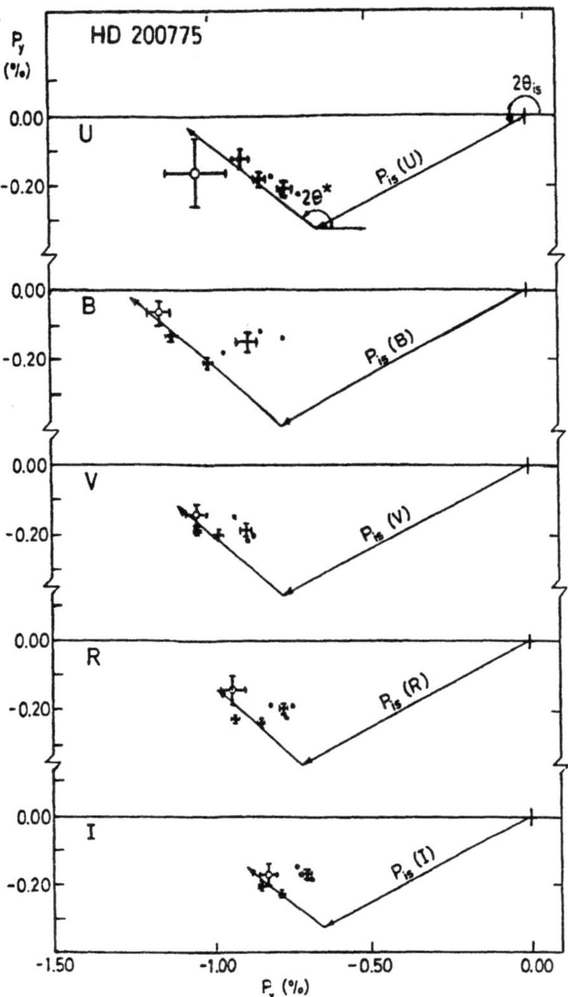

Figure 1 - UBVRI polarization observations of HD 200775 expressed in terms of the polarization vector components Px and Py. Our new data (August 1990) are given by the filled dots and the adjacent error bar crosses show their mean values with the standard error of the mean. Earlier data and the determined intrinsic and interstellar polarization components from Pfau et al. (1987) are also shown.

component PPR obtained for HD 200775 resembles that of classical Be-star polarization, produced by electron scattering in an extended circumstellar envelope.

Most of the interstellar-type component is apparently produced within the cloud, but shows a wavelength dependence and thus the grain size distribution rather similar to the surrounding interstellar space.

We show our new observations of HD 200775 in Fig. 1 together with the earlier data from PPR, and the intrinsic and interstellar polarization vector determinations referred above. Our observed points do not lie exactly on the intrinsic polarization line of PPR, which means that the position angle of the intrinsic polarization has not remained constant over long time intervals (several years). The shifts of the points corresponds to a position angle change of 10-15 degrees. Such changes are not uncommon in classical Be stars even on shorter time scales. Since the direction of polarization fluctuates around a mean value (perpendicular to the equatorial disk plane) the accumulation of more new data will make it possible to improve the determinations of the intrinsic and interstellar polarizations later.

We give in Table 1 the average polarization for HD 200775 from our three observing nights in August 1990, since no statistically significant variations from night to night are present. Attention should be paied to the different diaphragms used by the various authors in the literature; we can remind that HD 200775 is surrounded by the reflection nebula NGC 7023 which, in turn, is a part of the dark cloud Lynds 1172.

4. POLARIMETRY OF NGC 7129-S

BD+65°1637 belongs to the complex NGC 7129-S in which more than 15 objects are included according to Hartigan and Lada (1985). Adopting the identification by these authors, we have observed polarimetrically the objects I, II, and III (BD+65°1637).

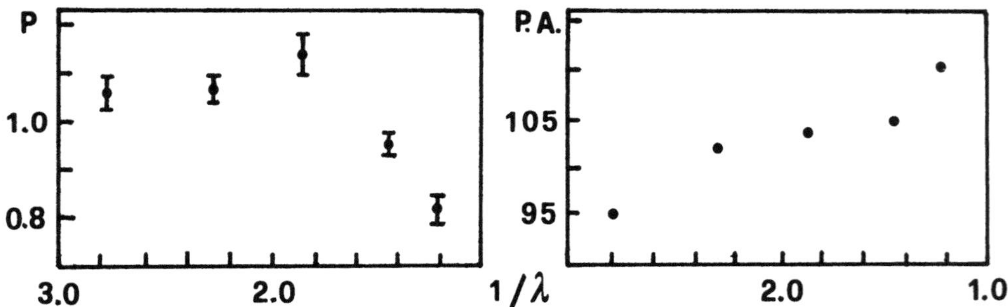

Figure 2 - Polarization data versus wavelength for object III (BD+65°1637).

In this case too the reflection nebulosity NGC 7129-S may supply a significant contribution to the polarization measured for BD+65°1637; at La Palma we have performed multiaperture observations (7.5 and 10 arcsec) that could help in estimating the strength of the nebular component. We have to remind that NGC 7129-S is illuminated by several stars located at different places in the nebula. The observations at different diaphragms gave similar results suggesting that the nebular light may not be a problem in that region.

The obtained weighted mean values are listed in Table 1 and shown in the Fig. 2 for the object III. At the 3-sigma level the object III shows definite polarization and position angle whereas for the objects I and II the results are preliminary as they were obtained in poor conditions (partially cloudy nights).

Also in the case of BD+65°1637, the comparison of the present observations with the previous ones (see the paper by Vrba et al. 1979) indicates the time variability in the amount of polarization. We find, as in Vrba et al. (1979), a slow rise towards the ultraviolet. The position angle turns out to be larger by about 10 degrees in all wavelengths.

However, we have to notice that probably the existing polarization measurements have been carried out with different sizes of diaphragms.

5. CCD IMAGING OF NGC 7129-S

Our main aim was to study the structure of the S-shaped source PP102 at different wavelengths. Fig. 3 shows the V,R,I images we obtained.

In order to show the faintest structures visible in our frames without overexposing the brighter emission arising from the reflection nebula NGC 7129-S, we used a logarithmic filter on the images. Some bright stars over the saturation level show clearly noticeable bleeding of the charge along columns of the detector. This effect avoids an accurate comparison of the polarization angle with the complex filamentary structures observed in the region.

REFERENCES

- Cohen, M., and Schwartz, R.D.: 1983, *Astrophys. J.* **265**,877.
- Garrison, L.M., and Anderson, C.M.: 1977, *Astrophys. J.* **218**,438.
- Gyul'budagyan, A.L.: 1982, *Pis'ma Astron. Zh.* **8**,232.
- Hartigan, P., and Lada, C.J.: 1985, *Astrophys. J. Suppl.* **59**,383.
- Herbig, G.H.: 1960, *Astrophys. J. Suppl.* **4**,337.
- Mundt, R.: 1987, in *Circumstellar Matter*, IAU Symposium no. 122, eds. I. Appenzeller and C. Jordan, Reidel, Dordrecht, p. 147.
- Origlia, L., Busso, M., Ferrari-Toniolo, M., Persi, P., Robberto, M., Scaltriti, F., and Silvestro, G.: 1990, *Mem. Soc. Astron. It.* **61**,95.
- Parsamian, E.S., and Petrosian, V.M.: 1979, *Sooshenia Biurakanskai Observatori Akad. Nauk. Armianskoi S.S.R.*, no. 51.
- Persi, P., Busso, M., Ferrari-Toniolo, M., Origlia, L., Robberto, M., Scaltriti, F., and Silvestro G.: 1988a, in *Mass Outflows from Stars and Galactic Nuclei*, eds. L. Bianchi and R. Gilmozzi, Kluwer Academic Publishers, p. 337.
- Persi, P., Ferrari-Toniolo, M., Busso, M., Robberto, M., Scaltriti, F., and Silvestro, G.: 1988b, *Astron. J.* **95**,1167.
- Pfau, W., Piirola, V., and Reimann, H.,-G.: 1987, Astron. Astrophys. **179**,134.
- Piirola, V.: 1973, *Astron. Astrophys.* **27**,383.
- Piirola, V.: 1988, in *Polarized Radiation of Circumstellar Origin*, eds. G.V. Coyne et al., University of Arizona Press, p. 735.
- Schmidt, G.D., and Vrba, F.J.: 1975, *Astrophys. J.* **201**,L33.
- Strom, S.E., Strom, K.M., Yost, J., Carrasco, L., and Grasdalen, G.L.: 1972, *Astrophys. J.* **173**,353.
- Strom, K.M., Strom, S.E., and Kinman, T.D.: 1974, *Astrophys. J.* **191**,L93.
- Vrba, F.J., Strom, S.E., and Strom, K.M.: 1975, *Publ. Astron. Soc. Pacific* **87**,337.
- Vrba, F.J., Schimdt, G.D., and Hintzen, P.M.: 1979, *Astrophys. J.* **227**,185.

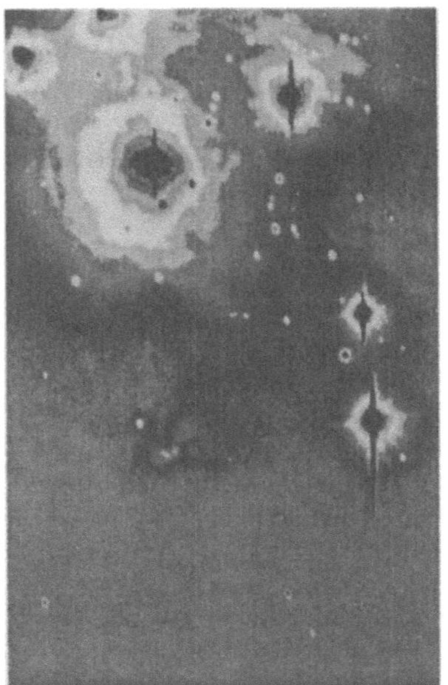

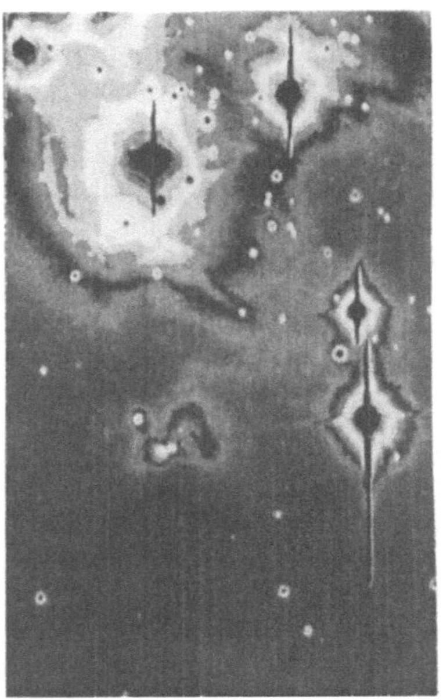

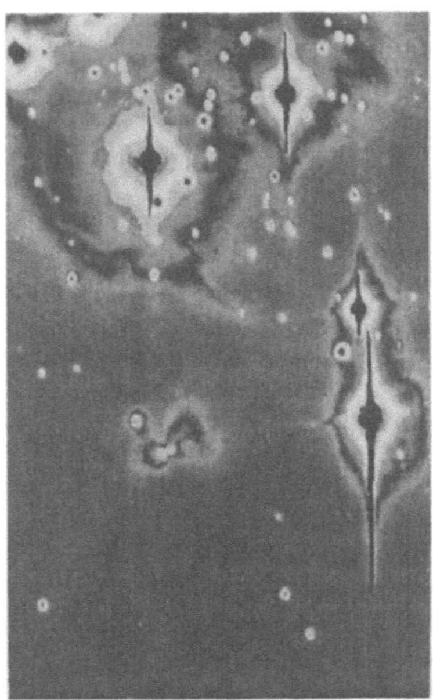

Figure 3 - V, R, and I (from top left clockwise) CCD images of the region NGC 7129-S. North is up, and east to the left. The S-shaped structure corresponds to PP 102 (see Parsamian and Petrosian (1979)). The objects I, II, and III (BD+65°1637) (from bottom) are the three black dots on the right part of the images (see also the Figure 22 in Hartigan and Lada (1985)).

DISCUSSION

Kraft: What are the units? Is it 1% polarization or 0.1% polarization?

Scaltriti: The units are as given - the typical polarization is 0.8%, but you have to take into account the interstellar polarization. If you compare our observations with others, you can find that the order of magnitude of the polarization is similar, but that you must be very careful about the subtraction of the interstellar polarization, because it can be a big effect.

ROTATIONAL VELOCITIES OF LOW MASS STARS IN YOUNG CLUSTERS

John R. Stauffer
Center for Astrophysics
60 Garden Street
Cambridge, MA 02138

ABSTRACT. Reviews of stellar rotation written prior to 1980 included essentially no information on the rotational velocities of stars later than G0 other than that they were generally slow rotators. Due to the advent of new instrumental capabilities, a large body of data on the rotational velocities of low mass stars is now available. In particular, several nearby clusters have good rotational velocity data for low mass stars, allowing us to estimate the evolution of rotation on and near the main sequence. Somewhat unexpectedly, many low mass stars have very large rotational velocities when they arrive on the main sequence. However, these stars spin down rapidly so that by the age of the Hyades essentially all of the low mass stars are relatively slow rotators. This paper provides a brief review of the "classical" rotation studies of the 1960's, an outline of the new discoveries of the 1980's and their current interpretation, and a description of the outstanding problems that remain.

1. **Introduction**

It is now twenty years since Bob Kraft published his review of Stellar Rotation (Kraft 1970), and it thus seems like a good time to take a new look at the subject. Despite a large amount of new data, most of which has been accumulated in just the last ten years, many of the questions we are still struggling with are those that Bob evoked in his review. A few quotations are in order:

"Theories of star formation customarily must predict the mass-frequency function $N(\mathcal{M})/d(\mathcal{M})$, but they can also be required to give the correct form of $<J(\mathcal{M})>$, the mean angular momentum for stars of mass $\mathcal{M}$. A leading task of spectroscopic observation is to provide this function, and though it is obviously convenient to study its form on the main sequence, the relationship between this and the desired initial form may be obscured by stellar angular momentum losses."

S. Catalano and J. R. Stauffer (eds.), Angular Momentum Evolution of Young Stars, 117–134.

"It is convenient to assume a star is a rigid rotator...but the physical validity of this assumption remains unknown. ... Does a star always rotate as a rigid body with concomitant radial exchange of angular momentum, or does each infinitesimal shell conserve its own angular momentum as the radius changes? Or does a star follow a course characterized by a more complicated radial redistribution of its angular momentum?"

"Since binaries constitute something like one half of all stars of types dK and earlier, it is of interest to ask whether the rotations of stars in binary systems are the same as single stars..."

Those were good questions then, and perhaps unfortunately (because it means we have not yet solved those problems), they are still good questons now. In this review, I will attempt to outline the progress that has been made since Bob's review paper, particularly with regard to open clusters.

2. Observations of Rotational Velocities in Open Clusters

Kraft (1970) summarized the observational data for stellar rotation up to that time. Very briefly, it was known that high mass stars are generally relatively rapid rotators ($v\sin i \sim 150$ km-s^{-1}) and that the distribution of rotational velocities for B and A stars can be roughly fit by a Maxwellian distribution; that the inferred mean angular momenta for high mass stars is well fit by a law of the form $J/\mathcal{M} \sim \mathcal{M}^{0.6}$ - the "Kraft Law"; that there is a break in the mean rotational velocities beginning about spectral type F0 - with later type stars showing progressively less rotation; that for the mid to late F type stars, rotational velocities decrease with time probably due to angular momentum loss by stellar winds or flares; and that the inclination axes of stars in open clusters are not aligned with the galactic pole and are probably randomly oriented. Due to their faintness, little was known about the rotational velocities of field stars later than G0 other than that they are ≤ 25 km-s^{-1}, and essentially nothing was known about the rotational velocities of low mass stars in open clusters.

From an examination of Kraft's open cluster data and Wilson's CaII K data, Skumanich (1972) concluded that both rotational velocities and CaII K emission for late F dwarfs decrease with time according to a $t^{-1/2}$ law. This has been taken as further evidence that stellar winds, generated by the same processes which produce the CaII K chromospheric emission, are responsible for angular momentum loss on the main sequence for low mass stars.

It was another decade before the next major revelation in the field. Based on an extensive photometric monitoring program, van Leeuwen and Alphenaar (1982) announced the discovery of a number of G and K dwarf members of the Pleiades with well-determined periodicities in their visual brightness. The periods for these variables were generally of order 10 hours. Their light curves were similar to those found for spotted stars, in which case the photometric period corresponds to the rotation period. For the Pleiades variables, this meant rotational velocities of order 100 km-s^{-1}, with the shortest period star having an indicated rotational velocity of about 170 km-s^{-1}. Such rotational velocities were unprecedented for single, main sequence low mass stars.

Spectra were soon obtained by Soderblom et al. (1983) and Stauffer et al. (1984) which showed that the short period, variable K dwarfs in the Pleiades indeed have very large rotational velocities. Not all of the late type Pleiades stars are rapid rotators, however - more than half of the K and M dwarf Pleiades members have $v\sin i < 10$ km-s^{-1}. Subsequent studies have now provided further Pleiades rotational velocities (Stauffer and Hartmann 1987; Soderblom, Jones and Stauffer 1990) as well as rotational velocities for stars in the Alpha Persei (Stauffer, Hartmann and Jones 1985, 1989; Prosser 1990), Hyades (Stauffer, Hartmann and Latham 1987; Lockwood et al. 1984) and IC 2391 (Stauffer et al. 1989) clusters. Figure 1 shows the new rotational velocity data for low mass stars in the Alpha Persei, Pleiades and Hyades clusters combined with data for the high mass stars in these clusters (Anderson, Stoeckly and Kraft 1966; Kraft 1965, 1967a,b). The nominal ages for the clusters (Mermilliod 1981) are 50 Myr, 70 Myr and 600 Myr, respectively. The primary empirical aspects of these plots are:

(1) the A and early F star rotational velocity distributions for the three clusters are roughly the same - no evolution with time is evident. The distributions have a mean of around 150 km-s^{-1}, and are grossly Gaussian or Maxwellian, in the sense that there is a broad distribution with few slow rotators;

(2) rapid rotators are present throughout the effective temperature range observed in the youngest cluster (Alpha Persei); there is a depression in the distribution in the next younger cluster (Pleiades) with the G stars appearing to have spun down more than the K stars; and in the Hyades, none of the late type stars are rapid rotators, though there are still some M dwarfs with moderate rotation; and

(3) perhaps most importantly, the rotational velocity distributions for the late type stars in these clusters are no longer "Maxwellian" due to the presence of a large number of slowly rotating stars (in many cases, the stars plotted at $v\sin i = 10$ km-s^{-1}are actually just upper limits). This latter fact is illustrated in Figure 2.

Assuming that the primary difference between the three clusters is age, it is possible to roughly estimate spindown timescales for low mass stars for the early, rapid rotation phase. For G dwarfs, the timescale is a few times 10^7 years, for K dwarfs, it is several times 10^7 years, and for M dwarfs it is a few times 10^8 years.

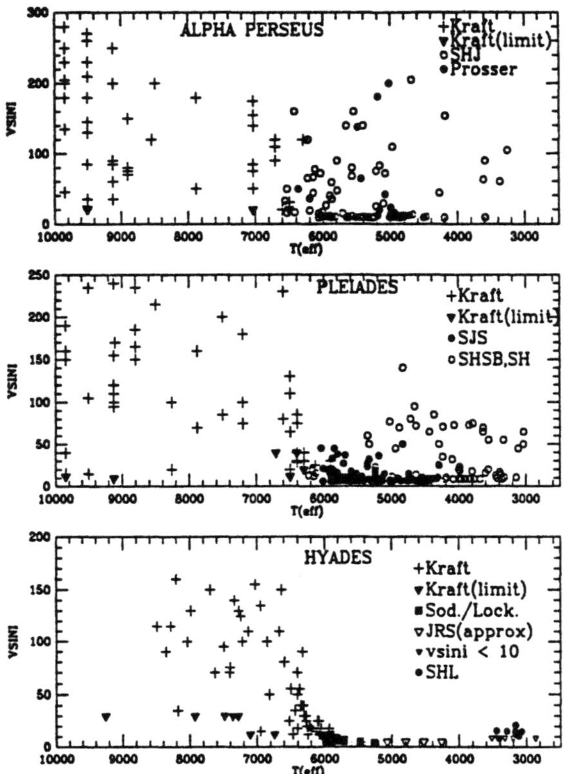

Figure 1: Spectroscopic rotational velocities for stars in the Alpha Persei, Pleiades and Hyades open clusters. Sources for the data are described in the text. The large inverted triangle symbols in the Hyades plot are just representative points indicating that all stars in this temperature range have spectroscopic rotational velocities less than 10 km-s^{-1}.

An important refinement of the observational data for the Hyades has recently been published by Radick et al. (1987). These authors obtained photometric periods for 23 F8-K8 members of the Hyades, and showed that not only do the periods increase (rotational velocities decrease) towards later spectral types, but that a smooth relation between B-V color and period fits the data within the errors of measurement ($\sim$ 1 km-s^{-1}) - that is, by the age of the Hyades, G and K dwarfs have spun down to rotational velocities that are a unique function of mass and age. The one exception to this relation was VA 500 (type=K7), the latest star in the sample, which presumably just marks the blue edge of the rapid rotator sequence in the Hyades (see Fig. 1).

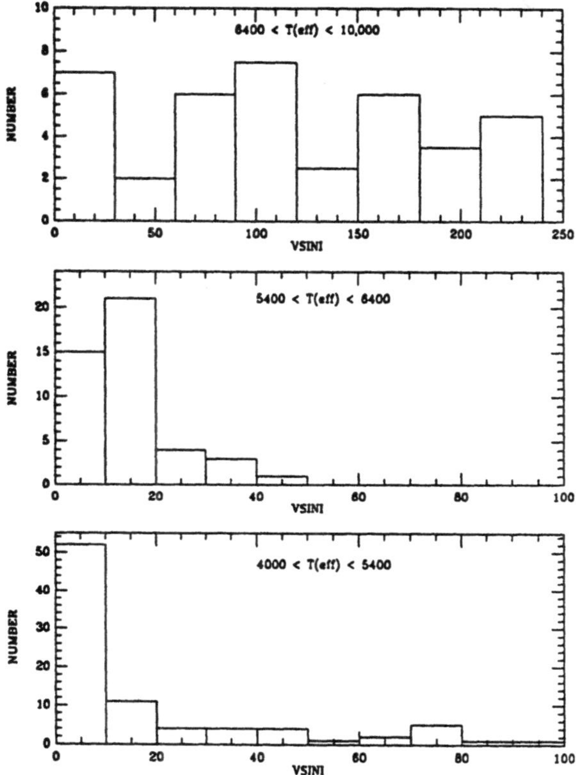

Figure 2: Distribution of rotational velocities for Pleiades stars in three mass ranges. Note the progressive increase in the number of very slow rotators for lower mass stars.

Two other surveys are especially noteworthy. Benz, Mayor and Mermilliod (1984) have obtained accurate spectroscopic rotational velocities with a resolution limit of only a few km-s^{-1} for F and G dwarfs in the Pleiades, Hyades, Praesepe and Coma clusters using CORAVEL. These data appear to show all three of the old clusters in this group have the same dependence of *vsini* on B-V color among the late F dwarfs - confirming the universality of the spin-down process (and also providing new evidence that rotation axes in open clusters are not aligned with the galactic pole because the Coma cluster is located near the pole and the other clusters are at low galactic latitudes). Smith, Beckers and Barden (1985= SBB) obtained rotational velocities for a large number of pre-main sequence stars in the very young Ori Ic cluster. The Orion star-forming complex provides a potentially rich source of data for studies of young star rotational evolution, however the value of the SBB survey was marred by the inclusion of a large number of non-members, as shown by Mermilliod and Mayor (1985) and McNamara (1990).

3. Related Observations

The arguments for randomly aligned axes in open clusters have been relatively indirect in the past. The rapidly rotating Pleiades star for which rotation periods have been derived make it possible to directly measure the distribution of inclination axes for the first time. In particular, there are now 14 Pleiades K dwarfs for which both $v sini$ and P_{rot} are known. From these data and published photometry for the Pleiades stars we can derive the rotational inclination axes for these stars. Figure 3 shows a plot of the inclination versus the derived rotational velocity. The solid curve denotes an effective selection limit appropriate to half of the stars. While no statistical test has been applied to these data, they appear consistent with a random distribution.

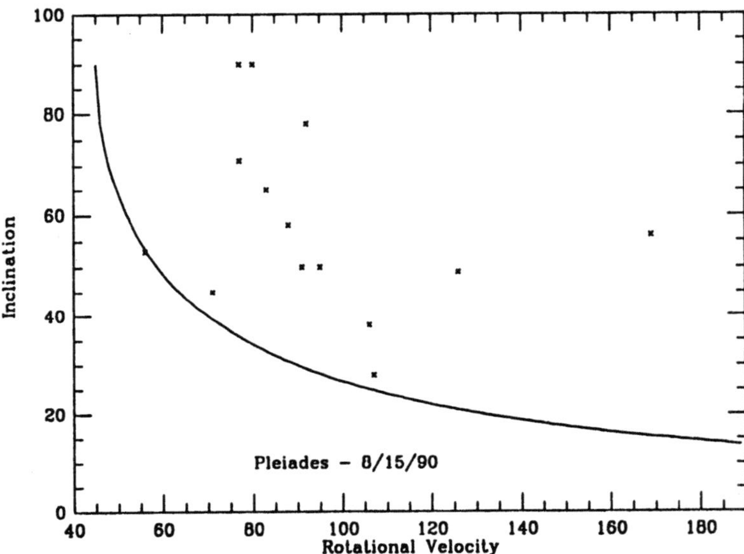

Figure 3: Estimates of the inclination axes of K dwarfs in the Pleiades for which both $v sini$ and P_{rot} known. Seven of the stars were first detected as short period variables, which implicitly selects against low inclinations but in a non-quantifiable manner. The other seven stars were identified as rapid rotators via spectroscopy and subsequently observed photometrically. The selection criterion ($v sini$ > 45 km-s^{-1}) is indicated by the curved line.

A few photometric periods have also been obtained for the Alpha Persei cluster, and it should eventually be possible to obtain inclination estimates for a similar sample of stars there. Rucinski (1989) has obtained CCD images of a portion of the NGC 3293 cluster, and has identified several faint, variable stars that are likely to be late type, spotted analogs of the Pleiades K dwarfs. The advent of 2048^2 CCD's and their use on wide-field telescopes should allow similar identification of late type rapid rotators in other moderately distant, young open clusters.

It would be useful to obtain periods for M dwarfs in the Hyades in order to extend the Radick et al. survey to lower mass stars. Because of the faintness of these stars, and their expected small amplitude of variability, that would be a difficult project. However, as shown by Duncan et al. (1984), a single measure of the CaII K flux plus the star's B-V color provide a very accurate estimate of the rotation rate of G and early K dwarfs in the Hyades. Therefore, a program to obtain chromospheric activity data for later type Hyades stars might suffice to estimate their rotational velocity distribution. The CaII K line would be difficult to observe for these stars, but Hα is much easier, and Stauffer et al. (1990) have now obtained Hα equivalent widths for a large number of Hyades K and M dwarfs. Figure 4 shows the distribution of Hα equivalent widths as a function of R-I color for these stars. We believe that Hα is acting as a proxy for rotation in this diagram. All the stars bluer than R-I = 0.6 have essentially the same Hα equivalent width at a given color, in accord with the Radick et al. CaII K data for stars in the same color range. One set of redder stars appears to continue this absorption sequence,

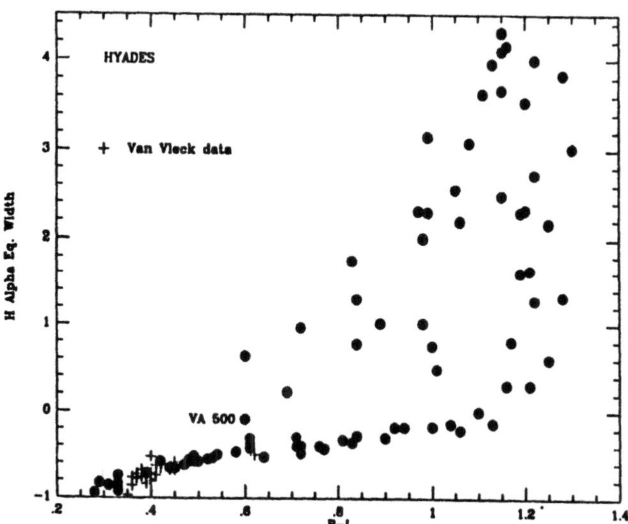

Figure 4: Hα equivalent widths for late type Hyades members versus Kron system R-I. The conversion between spectral type and color is: R-I=0.4 (K4), R-I=0.65 (M0), R-I=0.90 (M2), R-I=1.25 (M4).

and we assume that these stars have small rotational velocities that extend the relation between period and color established by Radick et al. A larger number of stars have Hα in emission, and we believe these stars are the more rapidly rotating population predicted by Radick et al. on the basis of the "discrepant" period for VA 500. The small amount of rotational velocity data which we have obtained confirms the general veracity of this picture. It will be useful to obtain periods or $v\sin i$'s for a reasonable sample of these late type Hyades members in order to produce a relation between rotational velocity and chromospheric emission for this presumably coeval sample of M dwarfs.

The open cluster data reviewed in § 2 indicate a very rapid spindown timescale for G and K dwarfs from their initial main sequence rotational velocities to the quite low rotational velocities found in the Hyades. This may be accomplished via a wind which is simply a scaled up version of the solar wind. However, observations by Cameron and Robinson (1989) suggest that angular momentum loss via discrete mass ejections may play an important role, as suggested by Schatzman (1962) many years ago. Cameron and Robinson have obtained high signal-to-noise, high resolution, time resolved spectra at Hα for the rapidly rotating, field K dwarf HD 36705 (AB Dor). Their spectra show weak Hα absorption features which move through the Hα emission profile on rapid timescales. They interpret these features as due to cool, dense clouds ejected from the tops of magnetic loops in the coronae of this star, and forced to corotate with its magnetic field out to several R_*. Taking reasonable estimates for the mass in these clouds, they derive plausible spindown timescales due solely to angular momentum loss via cloud ejection of order 10^7 to 10^8 years for HD 36705. A number of arguments suggest that HD 36705 may be an escaped Pleiades member (Rucinski 1982; Innis, Thompson and Coates 1986), so these observations may indicate the method by which all young, low mass stars lose the majority of their initial main sequence angular momentum. Continued observation of this star, and a search for other stars bright enough to perform this type of observation would be valuable.

Observational evidence suggesting the presence of disks around pre-main sequence stars has grown considerably in the past few years (Bertout 1989; Kenyon and Hartmann 1987; Strom et al. 1989). Many of the spectral peculiarities of T Tauri stars are explicable with such a model, in particular the UV excess and strong emission lines may arise as a result of accretion of disk matter onto the star. Because the disk matter arrives with Keplerian velocities, such accretion could clearly have a strong affect on the star's rotation. Furthermore, there appears to be a range of disk masses for PMS stars (Walter et al. 1988; Strom et al. 1989) which could then lead to a range in rotation. For this and other reasons, it is thus of importance to estimate the ubiquity of disks around young stars. The nearby open clusters may help answer this question. Backman et al. (1990) have recently shown that a significant fraction of the A stars in the Pleiades and Alpha Persei clusters may have weak but detectable IR excesses as derived from data from IRAS. Ground-based observations at 10μ can improve upon these data considerably, so it should soon be possible to produce a census of the remnant disk structures around the A and perhaps F stars in these clusters. A correlation of these data with the rotational velocities for these stars could help determine the influence of late accretion on the main sequence rotational velocities of young stars.

4. Impediments to Understanding the Open Cluster Data

4.1. AGES AND AGE SPREADS

Use of open clusters to calibrate any stellar evolutionary phenomenom requires an accurate knowledge of the age of the cluster but also requires the assumption that all of the members of the cluster are essentially coeval. It is possible that neither the requirement nor the assumption are valid. If so, this would seriously hamper the ability to use the observations to calibrate models of rotational velocity evolution.

The commonly quoted age for open clusters is that derived from fitting the upper main sequence turn-off to theoretical, post-main sequence evolution models. For the clusters of most interest for the present purposes, modern "standard" models have produced ages of order 70 Myr for the Pleiades and 600 Myr for the Hyades (Patenaude 1978; Mermilliod 1981). However, a new series of models are becoming available which include convective core overshoot (and which also include rotation, though that effect is secondary) - and these models generally have the effect of increasing the derived cluster age. Representative ages calculated for the Pleiades and Hyades with these new models are 150 Myr and 1.2 Gyr, respectively (Mazzei and Pigatto 1988, 1989), about twice the previously estimated ages. These new ages do not change the relative ordering of the ages of clusters, and so do not change the qualitative picture of the rotational velocity evolution on the main sequence. They become important, however, when comparing theoretical models of rotational evolution to the observational data.

A more troublesome problem is the question of age spreads within the stars of a given open cluster. The expected theoretical age spread among stars in an open cluster due solely to the sound-crossing time of the proto-cluster is only of order 1-10 Myr. The first detailed discussion of the possibility of a considerably larger age spread was by Herbig (1962), who noted that the apparent age of the low mass stars in the Pleiades (from fitting Henyey track isochrones) was > 220 Myr as compared to the nuclear age of 60 Myr. Since that time, a number of other authors have proposed various kinds of evidence for age spreads in open clusters. From fitting isochrones to high mass stars, Eggen and Iben (1988) proposed bi-modal (or many modal) age distributions within a number of clusters. Duncan and Jones (1983) proposed an age spread among the low mass stars in the Pleiades of order 200 Myr based on an observed spread in the inferred lithium abundances of those stars. Stauffer and Hartmann (1986) noted that an age spread of order 100 Myr among the low mass stars in the Pleiades could explain the difference between the distribution of rotational velocities among the Pleiades K dwarfs and the rotational velocity distribution of their progenitors (as exemplified by the low mass T Tauri stars).

However, none of these suggestions for possible age spreads should be regarded as observational proof for this phenomenon. Using improved observational data and modern pre-main sequence isochrones, Stauffer's (1984) best estimate of the apparent age difference between the low and high mass stars was only ~30 Myr, and this difference could just reflect errors in the observational data or the theoretical models. Soderblom et al. (1990) have shown that the apparent spread in lithium abundance at fixed effective temperature (and hence the apparent age spread) in the Pleiades may not be real because the KI $\lambda 7699 Å$ line - which is "chemically" similar to the $6707 Å$ lithium line - shows a similar, though smaller, spread. The rotational velocity distribution of the Pleiades K dwarfs can be approximately reproduced without an age spread if angular momentum loss during PMS evolution is essentially not a function of angular rotation rate (Stauffer and Hartmann 1987), a suggestion compatible with the observed lack of correlation between emission line strength and rotation for T Tauri stars (though see Bouvier (1989)).

Probably the best means to prove the existence of significant age spreads in young open clusters would be to show that single isochrones cannot fit the distribution of cluster members in a color-magnitude diagram. When applied to very young clusters, this method is not likely to be reliable due to the difficulty of deriving accurate temperatures and luminosities for T Tauri stars and the possible modification of the evolutionary tracks by mass outflow/inflow (Hartmann and Kenyon 1990); in addition, some previous claims for a time spread of star formation among the low mass stars in very young clusters were erroneous due to incorrect usage of the evolutionary track ages (Stahler 1985). The method is also unreliable for clusters older than about 100 Myr because the PMS isochrone becomes too close to the ZAMS. However, for a narrow range of nuclear ages (roughly 30 Myr to 100 Myr) the method should be viable, and there are several clusters in this age range which should eventually be amenable to this sort of test. The main difficulty with the test is the need to account for binary stars, because a star located above the main sequence can either be a younger star of similar mass or a binary. The simplest method to correct for the presence of binaries is to assume that the binary frequency and mass function is the same for high mass stars in the cluster (more massive than those expected to be still on PMS tracks) as for the stars in the magnitude range where the test is being made. One then fits a lower envelope to the cluster stars in a given color-magnitude diagram, calculates displacements above that fit, and compares the distribution of displacements in the two mass ranges to estimate if an age spread is present. Calibration of the test should be made by comparison to model open clusters constructed from theoretical evolutionary tracks, as illustrated in Figure 5.

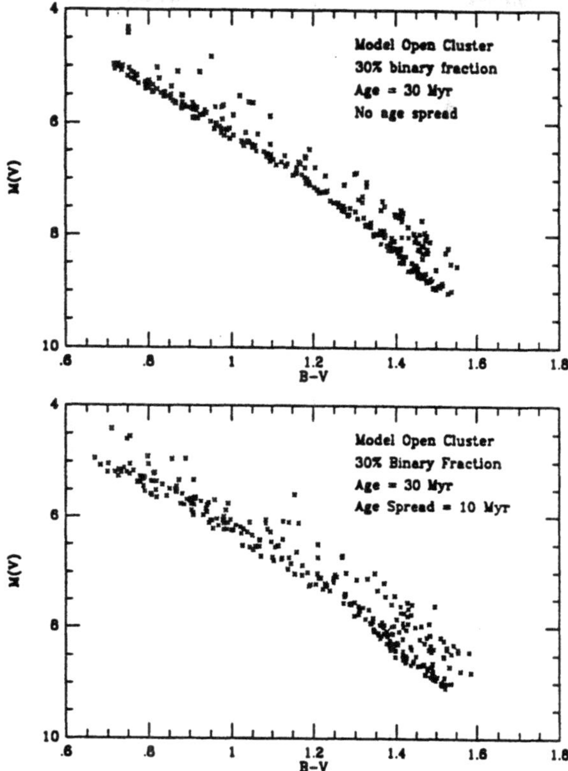

Figure 5: Model open star clusters for two different assumptions about the possible age spread in the cluster. The models assume a 30% binary fraction, for binary companions in the range from equal mass to 0.4 times the mass of the primary; 3% of the systems are assumed to be triples. A photometric error of 0.025 percent has been assumed for both axes.

While a definitive determination of the possible age spread within open clusters is not yet possible, one can place limits on the age spread which are useful for the present context. As shown in Figures 4 and 5 of Stauffer and Hartmann (1987), the K dwarf rapid rotators in the Pleiades are all essentially on the main sequence. Using Vandenberg's isochrones, this implies that they are older than 50 Myr. Therefore, models which attempt to explain the rotational velocity distribution among the low mass stars in the Pleiades or the lithium abundance spread should not rely on an age spread which requires ages younger than 50 Myr for the youngest stars. Similarly, observations of IC 2391 (Stauffer et al. 1989) show that a large rotational velocity spread is possible without an appreciable age spread.

The above arguments lead me to the belief that age spreads are not the explanation for the spread in rotational velocities or the lithium abundance spread among low mass stars in young clusters. It is more likely that the peculiar rotational velocity distribution for low mass stras in young clusters is induced during radiative track PMS evolution as a result of the angular momentum loss process or as a result of late accretion adding angular momentum to some of the stars. It is also possible, though less likely I believe, that the known Taurus population is not an accurate representation of the open cluster progenitors either because the most slowly rotating (least active?) Taurus members have not yet been found or because low mass stars born in associations have different rotational velocity histories from stars born in bound clusters. In any event, if I am correct, there is a real (and large) spread in rotation for low mass stars upon arrival on the ZAMS. The Sun could have had a rotational velocity of 100 km-s^{-1} or 10 km-s^{-1} at 50 Myr!

4.2. INTERNAL ROTATION

We observe only the surface rotational velocities for the open cluster stars. Interpretation of these data in terms of angular momentum loss rates requires an assumption about the internal rotation. This is important because whereas the total moment of inertia decreases by a factor of 10 from 1.0 $\mathcal{M}_\odot$ to 0.4 $\mathcal{M}_\odot$, the moment of inertia of the convective envelope is sensibly constant at 10% of the Solar main sequence moment of inertia in this mass range. Some support for the idea that only the outer convective envelope is initially spun down has been claimed: (1) the original Endal and Sofia (1979) models for spindown of the Sun derived spindown timescales similar to those inferred from the cluster data, and their models essentially decoupled the envelope from the core; (2) the cluster data show that G dwarfs spin down considerably faster than K dwarfs - implying much larger angular momentum loss rates for G stars than K stars if the whole star is being spun down (why should angular momentum loss rates be zero for early F stars, peak for the G stars, and then decrease sharply to later types?).

Figure 6 illustrates the quantitative differences required by the two extreme assumptions regarding internal rotation. The figure was derived by fitting an upper envelope to the Alpha Persei $v\sin i$ distribution, adopting very rough estimates of spindown time as a function of mass as indicated by the cluster data, and using Vandenberg's moments of inertia. It shows that if only the outer convective envelope is being spun down, the initial angular momentum loss rate is about 1000 times the current solar rate and is not a strong function of stellar mass; if the entire star is being spun down, then the initial loss rate for 1.0 $\mathcal{M}_\odot$ stars is about 10^4 times the current solar rate, decreasing to 1000 times solar for M dwarfs. If the angular momentum loss rate was 10^4 times solar upon arrival on the ZAMS, then it was also presumably very large just prior to the ZAMS - making it difficult to understand how the star managed to spin up to large rotational velocities during PMS evolution.

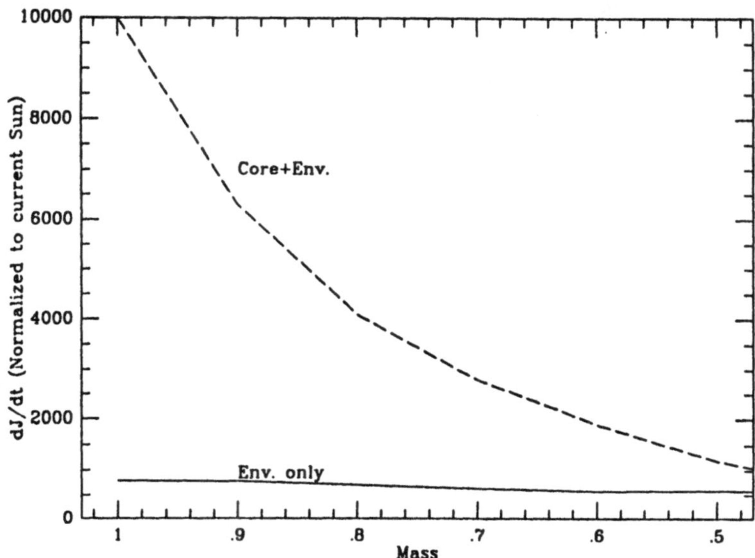

Figure 6: Rough estimates of the average angular momentum loss rate experienced by stars of different masses upon arrival on the main sequence if either the entire star is being spundown or if only the outer convective envelope is being spundown. Units are in terms of the current solar angular momentum loss rate.

Because internal rotation is the topic of several other papers at this meeting, I will not discuss it further. I hope that the theorists will solve this problem.

5. Summary

We still have not solved many of the problems noted by Kraft in his 1970 review. However, we are much closer now and we have made much progress. The most valuable observational contributions during the next few years should be: (1) to derive rotational velocities for a rich cluster slightly younger than Alpha Persei in order to extend the cluster time sequence; (2) to derive rotational velocities for a cluster slightly older than the Pleiades; (3) to obtain accurate photometry of high probability Alpha Persei members in order to place limits on the possible age spread in that cluster; and (4) to make sure that we have a complete list of Taurus cloud members (i.e. make an unbiased proper motion survey) and obtain more rotation periods in order to determine a good distribution of rotational velocities for the presumed precursors of the open cluster stars. All of these projects are technically feasible.

One of the most promising aspects of this field during the past few years has been the development of the first theoretical models which attempt to explain the young cluster rotational velocity and lithium observations (Kawaler 1988; Pinsonneault et al. 1990; MacGregor and Brenner 1990). These models offer the hope of taking the hard won observational data and eventually deriving answers to some of the questions posed by Kraft. It is the job of the observers to make sure that the theorists interpret the observational data correctly and understand the limitations of the data.

REFERENCES:

Anderson, C., Stoeckly, R., and Kraft, R. 1966, *Ap.J.*, **143**, 299.
Backman, D., Stauffer, J. and Witteborn, F. 1990, in preparation.
Bertout, C. 1989, *A.R.A.A.*, **27**, 351.
Bouvier, J. 1989, A.J. in press.
Cameron, A.C. and Robinson, R.D. 1989, *M.N.R.A.S.*, **236**, 57.
Duncan, D., Baliunas, S., Noyes, R., Vaughan, A., Frazer, J., and Lanning, H. 1984, *P.A.S.P.*, **96**, 707.
Duncan, D. and Jones, B.F. 1983, *Ap.J.*, **271**, 663.
Eggen, O. and Iben, I. 1988, *A.J.*, **96**, 635.
Hartmann, L. and Kenyon, S. 1990, *Ap.J.*, **349**, 190.
Herbig, G. 1962, *Ap.J.*, **135**, 736.
Innis, J., Thompson, K. and Coates, D. 1986, *M.N.R.A.S.*, **223**, 183.
Kawaler, S. 1988, *Ap.J.*, **333**, 236.
Kenyon, S. and Hartmann, L. 1987, *Ap.J.*, **323**, 314.
Kraft, R. 1965, *Ap.J.*, **142**, 681.
Kraft, R. 1967a, *Ap.J.*, **148**, 129.
Kraft, R. 1967b, *Ap.J.*, **150**, 551.
Kraft, R. 1970, in Spectroscopic Astrophysics, ed. G.H. Herbig (Berkeley: University of California Press), p. 385.
Lockwood, G.W., Thompson, D., Radick, R., Osborn, W., Baggett, W., Duncan, D. and Hartmann, L. 1984, *P.A.S.P.*, **96**, 714.
MacGregor, K. and Brenner, M. 1990, preprint.
Mazzei, P. and Pigatto, L. 1988, *A.A.*, **193**, 148.
Mazzei, P. and Pigatto, L. 1989, *A.A.*, **213**, L1.
McNamara, B. 1990, *Ap.J.*, **350**, 348.
Mermilliod, J-C. 1981, *A.A.*, **97**, 235.
Mermilliod, J-C. and Mayor, M. 1985, in IAU Colloq. 85, Stellar Radial Velocities, ed. A.G.D. Philip and D.W. Latham (Dordrecht: Reidel), p. 367.
Patenaude, M. 1978, *A.A.*, **66**, 225.
Pinnonneault, M., Kawaler, S. and Demarque, P. 1990, *Ap.J.Supp*, **74**, 501.

Prosser, C. 1990, priv. comm..
Rucinski, S. 1982, Inf.Bull.var.Stars, No. 2203.
Rucinski, S. 1989, priv. comm..
Schatzman, E. 1962, *Ann.d'Ap.*, **25**, 18.
Skumanich, A. 1972, *Ap.J.*, **171**, 265.
Smith, M., Beckers, J. and Barden, S. 1983, *Ap.J.*, **271**, 237.
Soderblom, D., Jones, B., and Walker, M. 1983, *Ap.J.Lett.*, **274**, L37.
Stahler, S. 1985, *Ap.J.*, **293**, 207.
Stauffer, J., Hartmann, L., Soderblom, D. and Burnham, J. 1984, *Ap.J.*, **280**, 202.
Stauffer, J., Hartmann, L. and Jones, B. 1985, *Ap.J.*, **289**, 247.
Stauffer, J. and Hartmann, L. 1986, *P.A.S.P.*, **98**, 1233.
Stauffer, J., Hartmann, L. 1987, *Ap.J.*, **318**, 337.
Stauffer, J., Hartmann, L., and Latham, D. 1987, *Ap.J.Lett.*, **320**, L51.
Stauffer, J., Hartmann, L., and Jones, B. 1989, *Ap.J.*, **346**, 160.
Stauffer, J., Hartmann, L., Jones, B. and McNamara, B. 1989, *Ap.J.*, **342**, 285.
Strom, K., Strom, S., Edwards, S., Cabrit, S. and Skrutskie, M. 1989, *A.J.*, **97**, 1451.
van Leeuwen, F. and Alphenaar, P. 1982, ESO Messenger, No. 28, p. 15.
Walter, F., Brown, A., Mathieu, R., Myers, P. and Vrba, F. 1988, *A.J.*, **86**, 297.

DISCUSSION

Strom: I usually agree with most of what you say. But, the issue of age spreads I think is fairly well established at least for clusters such as the Orion nebula cluster or NGC 2264 where if you focus your attention simply on those stars which lack evidence for disk related phenomena, looking at the proper motion members, you get an age spread of 20 or so million years, for both clusters. It seems to me that if they are representative of open clusters, which seems logical, that you would expect an age spread of order 10-30 Myr for the Pleiades. With all due respect to the data for IC 2391, I think the data on younger clusters suggests a considerably larger age spread.

Stauffer: I will give one answer, then I think Lee Hartmann wants to give his answer also. The main points I want to make is that the age spreads of order 10^8 years that can be found at various points in the literature should not be taken as gospel, and very probably are wrong. I am perfectly happy to accept a possible age spread of 10 Myr, and that is compatible with all the open cluster observations (the observations are not yet capable of detecting an age spread of this size). The limit on the possible age spread in IC 2391 is probably of order 10-15 Myr, while that for Alpha Per is probably of order 30 Myr. You can't use an age spread much larger than this to try to explain the lithium data or the rotational velocity distributions.

Strom: The question is whether 10-20 Myr is an interesting number from the point of view of the spindown.

Stauffer: An age spread of 10-20 Myr would not allow one to explain the Pleiades rotational velocity distribution without also still invoking one of the other mechanisms I don't think - in which case one might just as well not have an age spread.

Hartmann: The point of note about the Hyades and the Pleiades and other bound open clusters is that they are much smaller than the regions you are talking about. If you take a typical dimension of a parsec, the crossing time is about a million years, whereas in 2264 and the other regions you are talking about, the crossing time for the gas is about 10 Myr. So, it is not surprising that you have a 10 Myr age spread in those regions. The question is whether in a much smaller volume of space with a much smaller crossing time, whether you would expect to have an age spread that large. In Taurus, it is not clear that you even have an age spread of 10 Myr where we can do a very accurate census of the stellar population. So, my feeling is that age spreads are much less of an issue in these clusters than in the other regions such as Orion or NGC 2264.

Bouvier: Can you say something about the frequency of spectroscopic binaries in these clusters? Would you expect to find at least one or two close binaries in the Hyades, which would appear rapidly rotating due to tidal spin-up?

Stauffer: Seemingly, unlike the people who use CORAVEL, I usually only have enough observing time to observe a star once. If it was a single line binary with a relatively long period, I probably wouldn't know it at all. If the radial velocity were off by a few kilometers per second from what I expected for a cluster member, I might not care basically, if I was certain it was a cluster member from other considerations. Of the stars we have looked at, there are of order five double lined binaries. I compared that number with the number of double lined binaries for M dwarfs in the Gliese catalog, which we were observing with the same equipment, and the numbers are quite comparable. So, that small bit of data suggests that the frequency of SB2's in the clusters are similar to the frequency for field stars. There are two apparently spun-up binaries in the Hyades in our H alpha sample (I believe they are VB 119 and VA 677).

There is one interesting binary question which I would like to bring up with regard to your question. Photometric data for the Pleiades shows that at least 30 main sequence stars have binary companions whose mass is within about a factor of two of the primary. It seems possible at least that one could have a binary where one component is a slow rotator, and the other a rapid rotator. However, none have so far been observed. This could to some extent be a selection effect, because we often expose our spectra just enough to obtain enough counts for the cross-correlation program to work. If the star appears to be narrow lined, we can derive the cross-correlation with very few counts. However, to obtain a detection of a rapid rotator takes considerably higher S/N. Therefore, we might have obtained spectra of a narrow/rapid binary, but only noticed the lines from the slowly rotating component. It would be interesting to conduct a more thorough search for such stars. Theyr absence might would suggest that the binaries are not capture systems (since in that case there should be a random mixture of rotation types).

Roxburgh: A comment on the suggestion that the angular momentum loss rate could be independent of anything. While I couldn't make a model which would do exactly that, it does't follow that the angular momentum loss rate must be highest when the rotation is highest. The angular momentum loss rate also depends on the topology of the field, and you can imagine that for a rapid rotator the topology is very complicated with a lots of closed regions and therefore little space on the star from which wind could escape and transfer angular momentum. So it is possible in fact to have rapidly rotating stars with lower angular momentum loss rate than more slowly rotating stars, if you believe taht the magnetic field which the dynamo produces is more complicated with œOP rapid rotation.

Stauffer: I think we would be happy with that.

Cameron: I would like to comment on Roxburgh's comment. In fact there is some evidence that as you go to extreme rotation rates, we now have observational evidence that the closed field regions start to shrink. There are also reasons to believe that once the keplerian corotation radius moves inside the closed loop region, hydrostatic equilibrium inside the closed field region is actually going to cause the closed field lines to break open as you go to extreme rotation rates, and in fact the dead zone shrinks at very rapid rotation rates.

I also have one thing to say about the M dwarf observations in the Hyades. If you assume that the correlation between H alpha and rotation holds, you have a very peculiar situation where apparently after 500 Myr or so there is still a spread in rotation rates. Now, this implies that there is a braking time-scale of order 500 Myr at least which certainly must apply to the rapid rotators and if the slow rotators are are explained by an age spread then in order to have gotten slow rotators in this time implies an age spread in the cluster similar to the age of the cluster. Now if in fact the age spread is a great deal smaller than that, it suggests to me that what we are actually looking at in the M dwarfs is a snapshot of the initial angular momentum distribution.

Stauffer: First, there is a rotation spread among the M dwarfs in the Hyades because we do have rotational velocities for some of the H alpha emission stars, and their rotational velocities are of order 20 km/s, whereas those that have H alpha absorption are less than 10 km/sec, so that there is at least that much of a rotation spread. That is as much as we can say at this time with low S/N, moderate resolution sprectra.

Otherwise, I agree with what you said. I don't think that the Hyades M dwarf H alpha distribution is indicative of an age spread because the inferred age spread is too large, and what you really are seeing is an intrinsic rotation spread - either produced by the braking mechanism during PMS evolution or built in during the protostar phase.

ROTATION OF YOUNG STARS IN THE ORION NEBULA REGION

D.K. DUNCAN
Space Telescope Science Institute
3700 San Martin Dr.
Baltimore, MD 21218
USA

ABSTRACT. Approximately 50 stars in the Orion Nebula region with V = 12-14, B-V = 0.6-1.1 were observed with the intent of empirically determining the evolution of rotation on the approach to the main sequence. Most of the stars are of $0.8 - 1.0\ M_{\odot}$, with apparent ages ranging from a few times 10^6 to a few time 10^7 years, and all are proper motion members. With 20 of the stars fully analyzed, there is no indication of extremely rapid rotation such as is seen among the K stars in the Pleiades. Typical $v \sin i$ values are 15-30 km s^{-1}, and there is no apparent increase from younger to older stars as they approach the main sequence.

1. Introduction

One of the most remarkable discoveries concerning the rotation of young late-type stars was that of Van Leeuwen and Alphenaar (1982), which suggested that many of the K stars in the Pleiades were rotating extremely rapidly, up to 150 km s^{-1}or about 100 times as fast as the sun. Further work expanded and confirmed these results (Van Leeuwen, Alphenaar, and Meys 1987; Stauffer and Hartmann 1986) and showed that in the even younger α Perseus cluster many of the G stars showed very rapid rotation. In the older Hyades cluster neither G nor K stars exhibit rapid rotation. A tentative picture emerged which suggested that post T-Tauri stars speed up quite significantly on contraction to the main sequence, and then lose their rapid rotation quickly, in less than about 10^8 years, through angular momentum loss in a strong stellar wind (e.g. Marcy, Duncan, and Cohen 1985).

A significant problem with the above scenario occurs, however. On the main sequence there is a good correlation between stellar activity and rotation rate (Noyes *et al.* 1984), and it is expected that fast rotating stars would generate the strongest stellar winds. Furthermore, stars of greater convection zone depth show greater activity for a given rotation rate. If a star could lose angular momentum very quickly while on the main sequence, would it not lose enough angular momentum on the approach to the main sequence to prevent rapid rotation from ever building up; i.e. how does a star ever become an extremely rapid rotator? Where on the approach to the main sequence does rapid rotation begin to appear? Does the observed rapid rotation of the young main sequence stars mean that their cores are rotating at the same speeds? What are the effects of core-convection zone coupling in

S. Catalano and J. R. Stauffer (eds.), Angular Momentum Evolution of Young Stars, 135–141.

a fast rotating, contracting star with a strong stellar wind? The present investigation was begun in 1985 in order to help answer these questions.

2. Observations

The Orion Nebula region was chosen for this investigation because it contains a large number of late-type stars in various stages of contraction to the main sequence. Covering a wide range in color and magnitude insured that stars of a range of ages would be included. The stars observed rather uniformly span the range $0.5 \leq$ B-V ≤ 1.1, and $11.0 \leq V \leq 14.0$. The present study differs in two important way from the major previous study of rotation among Orion stars, that of Smith, Beckers, and Barden (1983). First, it reaches over two magnitudes fainter, thus including stars which are actually 1 $M_{\odot}$ and less. Second, all stars were selected from proper motion studies and are known to be cluster members; all are in the Orion Ic region but only a minority in the Trapezium region (McNamara 1976; McNamara and Huels, 1983). The Smith, Beckers, and Barden sample is known to include many nonmembers, and the masses of their stars are typically 1.5 – 2.0 $M_{\odot}$ (Rydgren and Vrba, 1984; McNamara 1990).

A sample of 44 Orion stars plus many standards was observed with the Cassegrain spectrograph of the Las Campanas Observatory in January 1985. All stars were observed in the blue, 3800-4600Å, with a resolution of 1.2Å, and in the red, from 5700-7000Å, with a resolution of 2.4Å. These data were used for spectral classification, reddening determination, and determining levels of Hα and Ca II H and K emission. Sixteen of the stars, plus three other Orion stars were observed in January 1987 with the Las Campanas 2.5 m. echelle. This fixed-format instrument covers most of the visible spectrum with a resolution of about 10-12 km s^{-1}. This is the primary sample reported here. The sixteen stars were chosen to include a mix of strong Hα emission stars, weak emission stars, and stars with Hα in absorption. An additional 50 stars were observed at similar high resolution but in only a few spectral orders with the Kitt Peak fibre-fed echelle in December 1988. These latter data will be reported in 1991.

3. Data Analysis

The Las Campanas echelle uses a 2-dimensional photon counting detector (2-D Frutti) behind an image-tube chain. The raw data was flat fielded, background subtracted, and extracted using standard IRAF routines. Spectra were then logarithmically binned and cross correlated with a narrow-lined reference spectrum. Both sky and slowly rotating standard stars were used as references, with little difference. Signal to noise of the data was typically 15-20 per resolution element and the wavelength range used, 3800-6600Å, contains hundreds of absorption lines and the cross correlation function was very clearly defined. It was fit with a sum of Gaussian plus quadratic (background), and the Gaussian FWHM taken as the measure of rotational broadening. The V sin i scale was calibrated by comparison with narrow (solar) spectra which had been convolved with Gaussians of known width and then analyzed. In fact, rotational broadening is not Gaussian, and the larger the rotation the greater the deviation from this shape (Soderblom 1988). The difference is small for most of our stars but could cause an error of perhaps 20% for the fastest-rotators.

The spectral range covered by the echelle data is so great that sensitvity to individual features is not large. Deleting individual broad features only caused changes at the 1% level. In the case of faint stars with very strong emission lines, the emission features were removed before cross correlation.

Color-color plots and color-spectral type plots were made for all stars which had been observed at Cassegrain. Spectral types agreed well with those of Walker (1983) for stars in common. Colors were in most cases photoelectric UBV observations of Walker (1969) or McNamara (1976), otherwise photographic colors derived from Parenago (1954) were used. The average reddening in Orion is not large, about 0.06 in B-V (Van Altena *et al.* 1988), although it varies from star to star. Reddening effects could be seen in the color-color and color-spectral type plots, but deviations from standard (unreddened) relations were generally not large. Two stars are exceptions: Par 2478 and Par 1617. In both cases the unusual colors are almost certainly intrinsic. Both are strong emission stars. Par 1617 is the well-known star YY Ori (Walker 1983) which shows strong hydrogen and Ca II H and K emission, inverse P Cygni profiles, and veiling of the blue part of the spectrum. Note that its position is intermediate in the color magnitude diagram, not near more luminous T-Tauri stars such as Par 1409 (classified as an SU Aur star by Herbig and Bell 1988).

Results for the primary sample of stars are tabulated in Table I (star numbers from Parenago 1954) and shown in Figure 1. Emission at Hα and in the Ca II H and K lines is noted if present as weak, medium, strong, or very strong. Numbers next to each point in the Figure give the measured rotational velocity in km s^{-1}. Also shown in Fig. 1 are isochrones of logarithmic age 6.5, 7.0, and 7.5 years, and evolutionary tracks for 1.0 and 0.8 $M_{\odot}$. The points are *not* corrected for reddening.

Table I. Measured rotational velocities.

Par #	V	B-V	$v \sin i$	Hα	H and K
1350	13.13	.76	20	(abs.)	weak
1409	11.60	.85	34	v. strong	v. strong
1426	12.54	.74	16		v. weak
1440	12.70	.95	16	weak	med.-strong
1477	14.0	.9	16:	strong	v. strong
1518	13.8	.8	24	weak	med.-strong
1554	12.30	.88	35	(flat)	med.
1617	13.5v	.88	30:	v.v. strong	v. strong
1647	12.53	.68	19	(abs.)	(abs.)
1699	13.04	.81	17		(abs.)
1732	12.47	.61	15		(abs.)
1774	12.75	.78	18	(abs.)	(abs.)
1827	13.4	.8	15	v. strong	v. strong
1904	14.10	.95	19		
1971	13.8	.9	18		
2048	13.9	1.1	14	weak	v. strong
2084	12.49	1.31	18	v. strong	v.v. strong
2152	14.02	1.39	21		
2576	12.00	.75	35	weak	med.

4. Discussion and Conclusions

Figure 1 is immediately interesting for what is not present: there are no stars with rotation rates comperable to the rapidly rotating Pleiades K stars. There is no clear tendency for $v \sin i$ to increase towards the main sequence along an evolutionary track. There is also no good correlation between chromospheric emission (Hα and Ca II H and K) and rotation rate.

Par 1409 is the most luminous star in the present sample and the one in common with Smith, Beckers, and Barden (1983) who found $v \sin i = 225$ km s^{-1}. Their value, based only on observations near the Na D lines, is not correct. SBB found a sharp feature of width $\sim$30 km s^{-1}and a much wider, shallow feature at the base of the D lines. The former was interpreted as circumstellar absorption and the latter as due to stellar rotation. We find $v \sin i = 34$ km s^{-1}and no sign of broad, shallow features. Using calculations of Gilliland (1986) and our value of 34 km s^{-1}we predict a rotational velocity of $\sim$170 km s^{-1}if Par 1409 evolved to the main sequence without angular momentum loss, slightly large for the young middle-F star it would become. This may give some idea of the angular momentum loss yet to occur on the approach to the main sequence.

The $v \sin i$ distribution of the present sample, which includes mostly post-T Tauri stars, is very similar to that found for low-mass T Tauri stars by Hartmann *et al.* 1986, except that in the mean velocities are about 50% higher. This is much less of an increase than would occur with conservation of angular momentum and solid-body rotation. The eight stars of the present sample clustered around the 7.5 isochrone have an average $v \sin i = 20\ (\pm\ 5)$ km s^{-1}. Even allowing for reddening, their average age should not be much less than 2×10^7 years. They can speed up by no more than about a factor of two when they reach the main sequence (Gilliland 1986). They will not produce a distribution such as that predicted in Fig. 11c of Hartmann *et al.* . This sample also seems to rule out the hypothesis that *all* late-type stars go through the phase of extremely rapid rapid rotation exhibited by some of the Pleiades K stars, and that the velocity differences among the Pleiades stars are primarily due to a dispersion of times since the stars arrived on the main sequence.

That there is not a good correlation between Hα or Ca II H and K emission and rotation is not necessarily surprising. The formation of chromospheric lines is much different in T Tauri stars than in main sequence stars (e.g. Edwards *et al.* 1987, Hartmann *et al.* 1990) and shows no correlation with rotation. The strongest emission seems to be associated with the presence of circumstellar disks. The present sample shows that the lack of rotation-emission correlation may extend to the post-T Tauri stars. The larger sample of the Kitt Peak observations now being analyzed will allow more quantitative tests of the above observations. It is interesting to note, however, that the present data shows examples of pairs of stars in almost the same position in the color-magnitude diagram but with very different levels of emission, and some stars apparently close to the main sequence with strong emission.

A caveat on which to end is that we of course only observe the rotation of the surface (convective) layers of these stars; the radiative core remains hidden. Through much of the pre-main sequence evolution the convection zone accounts for essentially all of the stellar moment of inertia; the radiative core passes it at a logarithmic age of 7.1-7.2 years (Gilliland 1986) and ends up with $\sim$90% of the total moment of inertia on the main sequence.

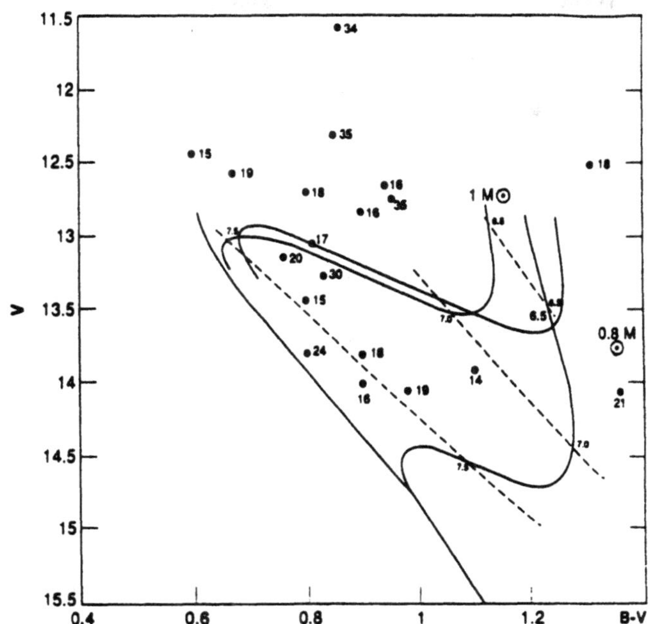

Figure 1. Rotational velocities of Orion stars.

5. References

Edwards, S., Cabrit, S., Strom, S.E., Heyer, I., Strom, K.M., and Anderson, E. 1987, *Ap. J.*, **321**, 473.

Gilland, R.L. 1986, *Ap. J.*, **300**, 339.

Hartmann, L., Calvet, N., Avrett, E.H., and Loeser, R. 1990, *Ap. J.*, **349**, 168.

Herbig, G.H., and Bell, K.R. 1988, *Lick Obs. Bull.*, **1111**.

Marcy, G.W., Duncan, D.K., and Cohen, R.D. 1985, *Ap. J.*, **288**, 259.

McNamara, B.J. 1976, *A. J.*, **81**, 375.

McNamara, B.J. 1990, *Ap. J.*, **350**, 348.

McNamara, B.J., and Huels, S. 1983, *Astr. Ap. Suppl.*, **54**, 221.

Noyes, R.W., Hartmann, L.W., Baliunas, S.L., Duncan, D.K., and Vaughan, A.H. 1984, *Ap. J.*, **279**, 763.

Parenago, P.P. 1954, *Trudy Sternberg Astr. Inst.*, **25**.

Rydgren, A.E., and Vrba, F.J. 1984, *A. J.*, **89**, 399.

Smith, M.A., Beckers, J.M., and Barden, S.C. 1983, *Ap. J.*, **271**, 237.

Soderblom, D.R. 1988 (private communication).

Stauffer, J.R., and Hartmann, L.W. 1986, *Pub. A.S.P.*, **98**, , 610.

Van Altena, W.F., Lee, J.T., Lee, J.-F., Lu, P.K., and Upgren, A.R. 1988, *A. J.*, **95**, , 1744.

Van Leeuwen, F., and Alphenaar, P. 1982, *ESO Messenger*, **28**, 15.

Van Leeuwen, F., Alphenar, P., and Meys, J.J.M. 1987, *Astr. Ap. Suppl.*, **67**, 483.

Walker, M.F. 1969, *Ap. J.*, **155**, 447.

Walker, M.F. 1983, *Ap. J.*, **271**, 642.

DISCUSSION

Pinsonneault: The Orion results, in combination with some of the things discussed by Stauffer, suggest an interesting possibility. Just as stars are born with a distribution of angular momenta, they might also be born with a distribution of magnetic field strengths and that they simply spin down at different rates. That might very well explain the pattern of slow rotators - which might just be the stars which spun down most efficiently - and a pattern of rapid rotators which might have had inefficient braking.

Soderblom: I will talk about that idea more tomorrow. I will just note one thing. There are various different ways of looking at the chromosphere, and you get a different impression depending on what you look at. One of the ways is H alpha, which is formed in a very different fashion from the various calcium lines, so you might get a misleading impression from just looking at H alpha.

Schatzmann: We had the Maunder minimum for the Sun which lasted 50 years. So we can not be sure in the case of weak activity if you are just not seeing a temporary period of inactivity.

Soderblom: Well, yes. But there are a couple of things you can say. First, the fact that the spots disappeared at the Maunder minimum tells you nothing about the global activity - there were no spectrographs operating then. The other thing you can say is that you could look at the Hyades - you don't see "drop outs" there - all the G stars fall in the same place, all neat and tidy, with no indication of any anomalously inactive stars.

Stauffer: Do you have lithium data or radial velocities for the Orion stars? Until you get that information, there is some chance that some of the apparently slow rotators might actually be non-members.

Soderblom: Well, yes that is possible. However, for these colors for these stars to have even this rotation, it is highly unlikely that these stars would be field stars. And the proper motions are now pretty good. So, I think it is unlikely that you would be getting much of a membership problem.

Hartmann: What are the rotational velocities for the stars that are going to be F stars on the main sequence? Are any of them rotating as fast as one would predict they should be?

Soderblom: Off-hand I don't know. The rotational velocities for those stars look like they average about 20 km/sec. I don't know how much spin-up you expect to see from there.

Stauffer: A factor of two more or less.

Soderblom: OK, then these velocities seem pretty low. You really expect to see higher rates than that.

Mayor: This is a question related to the duplicity of stars and the influence on the *vsini* distribution. We have some examples of the influence of duplicity and I will show a few examples during my talk.

Strom: Are these stars predominantly in the Ori Ic region, which is characterized by an admixture of slowly and rapidly rotating B stars or from the Id region which is dominated by rapidly rotating B stars?

Soderblom: They are from the Ic region, but not from the Trapezium cluster (that is, they are from a region about 1 degree from the Trapezium cluster).

ROTATIONAL VELOCITIES OF STARS IN OPEN CLUSTERS: THE TIME DEPENDENCE REVISITED

M. Mayor and J.–C. Mermilliod
Geneva Observatory and Institut d'Astronomie, Lausanne University
CH–1290 Sauverny, Switzerland

ABSTRACT. Using well defined rotational velocity distributions of G–type stars in the Hyades + Praesepe and Pleiades clusters, we have reexamined the age dependence of stellar rotation. The shape of the Vsini distribution is an extremely sensitive function of the α index of the braking law (dV/dt = $-AV^{\alpha}$). The comparison of Hyades + Praesepe and Pleiades observations favours the value $\alpha = 2$. The Vsini distributions could not be reproduced with $\alpha = 3$, the value corresponding to the so–called $t^{-1/2}$ law.

Key words: open clusters – rotation

1 INTRODUCTION

The angular momentum evolution of G–type or solar–type stars has been generally studied by using the mean rotational velocities of stellar samples of known ages. In that respect, open clusters offer a unique material. However, the time behaviour of the mean rotational velocities <Vsini(t)> is not equivalent to the time dependence for a single star, since V(t) depends on V_0 (formula (2)).

A parametrized formulation of the braking law can be formally written as:

$$dV/dt = -AV^{\alpha} \tag{1}$$

where α contains information on the magnetic field geometry and the relation between magnetic field and rotation. The coefficient A is itself a function of the stellar mass, radius, mass loss rate and physics related to the wind and magnetic field generation (Kawaler, 1988). Not only does the mean rotational velocity <Vsini(t)> depend on the braking law, but the shape of the Vsini distribution is also very sensitive to α.

Extensive Coravel observations of F and G type dwarfs in nearby clusters, which extend those used in a previous discussion by Benz, Mayor and Mermilliod (1984) have produced well–defined Vsini distributions which we use to constrain the values of A and α.

The new grids of stellar models with core overshooting and mass loss published by Maeder and Meynet (1989) provide new theoretical isochrones which reproduce very well

S. Catalano and J. R. Stauffer (eds.), Angular Momentum Evolution of Young Stars, 143–150.

the main-sequence morphology and allow a new determination of cluster ages. These are generally older by 0.2 – 0.3 dex than the ages previously estimated by Maeder and Mermilliod (1981), or from most standard models.

2 ROTATIONAL VELOCITIES OF LATE-TYPE STARS IN OPEN CLUSTERS

Rotational velocities have been measured by cross-correlation spectroscopy with the Coravel scanners (Baranne et al., 1979) both at the Haute-Provence and La Silla Observatories. The width of the cross-correlation function in terms of Vsini has been calibrated by Benz and Mayor (1981, 1984). Rotations as low as 2 km/s can be detected with this technique. The typical uncertainty on a Vsini determination is about 2 km/s. In fact, the cross-correlation function is so sensitive to rotational broadening that measurements of stars with Vsini larger than 40–50 km/s are difficult or impossible with Coravel.

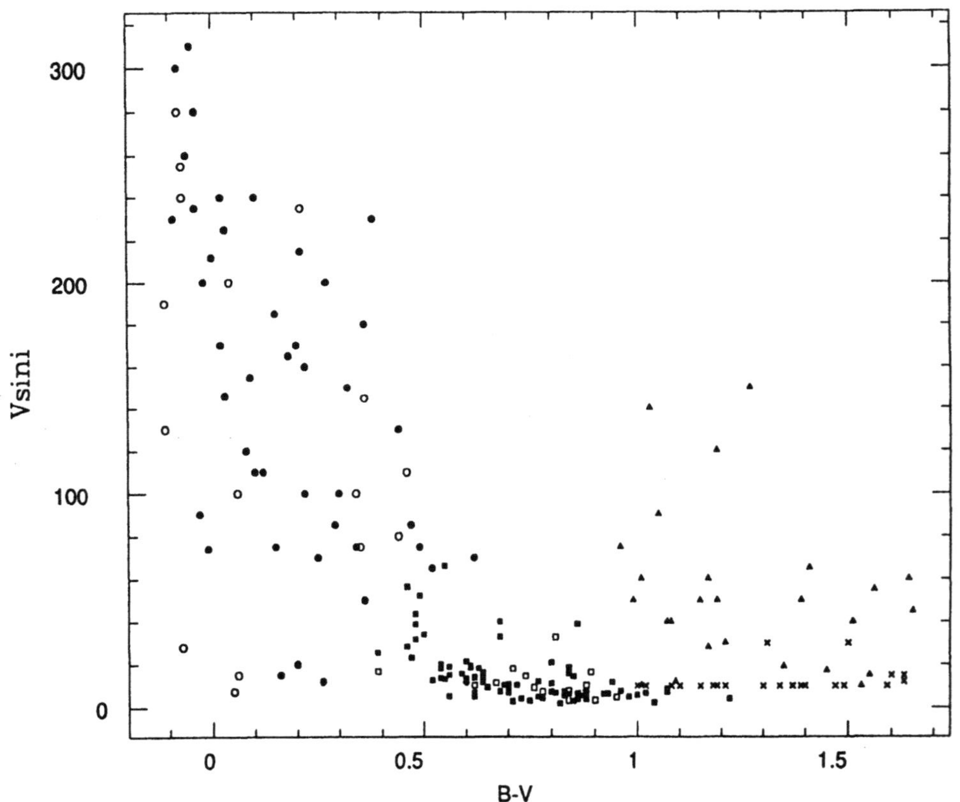

Figure 1: *Vsini vs B–V for all main-sequence stars in the Pleiades. Coravel observations are plotted as squares. Triangles represent observations by Stauffer and collaborators, (x) indicating upper limits (10 km/s). Filled circles are B–A–F stars observed by Kraft and collaborators (Anderson et al. 1965; Kraft 1967), open symbols denote spectroscopic binaries.*

Our survey includes about 700 red dwarf members of 11 open clusters (IC 2391, IC 2602, Alpha Persei, Pleiades, Blanco 1, NGC 6475, 7092, Coma, Hyades, Praesepe and NGC 752). Most stars have spectral types between F5 and K0. The age ratio of the oldest to the youngest clusters is about 50. Owing to the faintness of these stars, V reaching 13.5 mag, we were very careful in gathering well-integrated correlation profiles.

In the present context, the detection of low Vsini value is crucial because 50% of the G-type stars have rotational velocities smaller than 5 km/s already at the age of the Hyades. Rotational velocities for stars fainter than our instrumental limits, or for fast rotators, have been taken from the papers by Stauffer and Hartmann (1987) for the Pleiades, and Stauffer, Hartmann and Burnham (1985) and Stauffer, Hartmann and Jones (1989) for Alpha Persei. Figure 1 shows the Vsini vs B–V diagram for all main-sequence stars in the Pleiades as an example of the wealth of data now available.

3 THE TIME DEPENDENCE OF THE ROTATION OF THE G-TYPE STARS

We shall discuss here the angular momentum evolution of G-type stars only, concentrating on the determination of the decay law and leaving the complete discussion of the entire material to a forthcoming paper. For that purpose, we have selected all stars with 0.60 $<(B-V)_0 < 0.80$ in the Pleiades (29 stars) on the one hand, and in the Hyades and Praesepe (66 stars) on the other. We have rejected spectroscopic binaries, because the rotation may be altered by tidally induced synchronism.

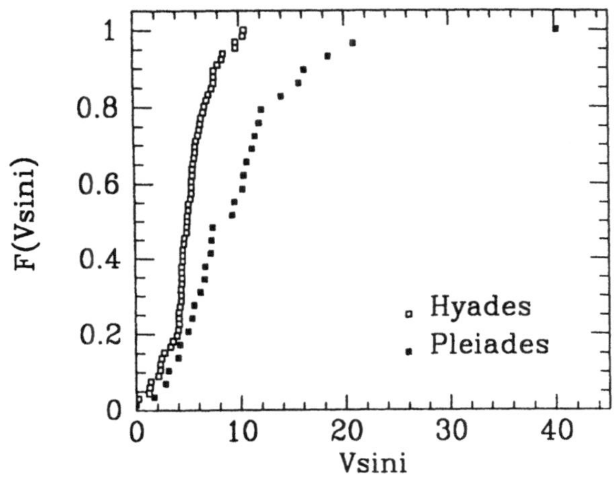

Figure 2: *Cumulative distributions of Vsini for G-type stars in the Pleiades (filled squares) and Hyades + Praesepe (open squares).*

The problem to be solved is illustrated by Fig. 2: which is the decay law which will bring the cumulative distribution of the rotational velocities of Pleiades stars taken at the Hyades age in agreement with that of the Hyades? We notice that the observed Vsini distribution for young clusters clearly exhibits a range of equatorial velocities at a given mass and age.

The general solution of equation (1) is given by

$$V(t) = \left[(\alpha - 1)A(t - t_0) + V_0^{1-\alpha}\right]^{1/(1-\alpha)} \tag{2}$$

A and α are the free parameters to be fixed ($\alpha \neq 1$).

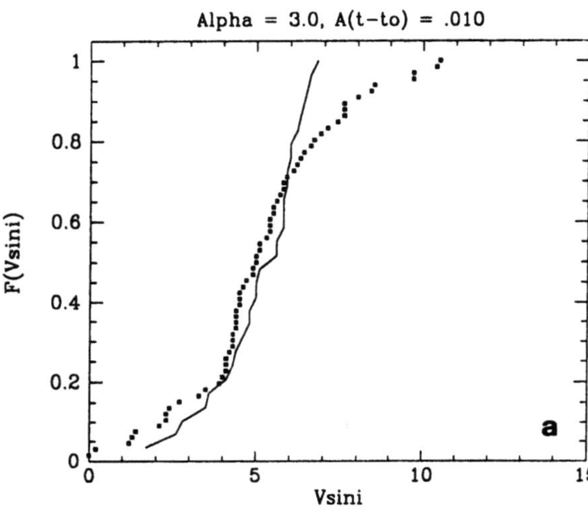

Figure 3: *The cumulative distribution of Vsini for G-type stars in the Pleiades F_P(Vsini, t_H) as computed for the age of the Hyades (t_H), assuming a braking law with index α (full line). For comparison, we show the observed distribution in the Hyades + Praesepe (squares). Despite the agreement of the mean velocity, the $t^{-1/2}$ law ($\alpha = 3$) cannot reconcile the two distributions (case a). The fit deduced from $\alpha = 2$ is very satisfactory (case b).*

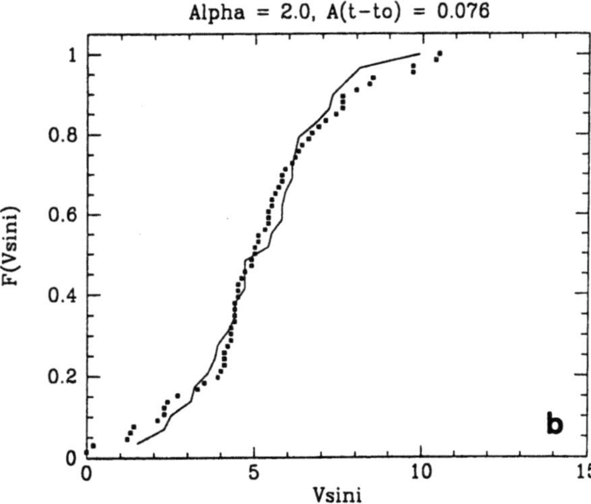

Let f(V,t) be the distribution of rotational velocities at a given age t. We shall assume that this function is the same for any cluster at the age t. Let us also define the observed distribution of the Vsini in the Pleiades as f_P(Vsini, t_P) and f_P(Vsini, t_H) that obtained after the evolution of each individual rotation velocity according to the law V(t). f_H(Vsini, t_H) is the observed function for the Hyades, where t_P and t_H denote respectively the ages of the Pleiades ($\log t_P = 8.0$) and Hyades ($\log t_H = 9.15$). The best fit between f_P(Vsini, t_H) and f_H(Vsini,t_H) will give the optimum pair of parameters (A, α) defining the braking law. In fact, the fit depends on the product A $\cdot$ (t–t_0), but not directly on A. The determination of α is therefore independent of the true age difference between the Pleiades and the Hyades. The comparison of the observed cumulative function of the Hyades and that computed for the Pleiades for different pairs (A,α) gives the best agreement for values of α in the range 1.6 to 2.3. The value $\alpha = 2$ is clearly favoured by further considerations of the mean rotational velocities. The value $\alpha = 3$, corresponding to the $t^{-1/2}$ time dependence (Skumanich, 1972) gives a much poorer agreement (Fig. 3)..

The mean observed rotation of the G type stars in the Pleiades is $<\text{Vsini}(t_P)> = 10.2 \pm 1.4$ km/s, and in the Hyades + Praesepe: $<\text{Vsini}(t_H)> = 5.1 \pm 0.3$ km/s. The mean rotational velocity <Vsini(t)> defined by a specific distribution at time t_0 is evidently a function of α and A. To constrain the range of possible values of α, we have required the mean velocity of the Pleiades computed for t = 5 Gyr to be the same as that observed for field G–type stars of that age. The ages have been taken from Barry (1988), the rotational velocities from Coravel observations. The mean velocity for field G–type stars with ages from 4 to 6 Gyr is <Vsini(5Gyr)> = 2.6 ± 0.2 km/s. This value restricts the choice of α from 1.9 to 2.3, and we adopt $\alpha = 2.0 \pm 0.3$.

With this value, the integration of the braking law dV/dt = –A V^α is especially simple:

$$V(t) = [A(t - t_0) + 1/V_0]^{-1} \tag{3}$$

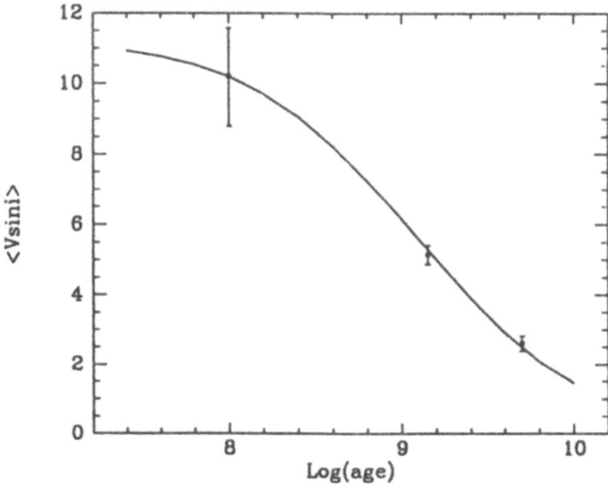

Figure 4: ***Time evolution of the mean rotation of Pleiades G–type stars < Vsini (t)> computed with the pair (A,α) from the best fit in Fig. 3b. The result is compared to the observed mean values for the Pleiades, Hyades + Praesepe, and field stars of mean age equal to 5 Gyr.***

It would be wrong to interpret this as a t^{-1} law, because the term $1/V_0$ is not negligible when compared to $A \cdot (t-t_0)$. The braking law resulting from $\alpha = 2$ and $A =$ const. correctly describes the surface rotational velocity evolution from 10^8 to 10^{10} yr. If we adopt an age difference $t_H - t_p = 1.3\ 10^9$yr, we find $A = 0.056(\mathrm{km/s})^{-1}\ \mathrm{Gyr}^{-1}$. When applied to the present observed rotations of the Pleiades stars, this braking law produces a superb agreement with the observed shape of the Vsini distribution of the Hyades + Praesepe stars (Fig. 3b). The agreement of the predicted mean rotational velocities with those observed in the Hyades + Praesepe as well as in field G stars of age $\sim$ 5 Gyr is also excellent.

We have carried out a similar analysis by comparing the Alpha Persei and Pleiades distributions. Once again, the evolution for G–type stars can be described with $\alpha = 2$, but now with a much larger value for the coefficient A. It does not seem possible to obtain a correct description of the rotational braking of very young stars by using the value of A required for older clusters.

4 REFERENCES

Anderson C.M., Stoeckly R., Kraft R.P. 1965, Astrophys. J. **143**, 299
Baranne A., Mayor M., Poncet J.–L. 1979, Vistas in Astron. **23**, 279
Barry Don C. 1988, Astrophys. J. **334**, 436
Benz W., Mayor M. 1981, Astron. Astrophys. **93**, 235
Benz W., Mayor M. 1984, Astron. Astrophys. **138**, 183
Benz W., Mayor M., Mermilliod J.–C. 1984, Astron. Astrophys. **138**, 93
Kawaler S.D. 1988, Astrophys. J. **333**, 236
Kraft R.P. 1967, Astrophys. J. **150**, 551
Maeder A., Mermilliod J.–C. 1981, Astron. Astrophys. **93**, 136
Maeder A., Meynet G. 1989, Astron. Astrophys. **210**, 155
Skumanich A. 1972, Astrophys. J. **171**, 565
Stauffer J.R., Hartmann L.W. 1987, Astrophys. J. **318**, 337
Stauffer J.R., Hartmann L., Burnham N. 1985, Astrophys. J. **289**, 247
Stauffer J.R., Hartmann L., Jones B.F. 1989, Astrophys. J. **346**, 160

DISCUSSION

van't Veer: This alpha exponent of the Skumanich relation has some magic history - everybody believes it, nobody believes it, and different people have gotten various values for it - bigger than 3, less than 3, I've seen 2.5 in the literature. But I am convinced that alpha does not exist - alpha depends on the rotation. When you go to much faster rotation, you cannot use alpha as 3. It becomes completely different. I think that we should be aware of the fact that while there is some mean value for alpha for some range of rotation we should not give alpha some magic significance.

Mayor: Do you have some reason to believe that alpha should be a function of the rotational velocity when the velocity is between 5 and 15 km/s, because our stars are in this relatively low rotation range?

Cameron: There is one very good reason to believe that alpha should drop considerably below 3 at around that sort of rotation rate (10-15 km/sec) and that is that all of the empirical evidence looking at the proxy indicators for magnetic field strength says that saturation in G stars occurs at something like 8 times the current solar rotation rate. Now this is getting very close to the sort of rotation rates which you are looking at. The alpha which you have been looking at is essentially the 1st power of the rotation rate and then it also depends on the power law dependence of the surface field strength on the rotation rate. If that is dropping from the linear dependence so that you no longer have a $B^2\omega$ which gives you your 3rd power ($B^2\omega$ being equal to ω^3 in that case) and if you then drop over to a saturated field where B is no longer dependent on ω, then you will drop down to a situation where $d\omega/dt$ is now proportional to ω, or if you take into account centrifugal driving and the effects of closed field lines, you get an even lower power law dependence and you can in fact end up with very high rotation rates with the angular momentum loss rate being almost independent of the rotation rate - so there is no theoretical objection to having a variable value of alpha as a function of rotation rate.

Schatzman: For the not too rapid rotators, a consistent dynamo theory leads to a value of alpha equal to 7/3, which is in fact in the range of values acceptable to your model. I agree with what Roxburgh has said before, that at very large rotation values, we have to take into account the actual surface of the stars where the lines of force are open, and this is a very complicated thing.

Soderblom: You mentioned, using averages among G dwarfs, but systematically the mass range which you are sampling is changing both with metallicity and with age. The Hyades are more metal rich and your are looking at higher masses for a given spectral type range and the field stars are higher masses because they are older. Rotation is very sensitive to mass. Have you applied a correction for that?

Mayor: Yes. But in our domain of B-V = 0.6 to 0.8, you are not very sensitive between solar type stars and the metallicity of the Hyades. If you compare to halo stars, you would have to take this into account.

Soderblom: I still think it would be important for your sample.

Gray: How much of a change in a color index are you talking about?

Soderblom: Perhaps 0.04 in B-V.

Stauffer: Perhaps 0.03 to 0.04 in B-V.

Soderblom: It is not inconsequential and the rotation function drops off very steeply with mass - you have a sliding window on a slanting line.

Mayor: It is actually not important for this, because you are almost in the flat part of the distribution for this color range - you are not in the steep part of the *vsini* distribution.

Catalano: I want to comment of the question of whether you should have the same value of alpha when you look at very young stars or old stars. When the 1/2 power law decay is applied to old stars, say ages greater than the Hyades, it is ok; but between the age of the Pleiades and the Hyades, and perhaps earlier, you have to take into account not only the shape or topology of the field, but also the transfer of angular momentum from the interior to the convective zone. We have to know whether you are slowing down the whole star or only the convection zone. So, things may be much more complicated. And we must look more closely at the data, and not only talk about surface rotation but angular momentum. If we can estimate the transfer from the interior to the exterior, then we can look for better understanding.

Pinsonneault: The absolute ages you assign to the clusters really do have an effect on the index or the time dependence which you derive. It looked to me like you adopted an age of about 1 Gyr for the Hyades, and you are bound to get a steeper relation than someone would get who used an age of say 500 Myr. So, the question of the absolute age scale does enter into this in a fairly sensitive way. Therefore, we do have to work out these theoretical disagreements of a factor of 2 or 3, which do exist, in the open cluster age scale.

Mayor: In fact, it is only if you compare the Pleiades and the Hyades that you can derive alpha without involving the absolute ages of the clusters. But afterwards, when you use the mean velocities of the field stars to derive a new alpha, then you need the ages of the clusters explicitly. (The point is that all the cluster ages would shift by about a factor of two together, so relative cluster ages would not change).

Pinsonneault: I disagree.

MAIN SEQUENCE ANGULAR MOMENTUM LOSS IN LOW-MASS STARS

David R. Soderblom
Space Telescope Science Institute
3700 San Martin Drive
Baltimore MD 21218 USA

ABSTRACT. Angular momentum loss for main sequence stars is sedate compared to the phenomena seen in young clusters. We believe we understand the basic processes that are responsible, yet rather few observations of good quality exist to delineate the rotation-age relation. The available data are here reviewed, including the existence of a rotation-age relation, its possible form, the statistics of rotation among solar-type stars, and other topics such as differential rotation and the relation of rotation to activity cycles. Finally, an observing program is outlined that could lead to significantly better understanding of angular momentum loss over the next few years.

1. GOALS, QUESTIONS, AND OBSERVATIONS

In studying the rotation of low-mass stars (which here means stars with convective envelopes, i.e., with $\mathcal{M} \lesssim 1.5\mathcal{M}_\odot$), our goal is to learn

$$J = f(\mathcal{M}, t, Z, J_0, R, \ell).$$

In other words, we want to determine how the angular momentum (AM) of a star depends on its evolutionary determinants (mass, age, composition, and initial AM), and how the AM is distributed with radius and latitude throughout the star. We would also like to know if other factors play a role – for example, planetary formation. (This discussion is predicated on the now-conventional paradigm of AM loss in stars like the Sun. This is discussed somewhat below.)

This is an ambitious goal, and we would get very close to approaching it if we could learn, say, how the AM loss rate, mass loss rate, and magnetic field strength depend on those quantities. More realistically, we would find satisfaction in just determining how the surface rotation depends on mass and age:

$$\Omega_{\rm surf} = f(\mathcal{M}, t).$$

S. Catalano and J. R. Stauffer (eds.), Angular Momentum Evolution of Young Stars, 151–165.

Several questions naturally raise themselves:

1. A single star like the Sun weaves a smooth, single-valued path through this multidimensional parameter space. Do similar stars proceed on similar paths (determinism) or do random factors play an important role (stochasticism)? Specifically, is the Sun typical for its mass and age?
2. Does a reasonably simple $\Omega(\mathcal{M}, t)$ relation exist that can be observationally characterized?
3. What is it that declines with age? Areal coverage of active regions? Overall magnetic field strength? The mass loss rate? The degree of differential rotation?
4. What conditions of formation and early evolution have later manifestations in the rotation of low-mass stars? Some possibilities are planets, duplicity, and cluster richness.
5. What additional observations are needed to clarify the situation?

I begin by considering the properties of low-mass stars that are pertinent to angular loss and that are observable. The most important and useful of these is the rotation period (P_{rot}), which is determined from photometry, spectrophotometry, or inference as a result of the presence of stellar surface inhomogeneities. Young stars (of the age of the Hyades or less) show perceptible brightness variations in intermediate-band filters (Radick *et al.* 1987), although the variations in color are much less. Thus extracting rotation periods for such stars is relatively easy, even if they are faint. On the other hand, the Mount Wilson group (Baliunas *et al.* 1985) has been able to detect rotation modulation in older stars with weak chromospheres by measuring the strength of the Ca II H and K line cores relative to the nearby continuum, thereby enhancing the contrast of the inhomogeneities. Finally, Soderblom (1985) has shown how to estimate rotation periods from knowledge of a star's HK emission strength (more on this below).

The advantages of determining rotation periods, compared to other observations, are considerable. First, the results are both precise and accurate, even for stars that rotate very slowly. There may be systematic effects that arise from measuring the rotation of the chromosphere instead of the photosphere, or from preferentially detecting the surface inhomogeneities at particular latitudes, but these must be much less that the errors inherent in measuring $v \sin i$. Moreover, P_{rot} measurements are aspect-independent and yield a more physically relevant number that an equatorial velocity. Finally, some observations of rotation periods suggest that we may be able to detect and measure stellar latitudinal differential rotation by that means.

The obvious drawback of measuring P_{rot} is that it is observationally intensive, requiring many data points. Thus much of the extant data on rotation, even for low-mass stars, is the result of $v \sin i$ determinations. This is quickly done if a high-resolution spectrum can be obtained, either directly from the analysis of line profiles, or by cross-correlation of the stellar spectrum with a narrow-lined star, typically the Sun (see Soderblom [1990] for a review of techniques involving high-resolution profiles). Line profile analysis is generally necessary unless rotation dominates the line broadening, although the Coravel group (Benz and Mayor 1981, 1984) have determined $v \sin i$ values down to the level of the Sun's from their cross-correlations. Line profile analysis requires data of high signal-to-noise, but cross-correlation can tolerate low S/N data and so is very efficient, especially for getting rotation rates of young stars in clusters. Cross-correlation can produce $v \sin i$ values that are precise to about ± 1 km s^{-1}. Line profile analysis can achieve precisions of about 0.3 km s^{-1} and accuracies of 0.5 km s^{-1} (Soderblom 1982), but only for the best cases. Even results that precise are of limited use for old solar-type stars because they rotate so slowly (note that the Sun's equatorial velocity is only about 1.8 km s^{-1}).

The domains of the different measurements techniques are shown schematically in Figure 1. Rotation periods can be measured only in the youngest stars at the high mass end but even in the oldest low-mass stars if spectrophotometric methods are used. Cross-correlation works best for relatively large $v \sin i$ values, which means almost all the higher mass stars but only the youngest low-mass stars. Line profile analysis will work over any portion of this diagram, but only for bright stars because of the need for high resolution and high S/N. As a result, you will note a regime of old, solar-type stars for which our only information on rotation comes from line profile analysis.

The mass of a star is arguably its most important property for determining its overall evolution. We know masses for few stars, of course, and those stars are in binaries where we suspect the rotation is affected by the duplicity. However, until we know much more about the details of rotation, it is sufficient to deduce rough masses from spectral types or colors.

The age, t, of a star is also a problematic, yet fundamental, quantity. The rotation-age relation is the essence of the work in this area, yet the ages of individual stars cannot now be determined with any accuracy or precision. Further discussion of this is presented below in discussing the relationship between chromospheric emission and age.

The initial AM of a star, J_0, is impossible to determine on a star-by-star basis, but we can study rotation in star formation regions and rotation distribution functions and thereby hope to infer something about initial conditions.

The actual AM loss rate and mass loss rate of low-mass main sequence stars have not been determined, except for the Sun. These are important quantities, but they will remain unobservable for some time to come.

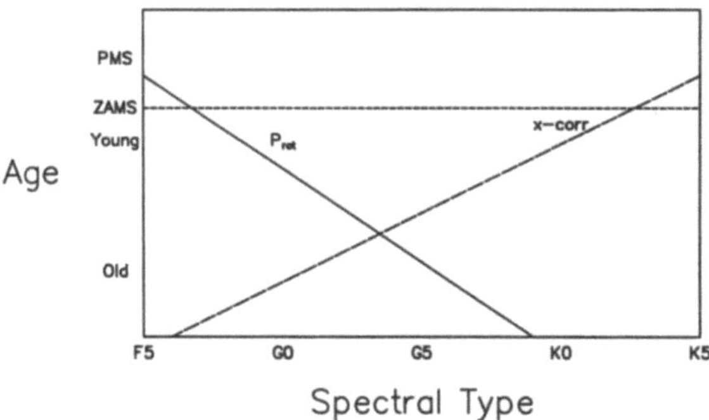

Figure 1. Schematic representation of the domains of observational techniques for measuring rotation in low-mass stars. The solid line shows the approximate region for which rotation periods are detectable, which is for all young stars and for older stars of lower mass. On the other hand, cross-correlation is useful only for relatively large $v \sin i$ values, which occur for all young stars and higher mass stars of all ages. There is a gap between these two techniques for the old G dwarfs, where line profile analysis has provided most of the available information.

The magnetic field strength, B, has in some cases been determined from direct measurements of Zeeman broadening, but the relevance of those measures to AM loss is arguable. We know of B mostly by inference from chromospheric activity and from some knowledge of the coverage of active regions for young stars.

2. *THE EXISTENCE OF A ROTATION-AGE RELATION*

Wilson suspected it, Kraft showed it, Skumanich quantified it, and Smith and Soderblom confirmed it: the rotation-age relation. Wilson (1966) appears to have been the first to suggest that stars like the Sun lose AM over their main sequence lifetimes as a magnetic field grips an ionized wind beyond the stellar surface, through the mechanism put forth by Schatzman (1962). Kraft (1967) provided a stronger observational basis for the relation by obtaining $v \sin i$ values from high resolution photographic spectra of late-F and early-G dwarfs in the nearby young clusters and the field; he also showed that field stars with strong Ca II H and K emission rotated much more rapidly that field stars that lacked that emission.

The underlying basis for this loss of AM as well as a decline of chromospheric emission (CE) with age has been assumed to be the presence of surface convective zones in stars like the Sun (Wilson 1963, 1966). This is dramatically illustrated (Figure 2) in the immortal diagram of Kraft (1967), which shows how precipitously rotation declines near 1.25 $\mathcal{M}_{\odot}$. All of this has led to our present paradigm for the main sequence evolution of AM and CE in solar-type stars: Rotation (especially differential rotation) interacts with convection, leading to a magnetic dynamo. The resultant magnetic field (especially those parts with open field lines) leads to AM loss if an ionized wind is present, and

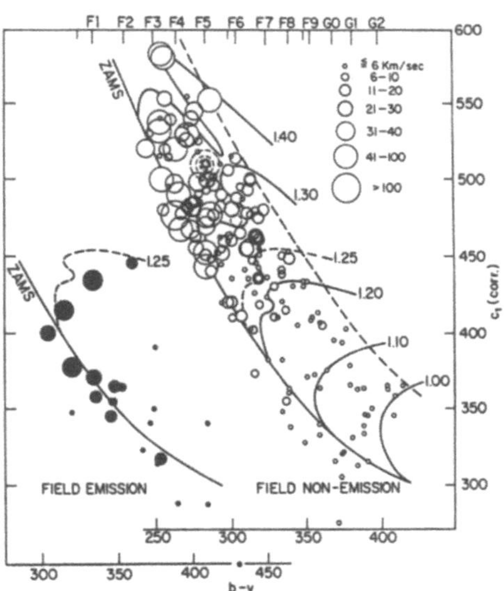

Figure 2. HR diagram of F dwarfs of the field (from Kraft 1967). The Strömgren $(b - y)$ color index is indicative of temperature and the $c_1(corr.)$ index indicates gravity. Circle sizes indicate the measured $v \sin i$ and a Zero-Age Main Sequence (ZAMS) has been shown. The evolutionary tracks are now obsolete but show how steeply rotation decreases with mass near 1.25 $\mathcal{M}_{\odot}$. The solid circles are field stars with strong Ca II H and K emission, and are displaced from the main diagram.

we now witness such AM loss on the Sun. Thus the loss of AM in a star like the Sun is an *inexorable* consequence of its main sequence evolution, not just a tendency, as long as that star exhibits solar-like phenomena over its lifetime.

Shortly after Kraft's work, Skumanich (1972) published his brief but pivotal paper that quantified the rotation-age relation. Skumanich took Kraft's data for the Pleiades and Hyades and used the rotation rate of the Sun to typify older stars. Similar data from Wilson for CE were also plotted on a log-log scale to illustrate the now-famous Skumanich power-law relations between rotation and age, and CE and age: $v \sin i \propto t^{-1/2}$; CE $\propto t^{-1/2}$. It should be noted that the cluster stars that Skumanich compared to the Sun were systematically and significantly more massive, but that error was fortuitously offset by the use of cluster ages that are significantly below those which are now believed to pertain.

After another interval of a few years, Smith (1978) added to Skumanich's data by determining $v \sin i$ values for some old field stars. This was then extended by Soderblom (1983), who determined many more $v \sin i$ values, especially near 1 $\mathcal{M}_{\odot}$. Soderblom used ages for the field stars that were estimated from their lithium abundances. That technique has since been questioned (Soderblom 1984, 1987), although those ages are probably at least qualitatively correct. Note, however, one critical point: the available observations are certainly *consistent* with a power law, but there are enough gaps and uncertainty to allow a variety of other relationships to explain the observations. For example, one could imagine drawing a step function through Soderblom's (1983) observations (shown in Figure 3), so that AM loss would be episodic with intervals of constant $v \sin i$.

3. *A STATISTICAL APPROACH TO STUDYING ANGULAR MOMENTUM LOSS*

Observations of a few field stars and young clusters show that a rotation-age relation does exist and that it is at least consistent with a power law, but we would like to learn more. One obvious question is whether solar-type stars of a range of masses obey the same rotation-age relation. We would also like to know the extremes of rotation seen in stars like the Sun and other general trends.

To do that it suffices to examine a large sample statistically, and such a sample exists through the survey of chromospheric emission undertaken by Vaughan and Preston (VP; 1980). VP obtained observations of the equivalent width of Ca II H and K emission

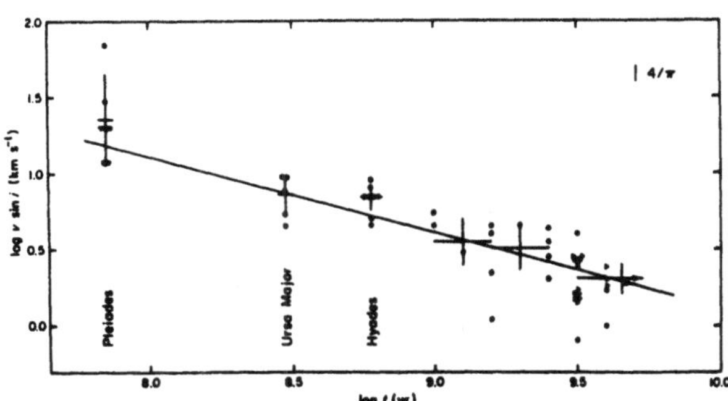

Figure 3. The $v \sin i$ observations of cluster and field stars of Soderblom (1983). For the field stars on the right side of the diagram, the indicated averages have horizontal lines to show the span of ages encompassed and vertical lines to show standard deviations. The slanted line has a slope of $-1/2$.

for an essentially complete sample of dwarfs later than F5, in the northern hemisphere and within 25 pc. They excluded binaries from their sample. Noyes *et al.* (1984) have shown how good the relation is between a normalized index of the HK emission and the Rossby number, which is the ratio of the rotation period to the convective turnover time. The latter number can be calculated from theory, which enables one to compute a rotation period for a star knowing only its HK emission strength. Soderblom (1985) applied this technique to the VP survey to derive the distribution of rotation rates shown in Figure 4. Note that this technique allows one to compute a star's rotation period to an accuracy of 20% from only one or two observations of Ca II H and K.

The results of this analysis illustrate the broad properties of rotation among stars like the Sun:

- The sample illustrated covers 0.50 to 1.00 in $(B-V)$ because the conversion of the HK emission index into physical quantities is best understood there. This color range corresponds to a mass range of about 0.8 to 1.2 $\mathcal{M}_{\odot}$.
- The maximum rotation seen is about ten times solar, but that is only for a few stars, and two to three times solar is more typical. The minimum is at about one-half solar.
- Most stars, as one would expect, are old and lie near the lower bound of the diagram. The oldest F dwarfs have already left the diagram and have evolved into subgiants and so are by definition no longer part of the sample. At the same time, the oldest G and K dwarfs have AM left to lose, so that the flat lower bound at $\sim 1/2\Omega_{\odot}$ is not a "floor" of rotation or activity.
- The range in Ω at any one color is independent of color, which indicates that the same AM loss law operates over this entire mass range.

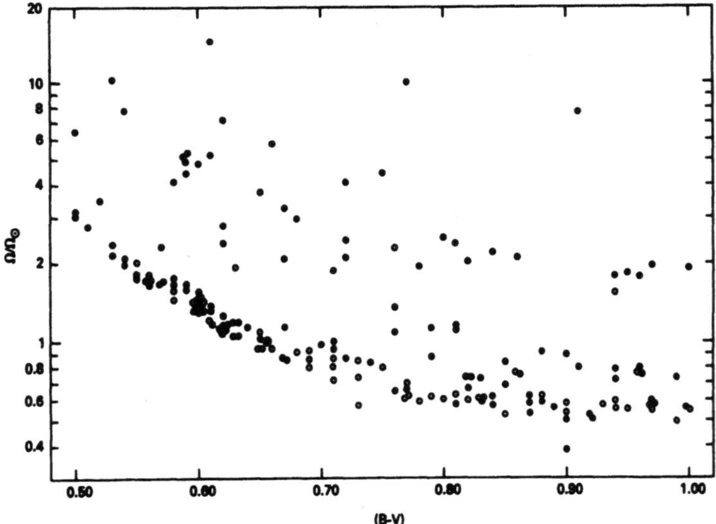

Figure 4. The distribution of stellar angular velocity with color for solar-type stars. The values of Ω are derived from observations of Ca II H and K emission strength, as described in Soderblom (1985). The solid points are stars of the young disk, while the open circles are high velocity stars of the old disk and halo.

- The high-velocity stars (open circles) almost all have low rotation rates, as expected for an old population.
- The Sun appears to be a very typical old 1 $\mathcal{M}_\odot$ star.

4. *A BACK DOOR APPROACH TO THE ROTATION-AGE RELATION*

The previous section noted the excellent relationship between CE and rotation found by Noyes *et al.* (1984). CE is much easier to observe than rotation: low-resolution spectra suffice, and CE is only weakly dependent on mass at any age (see, e.g., Soderblom [1985] for the distribution of CE with color for the stars of Figure 4). The intimate relationship between CE and rotation then allows the rotation-age relation to be studied if we can determine the CE-age relation.

Skumanich (1972) postulated a CE $\propto t^{-1/2}$ relation like the power law he put forth for rotation versus age. More recently, more complex CE-age relations have been suggested. For example, Barry (1988) has used low-resolution observations of solar-type stars in open clusters of many ages to calibrate the CE-age relation for stars of the VP survey. Barry uses a relation of the form CE $\propto$ exp($kt^{1/2}$) because CE $\propto$ exp($1/P_{\rm rot}$) (Noyes *et al.* 1984) and $\Omega \propto t^{-1/2}$. Barry then applies his CE-age relation to the VP survey, and concludes that star formation over the life of the Galactic disk has been highly non-uniform. In particular, he finds evidence for a recent burst of star formation.

As noted above, the Skumanich relation is certainly consistent with the observations of rotation in solar-type stars, but the data are also consistent with other functional forms. Thus the particular CE-age relationship of Barry is not compelling. More important, one can invert the process and ask what relation between Ce and age would have to hold if one assumed that the Star Formation Rate (SFR) is constant for the stars in the VP survey.

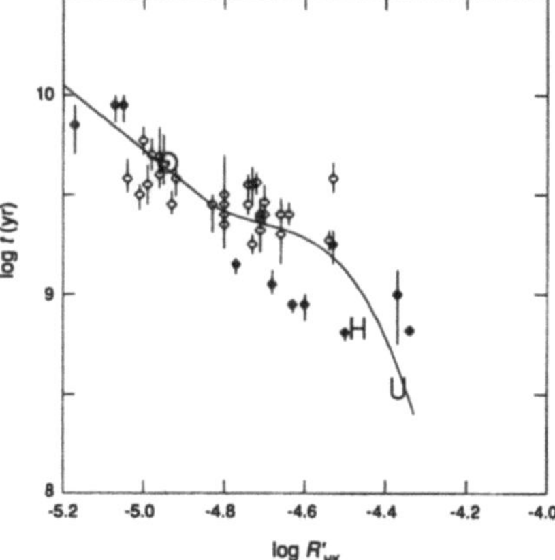

Figure 5. Observations of Ca II H and K emission in secondaries of wide binaries (filled diamonds) and slightly evolved F stars (open diamonds). Also shown are points for the Ursa Major Group (U), the Hyades (H) and the Sun ($\odot$). The ages of the binaries are determined from Strömgren photometry of the primary, while the G dwarf secondary is used to determine CE. The solid curve corresponds to a constant Star Formation Rate, determined by forcing the stars of the Vaughan-Preston survey to fall in equal numbers in bins of equal width in age.

Such a constant SFR curve is shown in Figure 5, together with some recent observations of Soderblom, Duncan, and Johnson (1991). The constant SFR curve is as consistent with these observations as is a power law, and leads to a much less radical conclusion. Until the CE-age relation is better defined, it is not possible to draw detailed conclusions about the SFR.

If the data of Noyes *et al.* (1984) define the rotation-CE relation, and the curve of Figure 5 defines CE versus age, they can be combined to derive rotation versus age. The result is shown in Figure 6, where the straight line is an example (not a fit) of a $t^{-1/2}$ power law. Clearly very modest errors in rotation or age would make it impossible to distinguish the points from the power law.

5. *OTHER TOPICS*

5.1 Differential Rotation

Differential rotation is a key ingredient of the dynamo models that are central to understanding magnetic activity on solar-type stars, yet we know virtually nothing about it for any star save the Sun. This is simply because differential rotation is devilishly difficult to observe. Its effects on line profiles are extremely subtle, and it would be difficult to attribute those effects solely to differential rotation with confidence, even if they were detected, given the presence of other broadening agents.

However, the Mount Wilson group has at least seen evidence for changing rotation periods in a few stars (Baliunas *et al.* 1985). An example is shown in Figure 7. However, these data, suggestive as they are of differential rotation, can also be explained in terms of a single rotation period and an active area on the star that grows and decays during the time covered by the observations.

Some evidence for stellar differential rotation has also been seen in period changes in the highly active RS CVn and BY Dra stars. In those systems the sense of the differential rotation can be the same as it is on the Sun, i.e., Ω increases with decreasing latitude. However, UX Ari shows differential rotation that is opposite to the Sun at one-tenth the degree (Vogt and Hatzes 1990). These stars are distinctly non-solar-like in several respects – particularly because they are close binaries – so that they may not be

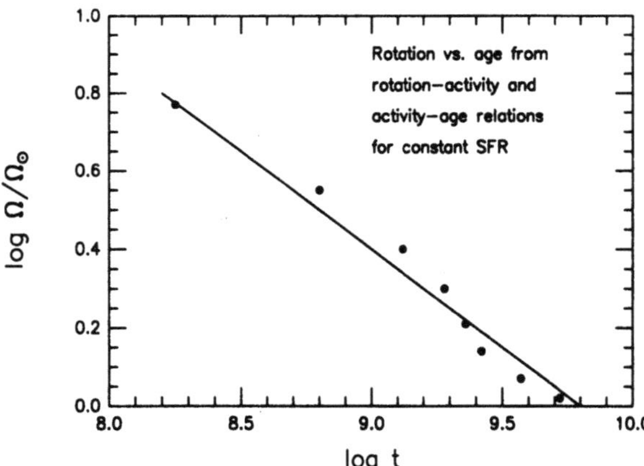

Figure 6. The rotation-age relation determined by combining the rotation-CE relation of Noyes *et al.* (1984) with the constant SFR curve of Figure 5. The straight line is a $t^{-1/2}$ power law drawn in for reference.

indicative of young, active stars in general. It is also true that these highly active stars consistently tend to have their spots congregate at the poles of the star, quite the opposite behavior to that of the Sun.

5.2 Rotation and Activity Cycles

We believe that rotation and its interaction with convection are at the heart of the dynamo mechanism. The eleven-year solar cycle is, we think, a manifestation of that dynamo, so it is natural to expect some relationship between rotation and activity cycles in stars. Based on experience from studying chromospheric emission, perhaps the Rossby number is better than rotation period.

The Mount Wilson group (Baliunas *et al.* 1985) have now obtained preliminary cycle periods for upwards of one hundred solar-type stars. The periods are preliminary because the solar experience shows that it is necessary to observe a star over several cycles before a reliable mean period can be established. Nevertheless, it is frustrating to see that there is a total lack of correlation between cycle period and either rotation period (Figure 8a) or Rossby number (Figure 8b). Perhaps better-defined cycle periods will help, but that may require 50 years of observation.

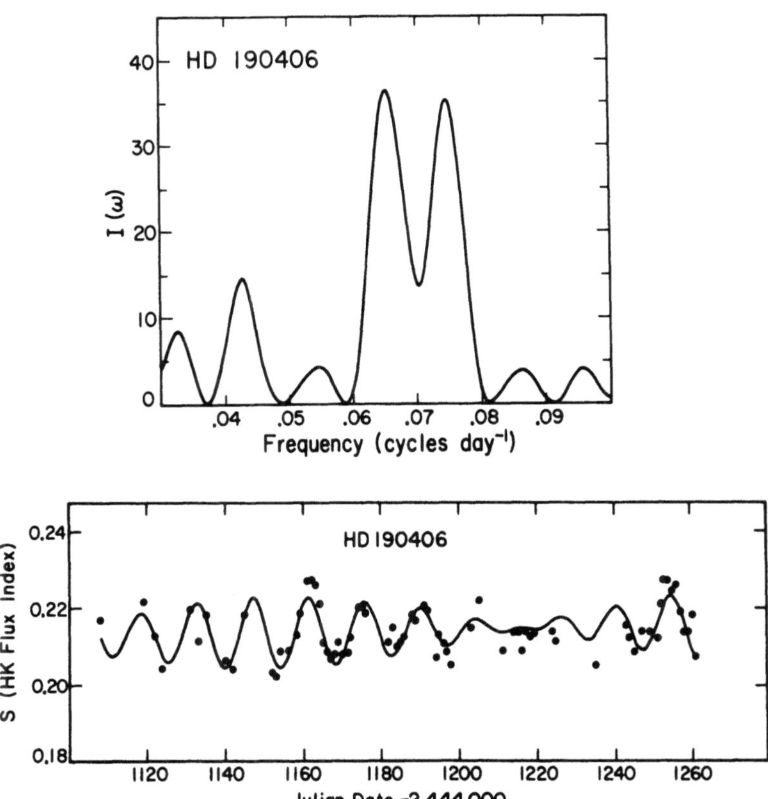

Figure 7. Periodogram of Mount Wilson H and K observations for the G dwarf HD 190406. Evidence for two periods is clearly present, but the individual observations (b) can also be fitted by a single period if an active region is added that grows and decays.

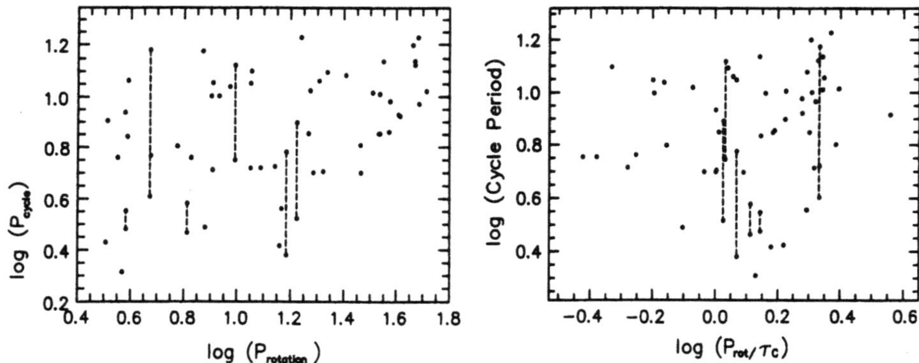

Figure 8. a. Observed long-term cycle periods of individual stars against their rotation periods. The vertical dashed lines join points for cases where two distinct cycle periods have been seen for one star. b. Cycle period against Rossby number.

5.3 Rotation in Old Disk and Halo Stars

So far we have discussed solely Population I stars, or stars no older than the Sun and with the same metallicity (to within 0.3 dex or so). It is of interest to consider rotation in stars of earlier populations as well: those of the halo and old disk. We would expect such stars to rotate very slowly because of their advanced age if they are like Pop I stars, but perhaps these stars may hold surprises as they did when examined for lithium abundances. For example, the thinner convective envelopes of these stars may inhibit the dynamo mechanism, reducing AM loss.

But few observations are available of old disk and halo stars at high spectroscopic resolution because such stars are generally very faint. As part of my thesis (Soderblom 1982) I observed μ Cas and τ Cet, and Smith (1978) has observed Gmb 1830. In all cases, no observable rotation is seen, the limits to detection of $v \sin i$ being at about the solar level or less, i.e., $\lesssim 2$ km s^{-1}. These stars have extraordinarily sharp lines, too sharp to be modeled with solar levels of broadening agents, even for zero rotation. In other words, it is necessary to use sub-solar values of macroturbulence to match the lines even with little or no rotation. Given the reach of new instruments, it would be of interest to observe these older stars more systematically, but it appears that only the highest levels of spectroscopic resolution will suffice.

5.4 Can Angular Momentum Hide Inside a Star?

It is unlikely that the Sun has a rapidly rotating core that will provide a reservoir of AM that it can tap during post-main sequence evolution, but it is not yet possible to rule that out. Observations of evolved stars may provide additional clues. For example, horizontal branch stars appear to rotate more rapidly that expected unless they have managed to bring AM to the surface from an unknown source.

As another example, consider Arcturus, a K giant. The mass of Arcturus is unknown since it is a single star, but spectroscopic determinations of its gravity, together with a well-determined radius, are consistent in indicating $\mathcal{M} < 1\ \mathcal{M}_\odot$. Now a 1 $\mathcal{M}_\odot$ star will leave the main sequence with $v \sin i \lesssim 2$ km s^{-1}, and by the time that star expands to

20 $\mathcal{R}_\odot$ (the radius of Arcturus) it will have an immeasurably small rotation rate. But the observed $v \sin i$ of Arcturus is about 2.5 km s^{-1}, and this has been measured by two observers.

There are several possible solutions to this apparent paradox. First, the gravities may be consistently wrong, and the real mass of Arcturus is closer to 2 $\mathcal{M}_\odot$, consistent with its current $v \sin i$. Second, the determinations of $v \sin i$ may have been misled by other broadening agents. Third, the observations are sound and the star left the main sequence with a rapidly rotating core that we now see as enhanced overall rotation. Fourth, the observations are sound and the star left the main sequence at 2 $\mathcal{M}_\odot$ and subsequently lost half its mass. And fifth, Arcturus started as a close binary, so that its current AM is not representative of a single star.

The first of these possibilities seems the most likely, despite the considerable effort that has gone into the analysis of the spectrum of Arcturus over the years. It is unlikely that other broadeners are fooling us. Some post-main sequence mass loss does occur in stars, but not at the levels needed to get rid of 1 $\mathcal{M}_\odot$ quickly. And supposing Arcturus used to be a binary is ad hoc. The detection of oscillations in Arcturus should help to determine the mass more accurately, and so resolve this.

6. Conclusions and a Plan for Future Observations

The existence of a rotation-age relation has been known for nearly 30 years, it characterization (in the $t^{-1/2}$ relation) for about 20 years, and its confirmation (via field stars) for about 10 years. Except for the dramatic and interesting work done in clusters recently, there has been little or no change in our observational understanding of AM loss in main sequence solar-type stars in some time. Fortunately the theory is drawing much more attention.

But the details of main sequence AM loss remain murky because the data in all three coordinates of interest (rotation, mass, and age) are "fuzzy." We can say at least that the observations are consistent with a power-law relation between rotation and age at any one mass, but other possibilities cannot be ruled out. The rotation-age relation also appears to be mass-independent near 1 $\mathcal{M}_\odot$. At any one age, rotation is strongly dependent on mass for $\mathcal{M} \gtrsim 1\ \mathcal{M}_\odot$, but isochrones flatten at lower masses. Finally, it is worth noting that essentially no field stars exhibit the phenomenon of ultra-fast rotation seen in the G and K dwarfs of the Pleiades and α Persei clusters, indicating that that phase is confined to the youngest stars.

In compiling this review, I was struck by how little things had changed in nearly a decade, at least for the issue of the rotation-age relation. Thus it seems appropriate to suggest new observational programs that could materially add to our knowledge and understanding.

First, we should endeavor to determine actual rotation periods for large and complete samples of stars in young clusters. Such a project is clearly feasible right now by using a CCD camera on a smallish ($\sim$ 1 m) telescope because Radick *et al.* (1987) detected the rotational modulation of Hyades G dwarfs in Strömgren bandpasses. The distance to which small-amplitude variations in young G dwarfs could be detected with a small telescope is substantial, which would add data for many clusters. Before such work could start it would be necessary to undertake studies of the membership of the low-mass stars in some cases, but that is worthwhile in itself.

The sheer quantities of data generated and the many nights of observing required are daunting, but from this kind of study we can first test if all clusters of the same age have essentially the same distribution of AM with mass. If they do not, that would suggest that we are seeing vestiges of the original distribution of AM within those clusters. Second, such observations will define isochrones in a $\Omega(\mathcal{M})$ diagram, helping to better determine the rotation-age relation. Finally, these observations would be critical for understanding the apparent convergence of rotation rates with age. By that I mean that in the Pleiades we see a huge range of rotation at any one color, yet in the Hyades there is little or no intrinsic spread. Clusters of intermediate age will presumably show an intermediate degree of convergence. The time scale for that convergence (and how that time scale may depend on mass) is surely an important clue to the physical processes taking place.

A second project of similar scope is to undertake the same kind of observations for older clusters, such as NGC 752, M67, or NGC 188. However, in these old stars conventional photometry will probably not suffice for detecting rotation periods because the contrast of stellar surface inhomogeneities is just too low. Instead one needs spectrophotometric data, such as is done at Mount Wilson in measuring the strength of the Ca II H and K lines relative to the continuum. The Mount Wilson instrument is far from sensitive enough to reach the faint stars in clusters, though, so one needs a large telescope. Additionally, liberal observing time may be needed because the rotation periods are expected to be long. There is, however, no other adequate means for studying rotation in older stars because $v \sin i$ values can never achieve the precision needed. Perhaps the proposed augmentation of the McMath feed to a 4 m diameter may make this project feasible.

A third project is to determine accurate rotation periods for a large number of field stars. In general ages are unknown for such stars but they are bright and easy to observe. A large enough sample could make the study of the statistics of the rotation of the sample meaningful. We could hope to confirm and extend the rotation-age relation and improve our ability to estimate rotation periods by observing chromospheric emission. It should be possible to detect differential rotation, given enough patience, and it may be particularly revealing to get rotation periods for all stars in a particular color range to see if they show the same gap seen in the Vaughan-Preston survey of H and K emission. Finally, some knowledge of stellar active regions (size, distribution, lifetimes) should result.

From this we may hope to see a substantially more complete picture of rotation in stars like the Sun about a decade hence.

REFERENCES

Baliunas, S.L. *et al.* 1985, *Astrophys. J.* ***294***, 310.

Barry, D.C. 1988, *Astrophys. J.* ***334***, 436.

Benz, W., and Mayor, M. 1981, *Astron. Astrophys.* ***93***, 235.

Benz, W., and Mayor, M. 1984, *Astron. Astrophys.* ***138***, 183.

Kraft, R.P. 1967, *Astrophys. J.* ***150***, 551.

Kraft, R.P. 1970, in G. Herbig (ed.), *Spectroscopic Astrophysics*, University of California Press, Berkeley, p. 385.

Noyes, R.W., Hartmann, L.W., Baliunas, S.L., Duncan, D.K., and Vaughan, A.H. 1984, *Astrophys. J.* ***279***, 763.

Radick, R.R., Thompson, D.T., Lockwood, G.W., Duncan, D.K., and Baggett, W.E. 1987, *Astrophys. J.* ***321***, 459.

Schatzman, E. 1962, *Ann. d'Ap.* ***25***, 18.

Skumanich, A. 1972, *Astrophys. J.* **171**, 565.
Smith, M.A. 1978, *Astrophys. J.* **224**, 584.
Soderblom, D.R. 1982, *Astrophys. J.* **263**, 239.
Soderblom, D.R. 1983, *Astrophys. J. Suppl.* **53**, 1.
Soderblom, D.R. 1984, in *Cool Stars, Stellar Systems, and the Sun*, eds. S.L. Baliunas and L.W. Hartmann (New York: Springer), p. 205.
Soderblom, D.R. 1985, *Astron. J.* **90**, 2103.
Soderblom, D.R. 1987, in *The Impact of Very High Signal-to-Noise Spectroscopy on Stellar Physics*, IAU Symp. 132, eds. G. Cayrel and M. Spite (Dordrecht: Reidel), p. 381.
Soderblom, D.R. 1990, in *Mem. Soc. Astr. Ital.*, in press.
Soderblom, D.R., Duncan, D.K., and Johnson, D.R.H. 1991, *Astrophys. J.* in press.
Vaughan, A.H., and Preston, G.W. 1980, *Pub. Astr. Soc. Pacific* **90**, 267.
Vogt, S.S., and Hatzes, A.P. 1990, in *The Sun and Cool Stars: Activity, Magnetism, and Dynamos*, IAU Coll. 130, in press.
Wilson, O.C. 1963, *Astrophys. J.* **138**, 832.
Wilson, O.C. 1966, *Astrophys. J.* **144**, 695.

DISCUSSION

Stauffer: Just two points. First, there is possibly some slight evidence that there are halo stars - or at least old disk stars - with moderate rotation still. There are a few old disk flare stars, which do have old disk motion and appear to be quite metal poor, but are flare stars and my guess is that they probably have rotational velocities of order 10 km/s or so based on data obtained by Hartmann and myself and published in the 1986 Ap.J. supplements.

Soderblom: What is that known from?

Stauffer: About one half of the dMe stars had detectable rotational velocities with the MMT, and those generally were of order 10 km/s. I think perhaps that one of the old disk flare stars had detectable rotation. But, in any case, just by comparison to the rotational velocities of the other dMe stars I would guess that the old disk dMe stars should have rotational velocities of order 10 km/s. Of course, being low metallicity probably matters less for them, because they have relatively thick convective envelopes anyway.

The other point is that Rucinski has looked at NGC 3293, which is a young southern cluster, with a CCD - hoping to find evidence of rapid rotators via their variability. He claims to have found a few good candidates. I agree with you that using large format CCD's to identify more rapid rotators in other clusters should be done.

Soderblom: What is the age of that cluster?

Stauffer: It is less than the Pleiades.

Rodonó: Can you clarify your statement that rotation depends on B-V and not mass?

Soderblom: I am sorry. I gave you the wrong impression. What I was trying to say was that simply inferring mass from the color was adequate for our purposes. Mass is presumably the real independent variable, but we can use color as a proxy.

Rodonó: I have a comment now on differential rotation. You said that there was no data on that, but actually from photometry there is quite a lot of data now. This comes from monitoring B white light variations and spot modelling in a very systematic way which can give you some hints on the differential rotation of these stars, especially in binaries where you have a time reference very well defined by eclipses, so you can monitor period variations which are very difficult to monitor on non-eclipsing stars where the time reference is uncertain.

Soderblom: Aside from the eclipsing stars, what other kinds of stars have you done this type of analysis for?

Rodonó: Also BY Dra stars.

Soderblom: For binary stars,there is always the nagging doubts about whether this would apply to single stars.

Gray: We know for the F stars from line profile analysis that differential rotation is not large, so we have some limits on differential rotation.

Soderblom: That is not a very severe limit though.

Gough: I was interested in your comment that you thought the mass of Arturus might be quite large. There has been some evidence recently that Arcturus is pulsating. So, my colleagues and I looked at the pulsation characteristics of models of Arcturus with the two canonical masses that people suggest - the high one not being as high as the 2 solar masses which is what you suggested. But, in any case, what is found is that we can get rough agreement with the high mass stars but not with the low mass ones. We don't yet know how sensitive this is to mass, and whether or not if it is as high as 2 solar masses we would get even better agreement.

Soderblom: What sort of mass do you use?

Gough: I've forgotten. I'm not good at remembering numbers. I think it was about 1.2 or 1.5 solar masses. The pulsation measurements are not very accurate at the moment, but that will improve with time. So there is a hope this issue can be cleared up.

Soderblom: That would be very interesting, to get masses of single stars.

Mermilliod: With respect to old clusters, I would like to note that we now have rotational velocities for the G dwarfs in NGC 752, a cluster thought to be older than the Hyades. So, you should be able to add another point onto your diagrams of rotation versus age.

THE ANGULAR MOMENTUM EVOLUTION OF YOUNG AND OLD BINARY COMPONENTS

FRANS VAN 'T VEER
Institut d' Astrophysique de Paris - CNRS
98 bis Boulevard Arago, 75014 Paris, France

CARLA MACERONI
Osservatorio Astronomico di Roma
Viale del Parco Mellini 84, I-00136 Roma, Italy

ABSTRACT: Young stars are losing great amounts of mass and angular momentum. It is important to know how these mass and angular momentum losses are related to a centrifugal wind excited by the high rotational velocity. The mechanism may be the same, in different physical conditions, independently of the age of the star. An important test in this respect is given by the components of late type close binaries rotating synchronously with the orbital motion. Even if they may be old objects their rotation may be at least as rapid as that of T Tauri stars. We discuss about what can be deduced concerning mass and angular momentum loss from the rapid rotation of old and young close binary components.

1 Introduction

The title of this workshop is the angular momentum (AM) evolution of young stars. Young stars are characterized by their tumultuous settling on the main sequence accompanied by violent stellar winds with various velocities in locally strong magnetic fields. These fields possess an unknown, but certainly complicated and variable geometrical structure and are at the origin of great amounts of angular momentum loss observed in magnetically active stars (Schatzman, 1962). The winds are able to clear away protoplanetary disks in less than 10^7 years. This time scale is so short that it is still a matter of debate how planets can possibly be formed before the removal of the gas.

From different talks of this workshop we know that during this stage of contraction to the MS great amounts of AM have often to be expelled in order to achieve the final stellar status.

The subject of the present contribution is to study how far these AML phenomena of young stars can be compared with the AML of older, but also rapidly

S. Catalano and J. R. Stauffer (eds.), Angular Momentum Evolution of Young Stars, 167–174.

rotating, components of synchronously rotating very close late-type binaries. The rotation of these components may exceed the measured equatorial velocities of the rapidly rotating young stars. So, with these systems, we possess an excellent material to study how far certain properties of young rotators can be understood as a consequence of their rapid rotation alone.

2 The rotation of very close binary components

The rotational velocity of T Tauri stars extends from 10 to 100 km/s and so their maximum equatorial velocity is still several times below the break-up velocity (Bouvier, these proceedings). For other young stars, looking at recently formed galactic clusters, somewhat higher maximum velocities are found. This is the case especially for IC 2391 (Stauffer, 1987) with $3 \cdot 10^7$ yr and α Persei ($5 \cdot 10^7$yr) and Pleiades ($7 \cdot 10^7$yr) (see Stauffer, 1987) which show upper limits of 150, 200 and 150 km/s respectively.

The rotation of late type very close binary components (VCBC) is controlled by the period and the degree of synchronization of the system. The latter depends on the timescales of AML from the components and angular momentum transfer (AMT) from the orbit to the components.

The great majority of solar type binaries with orbital periods $P_K \leq 3^d$ have synchronized rotations (see next paragraph) and hence from their Roche lobe configuration, combined with an adequate standard radius, we may compute the equatorial velocity v_{eq}. We find that from P_K=4 to 0.25 days (an estimate of the period corresponding to the contact stage) the rotation with respect to the center of the component varies from v_{eq}=12 to 195 km/s. Rotational velocities in this range are indeed measured by spectroscopic and photometric (i.e. spot modulations) observations. The local velocities with respect to the center of gravity are still higher (see fig.1). We conclude that there is a large range of periods for which VCBC's rotate at least as rapidly as the young stars, main topic of this workshop.

3 The age of the very close binary components

We all know that VCBC's may be young or old for there are no reasons to suppose that their formation has taken place during a limited timespan of the life of the galaxy. Star formation is considered as a continuous process and most stars are born in binary or multiple systems. It is suggested from observations (Abt and Levy, 1976) that the number of binaries of a given mass, formed per log P_K interval $\Delta \log P_K$ does not depend (or only weakly depends) on P_K. The main problem (van 't Veer and Maceroni, 1988) is to know where are the limits of the period interval of formation. We can only obtain some statistical knowledge of close binaries, namely their observed P_K distribution. To connect it to the initial distribution we need the period evolution function (PEF) of angular momentum losing orbit-spin coupled (AMLOSC) binaries. The convolution of IPD with the PEF then gives the

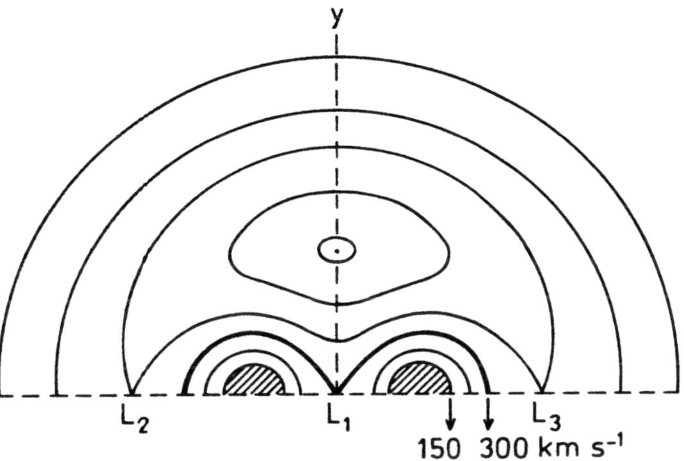

Fig.1: A close detached ($P_K = 1^d$) and a contact ($P_K = 0^d.25$) binary with their Roche equipotentials in the orbital plane. The maximum velocities with respect to the system barycentre for G5V components are also displayed.

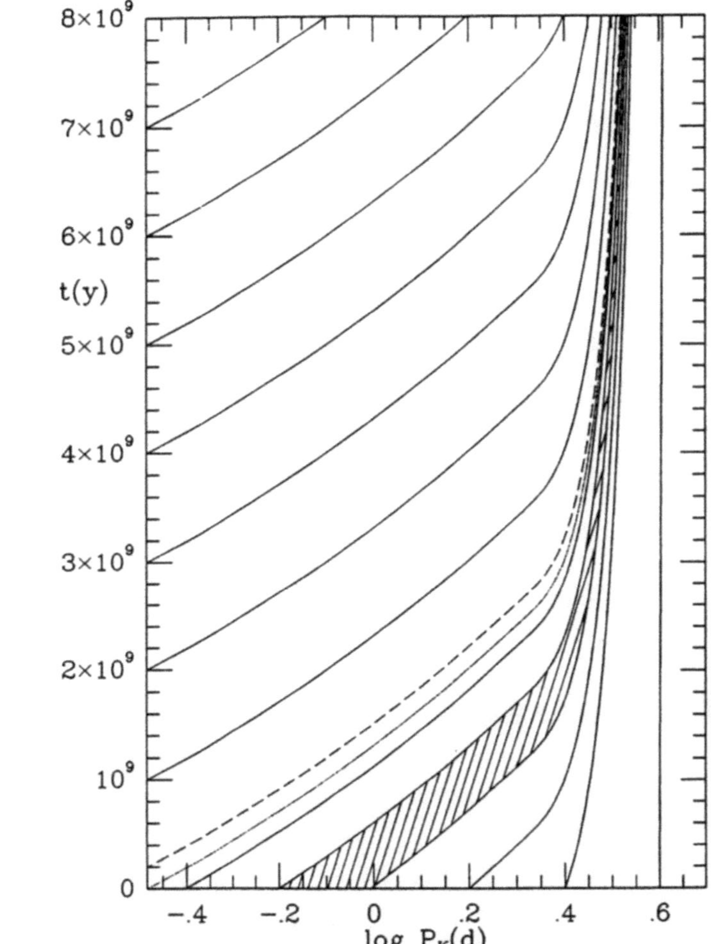

Fig.2: The PEF diagram. Every AMLOSC binary of the galactic disc was born at a location in the diagram determined by its present age (ordinate) and initial P_K. The binaries presently observed ($t = 0$) in a period bin ($0 < \log P_K < 0.2$ for example) were all formed in the (shadowed) area having the bin as base. Binaries formed on the dotted line are just coming into contact, binaries on the dashed line are just entering from the contact into the single star phase ($2 \cdot 10^8$ yr after first contact). Binaries formed in the region above and left of the dashed line have all become single stars.

computed present period distribution (PPD), to be compared with the observed one. This treatment of the dynamical evolution was developed in a series of papers (van 't Veer and Maceroni, 1988, 1989; Maceroni and van 't Veer, 1991; hereafter VM1, VM2 and MV); without entering into details we reproduce the PEF as we derived it for AMLOSC binaries made of two identical G5V components (fig.2).

The figure represents the variation of P_K with time. The variation is such that $dP_K/dt = f(P_K)$ depends only on the orbital period for a given set of AML and AMT parameters. Hence every curve is obtained by vertical shift of a standard PEF computed with the parameter values which best fit the observed PPD.

So to use the diagram of fig.2 we just have to plot the binary at his birth (t years ago) and its initial P_K. The consecutive evolution is then easily found by a vertical shift of the standard PEF to the birth point. In fig.2 time is started $8 \cdot 10^9$ years ago taken for the age of the oldest galactic clusters.

It is easy to see from fig.2 where all the binaries found now in a given velocity (= period) interval are coming from (see shadowed area of fig.2). We can also determine their mean age, provided we know (or can reasonably guess) the lower P_K limit of formation.

Taking all these points together we can now divide the shadowed area in two equal portions as it is done in fig.3. The separation between the two portions defines a median age t_m which has the property that half of the binaries formed in this area have age $t > t_m$. The result for t_m evidently depends on the lower period limit of binary formation. Two different possibilities are indicated in fig.3. Some median ages for different period bins and lower period limits are also presented.

In the preceding development we consider all AMLOSC binaries as the result of some fragmentation process with consecutive AML controlled approach of the components and we abandon the idea of young fission made contact binaries (Roxburgh, 1965). We hope to come back to this question in a forthcoming paper.

4 The dependence of angular momentum loss on rotation

In the preceding paragraphs (and much more explicitly in VM1, VM2, and MV) we described how the PEF of AMLOSC binaries can be computed from the most probable IPD and the observed PPD. The observed PPD must be corrected for different types of selection effects and the IPD can only be determined from an extrapolation of binaries without tidal interaction towards smaller periods. After a great number of unsuccessful attempts we were forced to conclude that the power law for spin down ($\dot{\omega} = b\omega^c$) first introduced by Skumanich (1972) and discussed (and criticised) by several speakers of this meeting cannot be extrapolated towards higher velocities than those for which it was derived ($v_{eq} < 20$ km/s).

Our conclusion, in spite of the uncertainties mentioned above, is without ambiguity: for ($v_{eq} > 20$ km/s) a sudden increase of the AML takes place and becomes approximately independent of the rotation (see fig.4).

We suggest that a parallel can be established between our results and those obtained for the rapid spin down of the members of very young clusters (Stauffer,

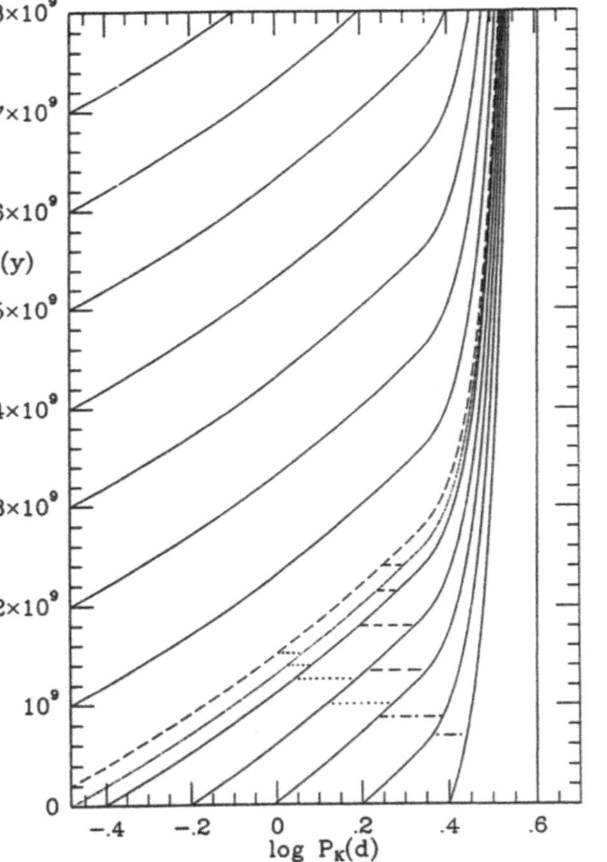

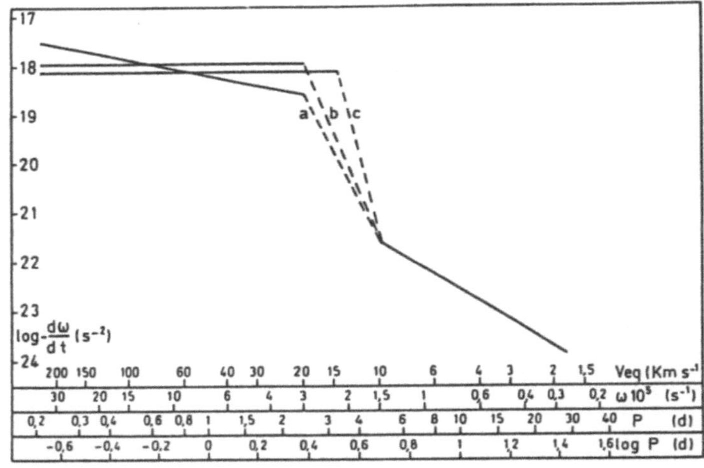

Fig.3: The age of the AMLOSC binaries. Every area (see fig.2) is divided by a median age t_m in two equal parts. The dotted lines are computed for the case when binaries are formed in the whole P_K-range. The dashed lines give the result when only binaries with $P_K > 1^d$ are formed.

Fig.4: Different outcomes for the spin-down function $\dot{\omega} = f(\omega)$. All these functions are compatible with the observed period distribution. They have in common a Skumanich-type relation for the longer periods, connected by a great jump to a more or less horizontal spin-down function for the short periods.

1987; Stauffer et al, 1989). The age of both groups of stars is not the same, but the high rotational velocity is their common property. It seems probable that the existence of different AML relations is principally, if not uniquely, connected with the rotational properties of the stars and not with their age.

From theoretical studies we know that the AML behaviour depends on the geometrical structure of the magnetic field, the mass loss rate and in a lesser measure on the degree of ionization of the stellar wind (Mestel 1984). The braking efficiency of the magnetic field decreases with increasing rotation, so from magnetic topography alone we should expect the opposite result (Roxburgh, 1983).A sudden increase of the mass loss rate seems to us a better candidate to explain the rotationally determined AML behaviour. At least four wind generation mechanisms are suggested in recent papers on this subject (see Evans et al., 1987) and we think that the question should be studied from this point of view. Moreover from their studies of pre-main-sequence winds of T Tauri stars Natta and Giovanardi (1990) conclude that the degree of ionization of stellar winds is extremely sensitive to changes of the temperature. This effect may also play a role in the braking efficiency of the winds.

Acknowledgements

This work has been partially supported by the italian Consiglio Nazionale delle Ricerche (FVV) and by the french Centre National de la Recherche Scientifique (CM).

References

- Evans, N.J., Levreault, R.M., Beckwith, S., Skrutskie, M.: 1987, *Astrophys. J.* **320**, 364
- Maceroni, C., van 't Veer, F.: 1991, *Astron. Astrophys.*
- Mestel, L.: 1984, *Proc. 3rd Cambridge Workshop on Cool Stars, Stellar Systems and the Sun*, S. Baliunas, L. Hartmann eds., Springer Verlag, New York
- Natta, A., Giovanardi, C.: 1990, *Astrophys. J.*, **356**, 646
- Roxburgh, I.: 1965, *Astrophys. J.*, **143**, 111
- Roxburgh, I.: 1983, *Solar and Magnetic Fields: Origins and Coronal Effects* ed. J.O. Stenflo, p.449
- Schatzman, E.: 1962, *Ann. Astrophys.*, **25**, 18
- Skumanich, A.: 1972, *Astrophys. J.*, **171**, 565
- Stauffer, J.: 1987, *Lecture Notes in Physics*, **291**, 182
- Stauffer, J., Hartmann, L.W., Jones, B.F.: 1989, *Astrophys. J.*, **342**, 285
- van 't Veer, F., Maceroni, C.: 1988, *Astron. Astrophys.*, **199**, 183
- van 't Veer, F., Maceroni, C.: 1989, *Astron. Astrophys.*, **220**, 128

DISCUSSION

Roxburgh: If you imagine that a dynamo operating in a rotating star is similar to any simple minded dynamical system which goes through period doubling, and therefore in an inverse cascade could go from one magnetic topology to another, that transition - other things being held constant - could lead to a substantial discontinuous change in the angular momentum loss rate.

van't Veer: Your model goes in the wrong sense. When the field becomes more complicated, the torque decreases. But I find an increasing angular momentum loss rate for more rapid rotation.

Roxburgh: I did not understand that is what you said.

van't Veer: As well as considering the magnetic topology, you also should consider the mass loss rate, which might increase substantially, perhaps by sudden coronal heating, and perhaps the degree of ionization of the wind might change.

Pinsonneault: You must have made some assumption about the initial distribution of periods, and you also had a diagram with ages. How were those ages derived?

van't Veer: You have an initial period distribution and a present period distribution. You have to go from the initial distribution to the present day one by a period evolution function. That is an integral of the period evolution function with the initial period distribution and it then gives you the present period distribution. And this present period distribution must be in agreement with what we observe. The initial period distribution was shown in one of the figures. But the latest figures were published by Abt in a review paper. He gives the distribution function of binaries as a function of period and it is a very large range. You can have periods from 1 day to a billion days - the latter being common proper motion binaries - and this distribution looks rather flat. Probably the initial distribution is flat also, stopping at perhaps about 1 day as a minimum period since I am not sure if you can make binaries whose initial periods are shorter than that.

AGES OF SPOTTED LATE-TYPE STARS

P.F. CHUGAINOV
Crimean Astrophysical Observatory
P.O. Nauchny
334413 Crimea
U.S.S.R.

ABSTRACT. It is shown that 87% of spotted single late-type field stars belong to the young disc and Local Association population with ages from 10^7 to 10^9 year. A comparison of photometric periods of T Tau-type, cluster, and single field stars confirms that the period difference is largest in the stars with ages of 10^7 - 10^8 year. Problems of spotted stars and their ages are discussed.

1 Introduction

At the present time the total number of late-type stars showing periodic or quasi periodic low-amplitude photometric variations, i.e. spotted stars, is about 200 including pre-main sequence, main-sequence and evolved objects. In recent studies by Stauffer et al. (1985), Innis et al. (1986), van Leeuwen et al. (1987) remarkable differences in photometric periods have been found in stars with quite similar physical characteristics. This interesting phenomenon has been interpreted as a short spinning-up of solar-mass stars in the stage of evolution which is near to the main sequence. We shall discuss the differences of rotational periods paying attention to the problems of age determination for field stars.

2 Observational Material

We have compiled lists of the following spotted stars: 50 single and 126 spectroscopic binary field stars, 37 T Tau-type stars, and 50 members of clusters. The relevant information for spectroscopic binary field stars and members of star clusters was taken from Strassmeier et al. (1988), Stauffer et al. (1985, 1988), Lockwood et al. (1987). Lists of single field stars, T Tau-type stars, and Pleiades members are given in Tables 1, 2, 3. Space velocity vectors U, V, W were determined by us or taken from the literature. Spectral types given in brackets were found from colors.

3 Age and Space Velocity

In order to establish the ages of stars which are not cluster members we need to know the constraints on the space velocity vectors depending on the age. We have adopted these

S. Catalano and J. R. Stauffer (eds.), Angular Momentum Evolution of Young Stars, 175–181.

constraints according to Eggen (1975, 1984) and some additional estimates have been supplemented according to recent data.

TABLE 1. Single field stars [1]

GCVS/HD	SP	Photometric ampl.	Photometric period	U	V km/s	W	Distance [2] (ps)	Remarks [3]
BE Cet	G2V	0.5^m	7.6^d	-36	-15	+2		YD,HS
OP And	gK1	0.1	76					
17144	G8III-IV	0.02	16.2	-34:	-8:	+10:	240	YD,HS
VZ Hor	K1V	0.1	2	-42	-18	-3		YD,HS
V491 Per	K2	0.03	7.37	-33	-19	-9		YD,HS
V891 Tau	G3	0.08	7	+15	0	-9	22	YD
V834 Tau	K3V	0.1	1.33	0	-4	-24		YD,LA
27536	G8III-IV	0.13	160					
31738	G5IV	0.04	?	0	+5	-20	23	YD
V1005 Ori	M1e	0.1	4.40	-29	-17	-23		YD
31993	(K3)	0.04	6.78	-11	-4	-6	13	YD,LA
YY Men	(F9)	0.2	9.55					
34198	KOIII	0.05	28.4	+30	-54	+27	206	OD
AB Dor	G8V	0.06	0.51	-11	-27	-11		YD,LA
37434	K2III	0.04	20.7	+8	-8	-22	151	YD
46291	(K1)	0.04	12.0					
VV Lyn	M3.5e	0.07	?	-9	-9	-20		YD,LA
YZ CMi	M4.5e	0.1	2.8	-13	-18	-12		YD,LA
DK Leo	dMO	0.2	8.0	-2	-17	-5		YD,LA
LQ Hya	dKO	0.1	1.66	-18	-11	-3		YD,LA
88230	K7Ve	0.01	6 ?	-7	-19	-35		YD,LA
94389	(K2)	0.15	78	0	-9	-1	22	YD,LA
DS Leo	dM2e	0.05	2.94	-15	+6	-15		YD
V838 Cen	K0/1p	0.08	1.84	-14	-20	-15	40	YD,LA
BF CVn	M1.5e	0.1	3.17	-11	-24	-8		YD,LA
DT Vir	dM2e	0.04	?	-34	-20	-8		YD,HS
115404	K2V	0.02	18.8	+37	+8	0		OD
FK Com	G2III-IV	0.1	2.4	-20	-27	-17	100	YD,LA
EQ Vir	K5V	0.02	4.0	-25	-14	-17		YD,LA
ξ Boo A	G8Ve	0.02	10.1	+7	+1	+1		YD
V343 Nor	KOV	0.02	4.4	+7	-2	-14	24	YD
V2133 Oph	KOV	0.04	11	+1	-1	-22		YD
152391	G8V	0.04	21	+79	-105	+8		OD
156026	K5Ve	0.02	21	-1	-31	-7		YD,LA
FK Ser	K5Ve	0.1	5.2	-13	-12	+1	46	YD,LA
PZ Tel	KOVp	0.04	0.94	-13	-24	-11		YD,LA
V1285 Aql	M2e	0.05	12 ?	-1	-3	-7		YD,LA
181943	KOVp	?	10.7	-30	-11	0	29	YD,LA-HS
CF Oct	(K1)	0.09	20.2	-6	+12	+8	21	OD
AU Mic	M1.6	0.3	4.9	-5	-13	-12		YD,LA
V1794 Cyg	G5III-IV	0.1	3.34	-8	-32	-1	69	YD,LA
201091	K5Ve	0.05	37	-90	-53	-8		OD
203251	(K2)	0.02	44.3	+10	+7	-13	17	YD
FL Aqr	M3Ve	0.1	1.95	-18	-9	-1		YD,LA
EV Lac	M4.5e	0.03	4.4	+20	-1	-1		YD
GT Peg	M3.5e	0.07	1.64	-34	-14	-23		YD
TW Psa	K5Ve	0.04	10.3	-4	-8	-14		YD,LA
218153	G8II	0.05	22	+89	-29	+95	955	OD
HK Aqr	dM1.5	0.08	0.43	-4	0	-3		YD,LA
OU And	G1III	0.05	23	+11	-5	-21	110	YD

(1) Data are taken mainly from Kholopov et al. (1985, 1987), Lloyd Evans and Koen (1987).

(2) Distances computed by us using spectroscopic parallaxes for stars not included in Gliese (1969).

(3) YD - young disc, OD - old disc, HS - Hyades Supercluster, LA - Local Association.

TABLE 2. T Tau-type stars (1)

Star designation	Sp	Photometric ampl.	period
V733 Tau	wt K2-3	0.1^m	3.4^d
V410 Tau	wt K3-7	0.2	1.9
BP Tau	tt K7	0.3	7.6
V819 Tau	wt K7	0.2	5.6
RY Tau	tt K1	0.1	5.6
HD 283572	su G6	0.1	1.5
T Tau	tt K1	0.04	2.8
DF Tau	tt M0.5	0.5	8.5
DH Tau	tt M0	0.35	7.0
DI Tau	wt M0	0.05	7.9
UX Tau A	wt K2	0.1	2.7
V827Tau	wt K7-M0	0.21	3.6
V826 Tau	wt K7-M0	0.06	4.05
V830 Tau	wt K7-M0	0.3	2.76
GI Tau	tt K5-7	1.0	7.2
GK Tau	tt K5-7	1.0	4.6
AA Tau	tt K7-M0	1.0	8.2
DN Tau	tt M0	0.15	6.6
HP Tau/G2	su G1	0.1	1.2
SU Aur	su G2	0.4	1.6 or 2.7
V836 Tau	wt K7-M0	0.29	6.99
GW Ori	tt G5	0.1	3.2
SY Cha	tt M0	0.1	6.2
TW Hya	tt K7	0.1	1.3 or 1.8
LHα332-20	tt K2	0.1	2.3
LHα332-21	tt K0	0.1	4.4
RU Lup	tt K		3.7
RY Lup	tt K4	1.0	3.9
Haro 1-1	? K5	0.3	3.3
Haro 1-4	tt K6	0.1	3.5
V2058 Oph	tt K5-7	0.2	6.9
Haro 1-8	tt K5	0.3	14.6
ROX 21	? M1	0.2	3.5
ROX 29	wt K6	0.2	6.3
V853 Oph	tt K-M1.5	0.1	23.8
Haro 1-14	tt K-M0	0.4	8.2
V2062 Oph	tt K2-3	0.1	3.4

(1) Data for 8 stars are from the paper by L.N. Berdnikov, K.N. Grankin, A.V. Chernyshiov, V.S. Shevchenko, S.D. Yakubov (in press). For the other stars data are taken mainly from Herbig and Bell (1988), Vrba et al. (1989), Herbst and Koret (1988).

The ages of solar neighbourhood stars are 10^7-10^8 years for the Local Association (the Pleiades group), 5×10^8-7×10^8 years for the young disc and the Hyades Supercluster, and 10^9-10^{10} years for the old disc and the halo (Eggen 1975, 1984). Eggen (1975) has concluded that the Sco-Cen group and the Tau-Aur association belong to the Local Association. However he has determined velocity vectors only for one T Tau-type star and their possible values for two others. Jones and Herbig (1979), Herbig and Bell (1988) have published the observational data which allow us to determine space velocity vectors for 36 T Tau-type stars belonging to the Tau-Aur association. Mean values of U, V, W for Hyades and Pleiades groups, Sco-Cen group and Tau-Aur association are compared in Table 4. The values for Sco-Cen and Tau-Aur groups are from our computations. The adopted distances are 170 ps for the Sco-Cen group and 140 ps for the Tau-Aur group. The errors given in Table 4 for these two groups are r.m.s errors for one star. Observational data for the

Sco-Cen group are mainly from Bertiau (1958).

TABLE 3. Pleiades members

Star designation		Sp	Photometric ampl.	period
HII	34	(K0)	0.05^{m}	6.55^{d}
	152	(G2)	0.07	4.12
	296	G8	0.11	2.53
	324	(K4)	0.17	0.41
	335	K5e	0.08	0.36
	625	K0	0.12	0.42
	686	K7e	0.10	0.40
	727	F7.5V	0.04	8.07
	739	G0V	0.04	2.70
	879	(K4)	0.07	7.39
	882	K3	0.12	0.58
	996	G2		1.26
	1124	K1	0.07	6.05
	1332	K4	0.05	8.3
	1531	K7-M0e	0.12	0.48
	1883	K2	0.20	0.24
	2034	K2.5e	0.07	0.55
	2244	(K2)	0.17	0.57
	2927	K4e	0.15	0.26
	3163	(K2)	0.10	0.42

Although some difference in velocity vectors exists between Pleiades, Sco-Cen and Tau-Aur groups, it is not large compared with errors. Accordingly we have taken extreme velocity values for stars belonging to these three groups as limits for solar neighbourhood stars with ages of 10^{7}-10^{8} years. They are given in the last line of Table 4. These values include errors, and by using them as a criterion for the age of 10^{7}-10^{8} years one may include some stars of a greater age.

TABLE 4. Space velocity vectors for different groups

Group	U	V	W
Hyades Supercluster (Eggen 1984)	-41	-18	-2
Pleiades Group (Eggen 1975)	-11 ±9	-25	-8 ±8
Sco-Cen Group	-7 ±7	-24 ±6	-10 ±5
T Tau-type stars of the Tau-Aur association	-17 ±2	-13 ±3	-10 ±4
The solar neighbourood population with ages $10^{7}-10^{8}$ y (limits)	+7 ÷ -30	-6 ÷ -38	+1 ÷ -22

4 Age and Photometric Periods

We are now able to summarize the results on ages of spotted stars. The young disc and the Local Association population is predominant or is the only population in all the groups considered with the exception of the group of spectroscopic binary field stars in which the proportion is lower, but still rather high (66%). The portion of this population in the group of single field stars is 87%. As it is well known, the ages of T Tau-type stars do not exceed 10^{7} yars, while the ages of α Per, Pleiades and Hyades clusters are 10^{7} years, 5-7$\times 10^{7}$ years, and 5×10^{8} years respectively. On the other hand we have found that the ages of majority of single field spotted stars are $10^{7} - 10^{8}$ years. We are, therefore, able to compare the photometric periods of stars of different ages.

Single stars with the shortest rotational periods of 0.2-0.9 days are present in Pleiades and α Per clusters but they are absent among Hyades and T Tau-type stars. These stars are believed to be in the spinning-phase. There are three field stars with periods of 0.43 d, 0.51 days, and 0.94 days, namely HK Aqr, AB Dor, and PZ Tel. All three probably belong to the Local Association. Thus ages of spinning-up spotted stars are $10^{7} - 10^{8}$ years. It is interesting to note also that there are stars with periods of 1-2 days. Their ages are about $10^{7} - 10^{8}$ years because they are met among T Tau-type stars, in α Per, Pleiades, and Local Association but they are absent in the Hyades. A part of them may represent stars evolving to the F spectral type. The others may be less massive stars in the stages of evolution just before and after the spinning-up phase.

There is a considerable difference in distribution of rotation periods between young disc G0-G6 and G8-K7 stars. In the first group, periods are 1.2-2.7 days for T Tau-type stars, 0.6 days for one star in the α Per cluster, 1.3-4.1 days in the Pleiades, 5.9-8.7 days in the Hyades, and 2.4-7.6 days in field stars. In the second group periods are 1.9-3.7 days for T Tau-type stars, 0.17-5.1 days in the α Per cluster, 0.24-8.3 days in the Pleiades, 3.7-12.6 days in the Hyades, and 0.51-44 days in field stars. In general, the difference between G0-G6 and G8-K7 groups is the absence of stars with periods less than 0.6 in the G0-G6 group. Thus one can suppose that in the G0-G6 interval either the spinning-up is not so large as in the G8-K7 interval or the pre-main-sequence spinning-up stars could not be observed because they possess opaque, more slowly rotating envelopes.

5 Problems

Let us remark here the following problems:

1) It is difficult to determine the photometric periods of T Tau-type stars because of irregular light variations of these stars. Besides the other difficulty is connected with the fact that T Tau-type stars constitute a class of objects which is not uniform. Vrba et al. (1986, 1989) have shown that the rotational modulation in the T Tau-type stars with strong emission lines is due to the presence of bright spots while in those with weak emission lines it is due to dark spots. Since, the bright spots may not be located on the star surface but on the circumstellar disc our comparison of T Tau-type and the other spotted stars may be no longer valid.

2) Studies of the spinning-up phenomenon should be restricted to single stars only because this phenomenon in spectroscopic binary systems may be hidden by the synchronization process. Fekel and Eitter (1989), Tassoul (1987) have shown that almost all spectroscopic binary systems with periods shorter than 30 days are synchronized. However there is a system, BY Dra with orbital period of 5.97 days, which is not synchronized and moreover it has the orbital eccentricity equal to 0.3. According to the computations made by Pettersen (1989) the synchronization time for BY Dra is 2 to 10^4 and the time of circularization of the orbit is 4 to 10^8 years. These values contradict the finding that BY Dra belongs to the old disc population (age of $10^9 - 10^{10}$ years). However it is possible that the last conclusion is due to uncertanties in the population attribution and/or errors in the parallax determination.

3) It is not clear if the subgiant spotted stars are undergoing the contraction phase, or if they are evolved stars. There exists a number of such stars between single field stars. One can consider FK Com and V1794 Cyg (HD 199181). Both are subgiants according to their spectral classification as well as from the comparison of their photometric periods and rotational velocities vsini. On the other hand our computations of components of space velocities have shown that both stars belong to the Local Association, i.e. their ages may be $10^7 - 10^8$ years. Thus these stars are rather unevolved.

4) A similar problem arises in the classification of spectroscopic binary systems with spotted components. Let us compare systems EI Eri (G5IV, P=1.94 ^{d}), V837 Tau (G2V, P=1.9 ^{d}), V815 Her (G8V, P=18 ^{d}), and V4778 Lyr (G8V, P=2.18 ^{d}). Their orbital periods and spectral types are nearly the same, but the difference between EI Eri and the other three systems is that the former is spectroscopically a subgiant. Therefore Fekel et al. (1988), Strassmeier t al. (1988) classified EI Eri as the evolved RS CVn class system and V837 Tau, V815 Her, and V478 Lyr as unevolved BY Dra class systems. One can find a confirmation of the classification of the EI Eri primary as subgiant in the fact that the lower limit of its radius, as obtained from the observed vsin$i$ and photometric period, is >1.9 $R_\odot$ but this may be an erroneous conclusion.

The system BY Dra itself is noteworthy in this respect. Vogt and Fekel (1979) adopting the spectral dM0e for its primary have concluded that it is an under-luminous and probably pre-main-sequence star. The recent most precise determination of the spectral type of the primary (Keenan, Mc Neil 1989) is K4Ve. The radius of 0.6-0.7 $R_\odot$, corresponding to this spectral type, is about the same as that deduced from th observed absolute bolometric magnitude and effective temperature of the primary, i.e. 0.77 $R_\odot$. But as Vogt and Fekel (1979) have shown, the mass of the BY Dra primary is 0.5-0.6 $M_\odot$. Combining this mass value with the observed mass function $M_1 \sin^3 i$=0.047 $M_\odot$, it is possible to deduce a value for the inclination angle i=28 °. The period of rotation is 3.83 days, and the projected velocity of rotation is vsin$i$=8 Km/s, then the radius of the primary turs out to be R=1.2-1.4 $R_\odot$, i.e. about twice the value obtained from the observed absolute bolometric magnitude and effective temperature. As Lucke and Mayor (1980) have pointed out, in order to remove this discrepancy one can suppose that the line broadening in the spectrum of the BY Dra primary is mainly determined by the macroturbulence. Of course, this hypothesis deserves further proofs and applications not only with respect to BY Dra but to other young stars. It seems also possible that outer layers of the atmosphere of the BY Dra primary rotate faster than layers in which the continuum radiation originates.

6 References

Bertiau, F.C., 1958, *Ap. J.* **128**, 533.
- Eggen, O.J., 1975, *Publ. Astron. Soc. Pacif.* **87**, 37. - Eggen, O.J., 1984, *Astron. J.* **89**, 1358.
- Fekel, F.C., Quigley, R., Gilles, K., Africano, J.L., 1988, *Astron. J.* **93**, 726.
- Fekel, F.C., Eitter, J.J., 1989 *Astron. J.* **97**, 1139.
- Gliese, W., 1969, *Veroff. Astr. Rechen-Inst. Heidelberg* **No. 22.**
- Jones, B.F., Herbig, G.H., 1979 *Astron. J.* **84**, 1872.
- Herbig, G.H., Bell, K.R., 1988, *Lick Obs. Bull.* **No. 1111.**
- Herbst, W., Koret, D.L., 1988, *Astron.J.* **96**, 1949.
- Innis, J.L., Thompson, K., Coates, D.W., 1986, *Month. Not. RAS* **223**, 183.
- Keenan, P.C., Mc Neil R.C., 1989, *Ap. J. Suppl.* **71**, 245.
- Kholopov, P.N., Samus, N.N., Kazarovets, E.V., Perova, N.B., 1985, *Inform. Bull. Var. Stars* **No. 2681.**
- Kholopov, P.N., Samus, N.N., Kazarovets, E.V., Kireeva, N.N., 1987, *Inform. Bull. Var. Stars* **No. 3058.**
- Lockwood, G.W., Thompson, D.T., Radick, R.R., Osborn, W.H., Baggett, W.E., Duncan, D.K., Hartmann, L.W., 1984, *Publ. Astron. Soc. Pacif.* **96**, 714.
- Lloyd Evans, T., Koen, M.C.J., 1987, *South. Afr. Obs. Circ.* **No. 11**, 21.
- Lucke, P.B., Mayor, M., 1980, *Astron. Ap.* **92**, 182.
- Magnitski, A.K., 1987, *Pis'ma Astr. Zh.* **13**, 1080.
- Stauffer, J.R., Hartmann, L.W., Burnham, J.N., 1985, *Ap. J.* **289**, 247.
- Strassmeier, K.J., Hall, D.S., Zeilik, M., Nelson, E., Eker, Z., Fekel, F.C., 1988, *Astron. Ap. Suppl.* **72**, 291.
- Tassoul, J.L., 1987, *Ap. J.*, **322**, 856.
- van Leeuwen, F., 1986, in *Flare Stars and Related Objects* (ed. L.V. Mirzoyan), Yerevan: Publ. House Acad. of Science, p. 289.
- van Leeuwen, F., Alphenaar, P., Meys, J.J.M., 1987, *Astron. Ap. Suppl.* **67**, 483.
- Vogt, S.S., Fekel, F.C., 1979, *Ap. J.* **234**, 958.
- Vrba, F.J., Rydgren, A.E., Chugainov, P.F., Shakovskaya, N.I., Zak, D.S., 1986, *Ap. J.* **306**, 199.
- Vrba, F.J., Rydgren, A.E., Chugainov, P.F., Shakovskaya, N.I., Weaver, W.B., 1989, *Astron. J.* **97**, 483.

ROTATION OF EVOLVED STARS

David F. Gray
Department of Astronomy
University of Western Ontario
London, Ontario N6A 3K7, Canada

ABSTRACT. Rotation is the driving mechanism for stellar activity, but in some parts of the H-R diagram, rotation itself seems to be closely regulated. Rotation of evolved stars is slowed by moment-of-inertia increases and by magnetic braking. The observations are used to map out two magnetic and two acoustic domains related to the powering of chromospheres and magnetic braking. Within the magnetic domains, rotation and convection are presumed to interact resulting in dynamo activity.

1. INTRODUCTION AND REVIEW

The age-old question is what happens to the rotation of a star as it evolves off the main sequence? The simple answer is "plenty!" Things start off peacefully enough, with simple increases in moment of inertia arising from the expansion of the star. Naturally, the surface rotation declines in response to this, although we may not yet understand the subtleties of the internal angular-momentum transfer. There are also some apparent inconsistencies in detail concerning the distributions of rotation rates. Specifically, an excess of slow rotators compared to a Maxwell-Boltzmann distribution is seen for A and B stars on the main sequence (e.g., Wolff et al. 1982), whereas F giants, which used to be A stars, do not show an excess.

Nevertheless, simple calculations of moment-of-inertia increases agree acceptably well for the initial stages of post-main-sequence evolution. Then comes the action. The action stems from non-thermal activity. For the giants (III) and subgiants (IV), at least some of it is magnetic. No, we can't see the magnetic fields directly in the sense of measuring the Zeeman effect because the spectral lines are too broad. What we do see is coronal and chromospheric activity, and they are indirect indicators of non-thermal activity, and for single stars we see rotational braking of the type readily associated with magnetic fields. Figure 1 shows the observations. The drop in rotation of the giants takes place between G0 III and G3 III. This is slightly earlier than the G5 III I had reported in 1982 (compare Gray 1989 with Gray 1982), a change resulting primarily from improved spectral types. The drop for subgiants occurs a little earlier, at F6 IV to F8 IV. the decline with spectral type for dwarfs takes place more slowly through the F-star region, and here the situation is different because we are dealing with a mass sequence instead of a time sequence. Connecting these points gives the rotation boundary shown on the last panel of Fig. 1. There is also a startling change in the distribution of rotation rates. The Maxwell-Boltzmann distribution of the F giants turns into a δ-function during the braking process.

The "action" may extend to luminosity class II, but the observed rotation of class Ib supergiants may be no smaller than one might expect from the simple moment-of-inertia calculations. I shall return to this point later for an important conclusion.

Also shown in the H-R diagram of Fig. 1 are the granulation and coronal boundaries. The

S. Catalano and J. R. Stauffer (eds.), Angular Momentum Evolution of Young Stars, 183–199.

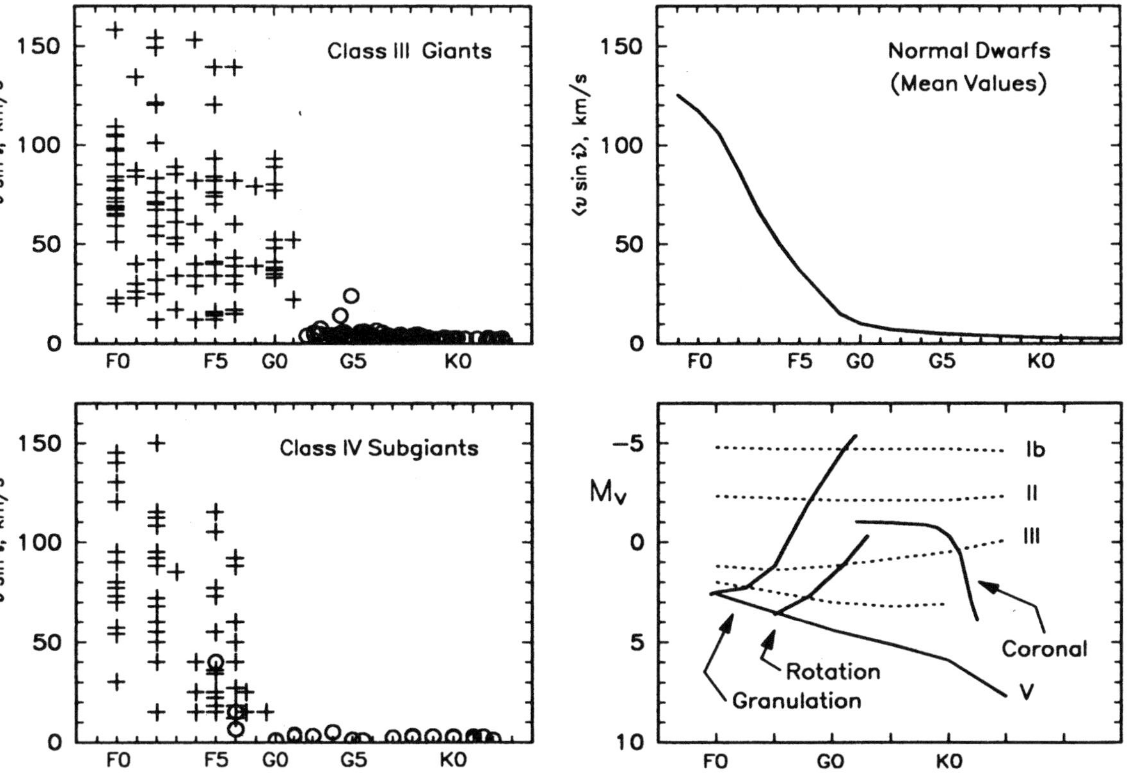

Fig. 1. Rotation of luminosity classes III, IV, and V are shown in three panels. The drop in rotation marks the rotation boundary. The H-R diagram in the fourth panel shows the position of the rotation boundary compared to the granulation and coronal boundaries. The data come from Gray (1989a), Gray and Nagar (1985), Gray (1984), and Fukuda (1982).

granulation boundary delineates the onset of solar-like granulation in stellar photospheres (Gray and Nagel 1989), and observationally identifies the start of deep-envelope convection. X-ray emission from hot ($T \gtrsim 10^6$ °K) plasma trapped in coronal magnetic loops is not seen for stars above or to the cool side of the coronal boundary (Linsky and Haisch 1979).

Let us clear up two points of potential confusion before going on. First is the question of evolutionary tracks, blue loops, and evolution up the asymptotic branch. Stars of luminosity classes I and II show extensive blue loops, and the transit times are shortest for first crossing, so we expect most of the observed supergiants and bright giants to be seen during the blue-loop phase. Luminosity class III stars do not show blue loops, but move once across the H-R diagram before evolving up the asymptotic branch near K2 III where their tracks join those of lower mass stars. Furthermore, essentially all class III giants come from the late B to early A-star portion of the main sequence, that is, the rapid-rotation domain. Class IV subgiants also have no blue-loop evolution, but some of the cooler members of this class originate from the slow-rotation domain on the main sequence. Clearly then, the large drop in rotation seen at the rotation boundary has nothing to do with either blue-loop evolution or with evolution up the asymptotic branch.

Second is the question of the "reality" of the rotation boundary (Gray 1981, 1982, 1988, 1989). Could the abrupt drop in rotation simply reflect a rapid rise in the moment of inertia as the star evolves across this region of the H-R diagram, or is braking involved? The question of angular momentum loss has been dealt with several times in the literature. Before the rotation discontinuity was discovered,

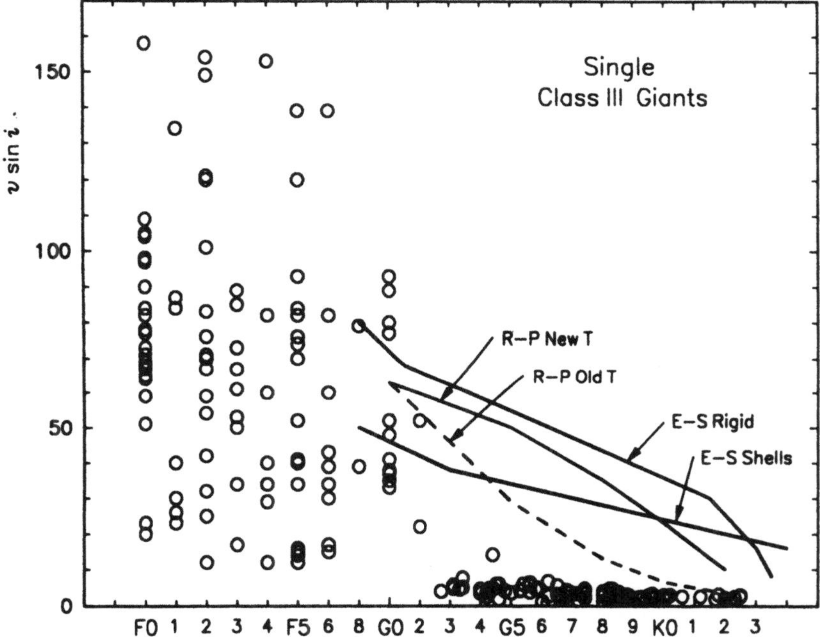

Fig. 2. Moment-of-inertia calculations predict rotation at the level of the curves. E-S is from Endal and Sofia (1979). Their detailed modeling fall between the rigid rotation and the case for conservation of angular moment in shells. R-P is from Rutten and Pylyser (1988). Their computations are shown for both older (Böhm-Vitense 1981, Flower 1977) and a newer (Gray 1991) effective temperature calibration. All of these curves are an order of magnitude higher than the observations.

Oke and Greenstein (1954), Abt (1957, 1958), Kraft (1970), and Endal and Sofia (1979) studied the problem. Andy Endal and I went over it once again (Gray and Endal 1982), as did Rutten and Pylyser (1988) some years later. As you can see in Fig. 2, the calculations show a smooth increase in moment of inertia, giving the curves shown in the figure. There is some range in the calculations owing to uncertainty in the effective temperature scale, but near ~ G3 III, the observed rates are only one tenth of the calculated mean, and there are numerous F giants rotating at twice this mean, increasing the discrepancy even more. We cannot escape the conclusion that an additional decline in rotation occurs beyond what is expected by moment-of-inertia increases. Indeed there is loss of angular momentum due to braking. Incidentally, Rutten and Pylyser (1988) used spurious data for their investigation and came erroneously to the opposite conclusion. The magnetic nature of the brake will become increasingly clear as we continue.

Now let us move on to the other activities of evolving stars.

2. THE ON-SET OF ACTIVITY

"Activity" for some people has come to mean the existence of temperature inversions, that is, the existence of emission from coronae and/or chromospheres. The familiar observational indicators are x-ray emission, HK line-core emission, H_α emission, and numerous ultraviolet emission lines. One can also tie in here the existence of starspots. Sometimes it is assumed that all such activity is magnetic in origin (getting back to the old solar terminology), but I will present evidence below that this is not the case.

Where do temperature inversions actually begin for these travelers across the H-R diagram? It begins at the granulation boundary, a demarcation separating "hot" stars from "cool" ones, and where deep-envelope convection begins (ref. Fig. 1). The granulation boundary is defined by the reversal of spectral line asymmetries (Gray and Nagel 1989). Stars on the cool side of this boundary have asymmetries qualitatively like the sun's, indicating the presence of granulation and deep-envelope convection. On the hot side of the boundary, the spectral-line asymmetry is reversed and larger, indicating some other kind of velocity field (Gray 1989b).

Chromospheric indicators show rapid growth across this region of the H-R diagram, for example, the C II and C IV shown in Fig. 3. Recall that evolution moves class III giants monotonically toward later spectral types, and owing to the increase of moment of inertia, the rotation is decreasing at a modest rate in the F0 III to G0 III interval. Even though rotation is declining, the chromospheric emission still rises steeply. It therefore seems obvious that the growth in chromospheric emission stems from the growth in the granulation and/or the convection zone more generally.

The question then naturally arises: might this initial growth in activity come from acoustic-wave dissipation rather than from some magnetic process? The answer is negative. To begin with, there are numerous x-ray emitters between the granulation and rotation boundaries, especially toward the subgiant region where the rapid evolution across the Herzsprung gap is less pronounced. It seems fairly well established that x-rays originate in magnetic coronal loops of trapped plasma, so x-rays are evidence for magnetic fields. Furthermore, you will already have noticed in Fig. 3 the drop in carbon emission at the rotation boundary. Give a rotating star some convection and it develops chromospheric emission. Give it a deeper convective envelope, and the chromospheric emission grows. But take away the rotation, as at the rotation boundary, and the emission drops precipitously. That looks convincingly like dynamo behavior. (That does not mean that acoustic waves contribute none of the power to chromospheric support, only that most of it is of magnetic origin.)

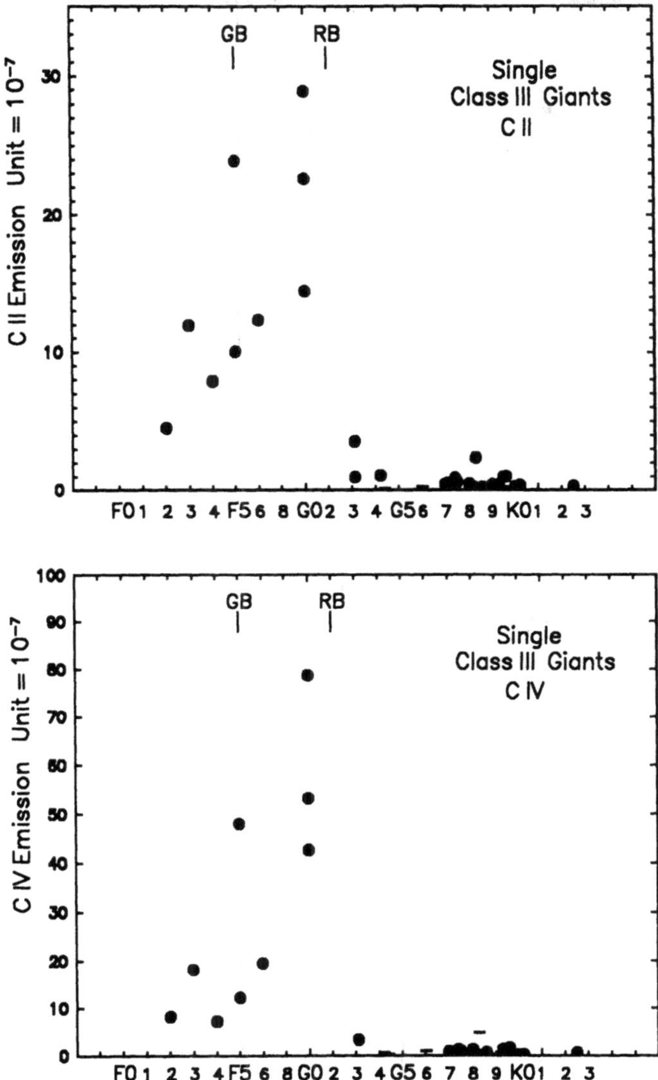

Fig. 3. The chromospheric carbon emission, expressed as a fraction of the total power output is shown against spectral type. The position of the granulation boundary, GB, and the rotation boundary, RB, are indicated. The emission grows rapidly across the granulation boundary and then drops abruptly at the rotation boundary. The carbon data is from Simon and Drake (1989).

3. THE MAGNETIC PHASE

The evidence above tells us the magnetic stage is already starting as a star crosses the granulation boundary. Magnetic fields are apparently eager to grow, and as soon as granulation puts structure into the envelope, the dynamo begins working. But then, why does the rotation boundary not coincide with the granulation boundary? In other words, why does the original magnetic-field growth not immediately lead to magnetic braking? I do not know the answer to this question, but let me point out that the rotation boundary parallels the granulation boundary. The important factor is not a time delay after granulation starts because the time to evolve between the granulation and rotation boundaries diminishes rapidly with increasing luminosity class. We must conclude that the position of the rotation boundary depends on a structural change, and the obvious one that parallels the onset of granulation is the deepening of the convection zone. Calculations of the convection-zone depth indicate that ℓ/R, i.e., depth normalized to the stellar radius, ranges from ~ 5% to 8% as the stars evolve from G0 III to G3 III. Perhaps a new dynamo mode is triggered at this ℓ/R. Or maybe the field simply gets strong enough to stretch long loops out from the star.

Now most F stars approaching the rotation boundary have more than enough rotation to activate a dynamo, but just how much rotation is really needed? Some guidance is given by the dimensional argument of the Rossby number, expressed here as

$$v_{lim} = v_{con}/(\ell/R), \qquad (1)$$

where v_{con} is the characteristic convective velocity, and v_{lim} is the minimum rotation needed to sustain the dynamo process (Durney and Latour 1978). Because this is a dimension argument, unknown constants may have been neglected, and it is only the shape and horizontal positioning that can be used.

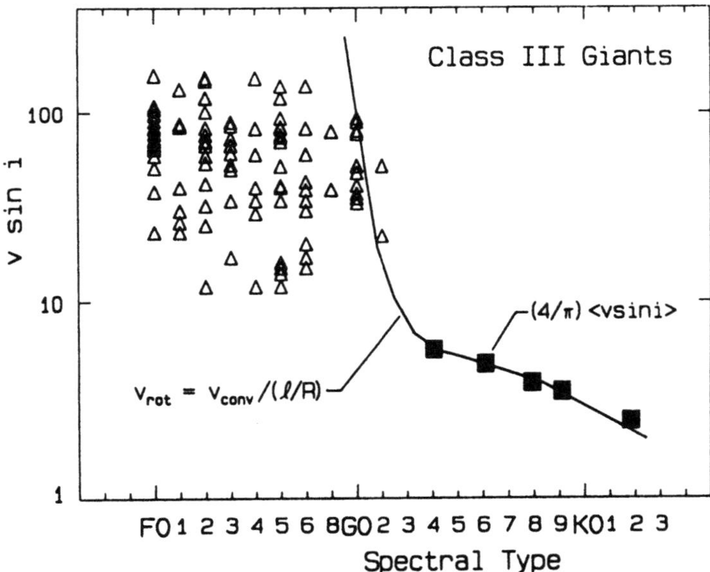

Fig. 4. Rotation of giant stars (symbols) is compared to the dynamo criterion (line) expressed as a limiting rotation rate needed to sustain dynamo action. Squares represent the actual rotation rates (without the sin i factor). The diagram is taken from Gray (1989a).

The calculated v_{lim} is compared to the observations in Fig. 4. In the first instance, it is the agreement of the position of the steep decline in v_{lim} with the observed decline in rotation that is important. It implies that for each F giant, there develops a deep enough convection zone to have $v_{con}\ /(\ell/R)$ fall below its rotation rate and, bang!, it has a magnetic brake. The v_{lim} curve in Fig. 4 is steep. The drop in rotation rates is steep, showing on average a tenfold drop while evolving a tenth or two of a spectral class. This implies a strong efficient brake. Presumably the whole convection zone looses its angular momentum over this same spectral interval. Some of you may even wonder if the whole star, core included, is not decelerated. If so, then the rate of angular momentum dissipation would be still more remarkable.

Equation 1 also predicts that the rotation will continue to decline under the influence of the magnetic brake until $v_{rot} \leq v_{lim}$, at which point the dynamo ceases to be driven and the brake is turned off. The observations also support this prediction. Stars after the rotation boundary show a drastically altered distribution of rotation rates. The original Maxwell-Boltzmann distribution is destroyed and compressed into a δ-function (Gray 1989a). If stars were simply increasing their moment of inertia, they would maintain their position within the distribution relative to their neighbors. They would all slow down, the Maxwell-Boltzmann distribution would be compressed, but it would still be there. The magnetic brake, by contrast, offers a natural explanation: rotation has dropped to $v_{rot} = v_{lim}$, and all G3 giants are therefore deposited at this limiting value independent of their initial rate.

Once we know v_{rot} is a single-valued function of spectral type, we can compute v_{rot} as a function of spectral type from the average observed $v \sin i$ in spectral intervals, assuming random orientation of rotation axes. Values of v_{rot} are shown as the filled squares in Fig. 4.

We cannot escape the magnetic-dynamo nature of stars on the cool side of the granulation boundary. But the magnetic phase does not simply start there and end at the rotation boundary. There is much more to the story.

4. THE ROTOSTAT MECHANISM

One might develop the picture above by proposing 1) the whole star was decelerated at the rotation boundary, 2) the dynamo turns off permanently at the end of the boundary experience. Unique rotation would be maintained from then on because it would simply be scaled by moment-of-inertia changes. But a serious problem arises here because the magnetic activity obviously continues as stars evolve through the G to early K types. In fact, chromospheric activity mimics the decline in rotation through this spectral interval. More specifically, the upper bound of chromospheric flux declines in step with the rotation (ref. Gray 1988, Fig. 5-26). Furthermore, activity indicators show a wide range of strength, implying a wide range in dynamo-activity among these post-rotation-boundary stars.

If we alter assumption 1), and instead presume only the envelope rather than the whole star is decelerated at the boundary, then the evolutionary deepening of the convective envelope should eat into the more rapidly spinning core and transfer angular momentum to the envelope. The star would spin up again as evolution progressed! But the observed rotation declines slowly.

The rotostat mechanism solves these problems and links the observations together in a natural way. The basic postulate is that the braking at the rotation boundary involves primarily the envelope and not the core, and when the rotation is reduced to $v_{lim} = v_{con}\ /(\ell/R)$, the braking ceases, but only until the star spins up again (slightly) as the deepening convective zone brings angular momentum from the core into the envelope. A small amount of spin-up makes the envelope rotation exceed the limiting value, and so the dynamo turns on once again. The new braking stage reduces the slight spin-up back to the dynamo-criterion value. The value of v_{lim} is now a little smaller however because $v_{con}\ /(\ell/R)$ is now smaller owing to a small increment of evolution. This process repeats itself many times as the evolution carries the star through its G to K stages. I call it flickering. In this way, the rotation of stars in the flickering mode is constrained or regulated (leading to the name rotostat) by the convection-zone parameters, v_{con} and ℓ/R, to follow v_{lim}. The initial unique rotation rate created at the rotation boundary in maintained dynamically by the rotostat.

Model calculations show ℓ/R to increase monotonically and v_{con} to decrease monotonically but more slowly with advancing spectral type (Gray 1988, Figs. 5-22 and 5-24). The calculated limiting value is compared to the observed rate of decline (squares) in Fig. 4. Only the slope is of interest, not the absolute level. I moved the calculated curve vertically to match the squares -- a degree of freedom implicit in the dimensional nature of eq. (1). The agreement is more than adequate. The main uncertainties lie in the physics of the calculations, where the mixing-length formulation is used to describe convection, and in the effective temperature calibration. Actually, the temperature scale does not have a strong effect here because the curve varies so slowly with temperature.

How closely stellar rotation actually follows the limiting value may depend on how much the rotation can exceed v_{lim} before the dynamo springs back into action, how strong the renewed magnetic field is, how long it takes to actually dissipate the new increment of angular momentum, and how long it takes for the magnetic field to decay after the rotation is brought down to the limiting value. But considering the fact that evolution is faster in the G0-G3 interval than in later stages, and the sharpness of the rotation boundary, I would guess the regulation is essentially instantaneous on an evolutionary time scale. Time scales comparable to the Maunder minimum of solar activity might be involved. The amount of angular momentum that is dissipated during a flicker is small compared to what took place at the rotation boundary. It is small because the spin-up above v_{lim} is small and it is small because the star's moment of inertia continues to increase.

Some estimate of the tolerance of excess rotation, i.e., rotation exceeding v_{lim}, can be made by considering how close the observed $v \sin i$ distribution is to the δ-function case. This involves matching the observed $v \sin i$ distribution to a $\sin i$ distribution convolved with the errors of observation. In a recent study of G and K giants (Gray 1989a), the errors of observation were estimated to be $\approx$ 20%, and the δ-function case was reproduced exactly within the statistics of the data. A conservative upper limit for the excess rotation allowed before triggering a dynamo flicker is then $\lesssim$ 20%. It is an upper limit because i) only a small fraction of the $\approx$ 20% observational error can be re-directed to the true range of rotation, and ii) there must be some cosmic scatter, i.e., giants span some range in mass across which v_{lim} may vary.

In principle we can find out something about the duration of the "on time" versus "off time" of the flickering by considering the statistics and distribution of chromospheric emission. The factors to consider here are differences in activity arising from rotational modulation, from activity cycles (which may be unrelated to flickering), and from effects arising from the inclination of the rotation-axis. Now we lack information on the size of rotational modulation in giants, but in dwarfs it amounts to $\lesssim$ 10%. Cyclic variations in dwarfs are $\lesssim$ 30% when they occur. The inclination effects can be studied by looking at the differences of chromospheric emission from the mean for the spectral type as a function of axial inclination. The axial inclination follows from the observed $v \sin i$ coupled with the unique v at each spectral type (Gray 1990). Figure 5 shows the combined C II, C IV, and Ca II emission normalized to the mean for each spectral type. The lines show the projected area of equatorial belts labeled with their latitude span. If the chromospheric emission is proportional to the projected activity belt, the maximum emission should lie near the curves. In practice, we see a wide range in emission, running from the curves down to the detection limit. So unless rotational modulation and/or cyclic variations are an order of magnitude larger for giants than for dwarfs, there is a large range that could be attributed to flickering. The data are few, but I have none-the-less constructed a histogram of the strength of chromospheric emission at constant "$\sin i$" in Fig. 5, as shown in Fig. 6. There appears to be a bimodal distribution which could be interpreted as the flicker-on and flicker-off modes. There are roughly equal numbers of stars with emission in each mode, but there are more observational upper limits in the off-mode (which are not included in the histogram). During the on-mode, the emission is approximately three times stronger than during the off-mode.

So, as you see, a great deal of information can be made into a coherent picture by the Rotostat Hypothesis.

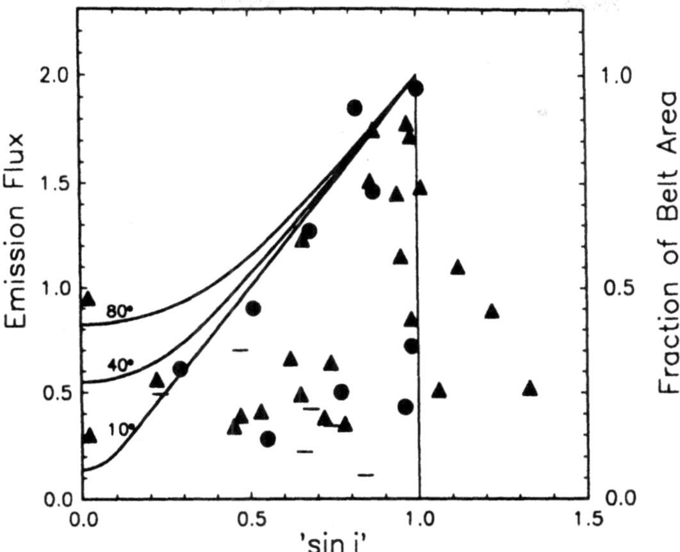

Fig. 5. Carbon II and carbon IV emission is denoted by filled circles, upper limits on carbon emission by dashes; calcium emission by triangles. All values are normalized to the mean for their spectral type. The abscissa is a computed rotation-axis projection factor: 'sin i' = 1 for the equator-on orientation. Stars seen near pole-on orientations show weak emission. The lines are projected equatorial belts extending over ±5°, ±20°, and ±40° of latitude. Stars show a wide range of emission strengths below the maximum represented by the curves. Adapted from Gray (1990).

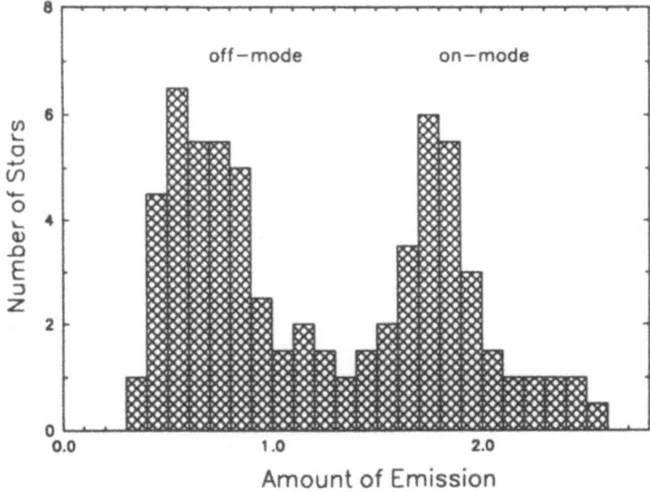

Fig. 6. The distribution of stars below the activity-belt lines of the previous figure shows two concentrations. Perhaps these peaks represent the on and off stages of dynamo activity when the rotostat is functioning.

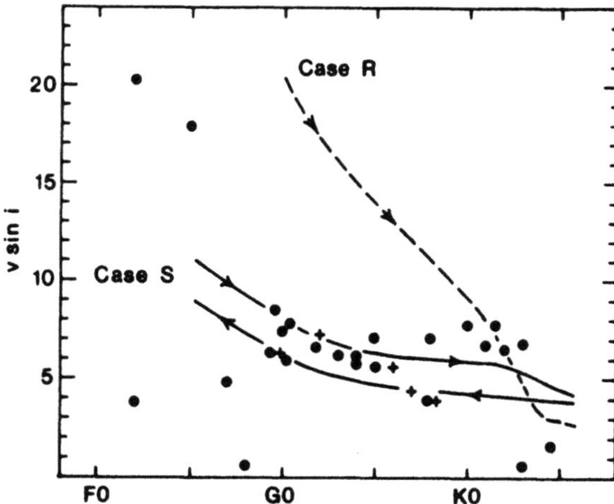

Fig. 7. Rotation of cool luminosity class Ib supergiants is shown as a function of spectral type. Models assuming conservation of angular momentum in shells (Case S) and for rigid rotation (Case R) are shown by the curves. From Gray and Toner (1987).

5. ROTATION IN HIGHER LUMINOSITY CLASSES

How much of the "action" do the more luminous stars get? Not as much as their smaller neighbors. Let us concentrate on the Ib supergiants. They certainly show systematic line-asymmetry changes with spectral type; the granulation boundary runs through G1 Ib. It would appear that convective envelopes and more-or-less normal granulation occur for Ib's on the cool side of the boundary, regardless of which phase of blue-loop evolution these stars are on. With granulation comes acoustic noise, and with the acoustic noise comes temperature inversions. Since both macroturbulence dispersion and line asymmetries increase with luminosity class (Gray 1988), we would expect acoustically-generated temperature inversions and their emission-line indicators to strengthen with luminosity class.

Do we expect dynamo activity? Remember, these supergiants have very large radii -- in the range of 100 times their main-sequence values. In fact, Fig. 7 shows a plot taken from an earlier paper (Gray and Toner 1987) showing the expected rates based on the $9M_{\odot}$ model of Becker (1981). As you can see, the Case S tracks (angular momentum conserved in shells) go right through the observations. The rigid-rotation case (Case R in the figure) is somewhat higher. Within the accuracy of such calculations, there is no compelling evidence for loss of angular momentum.

You might immediately ask about the distribution of the rates. If they show a Maxwell-Boltzmann shape, we'd have evidence for no braking, etc. Well, there aren't many Ib supergiants in the sky accessible to my 1.2 m telescope, but there are enough to show that the distribution is neither the Maxwell-Boltzmann nor the δ-function. So we do not immediately glean evidence for or against dynamo activity from the distribution of rates.

As another alternative, consider rotation as a function of luminosity. I have argued that the rates for classes III and IV are controlled by the rotostat, and if this is correct, we have some estimate of how much rotation it takes to run a dynamo. Fig. 8 makes an interesting comparison. The rotation for subgiants is a few km/s. The values for giants are a few km/s and slightly higher than for subgiants. And the values for supergiants are a few km/s and slightly higher than for giants. In other words, the simple increase in moment of inertia brings the rotation of Ib's down to, or within striking range of, the dynamo

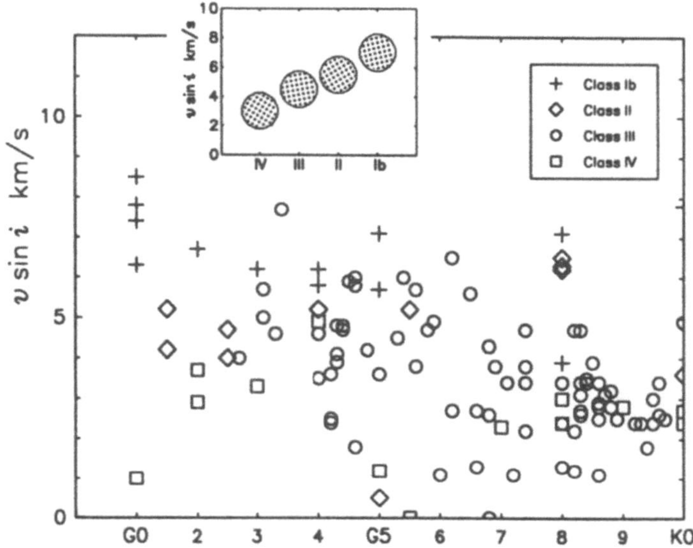

Fig. 8. Rotation of evolved stars is compared here. The rotation rates depend only weakly on luminosity class, but there is a systematic trend, especially in the G0 to G5 interval, shown schematically in the smaller graph. Data are from Gray and Nagar (1985), Gray and Toner (1986), Gray and Toner (1987), and Gray (1989a).

criterion v_{lim} for their internal physics. By the time convection runs deep enough to couple with the rotation and run a dynamo, the average Ib rotation is already at the limiting value. We conclude: on the average, dynamos will not turn on in Ib supergiants. Several implications then follow.

Implication a) those Ib's in the high-velocity tail of the Maxwell-Boltzmann distribution may rotate fast enough to trigger dynamo activity.

Implication b) Most Ib supergiants will be devoid of coronae because there is no magnetic field to form coronal loops. A few fast rotators from Implication a) may be exceptions.

Implication c) the chromospheres on most Ib's are acoustic-wave driven, or at least not magnetically driven.

Implication d) the rotation-velocity distribution of late-type Ib's should have a Maxwell-Boltzmann shape with the high-velocity tail grossly underpopulated because of magnetic braking.

There seem to be a few stars above the coronal boundary, such as β Dra, in agreement with Implication a). Implication b) explains the physical reason for the horizontal portion of the coronal boundary. The hot-loop/cold-loop dichotomy of Antiochos and Noci (1986), for example, is no longer needed. Should Implication c) be correct, we may be in a position to separate the physics of acoustic from magnetic heating. I have yet to study the numbers with regard to Implication d), but certainly the Ib supergiants do show an anomalous distribution of rotation rates, and Implication d) may just be the answer.

We might also make a more general empirical statement about characteristic convective velocities.

Roughly speaking (within about a factor of two), $v_{lim} \sim$ constant with luminosity (Fig. 8). The dynamo criterion of eq. (1) then implies that $v_{con} \propto \ell/R$. That is, the convective velocity grows as the fractional depth of the convection zone. Naturally differences with luminosity in internal structure, namely the density and temperature gradients, may alter the details, but within the factor of two or so, we are likely to find v_{con} varies roughly as ℓ/R. The trend in Fig. 8 actually implies an increase over direct proportionality of $\sim$ 2-3 times going from class IV to class Ib.

6. ROTATION OF EVOLVED K STARS

As stars evolve into the K2 to K5 spectral interval, they approach the Hayashi boundary where they evolve upward in luminosity and in moment of inertia. Rotation drops accordingly. Rotation falls below and stays below v_{lim}. If it existed at earlier evolutionary stages, dynamo action now ceases. Coronal loops disappear. We see no more x-rays. The right-hand portion of the coronal boundary is explained. Temperature inversions for stars on the cool side of the coronal boundary are acoustic driven.

7. ROTATION OF EVOLVED BINARIES

Although this paper is concerned with single stars, it is important to link in the synchronized binaries. Their orbital periods are one to two orders shorter than single star rotation periods, and so the synchronism enforces rotation well above v_{lim}. The stars respond by being hyper-active. Coronal and chromospheric emission are very strong. Gigantic dark starspots are seen. This type of response to rapid rotation within the magnetic domain is just what is expected.

Many fascinating details and valuable facts are obtained from synchronized binaries, and I refer you to the literature, for example, M. Rodono's contribution to this volume, Strassmeier et al. (1988), Hall and Henry (1990), Evans (1971), and numerous conference proceedings including Tuominen (1991), Wallerstein (1990), Linsky and Stencel (1987), and so on.

8. NON-CONFORMISTS

A few stars showing rapid rotation are seen within the magnetic domain. Examples are 7 Boo and the more extreme FK Com-type stars. How can these stars get around the usual magnetic braking and the rotostat regulation applied to their neighbors in the H-R diagram? One trivial possibility is that they really are synchronized binaries and the comparison star has escaped our detection. Another suggestion is that the non-conformists were binaries before the two components coalesced (Bopp and Stencel 1981, Bianchi et al. 1985). We would then be seeing the orbital angular momentum in action, but it would also be dissipating rapidly if the coalesced star subsequently follows the pattern of single stars.

Another possibility comes to mind. A dynamo functions by amplifying a "seed" field of some sort, possibly field running through the material from which it condensed. Could it be that occasionally a star forms with no magnetic field whatever? Then the dynamo would have nothing to amplify and the magnetic activity could not develop. Is the chromospheric activity in such stars powered without magnetic fields?

Or could it be that some of the too-rapid rotators got involved in a rotostat flicker in which the dynamo did not re-start for some reason, and so the star continued to spin up?

There are other non-conformist such as HR 1362 which appear to be slow rotators but have anomalously high chromospheric emission (Strassmeier et al. 1990).

Non-conformists are relatively few in number, and although they are curious and interesting, they should not deter us from understanding the pattern for the majority of evolved stars.

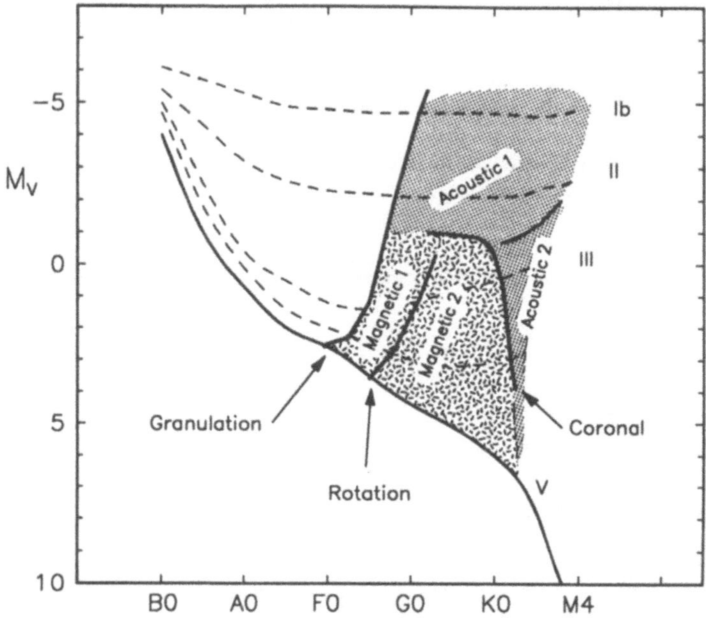

Fig. 9. The magnetic and acoustic domains are indicated on this H-R diagram. Magnetic activity in the first domain does not produce rotational braking. Magnetic activity in the second domain maintains the rotation at the dynamo rotation limit given in eq. (1), and is under the control of the convection-zone parameters. In the first acoustic domain, most stars rotate too slowly to drive a dynamo, so only the faster-than-average rotators will show magnetic activity. Stars in the second acoustic domain evolve out of the magnetic domain by rapidly increasing their moment of inertia.

9. SUMMARY

Initially rotation is altered only by moment-of-inertia increases as stars evolve across the H-R diagram. For stars of the lower luminosity classes, magnetic braking plays an important role in altering the initial angular momentum. The magnetic domain extends from the granulation boundary on the hot side to the coronal boundary on the cool side (see Fig. 9), and from the main sequence on the bottom up to the coronal boundary on the top. The rotation boundary divides the magnetic domain into two sub-domains: 1) the region between the granulation and rotation boundaries where magnetic braking seems to be absent and 2) the region between the rotation and coronal boundaries where rotostat magnetic braking occurs. Although the initial dissipation of angular momentum at the rotation boundary is large and abrupt, subsequent dissipation of angular momentum is metered out by the convection-zone through the rotostat mechanism. Synchronized binary stars do not follow this sequence of events because their orbital angular momentum is too large for the magnetic brake to dissipate on evolutionary time scales. They nevertheless behave as expected for rapid rotators caught in the magnetic domain.

Outside the magnetic domain, temperature inversions are driven by acoustic power. Two portions of the acoustic domain differ slightly. In the first one, a few stars - those in the high-velocity tail of the rotational velocity distribution - may rotate fast enough to engender dynamo action, but the rest will never have experienced dynamo action. The second acoustic domain is populated by lower mass stars, all of which previously had functioning dynamos. All stars in this second acoustic domain should now be devoid of dynamo activity because the evolutionary rise up the asymptotic branch forces the rotation below the limiting value of eq. (1). Some of the wide-ranging chemical patterns seen in asymptotic-branch stars may result from the encounter or avoidance of dynamo activity in earlier stages of evolution.

Rotation enters one way or another into most of the observable characteristics of evolved stars. Disentangling the details is a worthy challenge.

I am grateful to the Natural Sciences and Engineering Research Council of Canada for financial support.

References

Abt, H.A. 1957, Ap.J., 126, 503.
Abt, H.A. 1958, Ap.J., 127, 658.
Antiochos, S.K., and G. Noci 1986, Ap.J., 301, 440.
Becker, S.A. 1981. Ap.J. Suppl. 45, 475.
Bö hm-Vitense, E. 1981. Ann. Rev. Astron. Ap. 19, 295.
Bopp, B.W., and Stencel, R.E. 1981. Ap.J. 247, L131.
Bianchi, L. Grewing, M., and Kappelmann, N. 1985. Astron. Ap. 149, 41.
Durney, B.R., and Latour, J. 1978. Geophys. Ap. Fluid Dyn. 9, 241.
Endal, A. S., and Sofia, S. 1979. Ap.J. 232, 531.
Evans, D.S. 1971. Mon. Not. Roy. Astron. Soc. 154, 329.
Flower, P.J. 1977. Astron. Ap. 54, 31.
Fukuda, I. 1982. Pub. Astron. Soc. Pacific, 94, 271.
Gray, D.F. 1981. Ap.J. 251, 155.
Gray, D.F. 1982. Ap.J. 262, 682.
Gray, D.F. 1984. Ap.J. 281, 719.
Gray, D.F. 1988. Lectures on Spectral-Line Analysis: F, G, and K Star, (The Publisher: Arva, Ontario N0M 1C0, Canada).
Gray, D.F. 1989a. Ap.J. 347, 1021.
Gray, D.F. 1989b. Publ. Astron. Soc. Pacific 101, 832.
Gray, D.F. 1990. Publ. Astron. Soc. Pacific 101, 1126.
Gray, D.F. 1991. The Observation and Analysis of Stellar Photospheres, (Cambridge: Cambridge), 2nd ed., Chapter 15.
Gray, D.F., and Endal, A.S. 1982. Ap.J. 254, 162.
Gray, D.F., and Nagar, P. 1985. Ap.J. 298, 756.
Gray, D.F., and Nagel, T. 1989. Ap.J. 341, 421.
Gray, D.F., and Toner, C.G. 1986. Ap.J. 310, 277.
Gray, D.F., and Toner, C.G. 1987. Ap.J. 322, 360.
Hall, D.S., and Henry, G.W. 1990. Active Close Binaries, Proc. NATO Advanced Study Inst. (Kluwer: Dordrecht), C. Ibanoglu and I. Yavuz, eds.
Kraft, R.P. 1970, Spectroscopic Astrophysics, ed. G.H. Herbig, (Berkely: U. Cal. Press), p. 385.
Linsky, J.L., and Haisch, B. 1979. Ap.J. Letters 229, L27.
Linsky, J.L. and Stencel, R.E. (eds.) 1987. Cool Stars, Stellar Systems and the Sun, Lecture Notes in Physics 291.
Oke, J.B., and Greenstein, J.L. 1954. Ap.J. 120, 384.
Rutten, R.G.M., and Pylyser, E. 1988. Astron. Ap. 191, 227.
Simon, T., and Drake, S.A. 1989. Ap.J. 346, 303.

Strassmeier, K.G., Hall, D.S., Zeilik, M., Nelson, E., Eker, Z., and Fekel, F.C. 1988. Astron. Ap. Suppl. 72, 291.

Strassmeier, K.G., Hall, D.S., Barksdale, W.S. Jusick, A.T., and Gregory, W.H. 1990. Ap.J. 350, 367.

Tuominen, I. (ed.) 1991. The Sun and Cool Stars: Activity, Magnetism, and Dynamos, IAU Coll. 130, Lecture Notes in Physics, (Springer: Heidelberg). F.C. 1988. Wallerstein, G. (ed.) 1990. Cool Stars, Stellar Systems, and the Sun, (Astron. Soc. Pacific: San Francisco).

Wolff, S.C., Edwards, S., and Preston, G.W. 1982. Ap.J. 252, 322.

DISCUSSION

MacGregor: Several times during your talk, you alluded to the disappearance of magnetism upon the cessation of dynamo activity. Could you discuss how that might occur?

Gray: Probably not. The simple fact of the matter is that I reason in a very primitive way, because I don't want to get very far from the observations. I simply say that once the rotation is smaller than this dynamo criterion limit, the dynamo turns off. What the actual time-scale for this is, or how the field lines dissipate, I haven't a clue. And I don't intend to interpret that. It simply is not in the observations I have.

Roxburgh: Is there in fact any evidence that the dynamo number is a realistic parameter for the switching on and off of a dynamo?

Gray: I would like to think that my computed curve, compared to the observations, indicates that it works.

Gough: There are kinematic theories that make it look promising.

Pinsonneault: You can also get behavior of this type if you just have a balance between angular momentum loss that proceeds continuously and angular momentum transport from below the convective zone as rapidly rotating material is entrained. So, you can strike a sort of balance that reproduces that kind of pattern as well.

Schatzman: Some of these stars will end up as white dwarfs, and some of them rotate slowly with a large magnetic field. How can you explain their spindown?

Gray: Some of the white dwarfs have had polarization measurements and there are large fields indicated, and some show giant Zeeman splitting, so large magnetic fields are well established. What is wrong with expecting them to show magnetic braking when they have a strong field? The rotational velocities are all small - 2 km/sec. Do you agree?

Schatzman: Much smaller than that. The period of rotation is several days, and with their radii this corresponds to rotational velocities of order 10-100 m/s.

Gray: You are saying that if I have a dynamo criterion limit, then how come it doesn't turn off and the white dwarfs should be rotating more rapidly. Well, I would suppose that this criterion only works for stars which have some semblance of normal structure. That is, a deep convective envelope and so on. Once you solidify the star, I think there is quite different physics involved.

Belvedere: Your argument on the existence of a limiting velocity for the onset of dynamo activity is based on the assumption that the dynamo operates in the convection zone. But now we are learning from solar oscillations that probably we have some difficulties with the dynamo operating in the convection zone. Probably we are led to think that the working location for the dynamo is in the boundary layer beneath the convection zone.

Gray: I don't think that will affect my interpretation of the observations at all. It is basically a cosmetic difference, unless you insist that the boundary zone is extensive and decelerates the core as well so I don't have any source of angular momentum for this picture.

van't Veer: Do you think that the Maunder minimum of the Sun is also the result of the switching off and on of the solar dynamo?

Gray: I have suggested that in print. There are some problems with that, because at least for many dwarf stars, it is questionable whether the cores are still rotating rapidly. In the Sun, the observations are not conclusive, but I think it is pretty certain that the core is not rotating at 50 times the surface rate. That can be established just from the sphericity of the Sun. I don't insist on my model explaining the Maunder minimum or predicting it for other stars, however, if that fails, it would not affect the rest of my argument.

Roxburgh: When you appealed to the switching on of the magnetic field, an alternative it seemed to me was a rapidly increasing strength of the wind. So actually the mass loss rate increases.

Gray: As I understand it, the mass loss rate increasing affects the magnetic fields and does not necessarily lead to greater braking - you have a playback on the Alfven radius. Correct?

Roxburgh: Other things being equal, if you increase the mass loss rate, you increase the angular momentum loss rate.

Comeron: Not so. The Alfven radius moves in to compensate exactly in the simplest Weber-Davis type models.

Gough: Is that true if the mass loss rate goes to zero?

Cameron: I will have to think about that.

Paternó: It seems to me that the criterion to establish the dynamo activity is based not on the dynamo number but on something related to the dynamo number. Essentiallly, it is the turnover time of the convective cells, and when this is large the alpha effect is sufficient to give a big helicity. This works in the sense that even if the dynamo location was in the overshoot layer beneath the convection zone, the criterion would remain the same because the scale height of this overshoot depends on the bottom scale height of the convection zone. Therefore, it doesn't matter if the dynamo is seated in the boundary layer or the convection zone - it would be correct in any case (if it is in fact correct).

Gray: I agree with you, but I don't think Belvedere does.

ROTATIONAL DISCONTINUITY OF EVOLVED STARS: WHAT INTERPRETATION ?

J.R. de MEDEIROS and M. MAYOR
Geneva Observatory
51, ch. des Maillettes
CH- 1290 Sauverny
Switzerland

ABSTRACT: A systematic survey of about 2000 evolved stars has been carried out with the CORAVEL spectrometer to determine rotational velocities with a precision of about 1.0 km/s. As a result, the behaviour of the *Vsini* distribution, as a function of colours and luminosity, is now well established.

For the luminosity classes IV, III, II and Ib, the rotational discontinuity is now precisely defined and solid arguments inspire that the origin of this one is not the same for all classes. While for the subgiants the discontinuity seems to be the result of a magnetic braking, for the giants, that one results from a division between objects of significantly different ages. For the most luminous classes, the discontinuity results from the short duration of the first-crossing phase.

1. Introduction

One of the best known properties of rotating evolved stars is that these ones rotate slowly. Herbig and Spalding (1953, 1955) have shown on observational grounds a cutoff in the distribution of rotational velocities for evolved stars, which seems to be present near G0 in all luminosity classes from IV to II-III. As they pointed out, later than this spectral type stars with appreciable rotation rates are rare. In general, these authors could only determine modest upper limits because their low resolution photographic technique was larger than what we know today to be the average rotational velocity for evolved stars.

Most recently, in a series of papers, D. Gray and co-workers investigated the properties of the rotational velocity for several luminosity classes of evolved stars on the basis of very accurate *Vsini* values. For the subgiant stars, Gray and Nagar (1985) conclude that a rotational discontinuity occurs near G0IV. Combining his results with those of Alschuler (1975), who obtained *Vsini* for giants of earlier spectral types, with a too modest resolution technique, Gray (1981, 1988) claimed a rotational discontinuity to occur near G5III. In fact, for this luminosity class there is a sudden decline in rotation but located at G0III and not at G5III (de Medeiros and Mayor, 1989). This result was also independently obtained by Gray (1989). In all these cases, Gray interprets the cutoff in *Vsini* as a result of a magnetic braking. For the most luminous classes, no evidence for rotational discontinuity was found (Gray and Toner, 1986, 1987).

Admittedly, in all previous works the statistical analysis, and hence the interpretation, have been somewhat hampered by the paucity of the data sample. In order to improve this situation, we have obtained the projected rotational velocities for a large and homogeneous sample of evolved stars (de Medeiros and Mayor, 1989; de Medeiros, 1990). This study has

S. Catalano and J. R. Stauffer (eds.), Angular Momentum Evolution of Young Stars, 201–206.

enabled us to define clearly the location of the cutoff in *Vsini* distribution. Furthermore, we have found a same trend for the Ib supergiant stars.

In this work we attempt to provide a possible explanation for the origin of this "rotational dividing line".

2. The observational data

We have observed about 2000 stars covering the spectral range from middle *F* to middle *K* of luminosity classes IV, III, II and Ib. All stars with this characteristic and located north of declination -25°, listed in "The Bright star Catalogue" (Hoffleit and Jaschek, 1982; Hoffleit et al., 1983) and in the supergiants list of Egret (1980), were selected and observed. The observations were done with the CORAVEL spectrometer (Baranne et al., 1979) mounted on the Swiss 1.0 m telescope at the Haute-Provence Observatory, Saint-Michel (France). The results discussed in this paper are based on the observations collected from March 1986 until December 1989 and, in principle, each star was observed at least twice to search for spectroscopic variability.

By using the *Vsini* calibration of Benz and Mayor (1984) for the luminosity classes IV and III, and with an extension of this one for the bright giants and supergiants (de Medeiros, 1990), we obtain *Vsini* values as acccurate as the ones obtained by the Fourier transform technique. A comparison of our *Vsini* values with those published by Gray and co-workers for a sample of 103 stars from class IV to Ib, gives an excellent agreement with a typical r.m.s. of about 1.1 km/s.

3. The rotational discontinuity

In figures 1-4 we present the CORAVEL *Vsini* measurements for our sample of stars as a function of their (B-V). In order to avoid synchronization effects, we present here only single stars. The cutoff in the distribution of the rotational velocity is well defined for all luminosity classes. Its location in each luminosity sequence is F8IV, G0III, F9II and near F9Ib, which correspond to the (B-V) values 0.55, 0.70, 0.65 and about 0.70 respectively.

It is important to notice the wide range of *Vsini* values on the left side of the discontinuity for all luminosity classes. This behaviour seems to reflect the broad distribution of rotation rates for stars on the main sequence. This one is essentially similar all along the main sequence from spectral types O through A (Wolff et al., 1982).

The spread in *Vsini* values on the left of the cutoff decreases with increasing luminosity. Except for few giants and bright giants with moderate *Vsini*, all stars to the right of the discontinuity show a low rotation and in each luminosity sequence their mean value decreases slowly with the increasing (B-V) colours.

4. Discussion

For the subgiant and giant stars, the rotational discontinuity is currently interpreted as a result of a strong magnetic braking, due to the evolutionary deepening of the convective envelope. This couple with the rotation, generates a strong dynamo-generated magnetic field, and dissipates angular moment through mass loss along open field lines. Nevertheless, there are problems with this scenario. As mentioned above, a large fraction of stars in the spectral range *F* to *A* seems to leave the main sequence already with low rotation rates and, certainly, these ones have to arrive at the left side of the discontinuity, rotating very

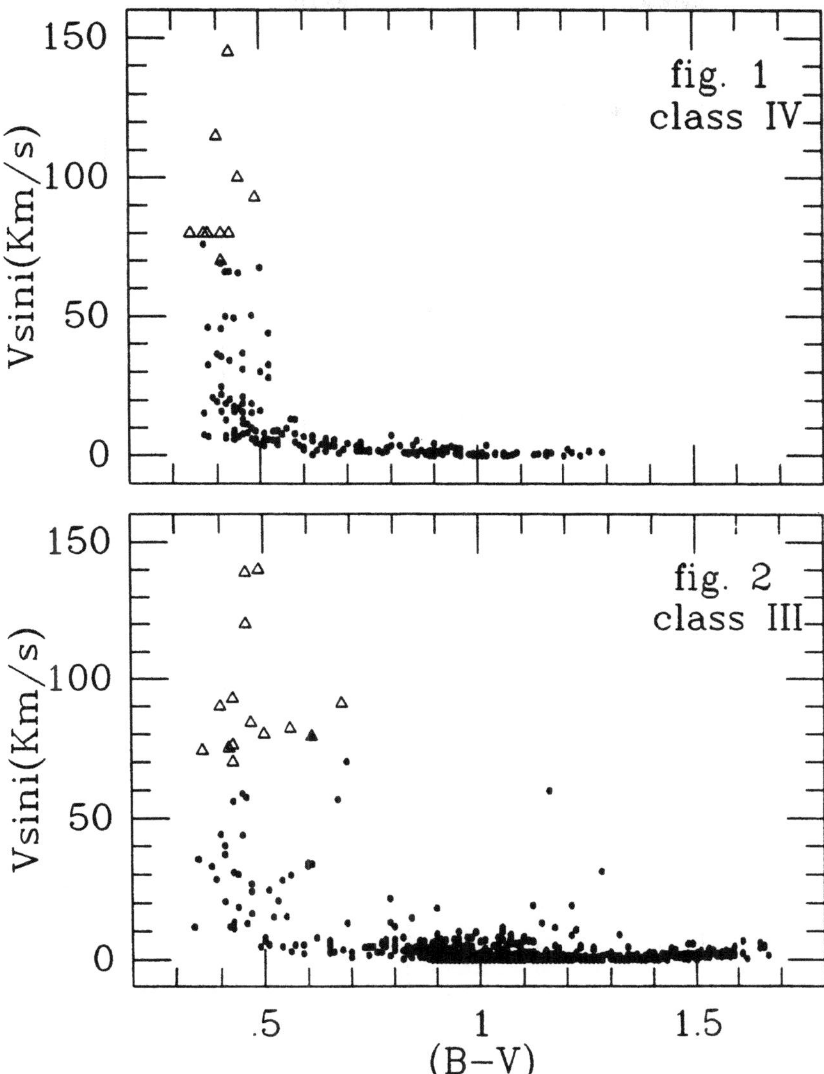

Figures 1 and 2. ***Vsini values as a function of colour (B-V). Triangles refer to values taken from "The Bright star catalogue".***

slowly. In this case, we do not need to invoke magnetic braking in the subgiant and giant domaine to explain the slow rotation for such stars.

The distributions of the rotation rates to the left of the discontinuity are shown in figures 5A - C. These distributions peak at low rotation values but they do not seem to be strikingly similar.

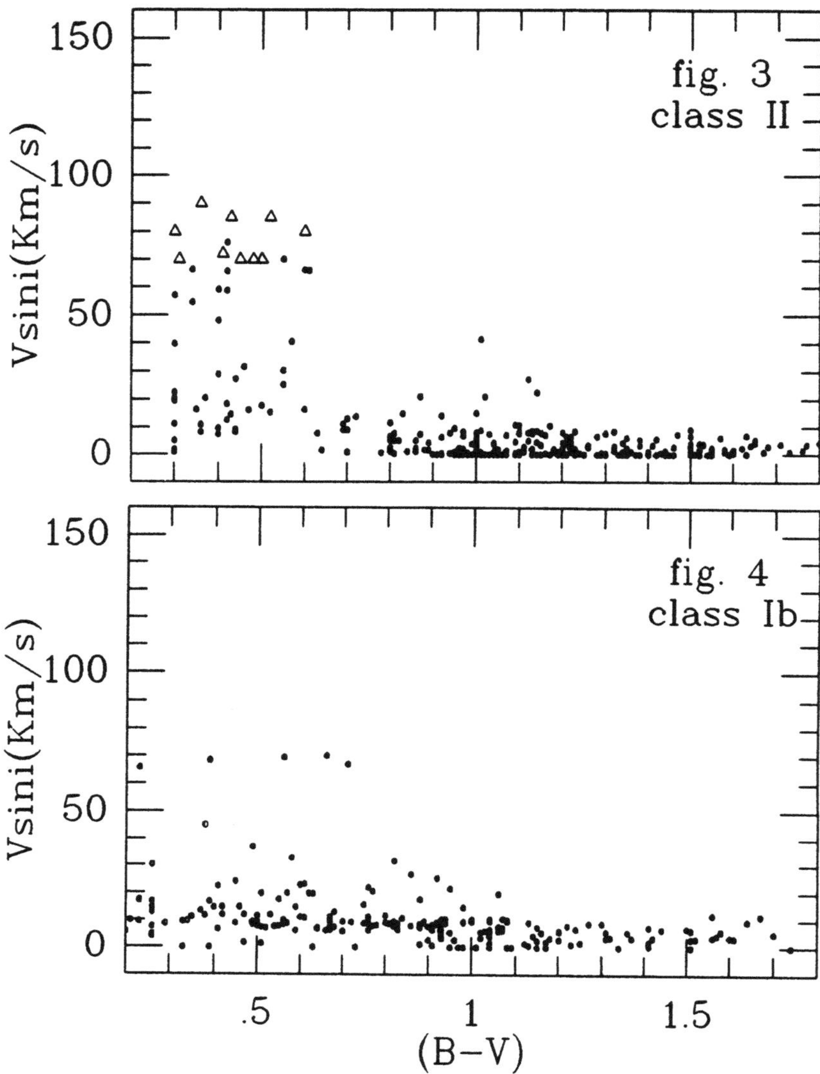

Figures 3 and 4. *Vsini values as a function of colour (B-V). Triangles refer to values taken from "The Bright star catalogue".*

The similarity of the distributions is expected if the same physical mechanism is involved determining the rotational velocity distributions. It is important to notice, for the supergiant stars, the large fraction of low rotators to the left of the discontinuity.

An examination of the kinematic-age relations between the stars located on the blue and red side of the discontinuity could clarify some aspects about the origin of this one. These relations may be deduced either from the observed radial velocities or space velocities.

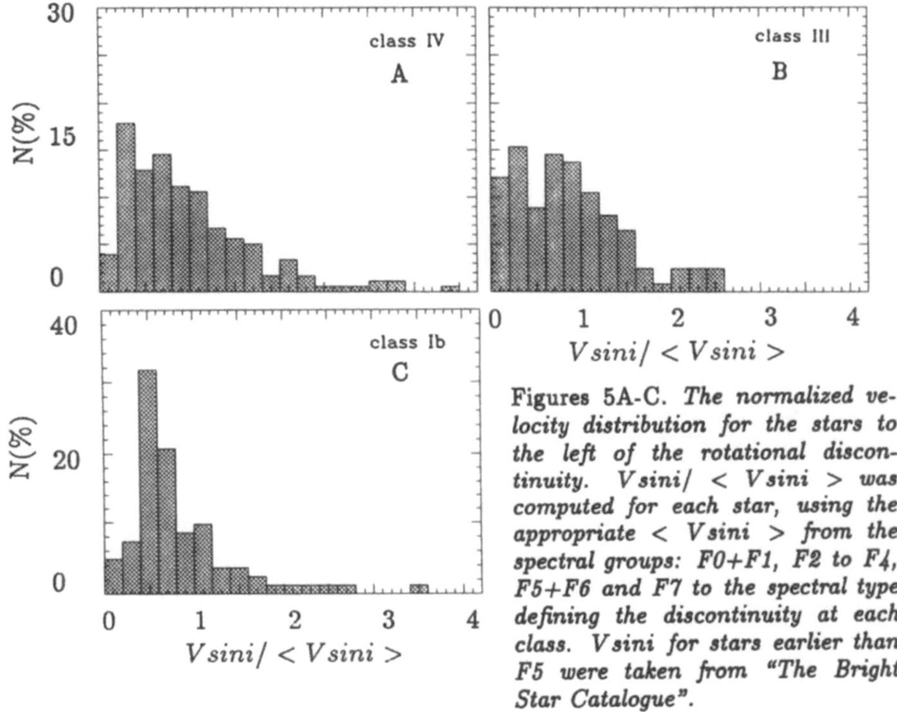

Figures 5A-C. *The normalized velocity distribution for the stars to the left of the rotational discontinuity. $Vsini/ < Vsini >$ was computed for each star, using the appropriate $< Vsini >$ from the spectral groups: F0+F1, F2 to F4, F5+F6 and F7 to the spectral type defining the discontinuity at each class. Vsini for stars earlier than F5 were taken from "The Bright Star Catalogue".*

To avoid to be influenced by the noticeable uncertainties on the distance determinations of giant stars we have considered here only radial velocities for our kinematical datation.

Table 1 gives, for the giant stars represented in figure 2, the velocity dispersion σ_{v_r} for several domains of colours, v_r corresponding to the radial velocity corrected for solar motion.

Table 1. *Radial velocity dispersion for the giant stars plotted in fig. 2.*

(B-V)	σ_{v_r}(km/s)	(B-V)	σ_{v_r}(km/s)
0.30/0.60	15.2 ±1.6	1.01/1.20	20.2 ±0.7
0.61/0.80	14.1 ±1.5	1.21/1.50	20.1 ±1.3
0.81/1.00	18.4 ±0.9		

It is well known that there is a progressive increase of the space velocities dispersion with stellar age (Mayor, 1974; Wielen, 1977). Therefore, on the basis of the result above, we can certainly conclude that giant stars on both sides of the discontinuity have significantly different ages. Stars on the blue side of this one are significantly younger than those on the red side, and in this case, it does not seem reasonable to interpret the cutoff in *Vsini* in the giants as the result of a magnetic brake.

In our view, this cutoff is the result of the age mixing associated with the very rapid evolution of giant stars in the Hertzsprung gap. In this case, the location of the cutoff should correspond with the blue edge of the gap.

For the subgiant stars, the magnetic braking proposed by Gray and Nagar (1985) seems to be the most plausible explanation for the cutoff at F8IV. From a similar analysis of radial velocity dispersion, there is no sign of an age division between stars earlier and later than F8IV. Moreover, this luminosity class undergoes a sudden decline in chromospheric activity near this spectral type (e.g., Simon and Drake, 1989; de Medeiros, 1990).

For the most luminous classes, we believe that the cutoff in the distribution of $V sini$ at F9II and near F9Ib is the result of the short duration of the first crossing phase. The spectral region, where the cutoff is observed, should also correspond to the blue edge of the Hertzsprung gap. Concerning the supergiants, about 90% of stars on the left side of the discontinuity show a $V sini < 25.0$ km/s, whereas about 10% present a $V sini$ larger than this value. This ratio low/large rotators present a good qualitative agreement with theoretical predictions from Maeder and Meynet (1988), if we assume the majority of low rotators evolving on the blue loops and the large rotators being still on the first crossing. According to these authors, the bluer extremity of the blue loops for some characteristic masses of bright giants and supergiants is located at (B-V) $\sim 0.90, 0.70, 0.30$ and 0.00 for $4, 5, 7$ and $9M_{\odot}$ respectively.

In this sense, Luck and Lambert (1985), based on a study on the relative CNO abundances of supergiants, pointed out several stars for which the relative mean value is consistent with the predicted value for stars observed following the first dredge-up. Among these stars, there are 2 *F type* and 4 early *G type* objects. For these stars we have a CORAVEL $V sini < 15.0$ km/s. This seems to be a supporting evidence that some, and perhaps all, low-rotators in the spectral region of the cutoff and on the left of this one are probably on the blue-loops.

References

Alschuler, W.,R., 1975: *Astrophys. J.*, **195**. 649

Baranne, A., Mayor, M. and Poncet, J. M., 1979: *Vistas in Astronomy*, **23**, 279

Benz, W. and Mayor, M., 1984: *Astron. Astrophys.*, **138**, 183

de Medeiros, J. R., 1990: *Ph. D. Thesis, Geneva Observatory*

de Medeiros, J. R. and Mayor, M., 1989: "Sixth Cambridge Workshop on cool stars, stellar systems and the Sun, *Astron. Soc. of the Pacific Conference Series, vol. 9, pp.404, Ed. G. Wallerstein.*

Egret, D., 1980: *Bull. Inf. CDS*, **18**, 82

Gray, D. F., 1981: *Astrophys. J.* **251**, 155

Gray, D. F., 1988: "Lectures on spectral-line Analysis: F, G and K Stars" *(Arva, Ontario; the Publisher)*

Gray, D. F., 1989: *Astrophys. J.* **347**, 1021

Gray, D. F. and Nagar, P., 1985: *Astrophys. J.* **298**, 756

Gray, D. F. and Toner, C. G., 1986: *Astrophys. J.* **310**, 277

Gray, D. F. and Toner, C. G., 1987 *Astrophys. J.*, **322**, 360

Herbig, G.H. and Spalding, J.F., 1953: *Publ. A.S.P.*, **65**, 192

Herbig, G. H. and Spalding, J. F., 1955: *Astrophys. J.* **121**, 118

Hoffleit, D. and Jaschek, C.: 1982, "The Bright star catalogue", *New Haven: Yale University Observatory*

Hoffleit, D., Saladyga, M. and Wlasuk, P., 1983: A supplement to "The Bright star catalogue" *New Haven: Yale University Observatory*

Luck, R. E. and Lambert, D. L., 1985: *Astrophys.J.*, **298**, 782

Maeder, A., and Meynet, G., 1988: *Astron. Astrophys. Supp. Ser.*, **76**, 411

Mayor, M., 1974: *Astron. Astrophys.*, **32**, 321

Simon, T., and Drake, S.A.: 1990, *Astrophys. J.*, **346**, 303

Wielen, R., 1977: *Astron. Astrophys.*, **60**, 263

MAGNETIC ACTIVITY AND ROTATION

MARCELLO RODONÒ
Institute of Astronomy, Catania University,
and Astrophysical Observatory
Viale A. Doria, 6
I-95125 Catania, Italy

ABSTRACT. Increasingly convincing observational evidence and theoretical arguments indicate the crucial role played by rotation, and convection in producing, amplifying and structuring stellar magnetic fields, which appear the ultimate responsible for the whole panoply of activity phenomena observed in stars.

A number of activity diagnostics are currently used to shed light into the physical characteristics of activity signatures and their intimate relation with global stellar parameters. In particular, the connection with those parameters that are relevant in determining the onset of the dynamo mechanism and the resulting magnetic field strength, geometry and variability. While the ultimate energy source of the observed phenomena is magnetic, the primary engine that appears capable of providing the required power is the mechanical flux originating in the convective zone, well inside the star. Therefore, the various forms of magnetic activity manifestation are intimately connected with the internal structure and evolution of stars, hence with their age and rotation regime.

The remarkable success achieved in understanding the general outline of magnetic activity requires, however, further observational and theoretical studies to be carried out in a systematic way and aimed at disentangling the role of specific parameters, such as rotation, and answering several still unresolved questions concerning the mechanisms that control the storage and conversion of magnetic energy into radiative, kinematic and other forms of energy outputs from magnetically active areas.

1 Introduction

The relation between rotation and magnetic activity, the so-called *rotation-activity connection*, has been studied very intensively in the last dozen years due to the combined efforts of new observation strategies and techniques, covering also the previously unexplored X-ray and UV spectral domains by using space-born instruments, and to the significant progress in the theoretical modelling of magnetic activity guided by the solar analogy. The development of dynamo models, first introduced by Parker (1955, 1979); and their application to stars (cf. Belvedere 1983, 1985; Gilman 1983; Schussler (1983)) has provided a qualitative but consistent picture on the generation of magnetic fields from the interaction of convection and rotation, on their appearance at the stellar surface and on their cyclic behaviour. From the observational side, after the early suggestion by Wilson (1963, 1966),

S. Catalano and J. R. Stauffer (eds.), Angular Momentum Evolution of Young Stars, 207–221.

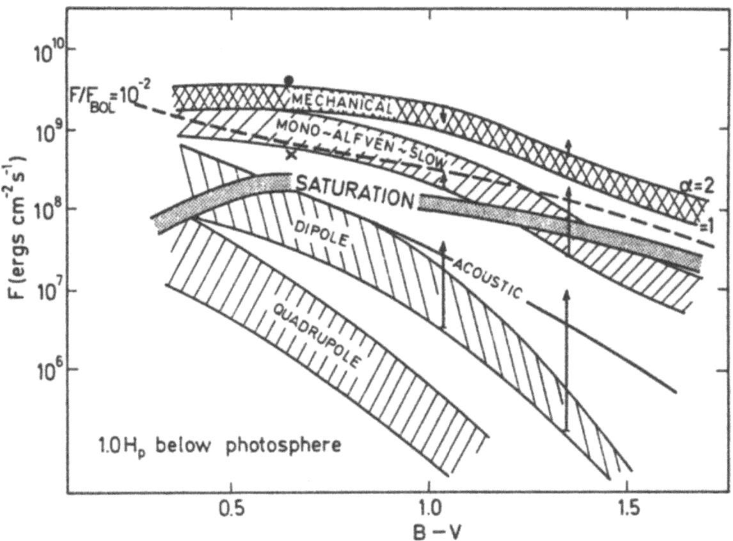

Figure 1: Observed saturated chromospheric-coronal radiative losses (SATURATION) and computed mechanical (MECHANICAL) and wave-fluxes. The dot and the cross are the maximal vertical and horizontal mechanic fluxes in solar granulation model (from Vilhu, 1987).

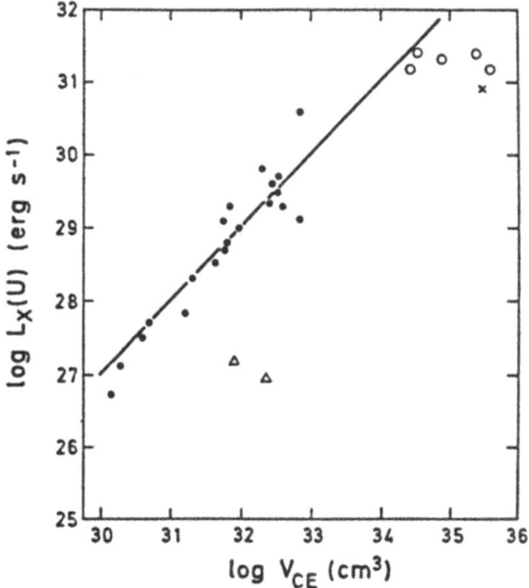

Figure 2: Relationship between coronal X-ray flux and the volume of the convective envelope, for dKe-dMe (filled circles), RS CVn (open circles) and a T Tauri (the cross)(from Pettersen 1989).

the primary role played by rotation was clearly identified through the discovery of tight empirical correlations between rotation and several activity indicators, such as chromospheric and transition region line fluxes (Skumanich 1972, Noyes et al. 1984, Schrijver 1986, Marilli and Catalano 1984) and coronal X-ray fluxes (Pallavicini et al. 1981). The rotation-activity connection has been extensively addressed in recent reviews (e.g. Baliunas and Vaughan 1985, Rodonò 1987, Hartmann and Noyes 1987, Linsky 1988, Catalano 1990a) and will not be duplicated here. The present review is aimed at identifying which are the basic facts and questions that deserve further attention in order to progress significantly in the understanding of magnetic activity and its connection with the star's rotation.

2 The energy source of magnetic activity

The remarkable role played by the magnetic field in activity phenomena is plainly demonstrated by solar images and magnetograms, where all observed phenomena, such as sunspots, plages, plasma network, flares and prominences, are directly connected with magnetic field strength and/or geometry. The role of convection, on the other hand, is not so visible, but the dynamo theory provides a coherent physical basis that requires the interaction of convection with rotation for the generation, amplification and evolution of magnetic fields through induction effects in conductive fluid masses. Both *kinetic or linear dynamos*, based on an externally assigned velocity field and disregarding the feedback of the magnetic field on the plasma motion, and *hydromagnetic or nonlinear dynamos*, where the magnetic field feedback is taken into account and the whole set of magnetohydrodynamical equations is simultaneously solved, highlight the important role of convection. Nevertheless, the actual energy output, attributable to magnetic activity, is observed from the star's atmosphere, where energy dissipation occurs following the occurrence of activity phenomena at photospheric up to coronal levels. From mixing length models of main sequence and giant stars, Vilhu (1987) computed the available mechanical, acoustic, and magnetic related (monopole, dipole, quadrupole) fluxes and, by comparison with total chromospheric - transition region - coronal losses from active atmospheres stars at saturation level (i.e. with filling factors $\simeq$ 0.8), showed that the mechanical energy source from convection supersedes all other possible form of energy output, independently from the spectral type (Figure 1). Therefore, according to Vilhu's (1987) calculations, the *heat engine*, or the *ultimate energy source* in the wording of Parker (1986), is provided by motions inside the stars, though for active stars of early spectral types the monopole or even the acoustic energy source might account for most or the total observed emission. It is interesting to note that saturated stars, as defined above, are rapid rotators, both single very young stars or tidally coupled members of close binaries. Another characteristic of Vilhu's plot is noteworthy, i.e., the level of total energy losses from saturated stars is always below the line $F/F_{bol} = 10^{-2}$, a limiting value approached towards the latest spectral types. Long-term observations of dMe flare stars show that the accumulated energy output due to flares, that constitute the most cospicuous energy loss from these *saturated* late-type dwarfs, attributable to magnetic activity, has an upper limit for L/L_{bol} of about 10^{-3}, which is consistent with the upper limit derived by Vilhu (1987) for the *quiescent* emission, i.e. outside of flares, from saturated active stars. Additional evidence on the relevance of convection in providing the necessary energy input for the development of stellar activity is provided by the tight correlation between the total luminosity of flare radiation in K- M dwarfs and the volume of the convection zone (Figure 2), as shown by Pettersen (1989)

3 The dynamo model and rotation

Dynamo models can be characterized by a dimensionless parameter, the dynamo number N_D, given by the squared ratio of the diffusive time (T_D) to the amplification time, (T_A), or the Rossby number N_R, given by the ratio of the rotation time of the convective zone (P_{conv}) to the convective turnover time (τ_c). These two parameters are very simply related by means of the following relation:

$$N_D = (T_D/T_A)^2 \approx N_R^{-2} = (P_{conv}/\tau_c)^{-2} \tag{1}$$

In fact, the dynamo number is defined by the relation:

$$\begin{aligned} N_D &= (T_D/T_A)^2 = (L_c^2/\eta)^2/(L_c/\alpha\Delta\Omega) \\ &= \alpha\Delta\Omega L_c^3/\eta^2 \end{aligned} \tag{2}$$

were α is the mean helicity $< v \cdot (\nabla \times v) >$ times the convective turnover time τ_c, $\Delta\Omega$ is the radial gradient of Ω, L_c is the typical length scale of convection, and $\eta = L_c \cdot v_c$ is the turbulent magnetic diffusivity, with v_c the convection velocity.

By assuming (Durney and Latour 1978)

$$\alpha \propto \Omega L_c \tag{3}$$

$$\Delta\Omega \propto \Omega \tag{4}$$

relation (2) becomes:

$$N_D \approx (\Omega L_c)^2 \times (\frac{L_c}{\eta})^2 \approx (\frac{\Omega L_c}{v_c})^2 \tag{5}$$

$$\approx (\Omega\tau_c)^2 \approx (\frac{\tau_c}{P_{conv}})^2 = N_R^{-2} \tag{6}$$

i.e., $N_D \approx N_R^{-2}$. Usually, instead of the rotation time of the convection zone (P_{conv}), the surface rotation time (P_{surf}) is considered to evaluate the Rossby number ($R_o = P_{surf}/\tau_c$), because surface rotation periods are readily available from several sources.

Mangeney and Praderie (1984) have introduced the following *effective* Rossby number

$$R_o^* = \frac{1}{2}(\frac{V_m}{\Omega L_c}) \tag{7}$$

where V_m is the maximum convective velocity and L_c is the total depth of the convection zone.

From the above definitions of the dynamo and Rossby numbers, it is evident that convection and rotation are the most significant parameters that preside over the dynamo action in stars and the associated activity phenomena. This straightforward statement, however, implies a dependence of dynamos on the internal structure and age of stars, so that it becomes quite a formidable task to try to isolate the effect of a given parameter on the observed activity and to identify accurate functional correlations. Moreover, whatever the type of dynamo model is assumed, a *shell dynamo* localized at the bottom of the convective zone or a *distributed dynamo* acting throughout the entire convection zone, the resulting activity scenario is largely controlled by nonlinear processes, that usually do not allow us to understand in detail the evolution of individual parameters from the end results of numerical integrations. On the other hand, the known empirical correlations between global stellar parameters and activity diagnostics generally involve mixed object samples, in the sense that any considered parameter of interest in the given star sample simultaneously depends on other parameters. As a matter of fact, the most significant dynamo parameters, rotation and convection, strongly depend on mass, age, chemical composition and on other circumstantial situations, e.g., membership in binary systems.
Therefore, in addressing the role of rotation in stellar activity we must bear in mind two basic facts: a) the evolution of stellar rotation with age and mass (Figure 3a and 3b), and b) the change of the effective Rossby number, or the associated characteristic velocity (Figure 4), versus mass and effective temperature. Catalano et al. (1988) have recently found that the rotation period depends not only on the square root of stellar age, as long ago discovered by Skumanich (1972), but also on stellar mass according to the following empirical relation:

$$P_{rot} = (1.32 - 0.92)10^{-3} M/M_{\odot} t^{1/2} + c(M/M_{\odot}) \tag{8}$$

where $c(M/M_{\odot})$ is constant wthin a narrow age range. Relation (8) and Figure 3 contain several interesting results. In particular, a) the lines of constant age converge at $M/M_{\odot} \approx 1.3$, corresponding to the spectral type F3, where it is likely that thick convection zones are absent, and b) the rate of spin-down is larger the lower the stellar mass is. This is suggestive of a more efficient spin-down in highly convective stars, because enhanced magnetic activity might increase the rate of momentum loss via magnetic wind braking. Actually, this behaviour is opposite to what observed for pre-main sequence stars, where magnetic activity is less important, if any. Figure 4 essentially reproduces the sizable variation of the Rossby number along the main sequence due to che change of the stellar structure and the relative importance of convection zones.

4 Activity diagnostics and rotation

Several diagnostic tools are available for magnetic activity studies. Firstly, the detection and measurements of stellar magnetic fields (B) and filling factors (*f*) from the comparison of observed and computed high-resolution profiles of magnetic sensitive and insensitive lines

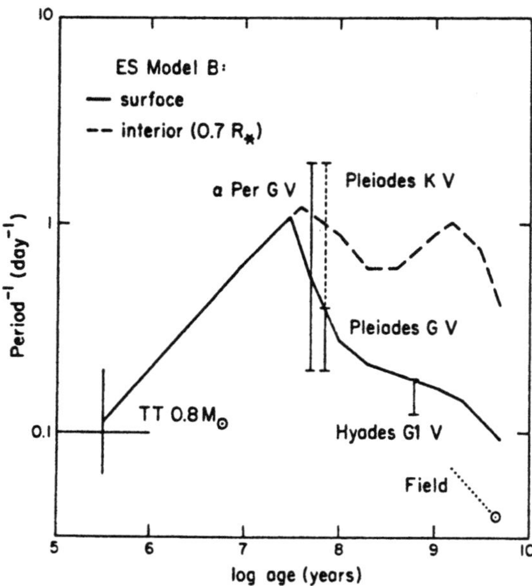

Figure 3a: Surface rotational velocities of solar-type stars as a function of age. The rotation evolution solar Model B of Endal and Sofia (1981) is displayed at the surface (solid line), and at the interior core-convective envelope interface (dashed line). Dotted line is the *Skumanich law* fit to the Sun (from Hartmann and Noyes 1987).

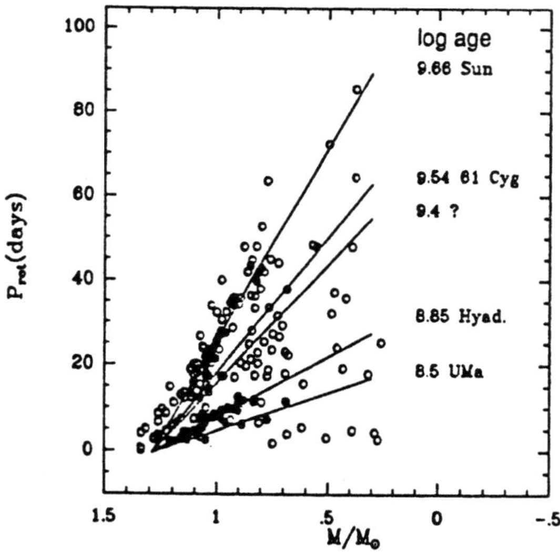

Figure 3b:Rotation periods, observed and computed from Ca II emission, versus stellar mass. Filled symbol refer to stars of known age. Straight lines are fit to stars of equal age (from Catalano 1990b).

that allows us to measure the relative Zeeman splitting (Robinson 1986).

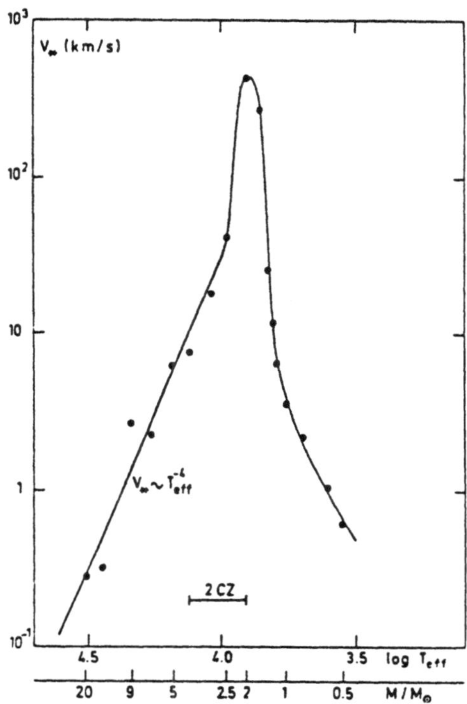

Figure 4: The characteristic velocity $v^* = 1/2(v_m R/L_c)$ as a function of the effective temperature for ZAMS models. v_m is the maximum convective velocity, R the photospheric radius, L_c the total depth of the convection zone (from Mangeney and Praderie 1984).

Saar and Linsky (1986), Saar et al. (1988) and Saar (1988) have progressively improved the Robinson's technique by taking into account the systematic errors, due to the presence of weak line blends in late-type star spectra and introducing the actual computation of radiative transfer effects. A sufficiently large body of data is now available to allow some important conclusions to be derived (Saar 1987). The total magnetic flux (B*f*) increases towards later spectral types and with rotation, and variability, both intrinsic and due to rotational modulation, has been detected. From multi-wavelength observations, Saar et al. (1988) were also able to derive a crude magnetic map of ξ Boo A. The result of Saar and Linsky (1986) that the magnetic field intensity for main sequence stars equates the equipartition field, i.e. $B_{obs} = B_{eq} = (8\pi P_{gas})^{0.5}$, suggests that the magnetic flux tubes are confined by gas pressure (P_{gas}). This result explains why the field strength increases, as the temperature and mass decrease along the main sequence towards late type stars. A very interesting and significant result was found by Saar and Schrijver (1987), who showed that the stellar Ca II emission flux scales with the magnetic field flux as $(Bf)^{0.6}$ giving the first direct evidence for the connection between photospheric magnetic fields and chromospheric emission in stars. Indirect evidence was already available from the spatial correlation of photospheric spots and chromospheric plages (Rodonò 1986, Rodonò et al. 1987), the former being assumed as good tracers of photospheric magnetic fields.

As summarized by Catalano (1984), chromospheric, transition region, and coronal global emission from active stars show well defined correlations with rotation (Figure 5). As indicated by the following empirical relations, higher temperature diagnostics decline steeper with rotation period than lower temperature ones:

$$L_X \cdot f_{cor}(M) \propto 10^{-(P/10.4)} \tag{9}$$

$$L_{CIV} \cdot f_{tr}(M) \propto 10^{-(P/23)} \tag{10}$$

$$L_{HK} \cdot f_{chr}(M) \propto 10^{-(P/32)} \tag{11}$$

The f(M) factors indicate some dependence on mass (see below). However, as noted by Simon (1986) and several others, these empirical correlations represent general trends that do not apply within individual luminosity classes or within specific stellar groups.

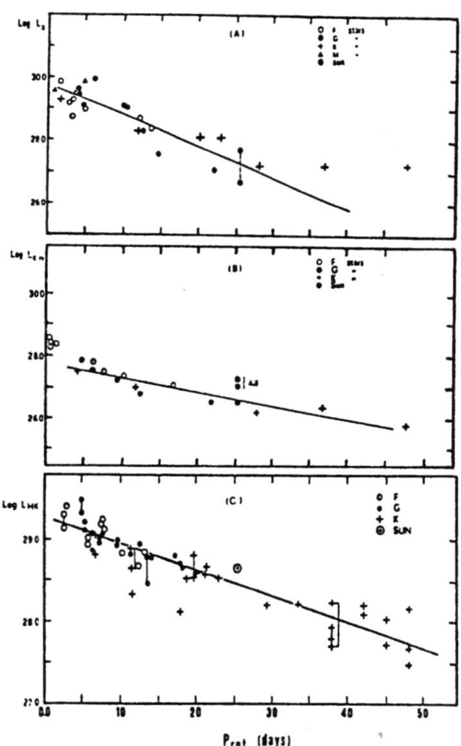

Figure 5: Emission luminosities of main sequence stars vs. rotation period, for chromospheric, transition region and coronal activity indicators. Straight lines are best-fits to the data (from Catalano 1984).

For example, the highly active RS CVn close binaries follow, as a group, the general rotation-activity relation for dwarfs and giants, but no rotation-activity correlation is found within the RS CVn group itself. This result might suggest that, in addition to rotation, at high emission levels other parameters or non-linear effects may dominate the mechanism that leads to the observed activity. Actually, quite different activity levels are observed on some late type flare stars, though they are characterized by very similar global parameters.

As implied by the existence of correlations (8) to (10), the activity levels at different atmospheric levels are strongly correlated one another. This indicates that activity phenomena in the atmospheres of stars derive their energy supply from a common reservoir that, however, can give rise to different phenomena characterized by different efficiencies, according to the physical characteristics of the plasma where they develop. This tight connection appears to hold not only at steady state active levels, but also on occasion of rapidly developing activity phenomena like flares. As shown in Figure 6 (Butler et al. 1988), integrated flare flux in a typical chromospheric diagnostic line (Hγ) is seen to correlate very tightly with coronal X-ray flare flux. In this context it is worth to caution about the solar activity analogy that should not be pushed too far. In fact, while from stellar activity data the mean correlations between fluxes in a range of temperature diagnostics spanning from $10^7 K$ to $\sim 10^4$K, are:

$$F_X \propto F_{CIV}^2 \propto F_{Ly\alpha}^{1.4} \propto F_{MgII}^3 \tag{12}$$

for the Sun these correlations are different:

$$F_X \approx F_{Ly\alpha} \ll F_{MgII} \tag{13}$$

Therefore, although the general trend of nonthermal energy deposition in the solar and stellar atmospheres indicates a definite increase with height, i.e. with temperature, the energy deposition in the highest atmospheric levels is up to two orders of magnitude larger on stars than on the Sun. Actually, the Sun deviates also badly from the empirical correlation between magnetic flux (Φ = B f), effective temperature (T_{eff}) and rotation velocity (V_{rot}), that was found by Saar (1987) for the stellar case:

$$\Phi \approx T_{eff}^{2.8} \cdot V_{rot}^{0.55} \tag{14}$$

As already anticipated, several authors have shown the tight correlation between activity and rotation. Because of the availability of rather extended data sets, the most complete studies have been carried out by using the Ca II emission flux (F_{HK}) or luminosity (L_{HK}), or the same parameters normalized to the bolometric ones. After the simple proportionality between emission and angular rotation velocity had been identified by Skumanich (1972), Noyes et al. (1984) evidenced the additional dependence on convection by showing that the normalized flux $R_{HK} = F_{HK}/(\sigma T_{eff}^4)$ is a smooth function of the Rossby number.

A similar relation was found by Hartmann et al. (1984) by using the Mg II h and k emission lines, which are the best chromospheric diagnostics in the ultraviolet spectral domain. These relations apply to main sequence stars and appear color or mass independent. The

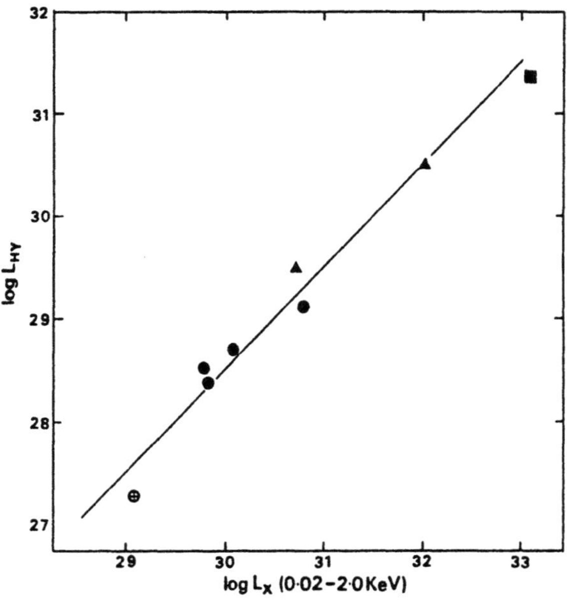

Figure 6: The integrated Hγ and soft X-ray fluxes for the stellar and solar flares. Symbols: triangles -YZ CMi, square - Gl 644B, filled circle - UV Cet and circle with central cross - the Sun (from Butler et al. 1988).

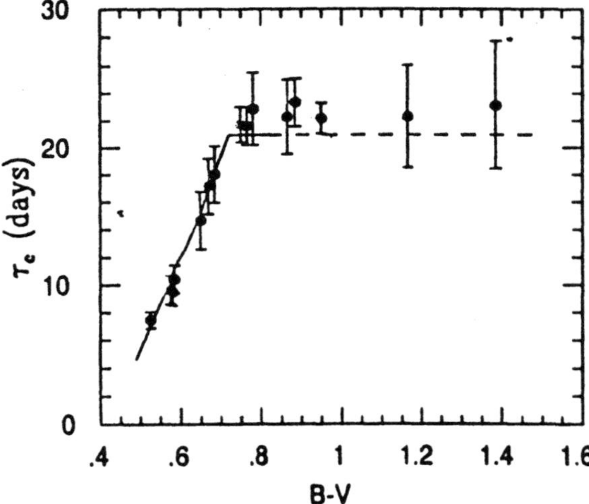

Figure 7: Convective turn-over time defined as the inverse of the a values of relation (16), as a function of B-V. The solid line is from Gilman (1980) for $\alpha = 2$; the dashed line is an extrapolation of Gilman's values for B-V >0.85 (from Catalano 1988).

Dutch astronomers (Zwan 1986, Schrijver 1986, Rutten 1987) used the same data corrected for the so-called *basal flux*, i.e. the quiescent chromosphere network emission, as identified from the observations of apparently inactive stars, and found color dependent correlations with the rotation period. The question of color dependence has been addressed more comprehensively by Marilli et al. (1986). They divided the available data into subclasses, each including a narrow range of B-V colors, and found that within each color interval the Ca II H and K line luminosity could be represented by the following exponential relation:

$$\log L_{HK} = -aP_{rot} + b \tag{15}$$

where both a and b are constant only within a narrow B-V range. This result quantitatively confirms the color or mass dependence of the rotation-activity relation. The b term represents the maximum activity level for $P_{rot} \rightarrow 0$, while the inverse of the coefficient a can be related to the convective turn over time τ_c. Actually, the remarkable agreement between the values $1/a$ and τ_c, the latter computed by Gilman (1980) for a ratio of the mixing length to the pressure scale height $\alpha = 2$, is shown in Figure 7 (Catalano 1988). Incidentally, we note that the minimum scatter in the correlation between activity index R_{HK} and Rossby number was found by Noyes et al. (1984) for $\alpha = 1.9$.

Coronal diagnostics, such as global X-ray fluxes, have given controversial results (Ayres and Linsky 1980; Stern et al. 1981, Pallavicini et al. 1981, 1990; Walter and Bowyer 1981; Walter 1982; Marilli and Catalano 1984; Caillault and Helfand 1985) on whether an exponential or power-law dependence holds between X-ray and rotation or Rossby number. Mangeney and Praderie (1984) have presented the most comprehensive study showing a well defined power-law correlation between the observed X-ray flux and the effective Rossby number defined by relation (6). The Mangeney and Praderie correlation is valid for active stars covering a wide range of spectral types, from O to M, irrespective of their luminosity class.

Having discussed at some length the question of rotation- activity convection, the principal conclusion that can be drawn is that, indeed, a *rotation-convection-activity connection* emerges both from observational evidence and theoretical arguments. The problem of disentangling the effect of rotation and convection, although still pursued by some authors, does not seem to be validated by our present understanding of stellar activity. Certainly, both rotation and convection enter the game of magnetic activity as the most important parameters, but the non linear character of the hydromagnetic dynamo, that seems the most adequate model to interpret magnetic activity, makes useless to inquire about the separate effect of rotation or convection alone, because the resulting activity level, as offered by observations, actually emerges from the non linear interplay of these two important parameters. As an extreme example, it is rather safe to anticipate that the extreme cases of highly-convective and very slowly rotating stars or non-convective rapidly-rotating stars, if magnetically active, would badly deviate from any *rotation-convection-activity correlation* derived from the characteristics of the majority of active stars.

5 Conclusion

By assuming that we are actually dealing with a *rotation-convection-activity* connection, other than simply with a *rotation-activity* connection, no further conclusion is worth to be put forward, except uselessly to reply the facts and arguments presented in the previous sections. Since this summary would not be of little use, given the limited length of the present review, I rather prefer to present a very short outline of the prospects of stellar activity studies.

The most important lesson we have learned from the study of stellar activity is that any future progress strongly depends on the systematic collection of long-term, multi-wavelength, and multi-site synoptic observations in order to provide much needed data on:

- magnetic field strengths, filling factors, surface mapping and variability, and their relations with activity signatures;
- physical characteristics, lifetimes and dynamics of active areas at photospheric, transition-region and coronal levels, and their spatial and/or physical connection;
- rotation periods, surface differential rotations and their evolution by monitoring the formation and migration of compact features on the stellar surfaces;
- long-term activity cycles, including null detection of cycles and Maunder-type minima of magnetic activity.

Concurrently, the following empirical/theoretical aspects deserve further attention:

- the improvement of existing correlations among activity signatures and with global stellar parameters, and the search for new significant correlations, bearing in mind that we are dealing with non-linear phenomena, that require extreme caution in identifying the most significant parameters; actually, as already emphasized in one of my past reviews (Rodonò 1987) instead of *macro-correlations*, involving activity and global stellar parameters spanning over a wide range of stectral types, we should search for *micro-correlations* within limited samples of stars that were similar in all other respect except the basic parameters under investigation for possible correlations;
- the overcoming of the present standstill in the theoretical development of non-linear dynamo models and on the mechanisms leading to the dissipation of magnetic energy in the outer stellar atmospheres;
- the further development of internal structure models for highly convective stars; these models, together with oscillation studies may through some light on radial differential rotation.

It is evident that to address the above listed questions and problems will not be a trivial task and reliable answers will depend on several circumstances. From the observational point of view, the implementation of international networks for multi-band, multi-site studies has proven to be very effective, so that their continued operation should be pursued strongly encouraged. From the theoretical point of view, the increasing availability of fast supercomputers and parallel computing techniques will certainly ease the enormous computational demands of nonlinear dynamo models in convective stars.

Acknowledgements. It is my pleasure to thanks Santo Catalano, Chairman of the Scientific Committee for organizing a very stimulating workshop and for invaluable help in finalizing the present review. Stellar activity research at Catania University and Astrophysical Observatory is supported by the Italian Ministry for University,and Scientific and Technological Research (*MURST*), the National Research Council (*CNR*), and the Italian Space Agency (*ASI*), whose financial support is gratefully acknowledged.

References

Ayres T.R., Linsky J.L.: 1980, *Astrophys. J.* **241** , 279.
Baliunas S.L. and Vaughan A.H.: 1985 *Ann. Rev. Astron. Astrophys.* **23**, 379.
Belvedere G.: 1983, in *IAU Coll. 71* , P.B.Byrne and M.Rodonò eds., Reidel Publ. Co., Dordrecht, p. 579.
Belvedere G.: 1985, *Solar Phys.* **100** , 363.
Butler C.J., Rodonò M., Foing B.H.: 1988, *Astron. Astrophys.* **206**, L1.
Caillault J.P., Helfand D.J.: 1985, *Astrophys. J.* **289** , 279.
Catalano S.: 1984, in *Space Research Prospects in Stellar Activity and Variability* , A.Mangeney and F.Praderie eds., Obs. Paris-Meudon Press, p. 243.
Catalano S.: 1988, *Irish Astron J.* **18** , 265.
Catalano S.: 1990a, NATO-ASI on *Active Close Binaries*, C. Ibanoglu and I. Yavuz eds., Kluwer Ac. Publ., p. 347.
Catalano S.: 1990b, in *New Window in the Universe; XI IAU ERAM*, F. Sanchez and M. Vazquez eds., Cambridge University Press, p. 161.
Catalano S., Marilli E., Trigilio C.: 1988, NATO-ASI on *Formation and Evolution of Low-Mass Stars*, A.K.Dupree and T.Lago eds. , Kluwer Ac. Publ., p. 377.
Durney B.R., Latour J.:1978, *Geophys. Astrophys. Fluid Dyn.* **9**, 241.
Endal A. and Sofia S.: 1981, *Astrophys. J.* **210**, 184.
Gilman P.A.: 1983, *IAU Symp. 102* , J.O.Stenflo ed., Reidel Publ. Co., Dordrecht, p. 247.
Hartmann L.W., Baliunas S.L., Duncan D.K., Noyes R.W.: 1984, *Astrophys. J.* **279** , 778.
Hartmann L.W., Noyes R.W.: 1987 *Ann. Rev. Astron. Astrophys.* **25**, p. 271.
Linsky, J.L.: 1988, in *Multiwavelength Astrophysics*, F. Cordova ed., Cambridge Univ. Press, p. 49.
Mangeney A, Praderie F.: 1984, *Astron. Astrophys.* **130**, 143.
Marilli E., Catalano S.: 1984, *Astron. Astrophys.* **133**, 57.
Marilli E., Catalano S., Trigilio C.: 1986, *Astron. Astrophys.* **167**, 297.
Noyes R.W., Hartmann L.W., Baliunas S.L., Duncan D.K., Vaughan A.H.: 1984, *Astrophys. J.* **279**, 763.
Pallavicini R., Golub L., Rosner R., Vaiana G.S., Ayres T.R., Linsky J.L.: 1981, *Astrophys. J.* **248**, 279.
Pallavicini, R., Tagliaferri G., Stella L.: 1990, *Astron. Astrophys.* **228**, 403.

Parker E.N.: 1955, *Astrophys. J.* **122**, 293.

Parker E.N.: 1979, *Cosmical Magnetic Fields: Their Origin and Their Activity*, Clarendon Press, Oxford.

Parker E.N.: 1986, in *Cool Stars, Stellar Systems, and the Sun*, M. Zeilik and D.M.Gibson eds., Springer-Verlag, Berlin, p. 341.

Pettersen B.O.: 1989, in *Solar and Stellar Flares,IAU Coll. 104,* B.M. Haisch and M. Rodonò eds., *Solar Phys.* **121**, p. 299. **224**, 182.

Robinson R.D.: 1986, in *Highlights of Astronomy*, J.P. Swings ed., Reidel Publ. Co., Dordrecht, p. 417.

Rodonò M.: 1986, in *Cool Stars, Stellar Systems, and the Sun*, M. Zeilik and D.M.Gibson eds., Springer-Verlag, Berlin, p. 475.

Rodonò M.: 1987, in *Proc. 5th EPS Meteting on Solar and Stellar Physics*, E.H.Schroter and M. Schussler eds., Springer-Verlag, Berlin, p. 39.

Rodonò M., Byrne P.B., Neff J.E., Linsky J.L., Simon T., Butler C.J., Catalano S., Cutispoto G., Doyle J.G., Andrews A.D., Gibson D.M.: 1987, *Astron. Astrophys.* **176**, 267.

Rutten R.G.M.: 1987, *Astron. Astrophys.* **177**, 131.

Saar S.H.: 1987 in *Cool Stars, Stellar Systems, and the Sun*, J.L.Linsky and R.E.Stencel eds, Springer-Verlag, Berlin, p. 10.

Saar S.H.: 1988, *Astrophys. J.* **324**, 441.

Saar S.H., Linsky J.L.: 1986, *Adv. Space Res.* **6**, No.8, 235.

Saar S.M., Schrijver C.J.: 1987, in *Cool Stars, Stellar Systems, and the Sun*, J.L.Linsky and R.E.Stencel eds, Springer-Verlag, Berlin, p. 38.

Saar S.H., Houvelin J., Giampapa M., Linsky J.L., Jordan C.: 1988 in *Cool Stars, Stellar Systems, and the Sun*, O. Havnes et al. eds, Kluwer Ac. Publ. Dordrecht, p. 45.

Schrijver C.J.: 1986, *Stellar Magnetic Activity*, Ph.D. Thesis, Utrecht University.

Schussler M.: 1983, *IAU Symp. 102*, J.O.Stenflo ed., Reidel Publ. Co., Dordrecht, p. 213.

Simon T.: 1986, *Joint NASA-ESA-SERC IUE Conf., London*, ESA SP-263, 53.

Skumanich A.: 1972, *Astrophys. J.*, **171** 565.

Stern R.A., Zolcinski M.C., Antiochos S.K., Underwood J.H.: 1981, *Astrophys. J.* **249**, 647.

Vilhu O.: 1987, in *Cool Stars, Stellar Systems, and the Sun*, J.L.Linsky and R.E.Stencel, Springer-Verlag, Berlin, p. 110.

Walter F.M.: 1982, *Astrophys. J.* **253**, 745.

Walter F.M., Bowyer S.: 1981, *Astrophys. J.* **245**, 671.

Wilson, O.C.: 1963, *Astrophys. J.*, **138** 382.

Wilson, O.C.: 1966, *Astrophys. J.*, **171** 695.

Zwan C.: 1986, in *Cool Stars, Stellar Systems, and the Sun*, M. Zeilik and D.M.Gibson eds., Springer-Verlag, Berlin, p. 19.

DISCUSSION

Paternó: The relationship linking the UV emission and the depth of the convection zone is not surprising - it has been predicted by theory for many years that the efficiency of the dynamo depends on the alpha effect, and this is sensitive to the cube of the depth of the convection zone.

Rodonó: Yes, the theory is not surprising. It is surprising to have the observations to show this.

Dziembowski: I didn't get your point on differential rotation. Do you have in your sample of stars with differential rotation any single stars or are they all binaries?

Rodonó: Mostly binaries, but there are some single stars. At least one single star that is similar to the sun shows differential rotation.

Vaiana: It is true that a decrease in scatter of a multivariate function is not necessarily an indication of the correctness of a given treatment, on the other hand, an increase in scatter is not indicative of correctness either. Many of these correlations are based on a small number of points, perhaps not chosen in a well-defined manner, and it is difficult to know whether the result is really significant. What I think we need to strive for is to conduct experiments in such a way that we can make well defined statistical tests and derive statistically supportable conclusions, as we have been trying to do in the X-ray field.

Rodonó: Yes, I agree.

LITHIUM, ROTATION AND AGE

Evry SCHATZMAN
Observatoire de Meudon
92195 Meudon Principal-Cedex, FRANCE

ABSTRACT. Lithium depletion as a function of time appears as a well established fact. After a review of the different physical processes which can be involved in this phenomenon, it appears that, very probably, stellar rotation does not play an important role. Stellar rotation is time dependant, lithium depletion can be used as a clock measuring stellar ages and can provide the relation between rotation and age, but this requires a calibration of the lithium-clock. This requests a well established model of lithium burning. Internal waves carry angular momentum and generate a mixing process which can be responsible for lithium depletion. The calibration of the lithium clock remains to be done.

1 Introduction

In an attempt to clarify the question of the relationship between lithium abundance and rotation it should be necessary to present the landscape of all physical processes which can play a role in the generation of lithium deficiency as it is observed in stars.

It should first be noticed (Cayrel de Strobel, 1990) that there are fluctuations in the metallicity: in the thick Galactic Disc population, the observed metal abundance has a dispersion by a factor 3 to (1/5), and in the thin Galactic Disk, the metal abundance has a dispersion by a factor 0.75 to 1.35. It should be recalled that the stellar models suggest in the Hyades a metal abundance which is 1.5 the solar abundance. In other words, when speaking of the cosmic abundance of Lithium, Log N(^{7}Li) = 3.3 (in the scale Log N(H) = 12.0) it should be remembered that we cannot exclude fluctuations in the initial stellar abundance of lithium. Therefore, we have to be careful in the analysis of the meaning of some of its abundance variations.

Mixing by macroscopic motions seemed to be an efficient tool, either slowing down gravitational separation (Schatzman, 1969), or increasing the rate of destruction of lithium (Schatzman, 1977). In both cases, it was suggested that the turbulent motion was generated in some way by rotation. The effect of turbulent motion has been considered in different aspects: turbulent viscosity and its effect on differential rotation (Tassoul and Tassoul, 1982, 1983, 1984a, b, 1986, 1989), turbulent viscosity and transport of angular momentum (Schatzman, 1987; Tassoul and Tassoul, 1989; Endal and Sofia, 1978; Pinsonneault et al., 1989); turbulent diffusion and lithium burning. One of the first difficulties which came out was that the turbulent diffusion

S. Catalano and J. R. Stauffer (eds.), Angular Momentum Evolution of Young Stars, 223–242.

coefficient which seemed fitted to explain the lithium deficiency in main sequence stars and especially in the Sun was too small by about an order of magnitude if it were used to explain the internal velocity of rotation of the Sun (Schatzman, 1987; Tassoul and Tassoul, 1989; Pinsonneault et al., 1989). It is naturally possible to accept the idea that the same dissipative process is working in both cases (transport of chemical elements and transport of angular momentum), and that the discrepancy between the two diffusion coefficients will be explained later. But it seems also that a better approach consists in trying to find the physical reason of this contradiction. The recent results concerning the internal rotation of the Sun (Brown et al., 1989; Thomson, 1990) are very important. They show in the equatorial plane a decrease of the angular velocity *inwards*, from the boundary of the convective zone (at 0.72 $R_\odot$) to about $0.4R_\odot$, and then an increasing angular velocity at smaller radii. The presence of a decrease *inwards* of the angular velocity is contradictory with the effect of a dissipative process. Assuming a flow of angular momentum *outwards*, transport of angular momentum by viscous effect due to shear flow, implies a negative gradient of the angular velocity, or to state things in the same way as above, an increase *inwards* of the angular velocity. Christensen-Dalsgaard (1990) shows the incompatibility of a negative gradient of angular momentum with the helioseismology data.

To this question of the physical process responsible for the transport of angular momentum it is necessary to join the question of the relation between rotation and lithium depletion . It has been already noticed by Balachandran et al. (1988) that fast rotators in αPer and in the Pleiades are more lithium-rich than slow rotators, and this is confirmed by Balachandran (1990). Schatzman (1990a, b) has suggested to explain the correlation between fast rotation and lithium abundance by the presence of an accretion disc which would bring angular momentum and lithium. It is difficult to keep this assumption because, as shown by Bouvier (1990), there is no indication of the presence of such a disc: this deserves to be discussed more carefully (Section 4). Schatzman and Baglin (1990) have shown, after a remark of Spruit (1990), that the velocity distribution, in the interval of the lithium gap of the Hyades, has an intrinsic dispersion, which rules out a biunivocal relation between rotation and lithium abundance: these results will be reported briefly in Section 2.

Then remains the question of the interaction between gravitational settling or radiative levitation and meridional circulation. Charbonneau and Michaud (1988a, b, 1990) have considered the problem. They have considered the meridional circulation as given by the model of Tassoul and Tassoul, but have ignored the differential circulation and the 2-D turbulence on equipotentials. If we consider a solar-like star, with an almost solid rotation in the radiative zone, it is difficult to give an estimate of the horizontal turbulent diffusion coefficient. From the estimates of Zahn (1983) the horizontal turbulent diffusion coefficient is of the order of the thermal diffusivity, of the order of 10^8 cm^2 s^{-1}. The time-scale of horizontal mixing is then of one million year, and is certainly shorter than the time-scale of the meridional circulation. Therefore, it seems necessary to combine the advection of chemical elements by circulation and horizontal mixing by turbulence.

There remains the relation between lithium deficiency and age and the relation between rotation and age. It is established that old clusters have a greater lithium deficiency than young clusters. If lithium deficiency is to be used as an age indicator, with the aim, for example, to establish properly the rate of spin-down, it is necessary to obtain a good calibration of the rate of lithium destruction.

We shall present first the reasons for which it seems reasonable to reject the assumption that rotation is at the origin of the processes which lead to destruction of lithium (Section 2). We shall then consider the properties of internal waves and show that they can provide the mechanism of transfer of angular momentum and the mixing process which is responsible for lithium burning in low mass stars (Section 3). Finally, we consider the situation in young clusters, and the question of calibrating the relation between lithium abundance and age.

2 Lithium depletion does not depend on rotation

2.1. VELOCITY DISTRIBUTION IN THE LITHIUM GAP

If the mechanism of lithium depletion is only rotation-dependant, the very small dispersion of the lithium depletion for a given mass in the Hyades should be associated with a very small dispersion of the rotational velocities. But in the earlier spectral types, the observed velocities do not seem to have such a small dispersion. The distribution of the velocities does not correspond to the $V \sin i$ distribution, an intrinsic dispersion of the rotational velocities is needed. This is especially true in the region of the gap (Boesgaard and Tripico, 1986). The cumulative distribution function (Figure 1),

$$\int f(V \sin i) d \cos i$$

looks more like representing $(1/C) \int\int dV d\cos i$ than $\int\int \delta(C-V) dV d\cos i$, where C is a constant and $\delta(C-V)$ is the delta function. The exact cumulative distribution in normalized variables, for a uniform velocity distribution from $v = 0$ to $v = V$, is:

$$\int_0^v f(v)dv = (\frac{v}{V}) \arctan \sqrt{(\frac{V}{v})^2 - 1} + 1 - \sqrt{(\frac{v}{V})^2 - 1}$$

It thus appears that there is an almost uniform distribution of the equatorial velocities between 0 and the maximum equatorial velocity C rather than a pure $\sin i$ effect. If the effect was due to rotation only, there would be a dispersion of the lithium abundances along the lithium gap and not a smooth distribution. Explaining the shape of the lithium gap by the envelope of the velocity distribution function, when the velocity dispersion presents an intrinsic distribution, is not reasonable. A physical process other than rotation is necessary to interpret lithium depletion (Spruit, 1990; Garcia-Lopez and Spruit, 1990).

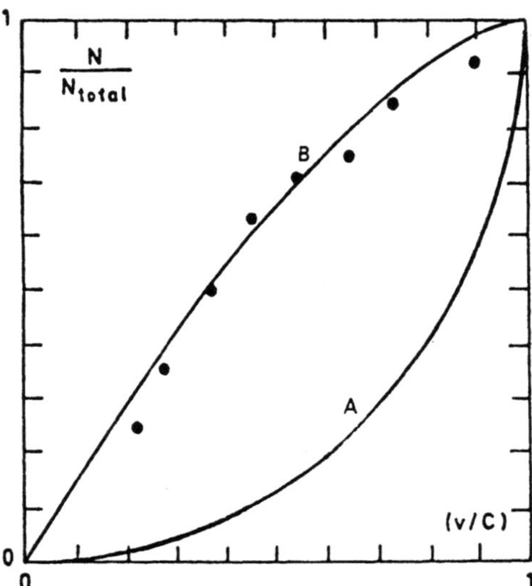

Figure 1: Cumulative velocity distribution function of the stars of the helium gap in the Hyades (from Schatzman and Baglin, 1990). Curve A corresponds to a dispersion due only to the view angle of the rotating star. Curve B includes the effect of an intrinsic uniform velocity distribution function between 0 and the maximum value of the velocity (here, 55 km/s).

2.2 GENERATION OF A TURBULENT FLOW

We shall now reconsider the instabilities which are supposed to generate the turbulent flow which brings lithium from the bottom of the convective zone to the burning level. We shall write again the expressions of the turbulent diffusion coefficients, given in the same form by Schatzman and Baglin (1990), being then easier to compare. We shall not rediscuss the theoretical problems concerning the instabilities which generate the diffusive processes, but we shall give estimates of their magnitude, and argue whether they can actually explain the observed lithium depletion and its mass dependence.

2.2.1 *Shear flow.* We assume here, with Zahn (1974), that the Richardson-Townsend condition is fulfilled, with a turbulent flow characterized by a Reynolds number $Re > Re_{critical}$, where, as usual, the critical Reynolds number is supposed to be of the order of one thousand. We then can write:

$$D_{sh} = \nu Re = \frac{3Lr^6\Omega^2}{G^2M^3}\frac{H_p}{r}\frac{\nabla_{ad}}{\nabla_{rad}}\frac{<\rho>}{\rho}\left(\frac{d\ln\Omega}{d\ln r}\right)^2\cos^2\theta \tag{1}$$

The different terms have their usual meaning: ν is the viscosity, with $\nu = \nu_{mol} + \nu_{rad}$

(Schatzman, 1977).
As we shall show now, the shear flow is probably not active. The turbulent flow generated by the shear flow instability will have to be taken into account if it exceeds the turbulent flow generated by the meridional circulation and if the quantity D_{shear} exceeds the critical value $D_{sh} > \nu Re_{crit}$.
Let us consider the condition

$$D_{shear} > \nu Re_{crit}$$

In the case of the Sun it corresponds to the condition:

$$\left|\frac{d\ln\Omega}{d\ln r}\right| > 43.18 \tag{2}$$

This has been obtained with the following values of the physical quantities, at the bottom of the convective zone : log T= 6.2826, log r = - 0.8387, $\Delta\nabla = 0.1$, log r = 10.70837, log $H_p = 9.67514$. Helioseismology (Brown et al., 1989; Thomson, 1990) provides an estimate of this quantity, certainly smaller than 0.2 and perhaps close to 0.
This leads naturally to the conclusion that in the present Sun, the shear instability is not active. The question can be raised whether the gradient of the angular velocity has ever reached, during stellar evolution, such a high value.

2.2.2 *GSF instability.* (Goldreich and Schubert, 1967; Fricke, 1968). It is possible to derive from the results of Kippenhahn et al. (1980) the turbulent diffusion coefficient due to GSF instability:

$$D_{GSF} = \frac{Lr^6\Omega^2}{G^2M^3}\frac{\nabla_{ad}}{\nabla_{rad}}\frac{H_p}{H}\frac{<\rho>}{\rho}\frac{1}{\Delta\nabla} \tag{3}$$

with

$$H = min(H_\Omega, H_\theta) \tag{4}$$

where H_Ω is the scale height of the angular velocity parallel to the axis of rotation and H_θ the scale, perpendicularly to the axis of rotation, of the specific angular momentum. $\Delta\nabla$, as usual, is the difference between the adiabatic and radiative logarithmic gradients (dlnT/dlnP) of the temperature. The efficiency of the turbulent flow generated by the GSF instability depends entirely on the scale H. In the present Sun, except perhaps in the boundary region below the convective zone, the scale H is large, of the order of several solar radii, and the Goldreich-Schubert-Fricke instability dominates only at great depths. One has also to compare the GSF turbulent flow to the turbulent flow induced by meridional circulation (see next paragraph). The turbulent flow induced by GSF instability is larger than the meridional flow, $D_{GSF} > D_Z$, if the logarithmic gradient is larger than 10 at the bottom of the convective zone; this is much larger than the observed value. With a solar model (Lebreton, 1990), the two turbulent diffusion coefficients become equal for $r \cong 4.10^9$ cm.

2.2.3 *Generation of the 3-D turbulence by the meridional circulation.* Zahn (1983, 1984) gives for the turbulent diffusion coefficient the average value on the sphere:

$$D_Z = \frac{L\Omega^2 r^6}{G^2 M^3} \frac{< \rho >}{\rho} \frac{1}{\Delta\nabla} \tag{5}$$

Let us first notice that some uncertainty remains on the value of the numerical coefficient in equation (4). When taking into account the exact value of the Rossby number for which the $2-D$ turbulence decays into a $3-D$ turbulence (Hopfinger et al., 1982, quoted by Zahn, 1983) it is necessary to introduce a factor $(0.4)^{-2} = 6.25$. Schatzman and Baglin (1990) introduce reference models where a turbulent diffusion coefficient $D = 6.25 D_Z$ has been used. In the discussion, Schatzman and Baglin (1990) introduce an efficiency parameter f, and write $D = f D_Z$.

At this point we come up with the fact that it is necessary to check whether the Richardson- Townsend condition is fulfilled when describing the generation of the $3-D$ turbulence by the $2-D$ turbulence on equipotentials. Zahn gives an estimate of the energy available for the feeding of the $3-D$ turbulence,

$$\epsilon_t = \eta D_{th} \Omega^2 \tag{6}$$

where η is the efficiency parameter,

$$\eta = \frac{\Omega^2 r}{g} \Delta\nabla^{-1} \tag{7}$$

The energy ϵ_t must exceed the work which is necessary to carry the turbulent cell vertically over the mixing distance l with a velocity ν. If there were no radiative exchange of heat with the surrounding medium, the work done per unit of mass and per second would be:

$$W = g\, l\, \nu\, \Delta\nabla\, (1/H_p) \tag{8}$$

where the product $l\nu$ will be considered later as representing the diffusion coefficient. The actual work W_{actual} to be supplied is obtained multiplying W by the ratio of the cooling time $t_{cool} = (l^2/D_{th})$ to the characteristic time of the turbulence $(1/\nu)$, which is the way of deriving the flux Richardson number (Townsend, 1958). Schatzman and

Baglin obtain the condition,

$$\frac{L}{4\pi G M_r \Delta\nabla} \frac{\nabla_{ad}}{\nabla_{rad}} \frac{\Omega^2 H_p^{1/2} r^{5/2}}{G M_r \Delta\nabla} > l\nu \tag{9}$$

The turbulent diffusion coefficient $l\nu$ is naturally identified to D_Z. Near the bottom of the convective zone of the present Sun, this gives $D_{turb} < 250$. Writing explicitly the expression of D_Z this gives the condition:

$$\left(\frac{H_p}{r}\right) > 1 \tag{10}$$

where the number 1 on the right hand side of the inequality is just an order of magnitude. Physically, it can be said that the power generating the turbulence is proportional to $(1/r)$, whereas the power necessary to overcome the gravity, which depends on the difference $(\nabla_{rad}\rho - \nabla_{ad}\rho)$, is proportional to $(1/H_p)$. The exact value depends (i) on the efficiency factor which can be adjusted to give the required turbulent coefficient $D = fD_Z$ and (ii) on the model for estimating the time-scale of the heat exchange by a turbulent element of size l by radiative diffusivity. The condition (10) is fulfilled inside the Sun, for $r < 0.11R_\odot$. In other words, in the presence of a density gradient, it is possible, that there is no turbulent flow generated by meridional circulation, except in regions close to the center of the star.

2.3 LITHIUM DEPLETION IN THE HYADES LOW MASS STARS

We are concerned here with stars of 1.1, 1.0, and 0.9 $M_\odot$. For these stars, the abundance of lithium decreases very regularly with mass. The difficulty of describing in a consistent manner the lithium deficiency of these stars and of the Sun is already mentioned by Baglin et al. (1985) and by Baglin and Morel (1990). Baglin et al. (1985) show that the calibration of the physical constants lead to a different result for the Hyades and for the Sun. Baglin and Morel (1990) show that the model computed with the diffusion coefficient of Zahn gives a lithium abundance which increases with mass instead of the important decrease which is observed (Figure 2).
The diffusion coefficient depends on the square of the angular velocity. Due to the spin-down effect the angular velocity decreases with time, and the diffusion coefficient is time dependant. Schatzman and Baglin (1990) have used the relation due to Schatzman (1989) between angular velocity and time,

$$\Omega = \Omega_0(1 + (t/t_0))^{(-3/4)}$$

The hope of explaining both the deficiency behaviour in the Hyades, and the solar deficiency by taking account of the time dependence of the diffusion coefficient turned out to vanish when carrying a more careful analysis of the problem (Schatzman and Baglin, 1990).

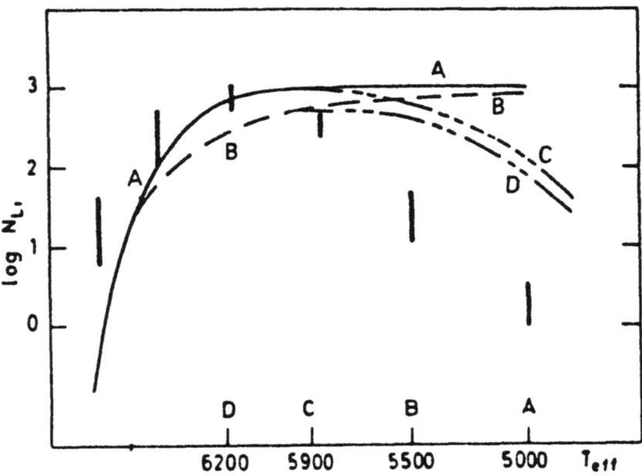

Figure 2: Lithium depletion in the Hyades (from Baglin and Lebreton, 1990). The observed values of lithium abundance are shown by the thick vertical lines. Theoretical values obtained with the diffusion coefficient proposed by Zahn (efficiency parameter : $f = 0.8$) are given by the curves AA and BB corresponding to extreme values of rotation velocity; the effect of overshooting over a depth of 0.7 H_p are given by the curves C and D.

This can be easily explained by looking at the expression (5) of the diffusion coefficient D_Z. The stellar radius decreases with mass and the bottom of the convective zone moves inwards for decreasing masses. Therefore the effect of the r^6 term is to decrease appreciably the diffusion coefficient. The mass of the convective zone increases for decreasing masses. Consequently, for decreasing masses, the lithium burning has to take place with a greater amount of lithium stored in the convective zone and a smaller diffusion coefficient. The fact that the bottom of the convective zone is nearer to the lithium burning region is not sufficient, and the efficiency of the diffusion mechanism decreases when going from 1.1 solar mass star to 0.9 solar mass star. The abundance of lithium, instead of decreasing with the stellar masses, increases. In order to compensate the effect of the diffusion coefficient and of the size of the convective zone by the time dependence of the diffusion coefficient it is necessary to introduce arbitrary values of the parameters: f efficiency parameter of the Zahn mechanism, V_0 initial velocity of rotation on the main sequence, t_0 characteristic time scale of the spin-down. These results are completely contradictory with the observed rate of spin-down, and with the observed dispersion of the initial velocities of rotation.

These results have been obtained assuming a quasi-solid body rotation below the convective zone. Compared to the situation a couple of years ago, we can say that the assumption of solid body rotation is supported by the results of helioseismology (Brown et al., 1989, Thomson, 1990). We can conclude this section in the following way: (i) from the velocity dispersion, it appears that there is no simple connection between lithium depletion and rotation. Let us just notice here that the case of young lithium rich fast rotators deserves special consideration; (ii) the role of turbulent flow generated by rotation has been overestimated, and this appears both in the analysis of the data on lithium depletion, and in the discussion of the validity of the theory of generation of turbulent flows.

3 Internal waves

3.1 NEED FOR ANOTHER MIXING PROCESS

The difficulty of explaining lithium depletion by turbulent diffusion mixing induced by rotation, leads to the consideration of another physical process and it appears that this has to be connected with other transport problems. The problem of transport and redistribution of angular momentum and the spin-down problem have taken a new aspect with recent results of helioseismology concerning the interpretation of the pressure modes splitting (Brown et al., 1989; Thomson, 1990). At the equator, where is the major fraction of the stellar angular momentum, the angular velocity ω in the radiative zone is smaller than the angular velocity ω_E in the convective zone, and decreases with the radius down to about 0.4 $R_\odot$. Inside the radiative zone, the angular velocity seems to be very little dependant on the latitude. The decrease of ω when going inside the Sun is contrary to what would be expected if angular momentum was carried outwards by a diffusive process only (Endal and Sofia, 1978; Schatzman, 1987; Pisonneault et al., 1989): these properties of the angular velocity cannot be explained by a flux of angular momentum proportional to the gradient of the angular velocity. In order to lift this contradiction, it seems necessary to take into account another physical process: the transport of angular momentum by internal waves, generated by the turbulent motion in the convective zone. It turns out that transport of angular momentum by internal waves seems to be a very efficient process (Gough, 1977; Spruit, 1987; Schatzman, 1990).

Press (1981) has estimated the flux of mechanical energy carried in the radiative zone by gravity waves excited at the boundary of the convective zone and has also given an estimate of their radiative damping A(r). Goldreich and Nicholson (1989), following the results derived from the theory of the mean Lagrangian (Dewar, 1970; Grimshaw, 1986), have shown how gravity waves carry angular momentum.

Furthermore, as gravity waves generate a random walk motion of the fluid elements (Press, 1981), they can also generate a diffusive process which transports chemical elements. It is necessary to study together the transport of angular momentum and the transport of chemical contaminants, as the macroscopic diffusivity D_M, and the flux of angular momentum depend on the same parameters. In the following, we shall summarize the recent results of Schatzman (1990).

3.2 ANGULAR MOMENTUM TRANSPORT

This is obtained in two steps, first by considering monochromatic internal waves, and then by integrating over the frequency spectrum generated by the turbulent flow of the convective zone.
In this oversimplified treatment, it is assumed that the waves are produced in a system rotating like a solid body, with a circular frequency ω_E, despite the fact that the convective zone does not present solid body rotation, showing a latitude dependence on the angular velocity. However, as the largest fraction of the angular momentum is concentrated near the equator, this can be considered as an acceptable approximation: it is like dealing with a cylindrical Sun. The choice of the circular frequency ω_E will be part of the boundary conditions. The gravity waves are propagating in a radiative zone rotating with a local value of the circular frequency $\omega(r)$, and are seen with a Doppler shift $\omega_E - \omega(r)$. There is a superimposition of waves with the frequencies $+\omega_c$ and $-\omega_c$, and the matter of the radiative zone sees waves with the frequencies $\omega_c + \omega_E - \omega$ and $-\omega_c + \omega_E + \omega$. The transport of angular momentum results from the effect of the superimposition of these two kinds of waves. Calling $\sigma+$ and $\sigma-$ the frequencies of these two waves, we follow Goldreich and Nicholson (1989), showing that the mechanical energy carried by the internal waves can be written

$$L_E = K\sigma^3 \tag{11}$$

where K is a constant determined by the mechanism of production of the internal waves. The corresponding flux of angular moment, neglecting any damping effect is:

$$L_H = \frac{L_E}{\omega_E} \tag{12}$$

We need now to obtain an estimate of L_E. The flux of mechanical energy is proportional to the product of the square of the horizontal velocity of the waves, multiplied by the radial group velocity. According to Schatzman (1990), if the frequency ω_c is much larger than the Brunt-Väissälä frequency, we can write for the flux of mechanical energy:

$$L_H = 4\pi r^2 \rho u_H^2 \frac{\omega_c^2}{k_H N} \tag{13}$$

3.3 SPECTRAL EFFECTS

The turbulent flow generates a spectrum of scales k and frequencies ω_c. We assume a Kolmogoroff spectrum with a cut-off at a certain wave number k_M and at a certain frequency ω_M. As the dispersion relation implies both k_H and k_r we shall assume with Press (1981) that the horizontal wave number is determined by the large scale flow pattern of the convective zone, but that the frequency spectrum and the velocity spectrum are determined by the effect of the pressure fluctuations in the convective zone. If the large scale flow pattern of the convective zone is so important, we have to take into account that it has the geometry of $2-D$ convective cells. The waves then should reflect this geometry with horizontal wave numbers of the same order of magnitude in two horizontal directions (Garcia-Lopez and Spruit, 1990).

It is the mean square horizontal velocity which contributes to the transport of angular momentum. If we call δu_{ij} the contribution to the velocity which is due to the pressure exerted on the boundary by the turbulent vortices 1, 2, ...i, ... V, in the wave number interval dk_i, we have to consider the quantity

$$\overline{\left(\frac{1}{V}\sum_{ij}\delta u_{ij}\right)^2}$$

Due to the fact that the action of the vortices is not in phase, we must take into account the fact that over a surface k_H^{-2} there are $(k/k_H)^2$ convective cells of a scale k^{-1}. Consequently, the amplitude of the motion which is generated at the frequency ω_c must be divided by the number of convective cells to the power one half and the square of the velocity must be divided by $(k/k_H)^2$.

In expression (13) we have to replace u_H^2 by the contribution to the spectral interval dk, $u_M^2 f(k)dk$, and include the reduction factor $(k_H/k)^2$. This gives the contribution to the flux of mechanical energy (the index I will be affected to all quantities at the boundary, or "close" to the boundary of the convective zone):

$$dL_E = 4\pi r_I^2 \rho_I u_M^4 \frac{k_M}{N_I}\frac{2}{3}\left(\frac{k_M}{k}\right)^{7/3} d\left(\frac{k}{k_M}\right) \tag{14}$$

It is suggested by Press to take for u_M the velocity u of the turbulent flow in the convective zone, given by the asymptotic formula of the quasi-adiabatic flow (Cox and Giuli, 1968) and to assume that:

$$4\pi r^2 \rho u^3 = (1/10)L_{star} = \varphi L_{star} \tag{15}$$

However, it should be noticed that angular momentum is mainly carried by waves produced by the inertial range of velocities of the turbulent spectrum. The observations of the solar granulation (Zahn, 1988) show that the velocity u_M which describes the inertial range of velocities is definitely larger than the velocity u given by equation

(15).

3.4 RADIATIVE DAMPING.

It is then necessary to take into account in the propagation the effect of the damping factor. Calling $A(r)$ the square of the radiative damping factor, we shall write in a different form the result of Press (1981)

$$A(r) = exp\left\{ - \int_r^{r_1} \frac{D_{th} k_H^2}{N} \left(\frac{N}{\omega_c}\right)^4 k_H dr \right\} \tag{16}$$

where D_{th} is the thermal diffusivity.

Superimposition of waves of the same amplitude, with circular frequencies $\pm\omega_c$ gives finally the contribution of the spectral interval dk to the flux of angular momentum:

$$dL_H = -4\pi r_I^2 \rho_I v_c^3 \frac{\omega_c}{N_I} \frac{1}{2} \frac{(\omega_c + \omega_E - \omega)^3 + (\omega_c + \omega_E - \omega)^3}{\omega_c^3} \frac{A}{\omega_E} f(k) dk \tag{17}$$

where N_I is the Brunt-Väissälä frequency close to the boundary of the convective zone, (N_I goes from zero to an almost constant value over a distance of 0.01 $R_\odot$); the term

$$4\pi r^2 \rho_I v_c^3$$

is an estimate of the mechanical energy flux from the convective zone to the radiative zone, taken near its boundary, ρ_I being the density at the boundary, given in order of magnitude (equ. 15). It is clear, with this expression, that the flux of angular momentum vanishes when $\omega = \omega_E$. If the frequency difference $\omega_E - \omega$ is small compared to ω_c, we can write

$$dL_H = -4\pi r_I^2 \rho_I v_c^3 \left(\frac{k_M}{k}\right)^2 \frac{\omega_c}{N_I} 3 \frac{(\omega_E - \omega)}{\omega_c} A(r) f(k) dk \tag{18}$$

We introduce here the Kolmogoroff spectrum of velocities and frequencies in order to carry the integration. We need only to express the damping factor A, which we write

$$A = exp\left\{ - \left(\frac{\omega_M}{\omega}\right)^4 F \right\} \tag{19}$$

with

$$F = \int_r^{r_1} \frac{\nabla_{ad}}{\nabla_{rad}} \frac{1}{\varphi} \frac{N^2}{g\omega_M} \left(\frac{r_I}{r}\right)^2 dr \tag{20}$$

The integration of equation 18 gives, when F is large (and this is true very close to the bottom of the convective zone),

$$L_H = \varphi L \frac{1}{N} \frac{9}{2} \frac{\frac{3}{4}!}{F^{3/4}} \frac{\omega_E - \omega}{\omega_E} \tag{21}$$

In stationary conditions, I being the momentum of inertia, we can write for the flux of angular momentum $L_H = (I\omega_E/\tau)$, where τ is the time-scale of the spin-down. The first aim being to obtain the order of magnitude, we shall write:

$$F = \frac{1}{\varphi} \frac{\nabla_{ad}}{\nabla_{rad}} \frac{N_I^3}{g_I \omega_M} r_I ln \left(\frac{r_I}{r}\right) \tag{22}$$

We finally obtain for the angular velocity, including in the expression of the flux of angular momentum L_H, a factor (2/3) which takes into account the average distance to the axis of rotation in the evaluation of the flux of angular momentum:

$$\omega_E - \omega = \frac{I N_I \omega_E^2}{\tau} \frac{F^{3/4}}{(3/4)!} \frac{3/2}{\varphi L} \tag{23}$$

where τ is the characteristic time-scale of the solar spin-down, of the order of the age of the Sun. With the following values of the different quantities, $r_I = 5 \cdot 10^{10}$ cm, $k_H = (\pi/H_p)$ with $H_p = 5.4 \cdot 10^9$ cm, $u_M = 3.8 \cdot 10^3$ cm s^{-1}, $\omega_M = 2.2 \cdot 10^{-6}\ s^{-1}$, $N_I = 1.26 \cdot 10^{-3} s^{-1}$, $g_I = 6 \cdot 10^4\ cm\ s^{-2}$ and the moment of inertia $I = 5.88 \cdot 10^{53} g\ cm^2$, the angular velocity $\omega_E = 2.85 \cdot 10^{-6}$, and the characteristic time $\tau = 5\ Gy$ we obtain

$$(\omega_E - \omega) = 312 \cdot 10^{-9} (ln(r_I/r))^{3/4} s^{-1}$$

which gives for $(r_I/r) = (0.72/0.567)$

$$(\omega_E - \omega) = 74.5\, 10^{-9} s^{-1}\ and\ (1/2\pi)(\omega_E - \omega) = 11.9\, 10^{-9} s^{-1}$$

instead of the results given by the analysis of the helioseismological data $(1/2\pi)\delta\omega = 16.6 \cdot 10^{-9} s^{-1}$ as given by Brown et al. (1989). These two values are so close to each other that we can consider that the transport of angular momentum actually takes place by internal waves.

3.5 MIXING PROCESS

Garcia-Lopez and Spruit (1990), following Press (1981) consider the turbulence induced by the shear flow due to the change of sign of the horizontal velocity over one half vertical wavelength. This process does not seem strong enough. But another process is present. Press (1981) has shown that, due to non linear dissipative effects, the r.m.s. of the displacement of a piece of fluid is finite; this implies the existence of a diffusion coefficient, which turns out to be proportional to ϵ^4 , where ϵ is the ratio of the horizontal displacement to the horizontal wavelength, the condition for applying the linear theory being that $\epsilon < 1$.

We have now to express the diffusion coefficient D_{mix}. The diffusion coefficient can be written as the product of a length l by a velocity ν. The length is the average distance l reached by a fluid element in an irreversible process after an overturn ω^{-1}; the velocity ν is the product of the distance l by the circular frequency ω, $\nu = l\omega$. For a gravity wave, with a vertical wave number k_r and a vertical velocity u_r, Press (1981) gives as an order of magnitude

$$l = \frac{k_r u_r^2}{\omega^2} \frac{D_{th} k_r^2}{\omega} \tag{24}$$

where D_{th} is the thermal diffusivity, $(D_{th} k_r^2/\omega)$ the fraction of the entropy memory which makes the process irreversible. It is assumed again that the horizontal wave number k_H is defined by the geometry of the convection and is of the order of $k_M = (2\pi/\alpha H_p)$ where α is the analog of the mixing length parameter and H_p the pressure scale height in the neighbourhood of the boundary of the convective zone. We have now to take into account the way in which the gravity waves are generated. For this purpose, we express k_r and u_r in terms of the horizontal components k_H and u_H, and write

$$l = \frac{k_H^3 u_H^2 N D_{th}}{\omega^4}$$

$$v = \frac{k_H^3 u_H^2 N D_{th}}{\omega^3}$$

The horizontal velocity is generated by the turbulent flow, the frequency ω corresponding to a wave number k and u_H being identified with the turbulent velocity u. The summation of the contributions of the wave number intervals dk must include the distribution function $f(k)dk$ and the number (k/k_M) of cells of size k which contribute to the horizontal displacement generating the wave (ω, k_H). We must also

take into account the expression given by Press (1981) of the horizontal velocity as a function of depth.
We also write $\omega = k_M u_M (k/k_M)^{2/3}$ and obtain, after integration over the Kolmogoroff spectrum (I is for the boundary of the convective zone) the average length and velocity and the diffusion coefficient:

$$D_M = \frac{\alpha}{2\pi}\frac{4}{35}\frac{7}{8}!\frac{5}{8}!(\Delta\nabla)_I \frac{L_r}{\varphi L_I}\left(\frac{M_I}{M_r}\right)^2 \frac{L_r}{4\pi G M_I \rho_I}\left(\frac{\rho_I}{\rho_r}\right)^4 \left(\frac{r_I}{r}\right)^{12}\left(\frac{\nabla_{ad}}{\nabla_{rad}}\right)_r^2 \frac{1}{F^{\frac{3}{2}}}$$

with

$$F = \int_r^{r_I} \frac{\alpha}{2\pi}\frac{GM_I}{r_I^2} r_I \frac{1}{v_M}\frac{L_r}{\varphi L_I}\frac{M_I}{M_r}\left(\frac{\rho_I}{\rho_r}\right)\left(\frac{r_I}{r}\right)^7 \left(\frac{\mu}{RT}\right)^{1/2}\left(\frac{\nabla_{ad}}{\nabla_{rad}}\right)_r^2 (\Delta\nabla)_r^{3/2}\frac{dr}{r_I}$$

The WKBJ solution of the eigenvalue problem is used here in the following way: the local rate of lithium burning q is neglected, down to the point where it becomes equal to the eigenvalue λ of the differential equation of diffusion.
An important parameter is the velocity u_M. It has been introduced in the definition of the energy spectrum. But the Kolmogoroff spectrum is valid only in the inertial range. The main contribution to the integrals comes from a narrow peak around $(k/k_M) \cong F^{3/8}$ which falls in the inertial range. Then, the parameter u_M^2 in the Kolmogoroff spectrum should be taken as the parameter which describes the inertial range, and not as the velocity coming from the definition (15) of the energy flow.

When applied to the problem of lithium depletion in the Hyades, the conclusions are the following:
(i) the rate of lithium burning in the 1 $M_\odot$ stars is definitely smaller than the rate of lithium burning in the 0.9 $M_\odot$ stars. The boundary of the convective zone is higher in the 1 $M_\odot$ stars than in the 0.9 $M_\odot$ stars. As the diffusion coefficient D_M decreases very quickly with depth, the lithium nuclei do not reach the burning level when the boundary of the convective zone is a little too high.
(ii) in stellar evolution, the boundary of the convective zone is receding upwards. This means that the rate of lithium burning becomes very small compared to what it is during the early phases of stellar evolution. The present rate of lithium burning in the Sun is then probably very small. This is an important conclusion, as an extrapolation of the lithium abundance in the $1 M_\odot$ stars of the Hyades to the present Sun would give a very great deficiency, with [Li] = -4, instead of -3, even when including the effect of the spin-down. A drop in the rate of lithium burning some time around 1 Gy would lift the difficulty.
(iii) With $\alpha = 2$, $\varphi = 0.1$, and according to the discussion above, $v_M \cong 2.5 \cdot 10^4$, the results are the following:

$$0.9 M_\odot \; stars, \; \lambda t_{diff} = 5.04, \; (\lambda t)_{observed} = 7$$

$$1.0M_\odot \ stars, \ \lambda t_{diff} = 1.26, \ (\lambda t)_{observed} = 3.1$$

These theoretical values are a little too small. However, the difference is not large, and it must be noticed that this result is based on a simplified solution of the eigenvalue problem and that the eigenvalue itself is very sensitive to the model. The time-scale of lithium depletion,λ^{-1}, varies approximately like $(\delta r)^{7/2}$, δr being the distance from the bottom of the convective zone. The consequence is that a small change of the distance of the bottom of the convective zone to the level of lithium burning can produce a relatively large change of the eigenvalue λ.

Notice also the difference between the velocity u_M which has been introduced in order to describe the inertial range of the Kolmogoroff spectrum, and the velocity u which is just a measure of the energy flux in the convective zone. If u_M instead of u is carried into the expression which gives the angular velocity inside the Sun, the difference $(\omega_E - \omega)$ is about four times smaller.

4 The lithium rich fast rotators

4.1 SPIN-DOWN

As already mentioned in the introduction, the lithium rich fast rotators raise an interesting problem. If we give up the idea of turbulent mixing induced by rotation, we nevertheless have to find the reason for which slow rotators in young open clusters are lithium poor. Furthermore, it has been mentioned in this meeting that fast rotators spin down slowly. It has been suggested that they are more spotty than slow rotators, and that the surface area of open lines of force being smaller, the mechanism of loss of angular momentum is less efficient. These assumptions can be approached quantitatively, if we take the estimates of Schatzman (1990) of the degree w (a dimensionless quantity) of the magnetic structure at the surface of a rotating star,

$$w \cong 0.15 \left(\omega \frac{\alpha H_p R}{\eta_{turb}} \right)^{2/3} \tag{25}$$

In the solar case, with $\alpha = 2$, this gives $w \cong 2.4$, or, expressed in degrees, $(\pi/2w) \cong 40°$, which is a reasonable value. For the fast rotators of αPer, this corresponds to an angle of about one degree. It is then quite possible that open magnetic field lines cover a small fraction of the disc. Let us call β the fraction of the solar surface covered by open magnetic field lines. The outgoing flux carried by these lines will be βB. In the differential equation describing the loss of angular momentum,

$$\frac{d}{dt} I_\omega = \frac{2}{3} \left(B * R^2\right)^2 \left(\frac{GM}{R}\right)^{-1/2} \omega \tag{26}$$

it is necessary to replace the mean square value of the magnetic field $(B*)^2$ by $(\beta B*)^2$, where β can be a small quantity if the angular velocity is large. This picture seems to fit qualitatively with the observations. When the angular velocity has already decreased, β takes a finite value, and we are back to the case described by Schatzman (1990a, b). The asymptotic value of ω does not depend on the initial conditions, if the time scale of the spin-down is short compared to the stellar age.

The picture could be the following. Some stars reach the main sequence with a high equatorial velocity. Due to the structure of their magnetic field, they experience a first phase of slow spin-down. Some other stars reach the main sequence with a smaller equatorial velocity, the spin-down process is immediately more efficient and, after 50 million years, they are on the asymptotic branch of spin-down.

4.2 LITHIUM

This picture of the evolution of stellar rotation in young clusters does not provide easily an explanation of the abundances of lithium. If the lithium burning does not depend on the velocity of rotation, all stars with the same spectral type and the same age should present the same abundance of lithium. The rate of lithium burning which has been obtained for the Hyades leads, for the Pleiades and for αPer, to a small depletion, except for the very late spectral types. It should be noticed that the depth of the convective zone is greater at the equator than for slow rotators. Lithium burning could take place faster near the equator and then the whole convective zone would be depleted faster. But this process should be inhibited for very fast rotators (the lithium rich stars). We can mention the possibility that for very fast rotators having a very strong magnetic field at the bottom of the convective zone, the internal waves are reflected by the magnetic field and do not enter in the radiative zone. This suggestion does not have to be taken too seriously, as it is not supported yet by any calculation.

5 Conclusions

We claim here that for low mass stars (around 1 solar mass), lithium depletion is governed by a transport process due to internal waves and not to a turbulent flow induced by rotation. Turbulent flow can overtake internal waves at great depths (about 0.4 $R_\odot$ in the case of the Sun), when the amplitude of internal waves, due to radiative damping, has decreased considerably. The place where this happens, with our present knowledge of the instability conditions, is not known exactly. As far as lithium burning is concerned, the present results are encouraging. Internal waves explain also the quasi-solid body rotation inside the Sun and the positive gradient of the angular velocity, in the solar equatorial plane.

Generalization of these results to fast rotators need a careful investigation. The properties of the non-linear dynamo suggest the presence, at the surface of fast rotators, of a very complex magnetic field, with a very small scale structure. The consequence is that a very small fraction of the magnetic field lines are open to space, with the result of a very low efficiency of the spin-down process. This can explain the presence of fast rotators in young clusters, and avoid the assumption of the presence of an accreting disc. But in the meantime, this does not lift the difficulty concerning lithium. If lithium burning is not dependant on rotation, this explains the presence of lithium in young stars, whatever their velocity of rotation is. But this is contradictory with the existence of lithium poor slow rotators in young clusters. We do not know which other effect, possibly dependant on rotation this time, has to be introduced. The quasi-discontinuity between slow and fast rotators suggests the existence of a sharp condition of existence of the mixing process, related to the velocity of rotation. No result has been obtained yet in this domain.

Acknowledgements. I would like to express my thanks to A. Baglin, R. Garcia-Lopez, Y. Lebreton (who gave me a useful solar model), H. Spruit, J.P. Zahn for lively and useful discussions on this problem.

References.

Baglin A., Morel P., Schatzman E., 1985, *Astron. Astrophys.*, **149**, 30
Baglin A., Lebreton Y., 1990, in *Inside the Sun*, G. Berthomieu and M. Cribier Eds., Kluywer Netherland, p.437
Baglin P., Morel P., 1990, in *Rotation and Mixing in stellar Interiors*. M.J. Goupil and J.P. Zahn Eds., Lecture Notes in Physics, N° 366 Springer-Verlag, p. 167
Balachandran S., 1990, *Astrophys. J.*, **354**, 310
Boesgaard A.M., and Tripico M.J., 1986, *Astrophys. J.*, 302, 249
Bouvier J., 1990, in *Rotation and Mixing in Stellar Interiors*, M.J. Goupil and J.P. Zahn Eds., Lecture Notes in Physic, N° 366, Springer-Verlag, p. 47
Boesgaard A.M., and Tripico M.J., 1986 *Astrophys. J.*, 303, 724
Brown T.M., Christensen-Dalsgaard J., Dziembowski W.A., Goode P., Gough D.O., Morrow C.A., *Astrophys. J.*, **343**, 526
Cayrel de Strobel G., 1990, in *Rotation and Mixing in Stellar Interiors*, M.J. Goupil and J.P. Zahn Eds., Lecture Notes in Physics, N° 366, Springer-Verlag, p.29
Charbonneau P., and Michaud G., 1988, *Astrophys. J.*, **327**, 809
Charbonneau P., and Michaud G., 1988, *Astrophys. J.*, **334**, 746
Charbonneau P., and Michaud G., 1990, *Astrophys. J.*, **352**, 681
Christensen-Dalsgaard J., 1990, *Reviews in Modern Astronomy*, **3**, 313
Cox ., Giuli R., 1968, *Principles of stellar evolution*, Gordon and Breach Publishers, New-York, chapter 14
Dewar R.L., 1970, *Phys. Fluids.*, 13, 2710
Endal A.S., and Sofia S., 1978, *Astrophys. J.*, **220**, 279
Fricke K.J., 1968, *Zeits. f. Astrophys.*, **68**, 317
Garcia-Lopez R., and Spruit H., 1990, (submitted to *Astrophys. J.*)

Goldreich P. and Nicholson P., 1989 a, *Astrophys. J.*, 342, 1075
Goldreich P. and Nicholson P., 1989 b, *Astrophys. J.*, 342, 1079
Goldreich P. and Schubert G., 1967, *Astrophys. J.*, 150, 571
Gough D., 1977, in *Proc. IAU Coll.* **36**, R. Bonnet and P. Delache Eds., p.3
Grimshaw R., 1984, *Ann. Rev. Fluid. Dynamics*, 16, 11
Hopfinger E.J., Browand F.K., and Gagne Y., 1982, *J. Fluid Mech.*, 125, 505
Kippenhahn R., and Thomas H.C., 1980, in *Fundamental Problems in the Theory of Stellar Evolution*, IAU Symposium No 93, D. Sugimoto, D.Q. Lamb, D.N. Schramm Eds., Reidel, Dordrecht, p.237
Lebreton Y., 1990, personal communication
Pinsonneault M., Kawaler S., Sofia S., and Demarque P., 1989, *Astrophys. J.*, **338**, 424
Press W.H., 1981, *Astrophys. J.*, **245**, 286
Scalo J.M., and Miller G.E., 1980, *Astrophys. J.*, **239**, 953
Schatzman E., 1969, Astron. Astrophys., 3, 331
Schatzman E., 1977, Astron. Astrophys., 56, 211
Schatzman E., 1984, in *Observational Tests of Stellar Evolution Theory*, IAU Symposium 105, A. Maeder and A. Renzini Eds., Reidel Publisher, Dordrecht, p. 491
Schatzman E., 1987, in *The Internal Solar Velocity*, B.R. Durney and S.Sofia, Eds., Reidel Publ. Co., Dordrecht, p. 159
Schatzman E. 1989, in *Turbulence and non-linear dynamics in magneto- hydrodynamic flows*, International workshop, July 4-9,1988, Cargèse (Corsica), France, M.Meneguzzi, A.Pouquet, P.L.Sulem Eds. North Holland, p.1
Schatzman E., 1990, in *Inside the Sun*, G. Berthomieu and M. Cribier Eds., Kluwer Publ. Co., Dordrecht, p. 5
Schatzman E., 1990, in *Rotation and Mixing in Stellar Interiors*, M.J.Goupil and J.P.Zahn Eds., Lecture Notes in Physics N° 366, p. 3
Schatzman E., and Baglin A., 1990, *Astron. Astrophys.*, (submitted to publication)
Spruit H., 1987, in *The Internal Angular Velocity of the Sun*, B.R. Durney and S.Sofia Eds., Reidel Publishing Company, Dordrecht-Holland, p. 185
Spruit H., 1990, in *Inside the Sun*, G. Berthomieu and M. Cribier Eds, Kluwer Publ. Co. Dordrecht, p. 415
Tassoul J.L., Tassoul M., 1982, *Astrophys. J. Suppl.*, **49**, 317
Tassoul J.L., Tassoul M., 1986, *Astrophys. J.*, **310**,786
Tassoul J.L., Tassoul M., 1989, *Astron. Astrophys.*, **213**, 397
Tassoul M., Tassoul J.L., 1983, *Astrophys. J.*, **271**, 315
Tassoul M., Tassoul J.L., 1984a, *Astrophys. J.*, **279**, 384
Tassoul M., Tassoul J.L., 1984b, *Astrophys. J.*, **286**, 350
Thomson M.F., 1990, *Solar Physics*, **125**, 1
Townsend A.A., 1958, *J. Fluid Mech.*, **4**, 39
Zahn J.P., 1974, in *Stellar instability and Rotation*, IAU Symposium N° 59, P. Ledoux, A. Noels and A.W. Rodgers Eds., Reidel Publ. Co., Dordrecht, p.18
Zahn J.P., 1983 in *Astrophysical Processes in Upper Main Sequence Stars*, 13th Advanced Course, Swiss Society of Astronomy and Astrophysics, Saas-Fee 1983, B.Hauck and A.Maeder Eds., p.253
Zahn J.P., 1984, in *Insabilités hydrodynamiques et applications Astrophysiques*, A. Baglin Ed., Société Française des Spécialistes d'Astronomie, p.389

DISCUSSION

Roxburgh: If you appeal to mixing by weak turbulence generated by internal gravity waves, that is in some sense the same as saying that there is a small amount of convective overshooting, isn't it?

Schatzman: I don't think so, because here these internal waves are gravity waves which are evanescent in the convective zone and propagate in the radiative zone. We still have to solve exactly the way in which the motions in the convective zone can generate motions in the radiative zone. And that has not been done.

Roxburgh: I understand. My point was just that convective over-shooting is a complicated thing, and I am not sure one can differentiate that from turbulence generated by gravity waves.

Geroyannis: You have shown a figure from a paper by Brown et al. which indicates that the solar rotation decreases beneath the convective zone. In your opinion, does this represent a general behavior that continues all the way into the center?

Schatzman: No. If you want to understand what is going on in the center, you have to solve completely the parabolic equation which describes the transfer of angular momentum and then you should include not only the internal wave transport of angular momentum which becomes very small around 0.4 solar radii, and then another process must take over - which must be a diffusive process. So, inside 0.4 solar radii, you may well have the angular rotation rate increasing towards the center.

Roxburgh: I understood the observations to be consistent with the statement that differential rotation disappears at the base of the convective envelope, and from there on in it is uniform. Is that not right?

Schatzman: I think there is probably a discontinuity in the angular rotation rate, but in the present situation of the theory, I cannot give the order of magnitude of the discontinuity.

Roxburgh: But, just from the observations, I believe they are consistent with a uniformly rotating core and a transition to differential rotation in the convective envelope.

ROTATION, CHROMOSPHERIC ACTIVITY, AND LITHIUM ABUNDANCES IN G AND K DWARFS OF THE PLEIADES

David R. Soderblom and J. Dan Hudon
Space Telescope Science Institute
3700 San Martin Drive
Baltimore MD 21218 USA

Burton F. Jones
Lick Observatory
University of California
Santa Cruz CA 95064 USA

John R. Stauffer
Harvard-Smithsonian Center for Astrophysics
60 Garden Street
Cambridge MA 02138 USA

ABSTRACT. The potential influence of rotation on lithium depletion is examined through a study of lithium in G and K dwarfs of the Pleiades. Both lithium abundance and rotation show an appreciable spread at any one color in that cluster, but excess Li is not well enough correlated with excess rotation for there to be a causal relationship. However, excess Li correlates very well with excess chromospheric activity and this is probably due in part to line formation conditions being altered by the activity. We also illustrate the distribution of $v \sin i$ for these stars.

1. *LITHIUM IN THE PLEIADES*

The study of the lithium abundances of stars like the Sun offers the promise of understanding processes related to surface convection for, we believe, it is only convection that can carry the surface Li to a depth where it can be destroyed in nuclear reactions. That depth is a little below the bottom of the Sun's convective zone (CZ), at a temperature of about 2.4 MK. Young solar-type stars have more Li than older stars, and the lower the mass the less Li at a given age.

Aside from those basic facts we are not confident of much about Li. But recent improvements in instrumentation allow us to explore the Li issue among much fainter stars than before, and at higher levels of precision. One of the problems worth looking at in more detail is the spread in Li seen among Pleiades G dwarfs by Duncan and Jones (1983). Some skepticism of their result is justified because it was based in part on data from photographic spectra, which can be unreliable. One of their conclusions was that there is probably an age spread present among stars of the Pleiades, in order to account for the Li spread. Further evidence for an age spread was presented by Butler *et al.*

S. Catalano and J. R. Stauffer (eds.), Angular Momentum Evolution of Young Stars, 243–249.

(1987), who showed a connection between excess rotation (the phenomenon of ultra-fast rotation) and excess Li.

To examine this problem further we have obtained high-resolution spectra of more than 100 Pleiades G and K dwarfs with the Hamilton spectrograph with a TI CCD, fed by the Lick Observatory 3m telescope. Details of these observations will be reported elsewhere; here we present some preliminary results that are pertinent to this meeting.

2. *LITHIUM AND ROTATION*

Do the findings of Butler *et al.* hold up in a larger sample? Figure 1 shows the distribution of observed equivalent width (W_λ) with color for the stars we observed. The colors are those observed, uncorrected for reddening, except that a few stars with anomalous reddening have had their colors shifted blueward to a value that would be seen if they experienced the same reddening as the rest of the cluster. This anomalous reddening is a problem for only a few stars in the southwest part of the cluster. We have presented the data in observational coordinates (as opposed to abundance versus temperature) because the effects of errors are much easier to understand and the global conclusions much easier to reach.

In Figure 1 different symbols have been used to denote different degrees of rotation. Two conclusions are obvious: First, there really is an intrinsic spread in Li at any given color in the Pleiades, a spread that cannot be removed by postulating, for example, any plausible degree of non-uniform reddening. The spread can be seen to grow with decreasing mass too. For the cool stars with strong Li lines, the 6708Å feature is saturated, so that the actual spread in abundance greatly exceeds the order-of-magnitude spread in W_λ that is seen. The second conclusion to be drawn from Figure 1 is that excess Li is correlated with excess rotation, but not in a very strict, one-to-one fashion. It is possible, of course, that some of the Li-rich stars with small $v \sin i$ values just happen to be seen pole-on, but there are too many of them for that to be plausible. An even more convincing argument is that Butler *et al.* themselves found two stars at $(B - V) \approx 1.2$ with $W_\lambda \lesssim 40$ mÅ, at the lower bound of the distribution shown. Those two stars are rapid rotators yet they have little or no Li.

Thus the connection between excess Li and excess rotation seems too weak to argue for a causal relationship. What about other quantities? The Pleiades G and K dwarfs also exhibit spreads in indices of chromospheric emission (CE). One such index can be formed by measuring the degree of filling in of the Hα line compared to chromospherically-weak field stars of similar color (some Pleiads even show overt Hα emission). Figure 2 shows the same data as Figure 1 except that degrees of CE are highlighted. In this case the correlation between excess Hα emission and excess Li is virtually perfect and is so strong that some explanation is called for – it cannot be merely coincidence.

Figures 1 and 2 taken together also imply that the vaunted rotation-activity relation breaks down in the Pleiades. This is true. All the Pleiades dwarfs have high levels of CE and most also rotate rapidly, but the correlation between the two is not strong on a star-by-star basis.

3. *LITHIUM AND CHROMOSPHERIC ACTIVITY*

What can give rise to the connection that Figure 2 illustrates? We think we understand the evolution of rotation in broad terms, especially since we believe that after reaching the Zero-Age Main Sequence stars only lose angular momentum, not gain it. Thus for a given mass a rotation sequence is an age sequence, unless some phenom-

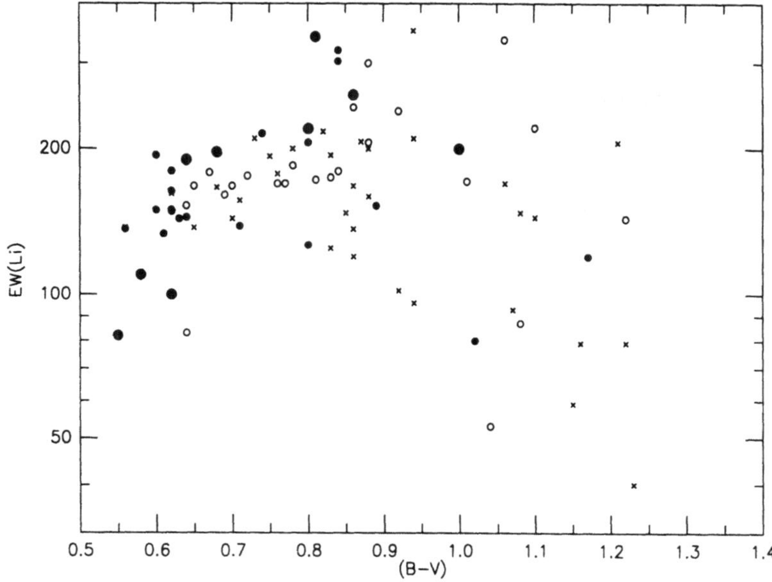

Figure 1. Equivalent width of the lithium 6708Å feature (in mÅ) versus observed $(B-V)$ color for G and K dwarfs of the Pleiades. The symbols indicate the relative degree of rotation for the stars, from large solid points for the most rapidly rotating, through smaller solid points, open circles, and then crosses for the slowest rotators. Note the presence of a number of stars with relatively large $v \sin i$ but low Li and the greater number of stars with small $v \sin i$ but strong Li.

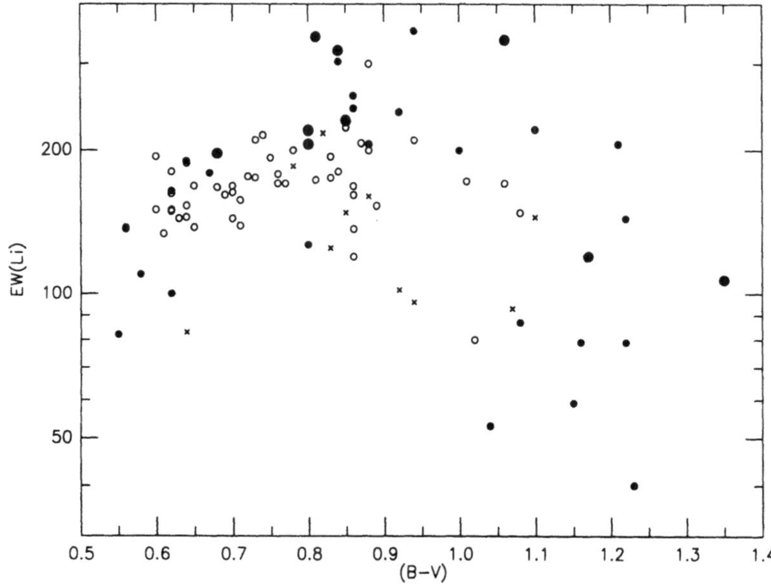

Figure 2. Same data as Figure 1, except that the points indicate the relative degree of filling in of the Hα line, in the same sense as for Figure 1.

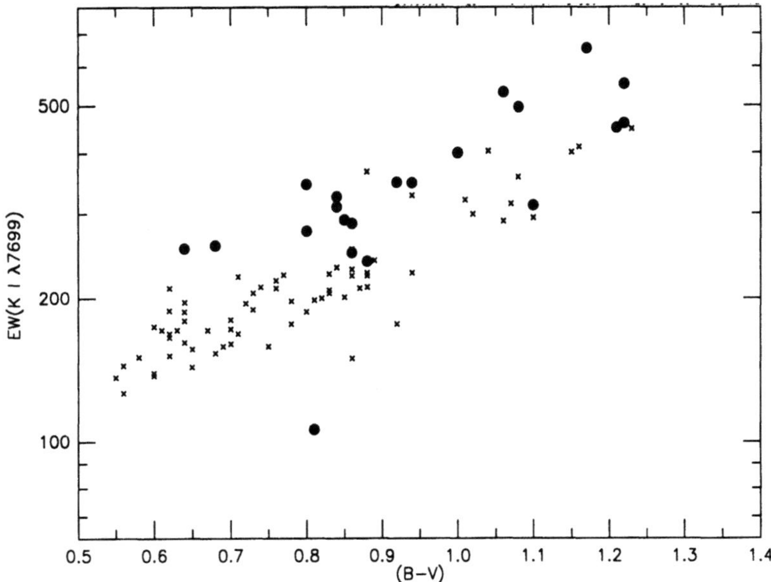

Figure 3. Equivalent width of the K I resonance line at 7699Å versus $(B - V)$ color for Pleiades stars. The solid points are the stars with the strongest Hα filling in.

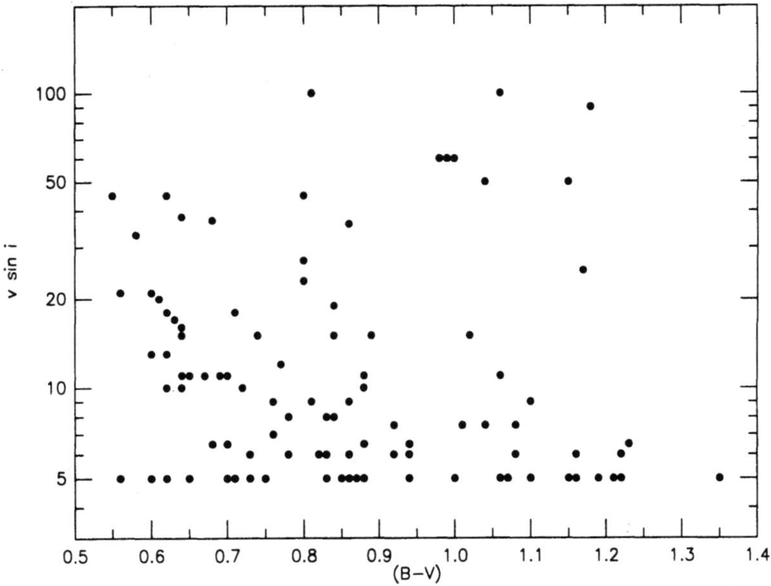

Figure 4. The distribution of rotation for the Pleiades stars we observed. Note the large number of stars with $v \sin i \lesssim 7$ km s^{-1}.

ena very new to us are going on. But the connection between rotation and CE may be subtler, especially in very young stars like these where the dynamo may be non-linear.

Is there a plausible link between CE and Li depletion? None comes to mind, but we should remember that the most Li-rich stars have Li lines that are quite strong. Much of those lines is formed in the highest layers of the star's atmosphere, the same layers that give rise to CE when modified slightly. Might CE be influencing line formation conditions in these stars? As a test of that hypothesis we have examined the strength of the K I feature at 7699 Å. This is also a resonance line that is formed under conditions similar to those of the Li doublet.

The result is shown in Figure 3. There indeed appears to be a tendency for the Hα excess stars to have stronger-than-average K I lines. The effect is only about 40% in W_λ, however. That is much much less than the spread seen in W_λ(Li) and suggests that the influence of CE may offer a partial explanation for the spread seen in Li, but is probably not enough to eliminate fully the Li spread. As this is written, observations are in progress to observe the secondary line of Li at 6103Å in several of these stars in the hopes of obtaining more accurate Li abundances.

4. THE DISTRIBUTION OF ROTATION IN THE PLEIADES

The resolving power of the Hamilton spectrograph has allowed us to determine $v \sin i$ for these stars down to the 6 km s^{-1} level. The result is shown in Figure 4. Note that this figure does not include other observations (such as some of the rapid rotators) that fall in this color range. These $v \sin i$ values were determined by cross-correlation.

Note that there is no such thing as a characteristic $v \sin i$ for a star of the Pleiades; there is just too great a range in rotation at any color. Note also the great abundance of slow rotators at all colors. This is a potential problem for stars near 1 $\mathcal{M}_\odot$ because such stars in the Hyades appear to have nearly uniform $v \sin i$ values of about 7 km s^{-1}. There appears to be too many Pleiads with $v \sin i$ less than that to allow for one distribution to evolve into the other whether one allows angular momentum loss or not.

One way out of this dilemma is to note that we have explained the rapid decline of surface rotation in Pleiades dwarfs (as indicated by the presence of both ultra-fast and slow rotators in the cluster) by postulating that the convective zone decouples from the radiative core. This allows a body with a significantly lower moment of inertia to be braked through magnetic torques. If that is correct then the angular momentum in the core must reemerge at some later date. Perhaps we are witnessing that reemergence as stars go from the age of the Pleiades to the age of the Hyades. Clearly observations of an intermediate-age cluster may be vital for understanding this.

REFERENCES

Butler, R.P., Cohen, R.D., Duncan, D.K., and Marcy, G.W. 1987, *Astrophys. J. 319*, L19.
Duncan, D.K., and Jones, B.F. 1983, *Astrophys. J. 271*, 663.

DISCUSSION

Hartmann: Were all of your chromospheric indicators absorption lines?

Soderblom: Mostly, especially for the IR triplet. The very reddest star in the sample had the IR triplet above the continuum. For H alpha, a number of the stars have overt emission, as blue as about B-V = 0.8.

Hartmann: The reason why I ask is that perhaps part of the reason you don't see as good a correlation is that if you start with a star which has no chromosphere, and then look at H alpha, first you increase the absorption equivalent width as you "turn up" the chromosphere, and then as you enhance the chromosphere further, you eventually get H alpha filling in and then going into emission. So, there is not a linear correlation between chromospheric "strength" and H alpha equivalent width.

Soderblom: At what level does that happen? For example, we have observations of Praesepe, where you also see filling in of H alpha though at more modest levels. So, where does the line start growing?

Hartmann: You have to have a detailed model - you could look at Cram and Mullan. Basically, there is a wide range chromospheric column density over which this non-linear behavior occurs.

Soderblom: That model really only applies for very red stars doesn't it?

Hartmann: No, I think something similar happens for K dwarfs too.

Bertout: We looked at the lithium abundance of about 30 stars in Taurus-Auriga, and we also found a very large spread going from an abundance of 4 down to about 1.5. We could not connect that lithium spread to any activity variation. The lithium spread also seemed to increase for lower mass stars.

Soderblom: Did you use the 6707 lithium line for your work?

Bertout: Yes.

Strom: First, in Debbie Paget's work, her stars showing the strongest lithium 6707 were also the stars showing the strongest potassium line for that spectral type. Second, with regard to the PMS stars, oddly enough for the WTT stars where there is seemingly little problem with veiling, there is apparently a correlation with rotation as well. That is, if you take Walter's *vsini* measurements and his lithium data, at a given spectral type, those stars with the most lithium also have the highest *vsini*. And, as Claude said, the inferred abundances can be quite large - the upper envelope for the WTT's (and the veiling-corrected CTT's) imply abundances as large as 4.0. So, either there is a real problem with the atmospheres - microturbulence, structure, or whatever - or much larger lithium abundances than we feel comfortable talking about in polite company.

Soderblom: One of the issues here is what condition stars get to the main sequence in at the age of the Pleiades. If they get there already with this spread built in, then things could get very confusing.

Strom: I think your point about the role a disk might play in replenishing lithium is an interesting one and is rather complex, particularly if you are adding a significant amount of material. The models that I thought I understood as of a year ago suggest that you could get some very significant spreads depending on exactly when the disk accretion occurs.

Soderblom: Shouldn't you at least expect some consistency. That is, a correlation between lithium and rotation?

Strom: Well, the distressing thing - as you know - is that the PMS stars with and without disks appear to have the same *vsini* distribution, and so if there is material hitting the surface it doesn't seem to be affecting the rotational velocities.

Bouvier: Just a further comment on PMS lithium. There is absolutely no evidence of any difference in the lithium abundance between the WTT's and the CTT's. So, this does not suggest that accretion from a disk is playing a role in lithium abundances.

Pinsonneault: Your Pleiades lithium data can be explained by some of the mixing models which I will describe later in this meeting, and you don't necessarily expect a correlation with the observed rotational velocity.

Soderblom: You will have to convince me.

Vaiana : Could you explain all of the lithium spread in the Pleiades as just due to an age spread?

Soderblom: No, I think the lithium spread is too large for that.

LITHIUM ABUNDANCE AND ROTATION IN SOUTHERN CHROMOSPHERICALLY ACTIVE STARS

S. RANDICH[1] and R. PALLAVICINI[2]
[1] *Department of Astronomy, University of Florence*
[2] *Arcetri Astrophysical Observatory*
Largo E. Fermi 5
50125 Firenze
Italy

ABSTRACT. We have carried out an extensive search for the Lithium I 6708 Å line in southern stars with active chromospheres. We found an anomalously high Li abundance for a large number of stars in the sample, *including many cool binaries of the RS CVn type*. These evolved stars are expected to have already depleted their Li while on the main-sequence and no appreciable Li should be detectable at spectral types later than ≈ G8. We have considered various possibilities to explain the observed Li excess in the cooler stars of the sample: 1) some stars may be young rather than post-main sequence objects; 2) some RS CVn binaries may have evolved from A-type progenitors with shallow convective zones; 3) tidally-enforced rotation in close binaries may have prevented an efficient depletion of Li by differential rotation mixing. In order to distinguish between these alternatives we have investigated the dependence of Li abundance upon stellar rotation. We show that case 1) probably holds only for a few stars in the sample while cases 2) and/or 3) appear the most plausible explanations for the majority of active binaries in the sample.

1. Introduction

The progressive depletion of Lithium and the decrease of rotational velocity with both age and spectral type in cool stars are two well known observational results. Tipically, a K-type star is observed to have a very low Li abundance and a small rotational velocity. The cooler components of RS CVn binaries can hardly be defined as "typical" since they have high rotational velocities and a related high degree of chromospheric activity. Nevertheless, if the cause of Li depletion in cool stars is the transport of surface Lithium to the base of the convective zone where Li is destroyed by nuclear reactions, very low Lithium should be detectable in these stars.

In contrast to these expectations, there have been occasional reports in the literature on the presence of the Li I 6708 Å line in active stars of very late spectral type, including stars classified as RS CVn binaries (see Pallavicini *et al.* 1987 and references therein). We decided therefore to carry out a systematic survey of chromospherically active stars in the Lithium region.

2. Data sample and observations

The sample stars were selected as follows. A primary group of objects comes from the list of southern active stars of Bidelman and MacConnell (1973). To these stars we have

S. Catalano and J. R. Stauffer (eds.), Angular Momentum Evolution of Young Stars, 251–257.

added several sources from the catalogue of Strassmeier *et al.* (1988) by selecting those that could be observed from the southern hemisphere. Finally other objects were taken from the lists of southern RS CVn candidates of Weiler and Stencel (1979) and Hearnshaw (1979). The list of Bidelman and MacConnell as well as those of Weiler & Stencel and Hearnshaw are based on the appearance of the Ca II H and K lines in low-resolution objective prism spectra. Since these lists use only one indicator of chromospheric activity, they are likely to be highly heterogeneous and to include objects that may be of different physical nature and in different evolutionary states. On the contrary, the RS CVn nature of the stars from the calalogue of active binaries of Strassmeier et al. is better established.

In total the sample comprises more than 60 southern stars of spectral types G and K and luminosity classes V, IV and III. Several inactive stars of various spectral types and luminosity classes were also observed for comparison. The observations were carried out at ESO, La Silla in several observing runs from Nov '86 to April '90. In all observing seasons we used the Coudè Echelle Spectrometer (CES) fed by the 1.4m CAT telescope. The short-camera and a CCD detector were used. The nominal resolving power was R=50,000 and the S/N ratio was in all cases greater than 100. The spectral range available in the Li region is $\approx$ 50 Å and was centered at 6708 Å. The Li I unresolved doublet at 6707.81 Å is close to a Fe I line at 6704.44 Å. However, since most stars in our sample are rapid rotators (with $V sini$ greater than $\approx$ 10 Km/s), the Li line is usually blended with the Fe line at 6707.44 Å. We must be careful therefore in correcting for the contribution of the Fe I line. We have done so by comparing our spectra with those of a number of narrow-line standard stars of different spectral types and luminosity classes. The available spectral range allows also the observation of the Ca I line at 6717.69 Å and of a number of other Fe I lines that can be used to estimate the metallicity of our stars. The Ca I line is also useful for a quick comparison with the Li line in stars that have similar spectral types and different Li abundances.

Examples of the acquired spectra are showed in Figs. 1, 2 and 3. Fig. 1 shows the spectrum of a catalogued RS CVn binary (HD 83442, Sp. K2 IIIp). No Li line is visible, the only prominent feature close to that wavelength being the Fe I line at 6707.4 Å. Fig. 2 shows another RS CVn binary of similar spectral type (HD 81410=IL Hya, Sp. K1 III) which shows instead a strong Li I+Fe I blend. In several active stars in our sample, the Li I line is extremely strong, even stronger than the Ca I line at 6718 Å. An example is the K2 IIIp star HD 219025 shown in Fig. 3.

Measured equivalent widths were converted to Li abundances using the curves of growth of Pallavicini et al. (1987), which were based on the atmospheric models of Bell and Gustafsson (1975, 1976). Although Li abundances derived from curves of growth are only approximate, we felt that this approach was sufficient for a survey such as the present one. A more detailed analysis using synthetic spectra will be done at a later stage. The major uncertainty on the derived Li abundances comes for the effective temperatures which were estimated from the colour index $(B-V)$. Even if effective temperatures derived from $(B-V)$ colours may be quite uncertain (especially for binaries), our main conclusions are not significantly affected by the uncertainty on the stellar parameters. Also the correction for the contribution of the Fe I line at 6707.4 Å is not critical in most cases, since the Li line, when present, is often much stronger than the estimated contribution of the nearby Fe I line.

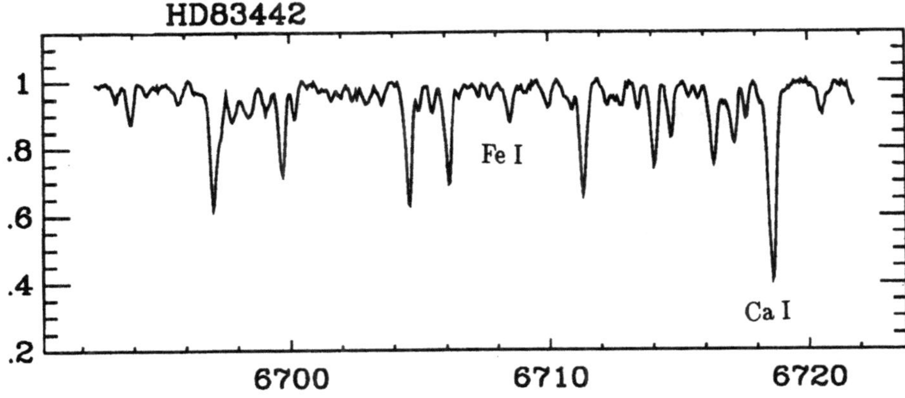

Fig.1 - CES spectrum of the RS CVn star HD 83442.

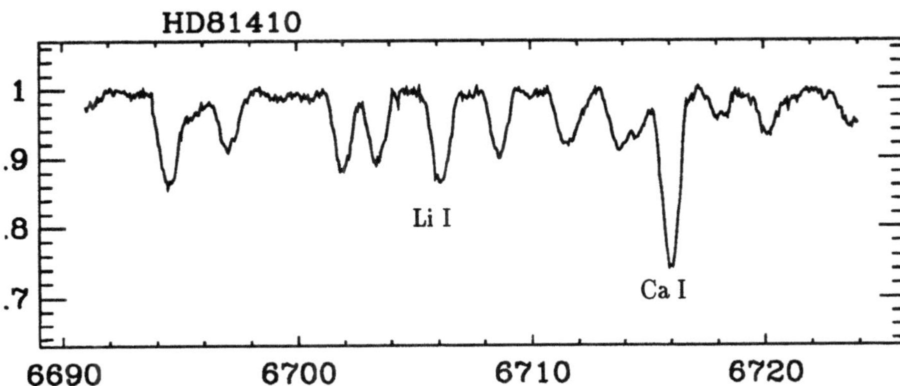

Fig.2 - CES spectrum of the RS CVn binary IL Hya. Note the strong Li line.

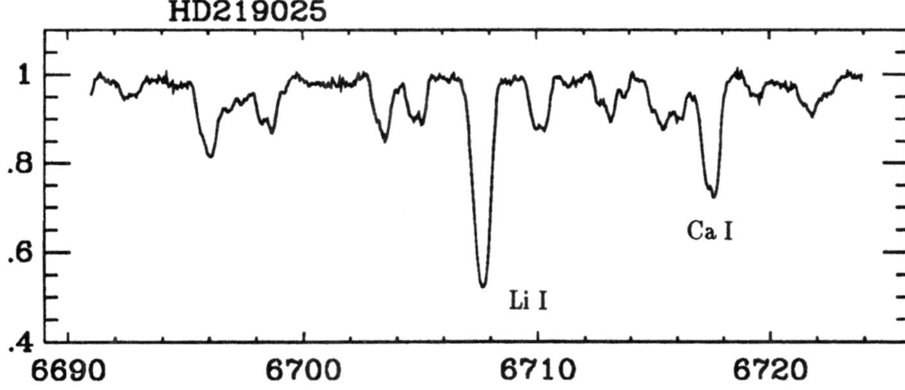

Fig.3 - CES spectrum of HD 219025. Note the extremely strong Li line.

3. Results

In Fig. 4 we show our results by plotting the derived Li abundances *vs.* effective temperature for both our sample stars (filled symbols) and for a sample of field stars observed with the same instrument (Soderblom 1985, Pallavicini *et al.* 1987) (open symbols). While chromospherically active stars with $logT_{eff} > 3.75$ do not show any significant departure from the typical behaviour of normal field stars, we have clearly an excess of Lithium for the cooler stars in our sample at $logT_{eff} \leq 3.74$. In particular, Lithium is present, with abundances ranging from ≈ 0.2 to ≈ 2.5, in most K-type stars in our sample, while only few of them do not show any detectable Lithium: this conflicts with the current belief that Li is completely depleted in K-type stars, except in the very young ones.

We also note that there is a small fraction of stars (the ones at the top of the diagram) which show a very large Lithium abundance, comparable to or even larger than the primordial Li abundance for population I stars.

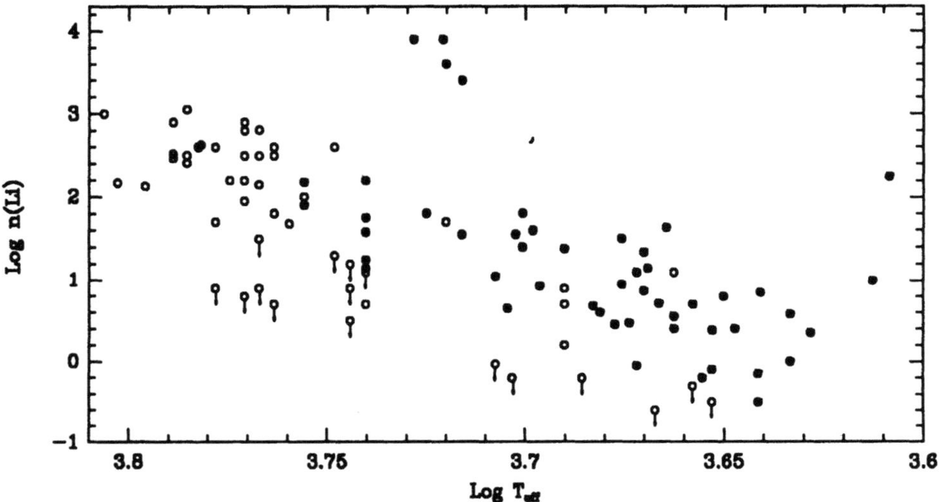

Fig.4 - Li abundance *vs.* effective temperature for the active stars in our sample (filled symbols) and for the stars in Pallavicini et al. 1987 (open symbols).

4. Discussion

First we should examine the possibility that the Lithium line observed in many active stars is not due to a genuine abundance effect. As Giampapa (1984) has shown for the Sun, surface activity may in fact affect the strength of the Li line. In particular Lithium may be enhanced by the presence of large cool spots over the stellar surface. If this is the case, monitoring "spotted" stars at different phases should reveal rotational modulation of the Li line.

This test was carried out at ESO in 1987. Four spotted stars (AB Dor, IL Hya, HR 1099 and YY Men) were observed simultaneously in the Li line and in broad-band UBV(RI)$_c$ filters. While strong photometric variations were found in the V-band (0.05-0.1 magnitudes), no significant variation in the Li line was observed (Pallavicini, Cutispoto and Randich

1990). This indicates that the strong Li line observed in many active stars in our sample is really due to a higher than normal Li abundance.

An explanation why Lithium is not completely depleted in RS CVn binaries has been proposed by Fekel *et al.* (1987). According to their suggestion, some of these stars may have evolved from late-A or early-F progenitors with very shallow convective zones; these stars should have suffered only negligible Lithium depletion while on the main-sequence. This explanation, though interesting, may hold only for the most massive stars in our sample (with masses $M \geq 1.5 M_{\odot}$). Lower mass stars (with masses in the range $\approx 1.0 - 1.5 M_{\odot}$) are likely to have been heavily depleted of Li as indicated by the presence of a Li "dip" in observations of open clusters (cf. Boesgaard 1990).

Two other possibilities to explain the high Li content of chromospherically active stars are suggested by their high rotational velocities. We find in fact that the majority of the stars in our sample are more rapidly rotating than inactive stars of similar spectral type.

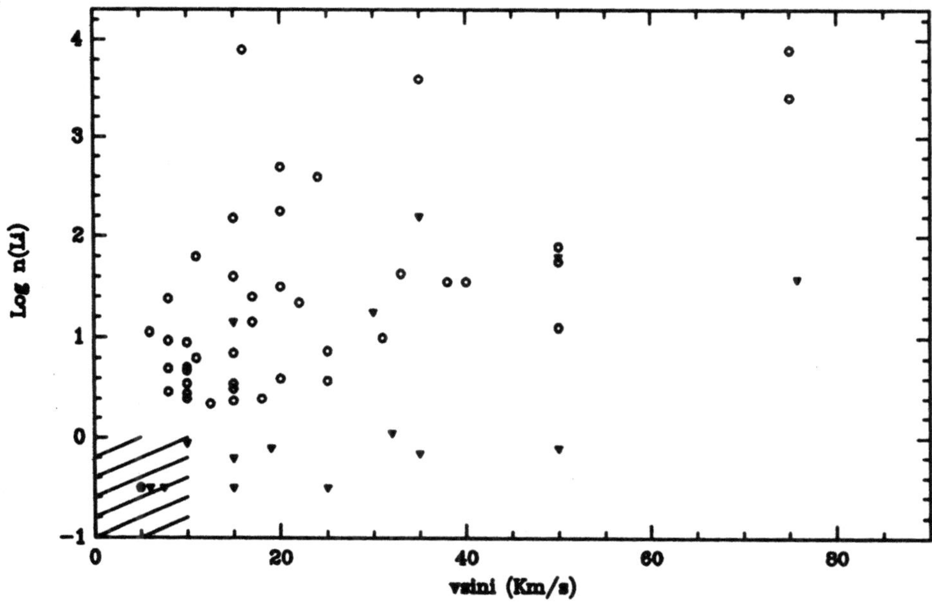

Fig.5 - Li abundance *vs.* projected rotational velocity for the stars in our sample. Filled triangles indicate upper limits on Li abundance. The dashed area in the lower left corner indicates the region of the diagram typically occupied by inactive K-type stars.

Rapid rotation may result either from the fact that some stars are very young or from tidal interaction in close binaries. In the first case the high Li abundance is simply a consequence of youth. In the second case, an explanation should be searched in the framework of the Li depletion mechanism recently proposed by Pinsonneault *et al.* (see Pinsonneault *et al.* 1990; Pinsonneault this conference; Sofia this conference). According to this mechanism, Li is depleted mainly by mixing due to radial differential rotation, when the outer layers of a star are spun down more rapidly than the interior. If the model of Pinsonneault et al. is correct, we can speculate that lack of an efficient braking in tidally coupled binaries may have prevented them from depleting Lithium by differential rotation mixing. Unfortunately it is difficult to distinguish between the various alternatives by simply looking at a correlation between Li and rotation since a spread in the distribution of initial angular momenta as well as differences in metallicity and age may affect the tightness of the correlation (we remind that in the Pinsonneault's model the Li depletion is related to the rotational history of the star, not to rotation itself).

As shown in Fig. 5, where we plot Li abundance *vs.* projected rotational velocity, the situation is indeed rather confusing. No definitive explanation can be given before determining other important parameters such as metallicity, degree of activity and space velocities.

However three different groups of stars are clearly present in the diagram: they include respectively, stars with no detectable Lithium, with moderate Lithium and with very strong Lithium. We suggest tentatively that the stars at the top of the diagram are very young objects, possibly PMS stars, still retaining all their original Lithium. The spread in *vsini* for these stars is not surprising since a similar spread is also observed in very young clusters (Hartmann and Noyes 1987, Stauffer this conference).

On the contrary, the stars at the bottom of the diagram may be the less massive and more evolved ones which have depleted their Lithium independently of rotation.

Finally the stars in the middle of the diagram, which have on average higher Li abundance and rotation rate than quiet stars of similar spectral type (dashed area at the lower left corner), may have preserved part of their Li either because their main-sequence progenitors had very shallow convective zones or because tidal interaction in close binaries prevented an efficient braking, and hence rotationally induced mixing.

References

Bell, R.A., Eriksson, K., Gustafsson, B., and Nordlund, A.: 1976, *Astron. Astrophys. Suppl.* **23**, 37.

Bidelman, W.P., and MacConnell, D.J.: 1973, *Astron. J.* **78**, 687.

Boesgaard, A.M.: 1990, in *Cool Stars, Stellar Systems and the Sun* (G. Wallerstein ed.), p. 318.

Fekel, F.C., Quigley, R., Gillies, K., and Africano, J.L.: 1987, *Astron. J.* **94**, 726.

Giampapa, M.S.: 1984, *Astrophys. J.* **277**, 235.

Gustafsson, B., Bell, R.A., Eriksson, K., and Nordlund, A.: 1975, *Astron. Astrophys.* **42**, 407.

Hartmann, L.W., and Noyes, R.W.: 1987, *Ann. Rev. Astron. Astrophys.* **25**, 271.

Hearnshaw, J.B.: 1979, in *Changing Trends in Variable Star Research* (F.M. Bateson, J. Smak, and I.H. Urch eds.), p. 371.

Pallavicini, R., Cerruti-Sola, M., and Duncan, D.K.: 1987, *Astron. Astrophys.* **174**, 116.

Pallavicini, R., Cutispoto, G., and Randich, S.: 1990, in *Surface Inhomogeneities in Late-type Stars* (P.B. Byrne and D.J. Mullan eds), in press.

Pinsonneault, M.H., Kawaler, S.D., Sofia, S., and Demarque, P.: 1990, *Astrophys. J.* **338**, 424.

Soderblom, D.R.: 1985, *Publ. Astron. Soc. Pacific* **97**, 54.

Strassmeier, K.G., Hall, D.S., Zeilik, M., Nelson, E., Eker, Z., and Fekel, F.C.: 1988, *Astron. Astrophys. Suppl.* **72**, 291.

Weiler, E.J., and Stencel, R.E.: 1979, *Astron. J.* **84**, 1372.

DISCUSSION

Kraft: I did not understand the sample. Were all of the K giants with fairly rapid rotation RS CVn stars?

Randich: No. Some are RS CVn stars, but a few are single. But, all are supposed to be evolved stars, and thus should not have lithium. They are not BY Dra stars.

Vaiana: There are also RS CVn stars on the main sequence. Does your sample include them?

Randich: There are just a few main sequence RS CVn stars, and in our sample there is only one. It is one of the stars with high lithium.

Cameron: Three years ago, I jointly published a full set of radial velocities for all of the stars in the Bidelman-MacConnell list. Two thirds of that sample of 55 stars are binaries, with periods up to about 100 days. The remaining 17 stars we could see no radial velocity variations at all over 4 years. So, they must be either single or at most very long period binaries. These stars showed *vsini*'s up to about 20 km/s.

LITHIUM DEPLETION INDUCED BY ROTATION IN YOUNG STARS

Corinne CHARBONNEL and Sylvie VAUCLAIR
Observatoire Midi-Pyrénées
14, avenue Edouard Belin
31400 Toulouse
France

ABSTRACT. Lithium depletion in Population I stars has been computed in the frame of Zahn's theory of rotation induced mixing. The implicit numerical code used to solve the turbulent and microscopic diffusion and nuclear destruction equations takes into account a Skumanich law of rotational braking during the stellar life.

The results are compared to the observations of lithium in galactic clusters. The depletion law follows the increase of the rotation velocities with T_{eff} in the F stars, and the lithium gap in the Hyades is well reproduced. The depletion curve for G stars can be accounted for if pre-main-sequence depletion is added.

1. Lithium Abundances in Young Stars.

Several extensive reviews of the observed lithium abundances in stars appeared recently in the literature. See, for example, Charbonneau and Michaud (1988) and Michaud and Charbonneau (1990). We shall focus here on the lithium observations in galactic clusters.

We can summarize these observations in the following way:

– Main-sequence cool stars (G and later spectral types) show a lithium depletion which increases for decreasing effective temperature. For a given effective temperature, the lithium abundance dispersion is larger in young clusters (one or two orders of magnitude) than in older clusters. The dispersion in the Hyades is exceptionally small. A general trend of lithium decrease with age is found.

– Main-sequence F stars show a lithium depletion for effective temperatures around 6600K in galactic clusters (Boesgaard and Tripicco (1986)), which increases with age. This dip also exists in field stars (Balachandran, Lambert and Stauffer (1988)), although more spread than in clusters. Recent observations of subgiants in the Hyades show that the lithium dip is still present when the stars evolve off the main-sequence (Balachandran (1990)).

S. Catalano and J. R. Stauffer (eds.), Angular Momentum Evolution of Young Stars, 259–264.

– On the hot side of the dip the observed lithium abundance do not vary, except in some peculiar A stars (Burkhart and Coupry (1989)).

– Galactic cluster giants show an average lithium depletion larger than expected with the simple dilution theory, and a large spread in lithium abundances is observed. This suggests that lithium has been more depleted in their main-sequence progenitors than expected by the standard nuclear destruction, even if this extra depletion is not seen at the surface of main-sequence stars with masses larger than 1.4 $M_\odot$ (Charbonneau, Michaud and Proffitt (1989), Gilroy (1989)).

Four different models have been proposed to account for the lithium dip in galactic clusters.

The first one is microscopic diffusion, as computed by Michaud (1986) (see also Vauclair et al (1978), Thevenin, Vauclair, Vauclair (1986)). This process gives the lithium dip at the right place, but it cannot account for all the observations. It needs some mass loss or other macroscopic motions to prevent lithium from becoming overabundant on the blue side of the dip.

In the second model, developped by Charbonneau and Michaud (1988), meridional circulation becomes the main process involved (lithium is then destroyed by nuclear reactions) and microscopic diffusion is now the secondary process which prevents lithium from being destroyed on the blue side, owing to the radiative acceleration which lifts it up (note that, in this model, there must be a fine tuning between the velocity of meridional circulation downwards and the microscopic diffusion velocity upwards for the lithium abundance to remain constant at the surface).

The third model is the rotation-induced turbulence model developped by Vauclair (1988) and Charbonnel, Vauclair and Zahn (1990). The lithium depletion is here supposed to be due to nuclear destruction and mixing induced by rotation. The mixing theory is developped in the framework of Zahn's 1987 theory. On the hot side of the gap, the normal lithium abundance is attributed to a separation between two different mixing zones related to two different loops of circulation. This model is discussed in more details below.

The fourth model involves mixing induced by gravity waves (Garcia Lopez and Spruit (1990)). This model can account for the dip if some special assumptions are made about the convective process. It cannot account for the depletion in G stars.

2. The Rotation Induced Turbulence Model

In the barotropic approximation, and with a small differential rotation, the turbulent diffusion coefficient induced by meridional circulation may be written:

$$D_T = \left| \gamma \cdot \frac{\Omega^2 r^6}{G^2(\nabla_{ad} - \nabla_{rad})} \cdot \frac{L}{M^3} \cdot \left(1 - \frac{\Omega^2}{2\pi G\rho}\right) \cdot P_2(cos\theta) \right|,$$

where Ω is the angular rotation velocity, r the radius and ρ the density of the considered shell, L and M the luminosity and the mass of the star, G the gravitational constant, $P_2(cos\theta)$ the 2-order Legendre polynomial and γ a factor of order unity.

The evolution of the lithium concentration c inside the star is obtained with the following diffusion equation:

$$\frac{\partial c}{\partial t} = D_T \frac{\partial^2 c}{\partial r^2} + \left[\frac{1}{\rho r^2}\frac{\partial(\rho . D_T . r^2)}{\partial r} - V_D\right]\frac{\partial c}{\partial r} - \left[\lambda(\rho, c) + \frac{1}{\rho r^2}\frac{\partial(\rho . V_D . r^2)}{\partial r}\right] c$$

where λ the nuclear reaction rate for lithium destruction by proton capture (Caughlan and Fowler (1988)),

and V_D is the microscopic diffusion velocity.

This equation has been solved with an implicit numerical code:

A spatial and temporal discretisation was done in the envelope of the star:

$$\frac{\partial \mathcal{C}^{\ell}}{\partial t} = \mathcal{M}\mathcal{C}^{\ell},$$

where $\mathcal{C}^{\ell}$ is the concentration vector at the given time $t = t_0 + \ell.dt$ (t_0 =initial time, dt=step of time), and $\mathcal{M}$ an implicit matrix.

Using the Cranck-Nicholson scheme ($\theta = 1/2$), the vector $\mathcal{C}^{\ell+1}$ at time $t' = t_0 + (\ell + 1).dt$ is given by:

$$\mathcal{C}^{\ell+1}(I - \theta.dt.\mathcal{M}) = \mathcal{C}^{\ell}\left[\frac{I}{\theta} + \left(\frac{\theta - 1}{\theta}\right)(I - \theta.dt.\mathcal{M}).\right]$$

This system was solved by Gauss's elimination method.

3. The Results

3.1. EFFECT OF TURBULENT DIFFUSION ON MICROSCOPIC DIFFUSION

We have first tested the effect of turbulence on microscopic diffusion in a static envelope model of 1.2 M $_\odot$, with an effective temperature corresponding to the effective temperature of the lithium dip. The convection zone is computed in the mixing length approximation; α, ratio of the mixing length to the pressure scale height, is here equal to 1.0. We used the Cox and Steward opacities. The evolution of the lithium abundance was followed over $6.1.10^9$ yrs. The rotation velocity is supposed to be constant during the stellar life time. The results are given in figure 1 for the ages of 0.6, 4.5 and 6.1 10^9 yrs.

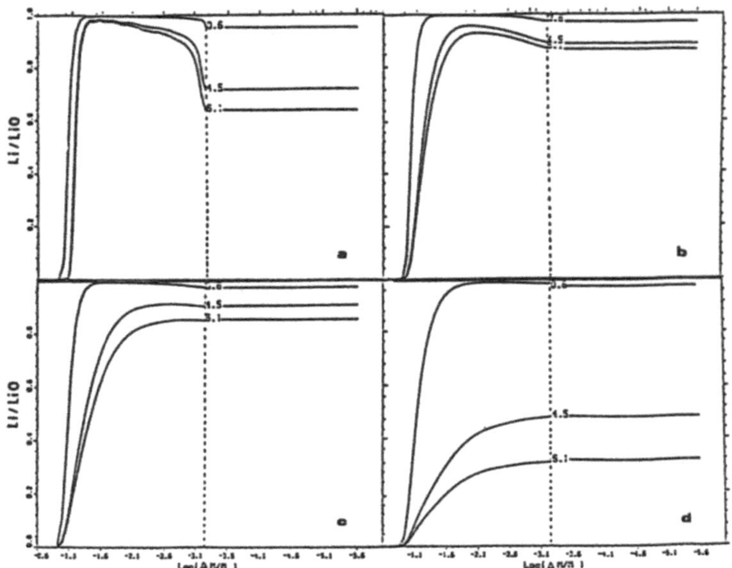

FIGURE 1. — Fractional lithium abundances versus depth ($\Delta M/M$ refers to the outer mass fraction) in a 1.2 $M_\odot$ star for different values of the rotation velocity: V=0, 1.8, 2.7 and 5.4 kms^{-1} respectively in fig. a, b, c and d. Theoretical lithium abundance curves are drawn for 0.6, 4.5 and 6.1 10^9 yrs. The vertical dashed line represents the boundary between the surface convective zone and the radiative zone.

As the turbulent diffusion coefficient increases, the gravitational settling is slowed down, and the surface abundance depletion is less pronounced for slow rotation velocities. Then turbulent diffusion becomes the main process involved for the lithium depletion. This occurs for a stellar rotation velocity smaller than the observed rotation velocity. The mixing induced diffusion process brings lithium to the region where it is destroyed by proton capture, and flattens the abundance gradients in the interior of the star.

Lithium is always depleted, whatever the turbulent diffusion coefficient: when it is large enough to prevent microscopic diffusion, nuclear destruction is no more negligible.

3.2. COMPUTATIONS FOR GALACTIC CLUSTERS

We have computed theoretical lithium abundance variations using static envelope models with $\alpha = 1.9$ and compared them to the observations of the Hyades. Microscopic diffusion is not taken into account in these computations. We assumed that all the stars began their life on the main sequence with a rotation velocity of $100 km.s^{-1}$, and were slowed down to their present observed velocities (given by the solid curve in figure 2a) with a Skumanich law:

$$\frac{1}{\Omega^{P-1}} - \frac{1}{\Omega_o^{P-1}} = ct,$$

where Ω_o and Ω are respectively the initial angular momentum and the angular momentum at time.

Pre-main-sequence depletion was introduced following the computation by Proffitt and Michaud (1989). Overshooting was added over one pressure scale height. Figure 2b shows three depletion curves: the first one for the age of the Pleiades (6.10^7yrs), the second one for the age of the Hyades (6.10^8yrs), and the last one for $1.5.10^9$ yrs.

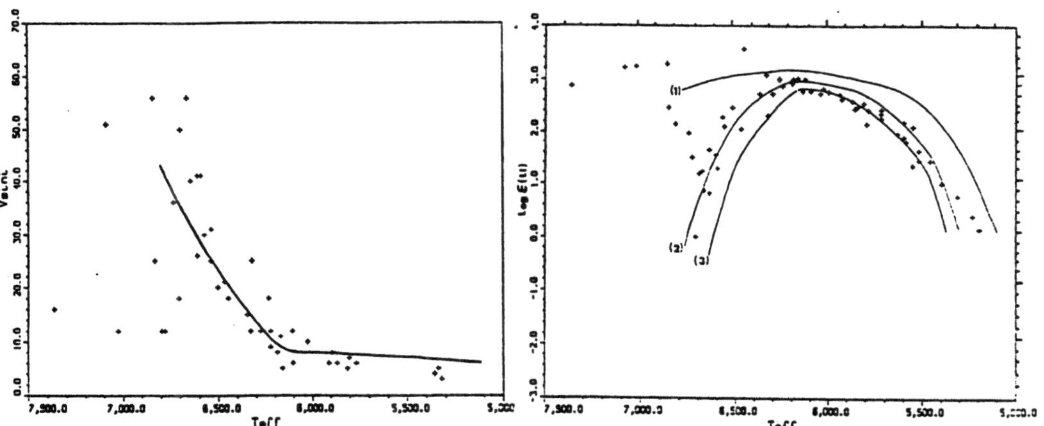

FIGURE 2. — **a.** Observed rotation velocities in the Hyades after Boesgaard (1987) and Radick et al (1987). The present rotation velocities used in our computations are given by the solid curve. **b.** Predicted Li abundances at the age of the Pleiades (1), of the Hyades (2), and at $1.5 * 10^9$yrs (3), vs. effective temperature. The assumed ZAMS abundance is log ϵ (Li)=3.3. The theoretical curves are compared with the lithium observations in the Hyades (Boesgaard and Tripicco (1986) and Cayrel et al (1984))

The hot side of the gap may be explained if two separate zones of meridional circulation develop in these stars. As shown by Vauclair (1988), the quiet zone lies inside the convection zone for cool stars, and gets out of it just for the effective temperature of the gap. This could also explain the "extra mixing" needed to account for the lithium observation in galactic cluster giants (Vauclair 1990).

References

Balachandran, S., 1990, Ap.J., preprint

Balachandran, S., Lambert, D.L., Stauffer, J.R., 1988, Ap.J., **333**, 267

Boesgaard, A.M., 1987, Pub. A.S.P., **99**, 1067

Boesgaard, A.M., Tripicco, M.J., 1966, Ap.J., **303**, 724

Burkhart, C., Coupry, M.F., 1989, Astr.Ap., **220**, 197

Caughlan, G.R., Fowler, W.A., 1988, *Atomic Data and Nuclear Data Tables* , **40**, 283

Cayrel, R., Cayrel de Stobel, G., Campbell, B., Däppen, W., 1984, Ap.J., **283**, 205

Chapman, S., Cowling, T.G., 1970, *The Mathematical Theory of Non Uniform Gases, Cambridge,* Cambridge Univ. Press.

Charbonneau, P., Michaud, G., 1988, Ap.J., **334**, 746

Charbonneau, P., Michaud, G., Proffitt, C.R., 1989, Ap.J., **347**, 821

Charbonnel, C., Vauclair, S., 1990, preprint

Charbonnel, C., Vauclair, S., Zahn, J.P., 1990, preprint

Däppen, W., 1989, private communication

Garcia Lopez, R.J., Spruit, H.P., 1990

Gilroy, K.K., 1989, Ap.J., **347**, 835

Michaud, G., 1986, Ap.J., **302**, 650

Michaud, G., Charbonneau, P., 1990, preprint

Radick, R.R., Thomson, D.T., Lockwood, G.W., Duncan, D.K., Bagett, W.E., 1987, Ap.J., **321**, 459

Thévenin, F., Vauclair, S., Vauclair, G., 1986, Astr.Ap., **166**, 216

Vauclair, S., 1988, Ap.J., **335**, 971

Vauclair, S., 1990, IAU, Symposium n° 145, Druzba, Bulgaria, G. Michaud ed.

Vauclair, G., Vauclair, S., Michaud, G., 1978, Ap.J., **223**, 920

Zahn, J.P., 1987, in *Astrophysical Fluid Dynamics* (Les Houches), ed J.P.Zahn and J.Zinn-Justin.

ROTATIONAL MIXING AND LITHIUM IN YOUNG STARS

M. H. Pinsonneault
Center for Solar and Space Research
Yale University
P. O. Box 6666
New Haven, CT 06511

ABSTRACT. The potential role of rotational mixing as an agent for lithium depletion is reviewed. For main sequence stars, mixing driven by angular momentum loss can cause lithium depletion for models with a wide range of masses. For young stars, however, the evolution of the structure inhibits lithium depletion in higher mass models. As a result, rotational mixing can cause a large dispersion in abundance for young low mass (late G and K) stars but not for young high mass (early G and F) stars. The properties of solar metallicity and metal-poor models are also compared.

1. Introduction.

One of the most important roles stellar rotation plays is its potential as a mixing agent. Explaining the surface abundances of stars requires more mixing than is predicted by standard stellar evolution theory; furthermore, evidence for additional mixing is seen for stars in a variety of evolutionary states, ages, masses, and metallicities. The surface abundances of light elements, especially lithium, are especially interesting as indicators of mixing.

Lithium is efficiently destroyed at a temperature ~ 2.5×10^6 K, which is modest by stellar interior standards. As a result, lithium can survive for an extended period of time only in the outermost layers of a star. If the surface convection zone of a star is sufficiently deep to burn lithium at its base, then the surface lithium abundance can be progressively reduced even in standard stellar models. Lithium depletion in standard models can be particularly effective in the pre-main sequence, when the surface convection zones of low mass stellar models are much deeper than on the main sequence. It is also effective on the main sequence in the lowest mass stars with the deepest surface convection zones. For solar metallicity models, main sequence lithium depletion is effective for masses less than 0.9 $M_{\odot}$. However, standard models fail to explain the observed depletion pattern in a variety of crucial respects (see §2).

Mixing can also cause surface lithium depletion as lithium-poor material is mixed upwards and lithium-rich material is mixed down (and destroyed in the interior). There are two classes of rotational mechanisms that can generate mixing. A rotating star is non-spherical, and the departure from sphericity causes meridional circulation currents to develop. Rotational instabilities driven by angular velocity gradients can also cause mixing; angular momentum loss in low mass stars is an effective agent for generating these angular velocity gradients, particularly in young rapid rotators. We have computed rotational

S. Catalano and J. R. Stauffer (eds.), Angular Momentum Evolution of Young Stars, 265–270.

models of the Sun (Pinsonneault *et al.* 1989), open cluster stars (Pinsonneault, Kawaler, and Demarque 1990), and halo stars (Deliyannis, Demarque, and Pinsonneault 1989; Deliyannis and Pinsonneault 1990; Pinsonneault, Deliyannis, and Demarque 1991a, b). See Pinsonneault *et al.* 1989 (and also Sofia, Pinsonneault, and Deliyannis 1991 in this proceeding) for a description of the method. We have found that rotational models have many of the properties required to explain lithium depletion in a variety of objects.

In this paper, I will concentrate on one aspect of this work : the degree to which the observed pattern of lithium depletion can be understood as naturally following from the stellar evolution. In §2 I briefly review some salient features of the lithium observations. In §3 the properties of high and low mass models are compared; mixing is found to be effective only in the lower mass young stars, while it is effective for a wider range of masses on the main sequence. In §3 I also compare the properties of solar metallicity and metal-poor models; the paper is summarized in §4.

2. Observed Lithium Depletion Pattern.

There now exists a substantial (and growing) database of lithium observations in low mass stars with a variety of ages, masses, and compositions. High lithium abundances are seen in young T Tauri stars, although the precise level is difficult to determine given the observational uncertainties (Strom *et al.* 1989). Higher mass stars have a relatively small range in abundance and little apparent depletion in the young Pleiades cluster, while lower mass stars are both substantially more depleted and exhibit a large dispersion in abundance at fixed T_{eff} (Soderblom *et al.* 1991). In the older open clusters, the depletion for the cool stars becomes very severe and strongly dependent on the effective temperature; in the Hyades, for example, the observed lithium abundance drops by three orders of magnitude over a range in effective temperature of less than 1000 K (Soderblom *et al.* 1990). By the age of the Hyades, there is also a population of F stars which is highly depleted with respect to both cooler and hotter stars (the F star lithium "gap", Boesgaard and Tripicco 1986), which leaves a narrow peak in abundance between the gap and the severe depletion for the coolest stars. In progressively older clusters, the peak narrows, the peak abundance declines, and there is an observed dispersion in lithium abundance. NGC 752 (Hobbs and Pilachowski 1986) is a particularly clear example.

This depletion pattern is exceedingly difficult to understand with standard stellar models. A dispersion in abundance at fixed T_{eff} is particularly problematical, given that standard stellar models with the same mass, age, and composition should have the same properties. It is difficult to cause steady main sequence depletion in G stars by deepening the surface convection zone (by overshoot, for example) without causing a drastic increase in pre-MS depletion, especially in cooler stars. As we will see below (in §3) rotational mixing naturally explains the observed depletion pattern.

The halo star depletion pattern is very different from that in the open clusters. There is a plateau over almost 800 K, with a nearly constant abundance and little dispersion. Cooler metal-poor stars exhibit progressively more depletion, but even there the abundance is a much weaker function of T_{eff} than for the open cluster stars (see Deliyannis, Demarque, and Kawaler 1990 for a list of observers and detailed discussion of the halo star lithium depletion pattern). Producing a model that reproduces this very different pattern is a theoretical challenge; once again, rotational models predict a depletion pattern for metal-poor stars that is qualitatively different from that for solar metallicity models and that matches the observations.

3. Lithium Depletion in Rotational Models.

3.1. MASS DEPENDENCE AND THE ABUNDANCE DISPERSION IN YOUNG LOW-MASS STARS.

The overall observed lithium depletion pattern can be understood by an analysis of the stellar evolution properties that drive the lithium depletion. In our models, instabilities driven by differential rotation with depth are primarily responsible for lithium depletion; angular momentum loss is the most efficient mechanism for generating differential rotation with depth. The timing and amount of angular momentum loss therefore strongly influences the degree of lithium depletion. Indeed, on the main sequence the two are strongly correlated in the models (Pinsonneault, Deliyannis, and Demarque 1991b). However, the amount of lithium depletion generated by rotational mixing also depends on the structure and evolution of the model; as we will see, early angular momentum loss is ineffective in driving mixing in higher mass models but highly effective in lower mass models.

Consider models with different mass but the same metallicity and total angular momentum. The lower mass models have longer pre-main sequence (pre-MS) lifetimes, so they have more time to experience angular momentum loss. They also have smaller moments of inertia, which implies higher surface angular velocities for the same total angular momentum. This causes more severe angular momentum loss in them because the angular momentum loss rate is a strong function of the surface rotation rate. Low mass models will therefore experience more pre-MS angular momentum loss than higher mass models.

The structure of the lower mass models also favors mixing when compared with the higher mass models. The surface convection is progressively deeper and hotter at the base for progressively lower mass stars. For sufficiently low mass stars, there is little or no lithium below the surface convection zone; as a result, any rotational mixing can easily transmit lithium-poor material to the convection zone. For higher mass stars (greater than about 0.9 $M_{\odot}$ for solar metallicity), however, there is a buffer zone below the surface convection zone where lithium is preserved. Any lithium-poor material must cross this region before it can reach the surface convection zone and modify the surface abundance; to make matters worse, the convection zone is rapidly retreating in the pre-MS phase. As a result, rotational mixing is strongly inhibited in higher mass stars.

What does this imply for lithium in young stars? *Cool (lower mass) young low mass stars are susceptible to mixing, while hot (higher mass) young low mass stars are not.* Different estimates for the time scale of mixing, of course, will produce different degrees of depletion in the lower mass stars. However. the overall pattern is not highly dependent on the particular set of parameters used in the rotational models.

Rotationally induced mixing has another property that is required to explain lithium observations in young clusters : models with the same mass and age but different initial angular momenta experience different degrees of mixing. This occurs because angular momentum loss forces models with different initial angular momentum (J_0)to the same surface rotation rate. Models with different J_0 then have different rates of core rotation with the same surface rotation rate. But rotational instabilities then force all of these models to have the same core rotation rate, which implies more angular momentum transport (and thus mixing) to occur in the models with the highest J_0. Because mixing is inhibited in young massive stars, this causes a small range in lithium depletion for the hotter young stars even given a range in angular momentum transport within them. For the cooler stars, however, angular momentum transport does cause lithium depletion; given a range in J_0 a range in depletion for models of a given mass and age will be produced. A dispersion in lithium abundance at fixed T_{eff} for young cool stars would then be expected; in fact, in the Pleiades exactly such a dispersion is seen (Soderblom *et al.* 1991, this proceeding). Later on, during the main sequence evolution, mixing is effective for both the lower and higher mass models. A dispersion in lithium abundance for a wide range of T_{eff} will then be produced during the main sequence; once again, this is seen in a variety of older open clusters.

3.2. METALLICITY DEPENDENCE AND THE LITHIUM PLATEAU IN HALO STARS.

The observational lithium depletion pattern in metal-poor stars is very different from that of solar metallicity stars (§2). Much of this difference can also be traced to the dependence of stellar structure and evolution on metallicity. Because the metal-poor stars are old, the ones we can observe today are relatively low mass (of order 0.8 $M_{\odot}$ or less). Because the total moment of inertia is only a weak function of metallicity, the halo star models lose as much angular momentum during the pre-MS as solar metallicity stars of the same mass. One might therefore expect substantial pre-MS mixing, in analogy with the solar metallicity models of the same mass. However, like the higher mass solar metallicity models the hottest halo star models have thin surface convection zones with a substantial lithium buffer zone below. Mixing is therefore relatively inefficient in the hot halo stars during the pre-MS phase. *This implies that halo star models will experience less rotationally induced lithium depletion than solar metallicity models.* Furthermore, most of the difference between halo models with different initial angular momenta is removed during the pre-MS phase when mixing is inefficient. As a result, *halo star models have a smaller dispersion in their lithium abundance than solar metallicity models.* This is also seen in the observations (Pinsonneault, Deliyannis, and Demarque 1991b).

4. Summary.

The observed pattern of lithium depletion is complex. This is not surprising given the fragility of lithium and the variety of potential mechanisms for depleting its surface abundance. What is both surprising and encouraging is the natural way that the observed depletion pattern follows naturally in stellar models with rotational mixing. Rotational mixing produces a dispersion in abundance for young cool stars, and a dispersion in abundance for older solar metallicity stars with a wide range in effective temperature. At the same time, when the same physical model is applied to metal-poor stars a very different depletion pattern emerges : there is a nearly flat plateau for hotter metal-poor models, with little dispersion in the plateau. It is worth noting that the observed CNO abundances in evolved stars also require additional mixing.

Larger samples of measured lithium abundances in clusters are needed to quantify the depletion and dispersion as a function of mass and time; once these samples are available they will provide a more quantitative test of the theory. More theoretical work is also needed, both in understanding angular momentum transport and mixing mechanisms and in understanding angular momentum loss (especially in rapidly rotating stars). It is already clear, however, that explaining the observed lithium depletion in stars requires some additional mixing mechanism, and that lithium provides a sensitive diagnostic of mixing in stars.

References

Boesgaard, A. M. and Tripicco, M.J. (1986), 'Lithium in the Hyades Cluster', *Ap. J. Lett.*, **302**, L49-L53.

Deliyannis, C.P., Demarque, P., and Kawaler, S.D. (1990), 'Lithium in Halo Stars from Standard Stellar Evolution', *Ap. J. Suppl.*, **73**, 21-65.

Deliyannis, C.P., Demarque, P., and Pinsonneault, M.H. (1989), 'The Ages of Globular Cluster Stars : Effects of Rotation on Pre-Main Sequence, Main Sequence, and Turnoff Evolution', *Ap. J. Lett.*, **347**, L73-L76.

Deliyannis, C.P. and Pinsonneault, M.H. (1990), 'Beryllium in the Galactic Halo : Surface Abundances from Standard, Diffusive, and Rotational Stellar Evolution, and Implications', *Ap. J. Lett.*, **365**, L67-L70.

Hobbs, L.M. and Pilachowski, C.P. (1986), 'Lithium in NGC 752', *Ap. J. Lett.*, **309**, L17-L21.

Pinsonneault, M.H., Deliyannis, C.P., and Demarque, P. (1991a), 'Evolutionary Models of Halo Stars with Rotation : I. Evidence for Differential Rotation with Depth in Stars', *Ap. J.*, in press (Jan 20 1991 issue).

Pinsonneault, M.H., Deliyannis, C.P., and Demarque, P. (1991b), 'Evolutionary Models of Halo Stars with Rotation : II. Effects of Metallicity on Lithium Depletion, and Possible Implications for the Primordial Lithium Abundance',.submitted *Ap.J.*.

Pinsonneault, M.H., Kawaler, S.D., Sofia, S., and Demarque, P. (1989), 'Evolutionary Models of the Rotating Sun', *Ap. J.*, **338**, 424-452.

Pinsonneault, M.H., Kawaler, S.D., and Demarque, P. (1990), 'Rotation of Low Mass Stars : A New Probe of Stellar Evolution', *Ap. J. Suppl.*, **74**, 501-550.

Sofia, S., Pinsonneault, M.H., and Deliyannis, C.P. (1991), 'Evolutionary Models of Rotating Stars', in S. Catalano and J.R. Stauffer (eds.), Angular Momentum Evolution of Young Stars, Kluwer Academic Publishers, Dordrecht, in press.

Soderblom, D.R., Oey, M.S., Johnson, D.R.H., and Stone, R.P.S. (1990), 'The Evolution of the Lithium Abundances of Solar-type Stars : I. The Hyades and Coma Berenices Clusters', *A. J.*, **99**, 595-607.

Soderblom, D.R., Hudson, J.D., Jones, B.F., and Stauffer, J.R. (1991), 'Rotation, Chromospheric Activity, and Lithium Abundances in G and K Dwarfs of the Pleiades', in S. Catalano and J.R. Stauffer (eds.), Angular Momentum Evolution of Young Stars, Kluwer Academic Publishers, Dordrecht, in press.

Strom, K.M., Wilkin, F.P., Strom, S.E., and Seaman, R.L. (1989), 'Lithium Abundances Among Solar-type Pre-main Sequence Stars', *A. J.*, **98**, 1444-1450.

DISCUSSION

Roxburgh: Can you tell us more about the physics which you have put in your model?

Pinsonneault: Yes, but Sofia will talk about that tomorrow, so perhaps it should wait until then.

Roxburgh: Just a hint now, please.

Pinsonneault: We include the shear instability - where we analyse the stability criterion provided by Zahn, and when it is violated we introduce a diffusion coefficient for it. We also consider the Goldreich-Schubert-Fricke mechanism, and we consider meridional circulation. We do not consider magnetic fields, and we do not consider angular momentum transport by them.

Roxburgh: Do you have the freedom to adjust angular momentum transport and material transport independently?

Pinsonneault: The angular momentum transport parameters were varied initially, but were not adjusted in our best case models which I presented. The chemical composition transport relative to angular momentum transport in the outer layers is empirically found to be substantially less. The transport of angular momentum must be much more efficient that the transport of material to explain the observational data.

Roxburgh: So, you do adjust the parameters to fit the data.

Pinsonneault: Just to get the sign.

Roxburgh: I don't understand what you are doing, unless you are just adjusting a number of free parameters to match the observations, because we don't theoretically understand any of these mechanisms.

Pinsonneault: You take one physical model, which admittedly has a number of uncertainties, and then apply this physical model to stars with a range of masses and evolutionary states and metallicities. You then get a consistent physical picture. What I am arguing is that the patterns of abundances and rotational velocities can be fit by this model.

Palla: Can you comment on the fact that during the PMS, you have evidence for mass accretion at a rate which is comparable to the mass loss rate which you need for your inferred angular momentum loss rates.

Pinsonneault: One hopes that we can start the analysis from a point in time after the disks stop being important.

INTERNAL SOLAR ROTATION

D. O. GOUGH
Institute of Astronomy and
Department of Applied Mathematics and Theoretical Physics
University of Cambridge
UK

ABSTRACT. The helioseismic determination of the angular velocity distribution within the sun is discussed. In particular, the bulk of the radiative interior of the sun is rotating at more-or-less the same rate as the photosphere. The latitudinal variation of angular velocity appears to change quite abruptly immediately beneath the base of the convection zone. Associated with that change is no doubt a meridional circulation which must be confined to only a thin layer; the photospheric lithium and beryllium abundances set constraints on the possible extent of substantial flow. There is slight evidence that the angular velocity changes with the solar cycle. The implications concerning a solar dynamo, if there is one, have not yet been extensively investigated.

Introduction

During the last half-decade it has been possible by helioseismological techniques to build up a picture of the spatial variation of the angular velocity $\Omega(r, \theta, t)$ throughout a substantial volume of the sun, and to make some tentative suggestions about is temporal variation. I describe briefly how the deductions are made. Then I summarize the results, in the light of what had been expected. Finally I discuss some immediate reactions to those results.

Solar Oscillations

Most observable acoustic modes of oscillation, from which the deductions about Ω have been made, are trapped essentially in a spherical shell whose upper surface is not far beneath the photosphere. The depth of the lower surface is a function of the order n and degree l of the mode. The modes are also concentrated outside a cone of semi-angle $\theta_0 \simeq \sin^{-1}\left[m/\left(l+\frac{1}{2}\right)\right]$, where m is the azimuthal order of the mode, whose axis coincides to the axis of rotation. Thus modes with different values of n, l and m sample different regions $\mathcal{R}$ of the solar interior. Although the outer surface of the region $\mathcal{R}$ within which acoustic modes are wave-like is usually

S. Catalano and J. R. Stauffer (eds.), Angular Momentum Evolution of Young Stars, 271–279.

beneath the photosphere, the modes can tunnel through the evanescent zone above $\mathcal{R}$ to the visible layers where they can be observed.

The sun is rotating slowly, in the sense that the centrifugal force is everywhere much less than gravity; therefore the sun is approximately spherically symmetrical, and consequently the oscillation eigenfrequencies are approximately degenerate with respect to m. It is basically for this reason that the radius r_t of the inner boundary of $\mathcal{R}$ depends only on n and l. However, the modes that have been observed to date have frequencies which lie in a relatively narrow range: 2mHz $\lesssim \omega/2\pi \lesssim$ 4 mHz. This provides a relation between n and l, so that roughly speaking r_t can be considered to be a function of l alone. (More precisely, r_t is a function of $\omega/\left(l+\frac{1}{2}\right)$ alone).

Rotational Splitting

The pattern of the wave motion that constitutes the modes is transported by the material flow within the sun. The main constituent of that flow is rotation, which is the topic of this lecture. To a first approximation the effect of Ω is simply to rotate the mode locally with angular velocity Ω. This both modifies the frequencies ω of nonaxisymmetric modes when viewed from an inertial frame, or from Earth, and, since Ω is nonuniform, distorts the wave pattern. Because ω/Ω is small, the global distortion of the wave patterns is also small, though locally the modification can be substantial when m is large. The frequency modification is simply $m\langle\Omega\rangle$, where the angular brackets $\langle\ \rangle$ denote an appropriately weighted average of Ω. The weighting is different for different modes. If a mode is regarded as the standing interference pattern of a propagating locally plane acoustic wave of frequency ω reflected at the boundary of $\mathcal{R}$ and resonantly interfering with itself, then the weighting is proportional to the time the wave spends in any given element of volume; it is proportional to the wave slowness k/ω, where $k(\boldsymbol{r})$ is the magnitude of the local wave number, and to the density of rays.

To date only the frequency modification, which is known as rotational splitting, has been measured. The measurement is accomplished by projecting the image of the sun onto the undistorted spatial eigenfunctions, and performing a frequency analysis of the temporal variation of the outcome. Of course there is some contamination from other modes, partly because even the undistorted modes are not orthogonal over the solar image (the eigenfunctions can be expressed in terms of spherical harmonics, which are orthogonal over the entire sphere, but of course are not orthogonal over a solar image, particularly the usable part which excludes an annulus to the limb) and partly from overlap with the rotational distortions of other modes. The latter, which has been discussed recently by Woodard (1989), is substantial only for modes of large azimuthal order, and is likely to be accounted for in imminent analyses of rotational splitting.

Inversion of Data

From the averages $\langle\Omega\rangle$ over different regions $\mathcal{R}$ of the star it is possible to infer the broad spatial variation of Ω by inverse techniques. Whatever method is used, the outcome is a localized average of Ω, the degree of localization having been determined predominantly by the number and range of modes whose frequencies have been measured. The first results were obtained from measurements of sectoral modes (for which $m = l$) alone (Duvall *et al.*, 1984), yielding averages which are apparently localized in radius r down to the edge of the energy-generating core and extending between latitudes $\pm\cos^{-1}\left[l/\left(l+\frac{1}{2}\right)\right]$. However, these averages are actually rather complicated to interpret, because each depends on a range of values of l, and therefore extends over a different range of low latitudes. More recently, measurements of rotational splitting of tesseral-mode frequencies has permitted the determination of the latitudinal variation of Ω in the region $r \gtrsim 0.5R$, where R is the radius of the sun (Hill *et al.*, 1988; Christensen-Dalsgaard and Schou, 1988; Brown *et al.*, 1989, Dziembowski, Goode and Libbrecht, 1989; Thompson, 1989, Korzennik *et al.*, 1990; Rhodes *et al.*, 1990; Sekii, 1990; Schou, 1991).

Theoretical Preconceptions

Before describing the results of the frequency inversions it might be worthwhile recalling some prior expectations. Because the sun is being spun down by the solar-wind torque, internal stresses are set up which redistribute the angular momentum within the sun and which drive motion with a meridional component. How well the deceleration of the interior keeps up with the deceleration of the photosphere depends on the magnitude of those stresses. It is about this that there is a wide diversity of opinion.

The response of the convection zone is fast. Reynolds stresses can transmit strain at the rms speed of the turbulent convective motion, and thus it takes the order of a year for the entire convection zone to readjust. Thus, on the spin-down timescale, adjustment is essentially instantaneous. That is not in doubt, though there has been some disagreement concerning to what state the convection zone adjusts. Notwithstanding the fact that deep in the convection zone the convective turnover time is comparable with the rotation period and that therefore the rotation is likely to impart a degree of anisotropy to the convective flow and consequently to the Reynolds stresses, thereby leading to differential rotation, many modellers have simply assumed that the Reynolds stresses can be parameterized with a scalar turbulent viscosity, which acts in the direction of rigid rotation. Numerical modellers such as Glatzmaier and Gilman (1982), on the other hand, have favoured rotation on cylinders, at least in low and mid latitudes where cylinders coaxial with the rotation are wholly within the convection zone. As I shall point out below, neither is correct.

Where disagreement has been most stark is with regard to the response of the radiative interior beneath the convection zone. There has been a wide spectrum of opinion, from those who believe that the response is slow and that therefore the entire radiative zone is rotating quite rapidly (though nowadays at least slowly enough to be consistent with the constraints imposed by the oblateness measurements), to those who believe that instabilities cause a relatively rapid response, except possibly in the energy-generating core, so that the rise in Ω with depth is gradual with perhaps a near-discontinuity at the core-envelope boundary, to those who believe that the solar interior is pervaded by a magnetic field the strength of whose poloidal component is at least 3μG or so, which is adequate to transmit the solar-wind torque on the timescale of the main-sequence lifetime and thereby establish rigid rotation throughout the entire interior beneath the convection zone.

Observed Internal Solar Rotation

Although the different representations of the internal angular velocity that have been published are not identical, it is probably true to say that broadly speaking none is inconsistent with the following picture:

In the convection zone the variation of Ω at the photosphere, where the rotation rate at the poles is some 25 – 30% lower than at the equator, is approximately maintained to the base of the convection zone. There is evidence that immediately beneath the photosphere, at a depth of about 2Mm, there is at least in the equatorial regions a maximum in Ω which is a few per cent higher than the photospheric angular velocity (Hill *et al.*, 1988), and another at a depth of about 50 Mm, some 8 per cent of the solar radius (Korzennik *et al.*, 1990). Immediately beneath the convection zone Ω falls at low latitudes and rises at high latitudes from the corresponding values in the convection zone, such that the dependence of Ω on colatitude θ at least in the outer layers of the radiative envelope is quite weak. The resolution of the data is insufficient to determine the distance over which the transition takes place; it is probably true to say that no-one at the moment can be sure that the transition isn't nearly discontinuous. Beneath the transition the latitudinal variation apparently continues to be weak. Indeed, Ω might be independent of θ down to $r \simeq 0.5R$, beneath which the inferences are unreliable.

The equatorial averages obtained solely from sectoral-mode splittings extend more deeply. Beneath the convection zone Ω appears to decrease with depth down to the edge of the energy-generating core (Duvall *et al.*, 1984; Korzennik *et al.*, 1990). The decrease is small, perhaps only a few per cent. Finally, there is some tentative evidence that the core rotates more rapidly, perhaps as much as or more than twice as fast as the photosphere. The evidence is based on the relatively high rotational splitting of the lowest-degree p modes reported by Duvall and Harvey (1984), whose data were inverted for the original determination of Ω, similar results presented by Isaak (1986; cf Claverie *et al.*, 1981) and Jefferies *et al.*, (1988), and some very tentative evidence of g-mode splitting reported, for example, by Delache and Scherrer (1983), by Fröhlich (1988) and by Pallé and Roca Cortés (1988).

Discussion of Observations

What can be said about these results? The first, and most obvious, is that they are quite different from prior expectation. The maintenance of the latitudinal dependence of Ω through the convection zone is inconsistent with rotation on cylinders. More interesting, perhaps, is the admittedly less secure inference of the radial dependence throughout most of the radiative interior obtained from sectoral-mode splittings: prior to and even after the first announcement of this result people had argued over how rapidly Ω increases with depth, or whether the rotation of the interior is rigid, but nobody had predicted the weak decline in Ω with depth that is suggested by observation. Why such a decline should be present is an extremely important theoretical issue. The only feature of Ω that had been discussed, indeed quite extensively, is the possibility of a rapidly rotating core, but it is very important to appreciate that the seismological evidence for this property is really quite weak.

The next remark one might make concerns the transition beneath the convection zone. Such a region cannot be in stable equilibrium in the absence of any motion other than rotation. Therefore there is bound to be a meridional flow, not unlike Eddington-Sweet and Ekman circulation, probably confined in the radiative region to a thin boundary layer by the convectively stable stratification, which might or might not be temporally varying and which presumably penetrates into the convection zone. The motion is likely to be dominated by a large-scale eddy in each hemisphere whose stagnation point is at the latitude at which the convection zone and the radiative interior rotate at the same rate: the dynamics of such motion is currently being studied by J.-P. Zahn and E.A. Spiegel. Evidently the motion provides a means of coupling the angular momentum in the outer layers of the radiative interior to the convection zone. In this regard, it is interesting to note that if it is assumed at the outset that in the radiative interior Ω is uniform, that in the convection zone Ω is independent of r, having the same θ dependence as is observed at the photosphere, and that the angular velocity in the rigidly rotating interior is determined so as to produce the best fit to the seismic data, then the mean angular momentum over spherical surfaces is continuous through the transition layer (Brown *et al.*, 1989). This implies that if the local stress is proportional to the discontinuity in Ω, as it might well approximately be, then there is no net torque between the convection zone and the radiative interior. The sun might therefore be in an approximately steady state.

The transition zone is pertinent to any discussion of the sun's magnetic field. Evidently, if the variation in Ω that has been reported persists in time, there cannot be a smooth radial component of the field through the transition zone unless the zone is in motion, for otherwise it would generate a continually increasing azimuthal component of the field. Either reconnexion is inevitable or the shear must reverse. Morrow *et al.*, (1988) have discussed the former briefly in terms of dynamo theory. Whether the shear reverses on the timescale of the solar cycle is a matter that will be resolved by future observations.

It is important to realize that the reliability of the inferences concerning Ω decreases substantially with depth. The decline in Ω with depth in the equatorial regions beneath the convection zone (yet outside the energy-generating core) was originally questioned because data reported by Brown (1984) were not precisely in accord with those reported originally by Duvall and Harvey (1984). Nevertheless, Brown's data too did show evidence for the decline, though rather more gradually. Since the data were obtained at different times perhaps this was simply evidence for a solar-cycle variation. After all, unlike the transition near the base of the convection zone it seems unlikely, though not wholly out of the question, that the angular-velocity gradient should be steady, unless it transpires that the polar angular velocity continues to rise with depth overtaking the value in the equatorial plane. The rapidly rotating core is yet more uncertain, because none of the observations of rotational splitting is totally convincing. Problems with interference from other modes and from noise, both in the sun and in the solar atmosphere, and difficulties caused by sidelobes from the observing window, have combined to prohibit a clear separation of the individual modes of low degree which are required for diagnosing the central regions of the sun. Nevertheless, in the wake of the initial inversions, Rosner and Weiss (1985) have discussed how such a rapidly spinning core might be a legacy of pre-main-sequence phases of the sun's evolution, a topic which is central to this conference. They failed to address, however, the relatively slow rotation in the radiative envelope immediately outside the core.

I conclude by mentioning briefly the possibility that the internal angular velocity varies with the solar cycle. Observations of the photospheric angular velocity show no substantial variation, and because the Reynolds stresses in the convection zone adjust so rapidly, one would expect that property to be indicative that the angular velocity throughout the entire convection zone is also more-or-less independent of time. However, the evidence of spatial variation in the radiative region beneath suggests that perhaps oscillatory motion may be present. This is borne out by recent seismic hints that from the base of the convection zone down to a radius of at least $0.4R$ the shear does indeed vary with time, being least at sunspot maximum (Goode *et al.*, 1990). Maybe this is evidence for a deep-seated magnetic oscillation, perhaps of the type discussed originally by Walén (1946) as being responsible for the cycle. Indeed, since the moment of inertia of the radiative interior is some ten times greater than that of the convection zone, one would therefore then naturally suppose that the cycle is controlled in the radiative interior, and not in the convection zone or at the interface between the convection zone and the radiative interior as most dynamo theorists suppose. If that were the case, however, one is left with the task of explaining the initially bewildering property that the interior manages to oscillate without it influencing the angular momentum of the imponderable convective envelope. It might require that the field strength be greatest in the outer layers of the radiative interior, imparting a greater rigidity where the contribution to the moment of inertia is highest.

What could one say about the restoring force, and the form of the oscillatory motion, should it exist? Since compression, which controls p modes, is restored in hours or minutes, buoyancy forces produce a response on a timescale of hours to give g modes, and vorticity stretching associated with inertial oscillations produces a response in a time comparable with the rotation period of the sun, the motion must be tightly constrained by these agents. There remains axisymmetric nearly azimuthal motion, that is to say rotational motion, which is restored by Lorentz forces resulting from magnetic-field distortion. To obtain an oscillation period of 22 years would require a large-scale kilogauss magnetic field pervading the radiative interior, which is not only similar in magnitude to sunspot fields but is also of the same order as some astronomical arguments have suggested for the internal zero-age main-sequence field (e.g. Mestel and Weiss, 1987; Gough, 1990). Associated with such motion there might also be a laminar spheroidal oscillation of the core, driven by the energy released from ^{3}He-consuming reactions, as has recently been discussed by Merryfield *et al.* (1990). That would certainly influence the solar neutrino flux. The asymmetry of such flow induced by the rotation might then react nonlinearly with the angular velocity and thus generate the long-period oscillation, the solar cycle, in a manner not dissimilar to the generation by gravity waves of the quasibiennial equatorial jet in the earth's atmosphere, and might in addition explain the apparent correlation with the solar cycle of Davis's measurements of the neutrino flux (Davis, 1988). But now I really am well into the arena of unfounded speculation; it is best at present to return to firmer ground, accepting what has been learned about the broad features of the convection zone and the layers immediately beneath, and awaiting future observations to tell us what is really happening at greater depth.

References

Brown, T.M. 1985, *Nature* **317**, 591-594

Brown, T.M., Christensen-Dalsgaard, J., Goode, P.R., Gough, D.O. and Morrow, C.A. 1989, *Astrophys. J.* , **343**, 526-546

Christensen-Dalsgaard, J. and Schou, J. 1989, *Seismology of the sun and sun-like stars* (ed. V. Domingo and E.J. Rolfe, ESA SP-286, Noordwijk) 149-153

Claverie, A., Isaak, G.R., McLeod, C.P., van der Raay, H.B. and Roca Cortés, T. 1981, *Nature* , **293**, 443-445

Davis Jr, R. 1988. Seventh workshop in Grand Unification/Icoban '86 (ed. J. Arafune, World Scientific) 237-276

Delache, P.H. and Scherrer, P.H. 1983, *Nature* , **306**, 651-653

Duvall Jr, T.L. and Harvey, J.W. 1984, *Nature* , **310**, 19-22

Duvall Jr, T.L., Dziembowski, W.A., Goode, P.R., Gough, D.O., Harvey, J.W. and Leibacher, J.W. 1984, *Nature* , **310**, 22-25

Dziembowski, W.A., Goode, P.R. and Libbrecht, K.G., 1989, *Astrophys. J.* , **337**, L53-L57

Fröhlich, C. 1988, *Advances in helio- and asteroseismology* (*Proc. IAU Symp.* **123**, ed. J. Christensen-Dalsgaard and S. Frandsen, Reidel, Dordrecht) 83-86

Glatzmaier, G.A. and Gilman, P.A. 1982, *Astrophys. J.* , **256**, 316-330

Goode, P.R., Dziembowski, W.A., Rhodes Jr, E.J. and Korzennik, S.G. 1990, *Progress of seismology of the sun and stars* (ed. Y. Osaki and H. Shibahashi, Springer, Berlin) 349-352

Gough, D.O. 1990, *Phil. Trans. R. Soc. Lond.*, A**330**, 627-640

Hill, F., Gough, D.O., Toomre, J. and Haber, H.A. 1988, *Advances in helio- and asteroseismology* (*Proc. IAU Symp.* **123**, ed. J. Christensen-Dalsgaard and S. Frandsen, Reidel, Dordrecht) 45-48

Isaak, G.R. 1986, *Seismology of the sun and the distant stars* (ed. D.O. Gough, NATO ASI-C **169**, Reidel, Dordrecht) 223-228

Jefferies, S.M., McLeod, C.P., van der Raay, H.B., Pallé, P.L. and Roca Cortés, T. 1988, *Advances in helio- and asteroseismology* (*Proc. IAU Symp.* **123**, ed. J. Christensen-Dalsgaard and S. Frandsen, Reidel, Dordrecht) 25-28

Korzennik, S.G., Cacciani, A., Rhodes Jr, E.J. and Ulrich, R.K., 1990 *Progress in seismology of the sun and stars* (ed. Y. Osaki and H. Shibahashi, Springer, Berlin) 341-347

Merryfield, W.J., Toomre, J. and Gough, D.O. 1990, *Astrophys. J.* , **353**, 678-697

Mestel, L. and Weiss, N.O. 1987, *Mon. Not. R. astr. Soc.* , **226**, 123-135

Morrow, C.A., Gilman, P.A. and De Luca, E.E. 1988, *Seismology of the sun and sun-like stars*, (ed. V. Domingo and E.J. Rolfe, ESA SP-286, Noordwijk) 109-115

Pallé, P.L. and Roca Cortés, T. 1988, *Advances in helio- and asteroseismology* (*Proc. IAU Symp.* **123**, ed. J. Christensen-Dalsgaard and S. Frandsen, Reidel, Dordrecht) 79-82

Rhodes Jr, E.J., Cacciani, A. and Korzennik, S.G. 1990, *Progess in seismology of the sun and stars* (ed. Y. Osaki and H. Shibahashi, Springer, Berlin) 163-172

Rosner, R. and Weiss, N.O. 1985, *Nature* , **317**, 790-792

Schou, J. 1991, *Challenges to theories of the structure of moderate-mass stars* (ed. D.O. Gough and J. Toomre, Springer, Berlin) in press

Sekii, T. *Progess in seismology of the sun and stars* (ed. Y. Osaki and H. Shibahashi, Springer, Berlin) 163-172
Thompson, M.J. 1989 *Solar Phys.*, **125**, 1-12
Walén, C. 1946 *Ark. Mat. astr. Fys.*, A**33** (18)
Woodard, M.F. 1989, *Astrophys. J.* , **347**, 1176-1182

DISCUSSION

Roxburgh: Could you indicate the magnitude of the error bars for the curve showing the internal angular rotation rate for the Sun?

Gough: The errors are small near the surface, and very large near the center. Also, the way in which the internal rotation is calculated is not entirely deterministic because the modeler has to discriminate between the real signal and aliases. This could lead to systematic errors in the calculated rotation curves.

Rodonó : Just a comment on how one estimates the *real* and the *observed* rotation of the Sun. You showed some new data from Stenflo, but we should also note that the rotation rate derived from sunspots changes with the age of the spots, in the sense that larger sunspots show a faster rotation than smaller ones. And this is linked with the age, and thus with the depth into the Sun where those spots are anchored.

Gough: I agree with everything you have said except possibly the last statement which I am not qualified to comment upon.

Sofia: I want to mention a success of dynamo theory. My group has been very successful recently in predicting the value of the maximum sunspot number during the next maximum by measuring the poloidal field of the previous minimum. This may just be a coincidence, but I believe that it is encouraging.

STELLAR ACTIVITY BELTS AS POTENTIAL INDICATORS OF INTERNAL ROTATION AND ANGULAR MOMENTUM DISTRIBUTION

G. BELVEDERE
Istituto di Astronomia, Università di Catania, Italy

ABSTRACT. The most recent helioseismological data suggest the location of dynamo action in the boundary layer at the bottom of the convection zone, where the radial gradient of angular velocity changes its sign at some intermediate latitude $\lambda_o = 30° \div 35°$. According to the dynamo criterion, this allows equatorial propagation of dynamo waves at low latitudes and poleward propagation at high latitudes, in agreement with the evidence shown by different tracers of solar activity.

This has been confirmed by a non-linear dynamo model in a very thin shell ($0.05R_\odot$), thus giving a substantial support to the idea that the latitude distribution of solar activity belts is a natural and direct consequence of the internal profile of angular velocity.

Extending the argument to solar type stars, we conversely suggest that observation of latitude distribution and migration of active regions, by photometric and spectroscopic methods, can in principle allow to infer the internal rotation and angular momentum distribution.

In practice, future improvement of observational techniques from space may make it possible only in the course of the next decade.

Comparison with future stellar oscillations data from space would be very useful to test the boundary layer dynamo hypothesis on a large sample of objects.

1. INTRODUCTION

The most recent solar oscillations inversion data (Harvey 1988; Brown et al. 1989; Dziembowski et al. 1989; Libbrecht 1989; 1990) seem to substantially agree as to the internal profile of angular velocity in the Sun: the surface latitudinal differential rotation persists throughout the convection zone, whereas, beneath the boundary layer at the bottom of the convection zone itself, rigid rotation dominates, with angular velocity ω_o = 436 nHz or $2.74 \cdot 10^{-6}$ rad s^{-1}, corresponding to the surface value at latitude $\lambda_o \approx 32° \div 33°$ (Morrow 1988; Libbrecht 1990; see also釀Paternò 1990).

S. Catalano and J. R. Stauffer (eds.), Angular Momentum Evolution of Young Stars, 281–288.

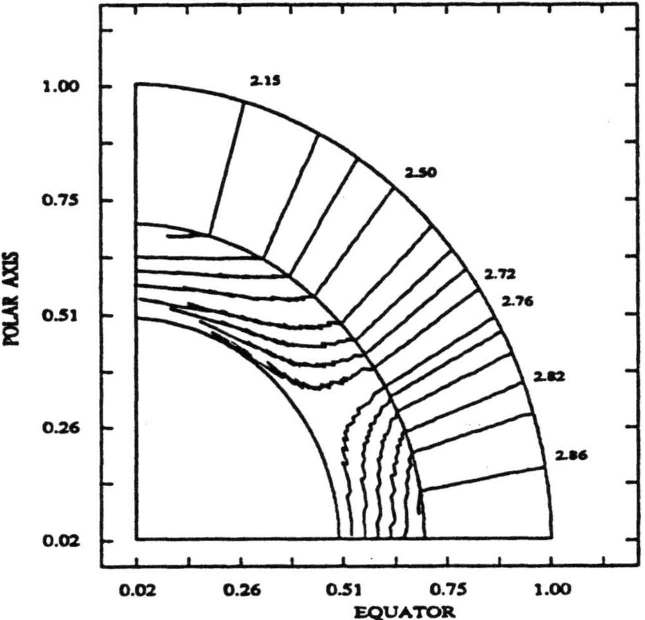

Figure 1: Isorotation surfaces in the solar convection zone and transition layer after Morrow (1988) and Paternò (1990). The internal rigid rotation, $2.74 \cdot 10^{-6}$ rad s $^{-1}$, corresponds to the surface one at latitude $\lambda = 32° - 33°$.

This implies that within the boundary layer between the radiative interior and the base of the convective zone ($0.65R_{\odot} - 0.70R_{\odot}$) the radial angular velocity gradient must change its sign at latitude λ_o (Fig. 1): in fact $\partial\omega/\partial r$ is positive for $\lambda < \lambda_o$ and negative for $\lambda > \lambda_o$. According to the $\alpha - \omega$ dynamo criterion (Parker 1955; Yoshimura 1975 b) based on the sign of the product $\Gamma = \alpha \cdot \partial\omega/\partial r$ and taking $\alpha < 0$ in the northern hemisphere boundary layer (Yoshimura 1975 a; Glatzmaier 1985 a,b), dynamo action is clearly supported by the helioseismological data (Belvedere 1990), as horizontal dynamo waves propagate polewards for $\lambda > \lambda_o$ and equatorwards for $\lambda < \lambda_o$, in agreement with the observational evidence shown by the latitude migration of different tracers of surface activity in the course of the solar cycle. Indeed, sunspots and most faculae, which belong to the equatorial activity belt ($\lambda < 35° \div 40°$), show equatorward migration, while polar faculae, filaments and large scale magnetic flux, which are characteristic of the polar activity belt ($\lambda > 40° \div 50°$) evidence poleward migration.

All this has a profound meaning as to the interpretation of solar activity (Belvedere et al. 1990 b):

(a) the latitude distribution of solar activity belts and the related migration are a natural and direct consequence of the internal profile of angular velocity, namely of the fact that the internal rigid rotation angular velocity has the same value as the surface one at $\lambda = \lambda_o$, which is just the fiducial latitude corresponding to the boundary between the equatorial and polar activity belts;

(b) mean field $\alpha - \omega$ dynamo in the boundary layer can still and better account for the observed phenomenological evolution in the course of the solar cycle, and keeps fit as the basic mechanism to understand solar and stellar magnetic activity.

2. THE BOUNDARY LAYER NON-LINEAR DYNAMO MODEL

The present picture has been given substantial support by the results of a non linear dynamo model in a very thin $(0.05R_\odot)$ spherical shell, representing the boundary layer, with full time and latitudinal resolution, and an integrated representation only in the radial direction (the so-called radial truncation).

The mean field dynamo equations for the time evolution of the magnetic field $\mathbf{B} = B(r,\theta)\phi + \nabla \times (A(r,\theta)\phi)$, where r, θ are spherical polar coordinates, ϕ is the unit vector in the azimuthal direction, Bϕ is the toroidal and $\mathbf{B}_p = \nabla \times A\phi$ the poloidal part of $\mathbf{B}$, are:

$$\frac{\partial A}{\partial t} = \alpha F(r,\theta)B + \eta_T\left[\nabla^2 - \frac{1}{r^2 \sin^2\theta}\right]A \tag{1}$$

$$\frac{\partial B}{\partial t} = r\sin\theta \mathbf{B}_p \cdot \nabla\left[\frac{U(r,\theta)}{r\sin\theta}\phi\right] + \eta_T\left[\nabla^2 - \frac{1}{r^2 \sin^2\theta}\right]B \tag{2}$$

Here αF is the usual α-effect, with F representing its spatial structure and α its magnitude. The quantity η_T is a turbulent diffusivity.

The dynamical influence of the magnetic field enters the model through its effect on the differential rotation $U(r,\theta) = u_o + u$, where u_o is a prescribed velocity field and u is a perturbation driven by the mean Lorentz force and subject to viscous damping.

The simplest equation that encompasses these features is:

$$\rho\frac{\partial u}{\partial t} = \frac{1}{\mu_o}[(\nabla \times \mathbf{B}) \times \mathbf{B}]_\phi + \rho\nu_T\left[\nabla^2 - \frac{1}{r^2 \sin^2\theta}\right]u \tag{3}$$

where ν_T is a turbulent viscosity.

The radially truncated model equations have been solved numerically with an explicit time stepping method of DuFort-Frankel type with suitable boundary conditions. For full details we address the reader to the extensive paper by Belvedere, Pidatella and Proctor (1990 a) who worked out a similar model in the solar convection zone.

Assuming the differential rotation profile u_o in the boundary layer as given by interpolating the most recent helioseismological data, the results show the existence of

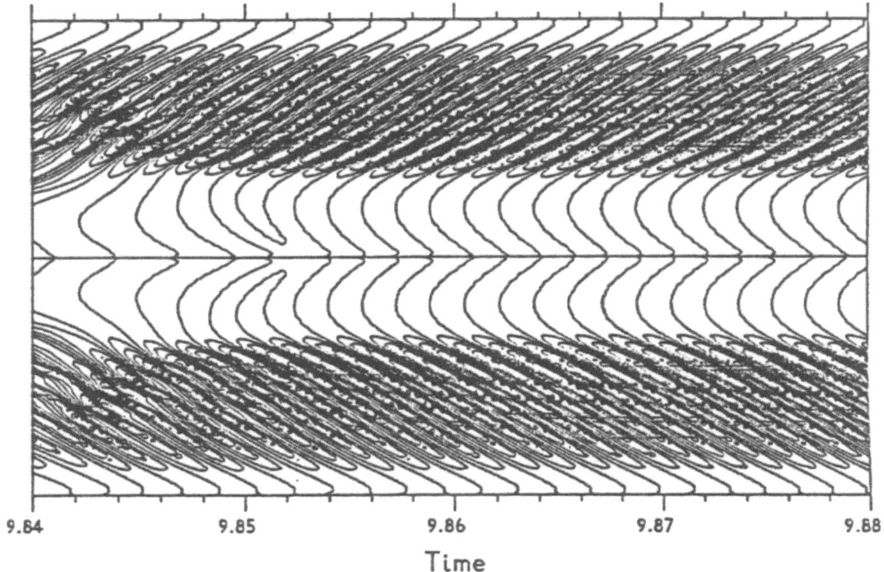

Figure 2: Butterfly diagram for $d = 0.05R_{odot}$ and D = -1 .

periodic dynamo wave-like stable solutions with both equatorward and poleward migrating branches, as shown by the related butterfly diagram (Fig. 2).

These solutions are found for dynamo number $D = \alpha\omega_o d^3/\eta^2{}_T \approx -1$, where d is the thickness of the boundary layer and ω_o is the surface angular velocity at latitude $\lambda = \lambda_o$. Incidentally, the relatively small value of $|D|$ is a consequence of the very steep radial angular velocity gradient in the thin layer, since for the onset of dynamo action $|\alpha \cdot \partial\omega/\partial r|$ must exceed a characteristic value.

The relative latitude extent and magnetic field intensity of the polar and equatorial branches shown in Fig. 2 is somewhat artificial, as it depends on the free choice of the latitudinal variation of the model parameter α, which has only to satisfy the requisite of antisymmetry with respect to the equator. Here we adopted $\alpha(\theta) \approx \sin\theta\cos\theta$, where θ is the polar angle, this giving more emphasis to the polar branch.

However, the really significant result of our numerical simulation is that horizontally propagating dynamo waves, with equatorial and polar branches, can exist even in a very thin layer, such as the boundary layer at the bottom of the convection zone, where the radial angular velocity gradient changes its sign at latitude $\lambda = \lambda_o$.

Of course it remains to be understood why the internal rotation is rigid and its angular velocity coincides with the surface one just at latitude $\lambda = \lambda_o$, and why do spots occur only for $\lambda < 35° - 40°$. A great amount of detailed study of the interaction of rotation, convection and magnetic field in the solar interior is clearly needed to try to explains these facts.

3. INVERTING THE ARGUMENT: STELLAR ACTIVITY BELTS AS A PROBE OF INTERNAL ROTATION

Although many uncertainties do still exist even in the solar case, nevertheless we are tempted to suggest that, conversely, observation of latitude distribution and migration of stellar activity complexes on late main sequence (G,K,M) slowly rotating stars, by photometric and spectroscopic methods, may in principle allow to infer the internal rotation profile and angular momentum distribution in a conceptually simple and direct way, offering a powerful tool to investigate their dependence on stellar parameters such as spectral type, average rate of rotation and age. For such a purpose we should need accurate measurements of: (i) the angular velocity at some fixed latitude (e.g. equatorial); (ii) surface differential rotation; (iii) latitude drift of activity belts over a suitable time span.

This way it should be possible to determine the surface latitude at which the direction of migration changes its sign and the corresponding angular velocity, that are the essential data in order to deduce the internal rotation profile.

Unfortunately, the present available observational data of surface distribution of stellar active regions, obtained by photometric and spectroscopic methods (e.g. Rodonò et al. 1986; Vogt 1983 (Doppler Imaging Method); Vogt and Hatzes 1990), only refer to binary, fast rotating, hyperactive stars of RS CVn and BY Dra type - whereas analogy to the Sun requires single, slowly rotating, mild activity stars - and show a clear tendency of active regions to be concentrated at high latitudes (polar belts: Fig. 3). Moreover, the sensitivity and resolution power of present observational capabilities are far from making it possible to perform high precision measurements such as those we suggest here.

However, future improvement of the observational techniques based on the analysis of both light curves and line profiles (to the latter regard, we also mention the Zeeman-Doppler Imaging Method (Donati et al. 1989) which allows surface magnetic cartography), and observation from space with large instruments (we refer expecially to the Spectrum Ultraviolet satellite project, SUV) may make it possible in the next decade.

Further, comparison between surface activity data and acoustic oscillations data for a large sample of stars (this may be performed in the framework of a single project, such as the PRISMA project) would be very useful in order to test the validity of the internal rotation probing method, whose basic principles are outlined here, and the reliability of the boundary layer hypothesis.

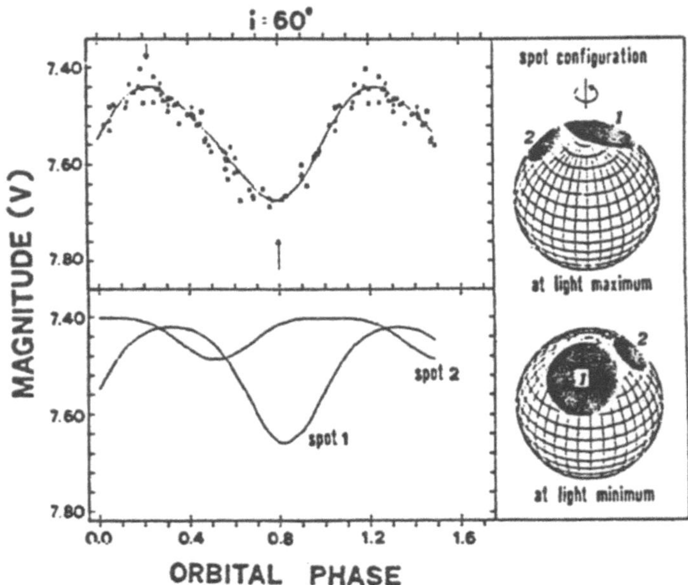

Figure 3: Spot configurations for II Peg (1981.8) as deduced from the analysis of the light curve (Rodonò et al. 1986). The bent of spots for being located at high latitudes seems to be fairly independent of the assumed inclination angle (here $i = 60°$).

4. REFERENCES

Belvedere, G.: 1990, in *Inside the Sun*, Proc. IAU Coll. **121**, G. Berthomieu and M. Cribier eds., Kluwer, Dordrecht, p. 371.

Belvedere, G., Pidatella, R.M., and Proctor, M.R.E.: 1990 a, *Geophys. Astrophys. Fluid Dyn.* **51**, 263.

Belvedere, G., Proctor, M.R.E., and Lanzafame, G.: 1990 b, in *The Sun and Cool Stars: activity, magnetism, dynamos*, Proc. IAU Coll. **130**, I. Tuominen ed., Springer-Verlag, Berlin, in press.

Brown, T.M., Christensen-Dalsgaard, J., Dziembowski, W., Goode, P.R., Gough, D.O. and Morrow, C.A.: 1989, *Astrophys. J.* **343**, 526.

Donati, J.F., Semel, M., and Praderie, F.: 1989, *Astron. Astrophys.* **225**, 467.

Dziembowski, W.A., Goode, P.R. and Libbrecht, K.G.: 1989, *Astrophys. J.* **337**,

Glatzmaier, G.A.: 1985a, *Geophys. Astrophys. Fluid Dyn.* **31**, 137.

Glatzmaier, G.A.: 1985b, *Astrophys. J.* **291**, 300.

Harvey, J.W.: 1988, in *Seismology of the Sun and Sun-like Stars*, E.J. Rolfe ed., ESA-SP **286**, 55.

Libbrecht, K.G.: 1989, *Astrophys. J.* **336**, 1092.

Libbrecht, K.G.: 1990, priv. comm.

Morrow, C.A.: 1988, in *Seismology of the Sun and Sun-like Stars*, E.J. Rolfe ed., ESA-SP **286**, 91.

Parker, E.N.: 1955, *Astrophys. J.* **122**, 293.

Paternò, L.: 1990, in *The Sun and Cool Stars: activity, magnetism, dynamos*, Proc. IAU Coll. **130**, I. Tuominen ed., Springer-Verlag, Berlin, in press.

Rodonò, M. et al.:1986, *Astron. Astrophys.* **165**, 135.

Vogt, S.S.: 1983, in *Activity in Red-Dwarf Stars*, Proc. IAU Coll. **71**, P.B. Byrne and M. Rodonò eds., Reidel, Dordrecht, p. 137.

Vogt, S.S. and Hatzes, A.P.: 1990, in *The Sun and Cool Stars: activity, magnetism, dynamos*, Proc. IAU Coll. **130**, I. Tuominen ed., Springer-Verlag, Berlin, in press.

Yoshimura, H.: 1975 a, *Astrophys. J. Suppl.* **29**, 467.

Yoshimura, H.: 1975 b, *Astrophys. J.* **201**, 740.

DISCUSSION

Dziembowski: What is the relative amplitude of your function U_o at the equator?

Belvedere: U_o is about 100, if I remember correctly.

Dziembowski: If so, there is another possibility to test your hypothesis, because measurement of the angular velocity variations at the equator and at the bottom of the convection zone are extremely accurate - the current limit is something like four nanohertz, comparing the solar data from various years. This is a very strict limit, but still consistent with your one part in 100.

Sofia: Do you think that in the near future you will be able to to use your formulation to forecast the strength of activity cycles.

Belvedere: I think it is possible, but it will probably be 5 to 10 years before it becomes technically feasible to measure the amplitude of magnetic field variations in stars other than the Sun.

Gough: What is the source of energy that drives your dynamo oscillations?

Belvedere: Rotation and convection.

Gough: So, it extracts energy from rotation?

Belvedere: Yes. This is the usual source for dynamo activity.

Cameron: Would you be happy to predict, using your model, what you might expect to see if the core of a K0 dwarf with a long rotation period had a core rotation rate five times faster than the surface rate? We have some very good Doppler imaging candidates where - if the theories about spindown in the Pleiades which involve differential rotation in the core are correct - the core should be rotating much faster than the surface. With Doppler imaging, we should be able to measure their spot sizes and distributions.

Belvedere: I think this model is independent of core velocity - it just doesn't appear in the equations. Anyway, who knows?

DYNAMICS OF SPOT GROUPS AND ROTATION OF THE EXTERNAL CONVECTIVE LAYERS IN THE SUN AND MAGNETICALLY ACTIVE STARS

A. F. Lanza [1], M. Rodonò [1,2], R. A. Zappalà [2]
[1] *Istituto di Astronomia dell'Università degli Studi, and*
[2] *Osservatorio Astrofisico, Città Universitaria*
Viale A.Doria 6, 95125 Catania, Italy

Abstract
The observed kinematics of sunspot groups and a simplified model, which describes the coupling between magnetized and unmagnetized plasmas, are used to derive some information on the rotation of the subphotospheric layers of the Sun. A comparison with present theoretical models and an extrapolation to some aspects of the phenomenology observed in the RS CVn binaries are presented.

1 Introduction

The study of the dynamics of the surface magnetic fields on the Sun and solar-like stars is of fundamental importance for the understanding of the structure and energetics of the outer atmospheres, including possibly processes which lead to mass loss and related angular momentum loss, as well as for studing the internal rotation and dynamo action. However, magnetic fields are not always passive as they interact with the surrounding plasma flows in a manner which is not at present satisfactorily understood. The study of the dynamical behaviour of magnetic structures in the atmospheres and in the outer convective layers of the Sun or of magnetically active stars, is therefore a very complicated task and many points need to be still clarified.

In this work we shall deal with sunspot groups dynamics and derive some hints on the rotation of the outer convective layers of the Sun, for which the presently available helioseismological data cannot give conclusive results (Hill, 1987, Woodard and Libbrecht, 1989).

The extension of these considerations to magnetically active stars, in particular to a sample of RS CVn binaries, is worth to be considered because they could help in interpreting the observed variations of the photospheric rotation period derived from spot rotational modulation data. We present the following ideas in order to stimulate further research and debate on this field.

S. Catalano and J. R. Stauffer (eds.), Angular Momentum Evolution of Young Stars, 289–296.

2 Observations on sunspot groups rotation

Sunspots and sunspot groups are commonly used as tracers of solar photospheric rotation. However, these magnetic structures exibit an age dependent angular velocity, which likely originates in the dynamical interaction between magnetized and unmagnetized plasmas in subphotospheric layers.

Using the Greenwich Photoheliographic Results for the years 1874-1976, Zappalà and Zuccarello (1989, 1990) studied the angular velocity of sunspot group barycentres. They found that, at each given latitude, the angular velocity of groups younger than about 10 days is higher, on the average, than the angular velocity of the photospheric plasma; the maximum difference is about 0.25 deg/day.

The average difference $\Delta\Omega_s$ is a monotonically decreasing function of group age; after about 5×10^6 s since its first appearence, the group angular velocity approaches that of the photospheric plasma at the same latitude.

$\Delta\Omega_s$ turns out to be dependent on group age only, with no significant dependence on latitude, group area or Zurich type.

3 Interpretation and hints on solar internal rotation

We assume Schussler's suggestion (1987) that the rotation of a spot group is dynamically coupled and is equal to that of the unmagnetized plasma at a depth z_a and that z_a becomes smaller and smaller (that is the group is coupled to more and more external layers) as it evolves, due to magnetic turbolent diffusion.

In this interpretation the dependence of group angular velocity on age gives a tool to probe the solar internal angular velocity.

Let us express the angular velocity profile in the external convective layers by the formula:

$$\Omega(r,\theta) = \Omega_0[\omega_0(r) + \omega_2(r)P_2(\cos\theta)] \tag{1}$$

where Ω_0 is a constant and P_2 is the second order Legendre polynomial, r is the radial distance from the centre of the Sun, θ the colatitude measured from the north pole.

In the light of the above interpretation the independence of $\Delta\Omega_s$ on latitude yields:

$$\omega_2 = cost. \tag{2}$$

$$\frac{d\omega_2}{dr} = 0 \tag{3}$$

while its decrease in time yields:

$$\frac{d\omega_0}{dr} < 0 \tag{4}$$

in the layers where spots are anchored.

4 Comparison with models of solar internal rotation

We now compare these hints on solar internal rotation with the theoretical studies of Durney (1985, 1987), Gilman and Foukal (1979), Gilman (1980). They found that in the convection zone of a star rotating with an average surface angular velocity Ω, two main dynamical domains can be distinguished, according to the values of the adimensional number Ro, which we define as $Ro = 2\Omega\tau$ (τ is the dominant eddy lifetime). In the lower part of the convective zone $Ro \gg 1$ and convective dynamics is strongly influenced by rotation, in the upper part $Ro \ll 1$ and rotationally induced perturbations on convective dynamics are much less significant.

For the study of spot group dynamics we are essentially interested in the upper domain where $Ro \leq 1$, thereinafter called the *external layer.*

The conclusions of the above quoted authors on the dynamics of the external layer can be summarized as follows:

- the latitudinal differential rotation generated within it is neglegible ($\frac{d\omega_2}{dr} = 0$);
- the tendency to a conservation of the angular momentum per unit mass during convectively driven motions and the action of the turbolent stresses lead to an angular velocity increasing inward:

$$\frac{d\omega_0}{dr} \simeq -\frac{1}{r} \tag{5}$$

In the framework of these models the latitudinal differential rotation observed at the surface is generated in the lower region $Ro \geq 1$ or possibly as the convective zone relaxes from the $Ro > 1$-state in the lower part to the $Ro < 1$-state in its upper part. Therefore, the latitudinal differential rotation observed in the external layer would merely reflect the rotation regime of the layers below it.

Considering the radial dependence of the angular velocity, we find that the order of magnitude of the maximum difference between the base and the top of the external layer is:

$$\frac{\Delta\Omega}{\Omega} \simeq \frac{d}{R} \tag{6}$$

where d is the total depth of the external layer. This approximation is valid for $d \ll R$.

Using a standard mixing length model (Belvedere et al., 1980), we estimated Ro as a function of depth. The condition $Ro = 1$ is attained at $d \simeq 2.1 \times 10^7$ m for the Sun, comparable with the depth of the supergranulation. This yields to a:

$$\left(\frac{\Delta\Omega}{\Omega}\right)_{com} = 3.0 \times 10^{-2} \tag{7}$$

to be compared with the observed maximum value for sunspot groups:

$$\left(\frac{\Delta\Omega}{\Omega}\right)_{obs} = 2.6 \times 10^{-2} \tag{8}$$

The orders of magnitude agree well giving credit to the proposed explanation.

5 Extension of the model to active stars: preliminary results

We applied the proposed model also to the magnetically active components of a sample of RS CVn systems (Lanza et al., 1990) determining the value of $\frac{\Delta\Omega}{\Omega}$ for each star. They are plotted versus the angular velocity Ω in Figure 1.

The rotation period of photospheric starspots on these stars differs slightly from the orbital period, the relative difference being in the range $10^{-2} - 10^{-4}$.

Moreover, in the best studied systems the photometric period turns out to be variable with variations of the same order, $\frac{\Delta P}{P} \sim 10^{-2} - 10^{-4}$, on various timescales (Catalano, 1983; Busso et al., 1984, 1985, 1986; Rodonò, 1986).

In Figure 2 we plotted $\frac{\Delta P}{P}$ versus the orbital angular velocity Ω. $\frac{\Delta P}{P}$ is the maximum relative difference between photometric periods for those stars for which at least two photometric periods were measured. Instead, for those stars for which only one photometric period was measured, the $\frac{\Delta P}{P}$ is the relative differences between photometric and orbital period.

We see that $\frac{\Delta P}{P}$ decreases with increasing Ω and at each Ω a significative dispersion in $\frac{\Delta P}{P}$ is present. The error of $\frac{\Delta P}{P}$ is of the order of 20%.

In Figure 3 the computed $\frac{\Delta\Omega}{\Omega}$ versus the observed $\frac{\Delta P}{P}$ for each star in our sample are plotted.

The most part of the points lie near the bisecting line. Only the group of very rapid rotators ($\log\Omega > -4.5$) presents a significative difference between the observed and computed $\frac{\Delta\Omega}{\Omega}$. The point distribution in Figure 3 can be interpreted considering that the computed $\frac{\Delta\Omega}{\Omega}$ are essentially upper limits for the observable $\frac{\Delta P}{P}$ because the range through which the coupling depth varies could be narrower than d, the total depth of the external layer.

These results stimulated us to propose the following *working hypothesis*:

- the difference between photometric and orbital period and the differences among photometric periods observed in RS CVn systems are due to differences in the angular velocity of the spot pattern, which reflect the radial angular velocity gradient in the external layer.

Starting from this hypothesis, we are presently developing a model which tries to explain the observed phenomenology. Some interesting preliminary results are the following:

- because of the latitude independence of $\Delta\Omega_s$ this angular velocity difference is detectable in disk integrated observations; the angular velocity determined by photometry reflects the average anchor depth or, equivalently, the average age of the spot pattern present in photosphere; therefore, the computed $\frac{\Delta\Omega}{\Omega}$ must be regarded as an upper limit for the observable photometric period variations $\frac{\Delta P}{P}$;

- large magnetic structures anchored on the external layer should be able to resist to the shearing action of latitudinal differential rotation: therefore, latitude independent

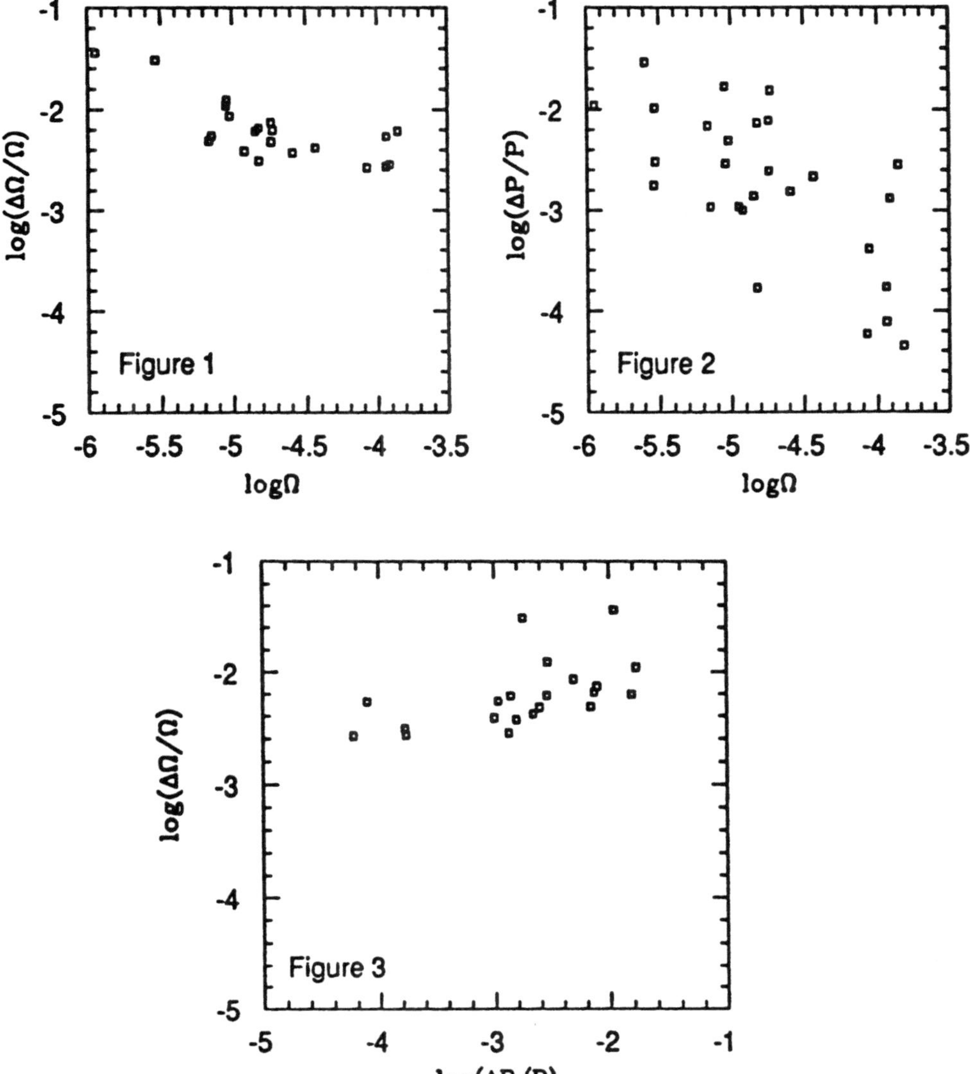

Figure 1: the computed values of the maximum $\frac{\Delta\Omega}{\Omega}$ plotted versus Ω for a sample of RS CVn systems.

Figure 2: the observed maximum values of $\frac{\Delta P}{P} \simeq \frac{\Delta\Omega}{\Omega}$ plotted versus the orbital angular velocity Ω.

Figure 3: the computed values of $\frac{\Delta\Omega}{\Omega}$ plotted versus the observed values of $\frac{\Delta P}{P}$.

variations in angular velocity could be the only ones showed by such a large magnetic structures;

- the coupling torque due to tidal interaction is sufficient to explain the nearly syncronous rotation of the active component in RS CVn systems, but may not be able to explain the extreme tightness of the coupling implied by the very long periods for the migration of the photometric waves (Scharlemann, 1981, 1982);
- the combined action of tidal interactions and of magnetic structures connecting the two stars in RS CVn systems (Simon et al., 1980; Mutel and Morris, 1988) might explain the presence of direct or reversal migration of the photometric waves and the reversal of migration direction observed in some systems.

6 References

- Belvedere, G., Paternò, L., Roxburgh, I. W.: 1980, *Astron. Astrophys.*, **91**, 356–359.
- Busso, M., Scaltriti, F., Blanco, C., Catalano, S., Marilli, E., Pazzani, V., Rodonò, M.: 1984, *Astron. Astrophys.*, **135**, 255–260.
- Busso, M., Scaltriti, F., Cellino, A.: 1985, *Astron. Astrophys.*, **148**, 29.
- Busso, M., Scaltriti, F., Cellino, A.: 1986, *Astron. Astrophys.*, **156**, 106-110.
- Catalano, S.: 1983, in *Activity in Red Dwarf Stars*, Eds. P. B. Byrne and M. Rodonò, Reidel, Dordrecht, p. 343–361.
- Durney, B. R.: 1985, *Astrophys. J.*, **297**, 787-798.
- Durney, B. R.: 1987, in *The Internal Solar Angular Velocity*, Eds. B. R. Durney and S. Sofia, Reidel, Dordrecht, p. 235–262.
- Gilman, P. A.: 1980, in *Stellar Turbolence*, Eds. D. F. Gray and J. L. Linsky, Springer-Verlag, Berlin, p. 19-37.
- Gilman, P. A. and Foukal, P. V.: 1979, *Astrophys. J.*, **229**, 1179–1185.
- Hill, F: 1987, in *The Internal Solar Angular Velocity*, Eds. B. R. Durney and S. Sofia, Reidel, Dordrecht, p. 45–50.
- Lanza, A. F., Rodonò, M., Zappalà, R. A.: 1990, Catania Preprint 16/1990.
- Mutel, R. L. and Morris, D. H.: 1988, in *Activity in Cool Star Envelopes*, Eds. O. Havnes et al., Kluwer, Dordrecht, p. 283–288.
- Rodonò, M.: 1981, in *Photometric and Spectroscopic Binary Systems*, Eds. E. B. Carling and Z. Kopal, Reidel, Dordrecht, p. 285–304.

- Rodonò, M.: 1986, in *Cool Stars, Stellar Systems and The Sun*, Eds. M. Zeilik and D. M. Gibson, Springer-Verlag, Berlin, p. 475–487.
- Scharlemann, E. T.: 1981, *Astrophys. J.*, **246**, 292–305.
- Scharlemann, E. T.: 1982, *Astrophys. J.*, **253**, 298–308.
- Schussler, M.: 1987, *The Internal Solar Angular Velocity*, Eds. B. R. Durney and S. Sofia, Reidel, Dordrecht, p. 303–320.
- Simon, T., Linsky, J. L., Schriffer, F. H.: 1980, *Astrophys. J.*, **239**, 911.
- Woodard, M. F., and Libbrecht, K. G.: 1989, in *Seismology of the Sun and Sun-like Stars*, Eds. V. Domingo and E. J. Rolfe, ESA-SP 286, p. 67-71.
- Zappalà, R. A. and Zuccarello, F.: 1989, *Astron. Astrophys.*, **214**, p. 369–374.
- Zappalà, R. A. and Zuccarello, F.: 1990, *Astron. Astrophys.*, in press.

DISCUSSION

Ternullo: I should like to give just a small comment. I suspect that the observational scenario upon which the theoretical models are constructed, as well as extrapolations from the Sun to other stars, is still poorly understood. We do not know the real fine structure of the butterfly diagram. When starting the last solar cycle, we observed the spots were relatively uniformly distributed and had an unusual configuration. The spots were not concentrated around one central activity center. Instead, there were about 7 activity centers, and their centers seemed to shift equatorward as the solar cycle proceeded. This complicates the phenomenological input to the theoretical models, and the attempt to extrapolate to other stars.

Lanza: In the case of stars, the latitude dependence of the spot location is not precisely known. But it could be possible that the enormous extension of stellar magnetic structures could give rise to structures which are rigid enough to overcome the shearing effect of the latitudinal differential rotation. This is a point which we are now studying, but I have no firm results.

THEORY OF MAGNETIC BRAKING OF LATE-TYPE STARS

A. COLLIER CAMERON, LI JIANKE and L. MESTEL
Astronomy Centre
University of Sussex
Falmer, Brighton BN1 9QH
England.

ABSTRACT. We review the basic theory of angular momentum loss via a magnetically channelled stellar wind. Various forms of power-law dependence of the instantaneous braking rate on the stellar rotation rate are derived. Observational constraints on the rotational evolution of young solar-type stars in open clusters and the field are used to calibrate and test the braking laws at rotation rates up to 100 times the solar rate.

1. Introduction

Since Kraft's (1967) pioneering studies of the rotational evolution of solar-type stars, it has been apparent that significant angular momentum loss occurs during the main sequence lifetimes of such stars. The magnetic fields generated by the interaction of convection and rotation in stars with outer convective envelopes are ultimately responsible for driving and channelling a hot, ionised stellar wind. At the same time, magnetic torques transfer angular momentum from the star to the stellar wind.

In the first part of this paper we review the basic theory of angular momentum loss via a steady stellar wind flowing outward along an axisymmetric, open magnetic field configuration rotating at the same rate as the stellar surface. In the simplest class of model, where the magnetic field in the thermally-driven wind is approximately radial and the field strength scales linearly with the rotation rate, the braking rate $\dot{\Omega} \propto -\Omega^3$, leading to the familiar $\Omega \propto t^{-1/2}$ braking law.

A more realistic model yields a somewhat reduced braking efficiency at low rotation rates, due to the trapping of some of the coronal gas within a closed-field "dead zone". At high rotation rates, centrifugal force both reduces the extent of the dead zone and leads to sporadic ejections of cooled condensed gas from the zone. Braking by the wind is affected by centrifugal driving, which brings the Alfvén surface nearer to the star, and by ill-understood rotationally-dependent changes in the base temperature and density of the wind zone.

These models give reasonable fits to the rotational evolution of older solar-type stars, but do not explain satisfactorily the angular momentum evolution of the very rapidly rotating young low-mass stars in clusters with ages less than a few times 10^7 y. Although the majority

S. Catalano and J. R. Stauffer (eds.), Angular Momentum Evolution of Young Stars, 297–314.

of stars in these clusters appear to arrive on the zero-age main sequence with relatively low rotation rates, the statistical "tail" of rapidly-rotating G and K stars somehow manage to reduce their surface equatorial rotation speeds from over $140\,\mathrm{km\,s^{-1}}$ to less than $40\,\mathrm{km\,s^{-1}}$ on a timescale of 2×10^7 y or so. This is an order of magnitude more rapid than conventional wind braking models can explain. Moreover, the G stars appear to spin down more rapidly than the K stars, which is inconsistent with their behaviour at lower rotation rates. Here we use a simple parametrised model to investigate the possibility that a rapid initial reduction in surface rotation rate may be achieved through a weak coupling between the radiative stellar interior and the convective envelope, so that only the envelope is braked initially.

2. The Wind Equations

2.1. THE INDUCTION EQUATION

A stellar wind blowing outward in the presence of an axisymmetric magnetic field configuration will to a good approximation interact with the field according to the perfect conductivity induction equation:

$$\frac{\partial \boldsymbol{B}}{\partial t} = \nabla \times (\boldsymbol{v} \times \boldsymbol{B}). \tag{1}$$

In a steady state this becomes

$$\nabla \times (\boldsymbol{v} \times \boldsymbol{B}) = 0 \Rightarrow \boldsymbol{v} \times \boldsymbol{B} = \nabla\Phi \tag{2}$$

where Φ is the single valued electric potential.

For convenience we separate the velocity $\boldsymbol{v}$ and the field $\boldsymbol{B}$ into poloidal vectors lying in the meridional plane, and toroidal components in the azimuthal direction:

$$\boldsymbol{v} = \boldsymbol{v}_{\mathrm{p}} + v_\phi \boldsymbol{i}_\phi, \qquad \boldsymbol{B} = \boldsymbol{B}_{\mathrm{p}} + B_\phi \boldsymbol{i}_\phi.$$

If the system is axisymmetric, $\nabla\Phi$ and so also $\boldsymbol{v} \times \boldsymbol{B}$ has no ϕ-component, whence the toroidal part $\boldsymbol{v}_{\mathrm{p}} \times \boldsymbol{B}_{\mathrm{p}}$ of $\boldsymbol{v} \times \boldsymbol{B}$ must be zero: the poloidal component $\boldsymbol{v}_{\mathrm{p}}$ (the "wind velocity") must be parallel to $\boldsymbol{B}_{\mathrm{p}}$, i.e.

$$\boldsymbol{v}_{\mathrm{p}} = \kappa \boldsymbol{B}_{\mathrm{p}}. \tag{3}$$

where κ is a scalar function of position.

The azimuthal component of the local fluid velocity can be expressed as $v_\phi = R\omega$, where ω is the local angular velocity of the fluid, and R is the distance from the rotation axis. If we use this and expand the induction equation (2) in cylindrical polar coordinates (R, ϕ, z) it can be shown (Chandrasekhar 1956; Mestel 1961) that

$$\begin{aligned} \nabla \times (\boldsymbol{v} \times \boldsymbol{B}) &= \frac{\partial}{\partial R}(R\omega B_R - \kappa B_R B_\phi) - \frac{\partial}{\partial z}(\kappa B_z B_\phi - R\omega B_z) \\ &= R(\nabla \cdot \boldsymbol{B}_{\mathrm{p}} + \boldsymbol{B}_{\mathrm{p}} \cdot \nabla)\left(\omega - \frac{\kappa B_\phi}{R}\right). \end{aligned} \tag{4}$$

Under axial symmetry $\nabla \cdot \boldsymbol{B} = \nabla \cdot \boldsymbol{B}_{\mathrm{p}} = 0$ and $\boldsymbol{B} \cdot \nabla \equiv \boldsymbol{B}_{\mathrm{p}} \cdot \nabla$; (2) and (4) then yield

$$\boldsymbol{B} \cdot \nabla \left(\omega - \frac{\kappa B_\phi}{R}\right) = 0 \tag{5}$$

so that $\omega - \kappa B_\phi/R$ is constant along field lines. In the simplest model the star is taken as rotating with uniform angular velocity Ω. Since below the slow magnetosonic point (see below) v_p given by (3) exponentiates down to small values, (5) reduces to

$$\omega - \frac{\kappa B_\phi}{R} = \Omega. \tag{6}$$

In a strictly axisymmetric model one can easily incorporate the expected equatorial acceleration of the stellar surface by allowing Ω to vary from one field line to another, but the modification to the braking rate will be modest. With a more realistic, non-axisymmetric field structure on the stellar surface it is questionable whether one can extrapolate the perfect conductivity equation (2) into the turbulent sub-photospheric region. One rather expects the whole magnetic structure to be moving round at a uniform mean rotation rate Ω – identified in the wind theory as the "stellar rotation" – with turbulent resistivity allowing the persistence of equatorial acceleration in a state that is steady in the frame rotating with Ω. From now on we shall therefore adopt (6) in the wind zone with Ω a constant.

2.2. MASS CONTINUITY

In the steady state, the continuity equation yields

$$0 = \nabla \cdot (\rho \boldsymbol{v}) = \nabla \cdot (\rho \boldsymbol{v}_\mathrm{p}) = \nabla \cdot (\rho\kappa \boldsymbol{B}_\mathrm{p}) = (\boldsymbol{B}_\mathrm{p} \cdot \nabla)\rho\kappa, \tag{7}$$

so that $\rho\kappa \equiv \eta$ is constant along field-streamlines; as the cross-sectional area A of an infinitesimal poloidal flux tube varies with distance from the star, both the material flux $\rho v_\mathrm{p} A$ and the magnetic flux $B_\mathrm{p} A$ stay constant, yielding

$$\frac{\rho v_\mathrm{p} A}{B_\mathrm{p} A} = \rho\kappa = \eta. \tag{8}$$

2.3. EQUATION OF MOTION

The rate at which angular momentum is carried by the wind out of unit volume is

$$\rho \boldsymbol{v} \cdot \nabla(\omega R^2) = \rho\kappa \boldsymbol{B} \cdot \nabla(\omega R^2) \tag{9}$$

from Eq. (3). In a steady state this must be made up by the moment of the toroidal component of the Lorentz forces operating on the unit volume:

$$\begin{aligned} \frac{1}{4\pi} R[(\nabla \times \boldsymbol{B}) \times \boldsymbol{B}]_\phi &= \frac{1}{4\pi} R[-(\nabla \times \boldsymbol{B})_\mathrm{R} B_z + (\nabla \times \boldsymbol{B})_z B_\mathrm{R}] \\ &= \frac{1}{4\pi}\left[B_\mathrm{R}\frac{\partial}{\partial R} + B_z \frac{\partial}{\partial z}\right](RB_\phi) \\ &= \frac{1}{4\pi}(\boldsymbol{B} \cdot \nabla(RB_\phi)). \end{aligned} \tag{10}$$

The ϕ-component of the equation of motion can thus be written with the help of (7) as

$$\boldsymbol{B} \cdot \nabla \left(\frac{RB_\phi}{4\pi} - \rho\kappa\omega R^2\right) = 0, \tag{11}$$

so that

$$-\frac{RB_\phi}{4\pi} + \rho\kappa\omega R^2 \equiv -\frac{\beta}{4\pi}. \tag{12}$$

Here $-\beta/4\pi$, constant on field-streamlines, is the steady flow of angular momentum along a unit flux tube, carried jointly by the gas flow and the moment of the Maxwell stresses. In other words, the flow of gas parallel to the poloidal field carries angular momentum $\rho v_p \omega R^2 A = (\rho\kappa\omega R^2)B_p A$ per second along an infinitesimal flux tube with local cross-sectional area A. This quantity changes with distance along the flux tube, but a steady distribution of angular momentum is maintained by the magnetic stresses transporting angular momentum along the tube at a rate $(-RB_\phi/4\pi)B_p A$ per second (Lüst & Schlüter 1955).

The essential physics of the situation can be likened to the flow of water through an elastic hose attached to a garden sprinkler head rotating at constant angular velocity. The water must flow parallel to the local direction of the hose, and the system must obey mass continuity. As it flows outward, the water in the hose has insufficient angular momentum to maintain co-rotation: the outer parts of the hose will tend to lag behind the inner parts, and the hose assumes a spiral form. Note that the local angular velocity of the fluid now depends on the flow velocity through the hose and on the local angle of the hose to the radial direction: it is *not* the same as the local angular velocity of the hose itself, which must be the same at all points along the hose. The tension in the hose due to its local curvature will now exert a torque on the water flowing in its interior. In the steady state, the local angle of the hose to the radial direction is related to its local curvature by the torque balance condition: the need to maintain a constant total angular momentum flux defines the shape of the hose.

2.4. EFFECTIVE CO-ROTATION

We now eliminate B_ϕ between (6), (12) and (8) to obtain

$$\omega\left(1 - \frac{4\pi\eta^2}{\rho}\right) = \Omega + \frac{\eta\beta}{\rho R^2}. \tag{13}$$

Note that

$$\frac{4\pi\eta^2}{\rho} = \frac{4\pi\rho v_p^2}{B_p^2} = \frac{v_p^2}{v_{Alf}^2},$$

where v_p is the poloidal component of the wind velocity and v_{Alf} is the local Alfvén speed defined by the poloidal field. Close to the star, $v_p \ll v_{Alf}$, but as the wind density ρ decreases outward, we eventually reach the critical *Alfvénic point* $R = R_A$ where $v_p \equiv v_A = v_{Alf}$ and so $\rho \equiv \rho_A = 4\pi\eta^2$. At this point Eq. (13) becomes singular unless

$$\Omega = -\frac{\eta\beta}{\rho_A R_A^2} = \left(-\frac{\beta}{4\pi}\right)\frac{1}{\eta R_A^2}. \tag{14}$$

This means that the total angular momentum flux along a unit flux tube is

$$-\frac{\beta}{4\pi} = \eta\Omega R_A^2 = \rho\kappa\Omega R_A^2. \tag{15}$$

It is apparent from this expression that the *total* angular momentum flux carried along the tube is *equivalent* to that which would be carried by the steady matter flux if the entire flow were kept in strict co-rotation at the stellar angular velocity Ω all the way out to the Alfvénic point (Mestel 1967).

In reality, however, it must be remembered that as in the garden hose analogy, the fluid does *not* co-rotate with the star, even within the Alfvénic surface. The local angular velocity ω of the fluid decreases smoothly through the Alfvénic point with distance from the star, and the remainder of the angular momentum flux is supplied by the magnetic stresses due to the curvature of the field lines.

The actual local values of ω and B_ϕ are found from Eqs. (13) and (15)

$$\frac{\omega}{\Omega} = \frac{1 - \left(\frac{v_{\rm p}}{v_{\rm A}}\right)\left(\frac{B_{\rm A} R_{\rm A}^2}{B_{\rm p} R^2}\right)}{1 - \frac{\rho_{\rm A}}{\rho}} \tag{16}$$

and from Eqs. (3) to (12)

$$R B_\phi = -4\pi\eta\Omega R_{\rm A}^2 \left(\frac{1 - \frac{R^2}{R_{\rm A}^2}}{1 - \frac{\rho_{\rm A}}{\rho}}\right). \tag{17}$$

From these we can see that close to the star, where $\rho_{\rm A}/\rho$ and $v_{\rm p}/v_{\rm A}$ are small, $\omega/\Omega \simeq 1$, since $B_{\rm p}R^2$ is roughly constant along field lines for a more or less radial field or for a dipole field. At very large distances from the star, however, the material transport term in Eq. (12) dominates, and $\omega R^2 \to \Omega R_{\rm A}^2$. Note, however, that near the Alfvén surface (and even well inside it) the local angular velocity ω of the material flow lags markedly behind the angular velocity Ω of the star and the magnetic field lines.

2.5. WIND ENERGETICS

In order to determine the location of the Alfvénic surface, the wind density and velocity laws and hence the rate of angular momentum loss, we need to incorporate the energetics of the wind flow in the model. In the frame corotating with the star, the total velocity $\boldsymbol{V}$ is parallel to $\boldsymbol{B}$, and so the magnetic body force $\boldsymbol{j} \times \boldsymbol{B}/c$ does no work on the gas. In this frame the equation of motion contains the Coriolis force (which also does no work) and the centrifugal force $\nabla(\frac{1}{2}\Omega^2 R^2)$. Thus the Bernoulli integral in this frame is simply

$$H \equiv \tfrac{1}{2}\boldsymbol{V}^2 - \tfrac{1}{2}\Omega^2 R^2 - \frac{GM}{r} + a_{\rm w}^2 \log\rho = E = \text{const} \tag{18}$$

where $a_{\rm w}$ is the sound speed in the wind zone (assumed isothermal). Transforming back to the velocities defined in the inertial frame

$$(V_{\rm R}, V_{\rm z}) = (v_{\rm R}, v_{\rm z}) = \boldsymbol{v}_{\rm p}, \qquad V_\phi = \omega R - \Omega R, \tag{19}$$

equation (18) becomes

$$\tfrac{1}{2}(v_{\rm p}^2 + \omega^2 R^2) - \omega\Omega R^2 - \frac{GM}{r} + a_{\rm w}^2 \log\rho = E. \tag{20}$$

This is the usual form of Bernoulli's integral, modified by the inclusion of the rotational kinetic energy $\frac{1}{2}\omega^2 R^2$ per gram and the work $\omega\Omega R^2$ done on the wind material per gram by the magnetic force $(\boldsymbol{j} \times \boldsymbol{B}/c) \cdot \boldsymbol{v} = (\boldsymbol{j}_\mathrm{p} \times \boldsymbol{B}_\mathrm{p}/c) \cdot \Omega R \boldsymbol{i}_\phi$ (Freeman & Mestel 1966).

For a known poloidal field configuration, equations (8), (16) and 20 combine to give a relationship of the form $H(\rho, r) = E$, thus defining ρ as a function of the coordinate r. The additional condition (14) ensures that all non-singular solutions for ρ pass through the Alfvénic point P_A. The critical points of the density-radius relation $H = E$ lie at the intersections $(\rho_\mathrm{s}, r_\mathrm{s})$ and $(\rho_\mathrm{f}, r_\mathrm{f})$ of $\partial H/\partial r = 0$ and $\partial H/\partial \rho = 0$, where the wind speed passes through the slow and fast magnetosonic wave speeds v_s and v_f (Weber & Davis 1967; Goldreich & Julian 1970; Sakurai 1985). The two conditions

$$H(\rho_\mathrm{s}, r_\mathrm{s}) = H(\rho_\mathrm{f}, r_\mathrm{f}) = E \tag{21}$$

ensure a smooth flow through these points and through P_A, so defining the solution along each field-streamline. This solution can be expressed in terms of the ratios ℓ_w of gravitational to thermal energy, k of rotational to gravitational energy and β_w of thermal to magnetic energy at the base of the wind:

$$\ell_\mathrm{w} = \frac{GM}{R_* a_\mathrm{w}^2}, \qquad k = \frac{\Omega^2 R_*^3}{GM}, \qquad \beta_\mathrm{w} = \frac{8\pi(\rho_0)_\mathrm{w} a_\mathrm{w}^2}{B_0^2}. \tag{22}$$

Here R_* is the radius of the star and B_0, $(\rho_0)_\mathrm{w}$ are reference values of the field strength and plasma density at the point where the field line considered emerges from the coronal base.

3. Braking Laws

The outflowing gas in the wind zone is responsible for the gradual loss of angular momentum from the star. In time, this will lead to a decrease in the rotation rate of the star. The precise form of the braking law relating the rate of angular momentum loss to the rotation rate of the star depends in turn on the field strength at the stellar surface, the sound speed in the wind zone, and the importance of centrifugal driving terms in the wind energy equation. There is some empirical evidence that the wind temperature increases with increasing rotation rate, and that the dynamo-generated field strength saturates at an upper limiting value for very high stellar rotation rates. For these reasons we shall consider the braking laws for slow and rapid rotators separately.

3.1. SLOWLY-ROTATING STARS

In a slowly-rotating star such as the Sun, the terms in ω in the Bernoulli equation (20) are small compared with the thermal pressure term. The wind is essentially described by Parker's (1963) thermally-driven wind model, in which the gas is accelerated rapidly through the slow magnetosonic point, and subsequently undergoes a much slower acceleration.

Studies of Zeeman broadening of photospheric lines in relatively slowly-rotating K and M dwarfs have shown that the observed field strength in stellar active regions is roughly equal to the photospheric gas pressure, and is thus a function of spectral type rather than rotation rate (Saar 1987). The *fraction* of the stellar surface covered by active regions does, however, appear to be a linear function of rotation rate (Linsky & Saar 1987). Other proxy indicators of active region covering fraction (such as CaII *H* & *K* emission strength) also scale linearly with rotation rate (Noyes *et al.* 1984) for relatively slowly-rotating dwarf stars of spectral types G, K and M. Since the average field strength in the corona and wind depends on the total magnetic flux emerging through the stellar surface, these observations suggest that a linear relationship $B_0 \propto \Omega$ between the average surface field strength and the stellar rotation rate is appropriate, at least for rotation rates up to roughly ten times solar.

3.1.1. Radial Field Models. In the simplest field models, the poloidal field component is assumed to be radial from the surface of the star outward. If the Alfvénic surface S_A is approximated by a sphere, the effective co-rotation prescription gives the angular momentum loss rate as approximately

$$-\frac{dJ}{dt} \simeq \frac{8\pi}{3}(\rho_A v_A r_A^2)\Omega r_A^2. \tag{23}$$

Along each field-streamline by (8)

$$\frac{4\pi\rho_0 v_0}{B_0} = \frac{4\pi\rho_A v_A}{B_A} = \frac{B_A}{v_A}, \tag{24}$$

so that

$$-\frac{dJ}{dt} \simeq \frac{2}{3}(B_A r_A^2)^2 \frac{\Omega}{v_A}. \tag{25}$$

The assumption of a radial field gives $B_A = B_0(r_*/r_A)^2$, so that

$$-\frac{dJ}{dt} \simeq \frac{2}{3}(B_0 r_*^2)^2 \frac{\Omega}{v_A}. \tag{26}$$

Since in slowly-rotating stars we expect the wind to be mainly thermally driven, little acceleration takes place beyond the slow magnetosonic point. Thus the wind velocity at the Alfvénic point will be $v_A \simeq (2-3)a_w$, more or less independently of the stellar rotation rate. In this case, the braking rate $-\dot{J}$ depends on B_0, a_w and Ω alone. Note that there is no dependence on ρ_0: if the the wind base density (and hence the mass loss rate) is increased, Eqs. (23) and (24) show that r_A must decrease to compensate exactly, leaving the angular momentum loss rate unaltered.

If a linear relationship $B_0 \propto \Omega$ is assumed for the dynamo-generated field, then with $J = k^2 M r_*^2 \Omega$, (26) yields

$$\frac{d\Omega}{dt} \propto -\Omega^3 \tag{27}$$

which integrates to $\Omega \propto t^{-1/2}$. This is the same power law index inferred by Skumanich (1972) from empirical comparison of the solar rotation rate with those observed among late-type stars in the Pleiades and Hyades clusters.

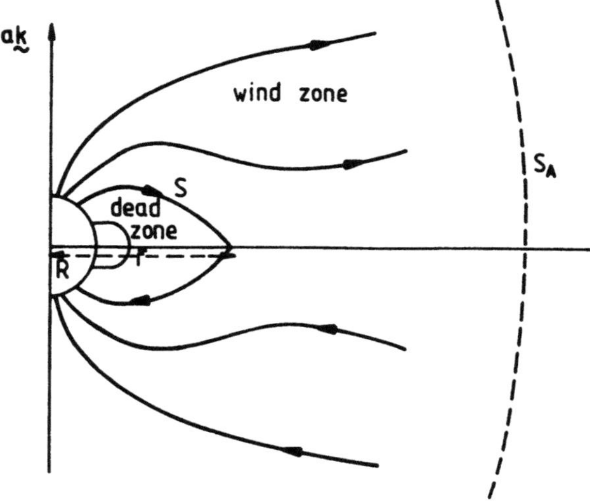

Figure 1. Schematic magnetic field model

3.1.2. Closed-Field Models. The radial field model described above is somewhat over-idealised. In reality, we expect that the region close to the stellar surface will be dominated by a mixture of closed coronal magnetic loop structures and open wind regions. Deep within the Alfvénic surface S_A, where the field is strong and $v_p \ll v_A$, the thermal and centrifugal forces that drive the wind are insufficient to distort the field significantly. As a result, $\boldsymbol{B}_p$ is nearly curl-free close to the star. Mestel (1968) and Mestel & Spruit (1987) assumed a dipolar flux distribution over the stellar surface, with the dipole aligned with the stellar rotation axis. In this case, flux tubes leaving the surface close to the equator form closed coronal loops which serve to confine hot, dense coronal plasma whose pressure is too weak to overcome the magnetic pressure. This closed-field region has been dubbed the "dead zone". This type of field configuration is illustrated schematically in Figure 1.

Field lines leaving the star nearer to the magnetic poles close at progressively greater distances above the stellar surface until the magnetic pressure is no longer sufficient to withstand the thermal and centrifugal pressures. Well within S_A the field structure thus changes from a static, nearly curl-free configuration to an open one with a wind blowing outward along the field lines (Mestel 1968; Pneuman & Kopp 1971; Mestel & Spruit 1987). Since the wind acts as an energy sink, the plasma temperature in the wind zone should be significantly cooler than that in the dead zone. X-ray images of the solar corona lend support to this picture, with the closed-field regions emitting brightly and open regions appearing as dark coronal holes.

The presence of the dead zone ensures that only part of the flux emerging from the star is connected to the open wind region. This means that $B_A r_A^2 < B_0 r_*^2$, so that by (25) the braking rate is less than would be expected for a radial-field model rotating at the same rate. In the model of Mestel & Spruit (1987) the approximately dipolar field has $B \simeq B_0(r_*/r)^3$ out to the radius $\bar{r} < r_A$ where the field changes from a closed to an open configuration. Beyond this point $B \simeq B_0(r_*/\bar{r})^3(\bar{r}/r)^2$, since along dipolar field lines $R^2 \propto r^3$. In this case equation (25) becomes

$$-\frac{dJ}{dt} \simeq \frac{2}{3}\frac{\Omega(B_0 r_*^2)^2}{(\bar{r}/r_*)^2 v_A}. \tag{28}$$

Thus the braking rate is modified by the dependence of the dead zone extent $\bar{r}$ on the rotation rate. The extent of the dead zone can be estimated from the condition that pressure balance must be maintained across the boundary that separates the wind zone from the dead zone. In the dead zone, the field lines are purely poloidal and cross the equator normally. Beyond $\bar{r}$ the field has a toroidal component, and the poloidal part is directed more or less radially. This implies a discontinuity in B_p and p along the separatrix S (shown as CXX' in Figure 1), which satisfies

$$\left(\frac{B_p^2}{8\pi} + \frac{B_\phi^2}{8\pi} + p\right)_w = \left(\frac{B_p^2}{8\pi} + p\right)_d \tag{29}$$

The plasma in the dead zone must obey hydrostatic equilibrium along each closed field line, which allows us to calculate p_d at any point in the dead zone. At the point C, where $(B_p)_d = 0$), the RHS of Eq. (29) consists of the gas pressure alone. Since C lies well within S_A, the contribution of B_ϕ to the magnetic pressure in the wind zone can be neglected. The gas pressure in the wind zone near C is reduced by the Bernoulli effect to a value which is very small in comparison to that in the dead zone. We are left with an approximate pressure balance condition at C:

$$\left(\frac{B_p^2}{8\pi}\right)_w = p_d. \tag{30}$$

This allows us to calculate the extent of the dead zone as a function of rotation rate, given an expression for hydrostatic equilibrium in the dead zone which takes account of centrifugal as well as gravitational forces, and an expression for the field strength in the wind zone as a function of rotation rate and distance from the star (Mestel & Spruit 1987).

If the thermal pressure in the dead zone is maintained by heating due to magnetic processes, we may assume that the coronal base values scale as $(\rho_0)_d \propto B_0 \propto \Omega$. At low to moderate rotation rates the condition (30) predicts an increase in the extent of the dead zone as the rotation rate increases. As a result, the covering fraction of open-field regions on the stellar surface which can contribute to the wind *decreases*. The ratio of the braking efficiency to that expected from a pure radial field model thus decreases toward higher rotation rates.

The detailed field structure is in fact determined by the magnetospheric currents, according to Ampère's law. Iterative numerical solutions have been calculated by Fitzpatrick (1988) and Barker (1990), for current distributions which take some account of the dynamics of the dead zone (see Section 3.2.2). The predicted fields and in particular the extent of $\bar{r}$ are in good agreement with the earlier estimates.

3.2. RAPIDLY-ROTATING STARS

So far we have seen that in the limit where centrifugal forces are relatively unimportant in either the wind or the dead zone, the braking rate varies with rotation rate as $\dot{\Omega} \propto -\Omega^p$, where the power law braking index p is 3 for a purely radial field, but is reduced somewhat below this value when a closed-field region is included in the model.

In more rapidly-rotating stars this picture is complicated by the increasing importance of centrifugal driving terms in the wind zone, and by shrinkage of the dead zone due to the contribution of centrifugal forces in the equation of hydrostatic equilibrium. At sufficiently high rotation rates, our earlier assumption that $B_0 \propto \Omega$ also breaks down: there is now a great deal of empirical evidence that magnetic activity levels in main sequence stars ***saturate*** at rotation rates roughly ten times faster than solar. Coronal temperatures (and hence possibly wind temperatures) also tend to be significantly greater in rapid rotators. Since young low-mass stars tend to be rapid rotators, the changes in braking behaviour at high rotation rates merit close attention. Here we consider them in the order in which they become significant as the stellar rotation rate increases.

3.2.1. Centrifugal Driving of the Wind. In a moderately rapidly rotating star, the terms in ω become dominant in equation (20) well before the Alfvénic surface S_A is reached (Belcher & MacGregor 1976). This net extra driving term has an energy per unit mass $\Omega^2 R^2(\omega/\Omega)(1-\omega/2\Omega)$. Well within S_A where $\omega \simeq \Omega$ this term is of order $\Omega^2 R^2/2$. The details of the wind speed (and in particular its value at the Alfvénic point) depend on the ratio $a_w/\Omega^2 r_A^2$. We have already seen that $v_{th}(r_A)$ (the wind speed at the Alfvénic surface for purely thermal driving) is typically 2 to 3 times a_w. Conversely, for purely *centrifugal* driving, the Alfvénic surface can be shown to lie at a distance r_A from the star where the local Alfvén speed has a value of order $v_A \simeq \Omega r_A/\sqrt{3}$ (Mestel & Spruit 1987). A reasonable approximation for intermediate cases is given by the RMS average

$$\left(\frac{v_A}{a_w}\right) = \left[\left(\frac{v_{th}(r_A)}{a_w}\right)^2 + \frac{\Omega^2 r_A^2}{3a_w^2}\right]^{1/2} . \tag{31}$$

In physical terms, the additional acceleration of the wind material due to centrifugal driving forces the gas to reach the local Alfvén speed at a smaller value of r_A than would otherwise be the case. This is because the local Alfvén speed is a monotonically *decreasing* function of r. The braking efficiency is thus *reduced* by centrifugal driving at high rotation rates.

3.2.2. Dead Zone Shrinkage. At higher rotation rates still, the hydrostatic equilibrium of the plasma inside the dead zone will be modified by centrifugal forces. Ultimately this leads to a situation where the component of centrifugal force along the field exceeds gravity in the outer parts of the dead zone. In this centrifugally-supported region, the pressure (and in an isothermal dead zone, the density) will thus increase exponentially outwards along the field. In the simple approximation where the gas pressure at the outer boundary of the dead zone matches the magnetic pressure in the adjacent wind region, it is clear that this rotationally-driven increase in dead zone gas pressure will lead to shrinkage of the dead zone.

In practice, the physics of the closed-field region in rapid rotators may be somewhat more complicated than this. The young southern field K dwarf AB Doradus, which has an axial rotation rate some fifty times the solar rate, has an X-ray corona with a mean plasma temperature of 1.7×10^7 K and an emission measure close to $1.5\times10^{53}\,\mathrm{cm}^{-3}$. This, combined with the observed lack of rotational modulation of the quiescent X-ray emission in this star implies that the X-ray emitting closed field region extends one or two stellar radii above the stellar surface (Collier Cameron *et al.* 1988). At this rotation rate, however, the point of centrifugal balance lies some $1.7\,R_*$ above the equator, within the inferred limits of the dead zone.

Time-series optical spectroscopic studies of this star recently revealed transient Hα absorption features, caused by scattering of chromospheric Hα photons by prominence-like clouds of mainly neutral material transiting the stellar disc (Robinson & Collier Cameron 1986). Significantly, the absorption transients always appear and disappear at the radial velocities of the approaching and receding limbs of the star respectively, and recur at intervals of one stellar rotation period. This implies that the clouds are in solid-body rotation with the star. Moreover, their apparent radial acceleration implies distances ranging between 2.5 and $9\,R_*$ from the stellar rotation axis: that is, they form *outside* the point of centrifugal balance, and so must be confined in the summits of co-rotating, closed loop structures (Collier Cameron & Robinson 1989a). New clouds have been observed to form in this region on timescales of order 2 days, possibly via radiative cooling instability in the centrifugally-compressed loop plasma. There is some evidence that during the lifetime of an individual cloud (of order a few days) the confining loop structure tends to evolve towards larger sizes, and that the clouds eventually break out of the confining field at least $10\,R_*$ from the rotation axis (Collier Cameron & Robinson 1989b).

These observations provide useful confirmation that the inner parts of the corona-wind structure in rapidly-rotating stars do indeed seem to be characterised by large, closed, co-rotating loop structures. They also suggest, however, that the dead zone is not completely dead: through the cloud formation and ejection process it too may contribute significantly to the overall mass and angular momentum loss rate from the star. Recent observational estimates indicate that the masses of individual clouds are of order $2\text{-}6\times10^{17}$g (Collier Cameron *et al.* 1990). Since the cloud formation/ejection rate is of order 2 clouds per day in the observable slice of the corona, the mass loss rate from the dead zone due to this process alone is at least $10^{-13}\,M_\odot\,\mathrm{yr}^{-1}$.

If we assume that the material ejected from the dead zone is incorporated in the overall wind flow, it seems that the combined effects of dead zone shrinkage and leakage may be such that the braking rate approaches the value obtained with a radial field model (plus centrifugal driving) at high rotation rates.

3.2.3. Dynamo Saturation. The braking laws derived above rest on the important assumption that the strength of the stellar magnetic field strength increases linearly with rotation rate. Vilhu (1984) found that this assumption appears to break down for rapid rotators. The surface-averaged fluxes of emission lines formed in the upper chromosphere and transition region increase strongly with increasing rotation rate for rotation rates less than ten times solar. At higher rotation rates, no further increase in line surface flux is seen: the fraction of the stellar surface covered by magnetic active regions appears to be saturated.

The soft X-ray emission from closed loop structures does not saturate completely. The power law index of the relation between X-ray emission flux and rotation rate does, however, decrease from a value close to 4 to a nearly linear dependence at high rotation rates. Vilhu attributed this change in behaviour to a saturation of the surface magnetic field strength, but with coronal loop heating (via twisting of loop footpoints by surface differential rotation) continuing to increase with increasing rotation rate.

In early G stars, this change in behaviour occurs at a rotation rate $\tilde{\Omega}$ roughly ten times greater than solar. If indeed B_0 is independent of Ω in stars with saturated dynamos, equation (28) shows that a braking law which takes the form $\dot{\Omega} \propto -\Omega^p$ becomes $\dot{\Omega} \propto -\tilde{\Omega}^2 \, \Omega^{p-2}$ for stars rotating more rapidly than the saturation limit.

4. Observational Constraints on Braking Laws

At present, the physics governing the relationships between field strength, wind density, wind temperature and the stellar rotation rate are poorly understood. Fortunately, observational advances during the last two decades have provided several important empirical constraints on the magnetic braking problem.

The work of Kraft (1967), Skumanich (1972) and later Soderblom (1983) showed that the rotational evolution of the F and G stars from the age of the Pleiades to that of the Sun could be traced by studying the rotation distributions in young and intermediate-age open clusters. Even in a cluster as young as the Pleiades, the rotation rates of the G dwarfs do not in general exceed the dynamo saturation limit, and the subsequent rotational history can be described adequately by the $\Omega \propto t^{-1/2}$ relation expected for the Weber-Davis model (Skumanich 1972).

The discovery of a population of ultra-fast rotators among the K dwarfs in the Pleiades (age $\simeq$ 70 Myr) and the G and K dwarfs in the α Persei cluster (age $\simeq$ 50 Myr) complicated the picture considerably (see Stauffer 1990 and references therein). The distribution of rotation rates among the G and K stars in both clusters shows a strong peak at roughly five times the solar rotation rate (Benz, Mayor & Mermilliod 1984) with a high-velocity tail in which the most rapid rotators have surface rotation rates up 100 times the solar rate.

This is a very different picture from that seen among the Hyades G and K dwarfs (age $\simeq$ 700 Myr). In the Hyades, the distribution of axial rotation periods has a single narrow peak, with no high-velocity tail (Radick *et al.* 1987). If we assume that the initial angular momentum distributions were similar for the G-K stars in all three clusters, the nature of the braking mechanism must be such as to erase all traces of the initial spread in rotation rates by the age of the Hyades.

The studies by Noyes *et al.* (1984) and Radick *et al.* (1987) of rotation periods among dwarfs in the field and the Hyades cluster respectively indicate that at any given age greater than that of the Hyades, rotation rates decrease towards later spectral types. In a simple braking model of the form $\dot{\Omega} = -\kappa\Omega^p$, the constant κ is expected to be a function of spectral type. The rotation rate at age t is found by integrating this expression to get

$$\Omega(t) = [(p-1)\kappa(t-t_0) + \Omega_0^{1-p}]^{1/(1-p)}. \tag{32}$$

After the end of the rapid rotation phase, the ratio of the rotation rate of a G2 dwarf to that of a K2 dwarf will be $\Omega_{G2}/\Omega_{K2} \simeq (\kappa_{G2}/\kappa_{K2})^{1/(1-p)}$. Since the value of p is close to 3, κ must increase towards later spectral types. This implies that at any given rotation rate, *the instantaneous braking rate should increase towards later spectral types*, if a single braking power law is applicable at all rotation rates.

While this appears to be the case for the slower rotators, the reverse is true of the rapid rotators in the young clusters. While rapid rotation is common among the K stars in both the Pleiades and α Per clusters, it is only seen among the G stars in α Per, which is the younger of the two clusters. It thus appears that the timescale for braking of G stars is comparable to the difference in ages ($\sim$ 20 Myr) of the two clusters, while that for K stars is somewhat longer. The timescale for M dwarfs appears to be longer still. Among the Hyades M dwarfs, the distribution of Hα equivalent widths as a function of spectral type reported by Stauffer (1990) shows a strong peak of stars with low activity levels and a more sparsely-populated high-activity tail.

In attempting to determine the rotational evolution of low-mass stars from the rotation distributions in these clusters, it is important that we should not confuse the shape of the distribution of initial angular momenta with genuine evolutionary effects taking place within the age spread of the cluster. The Hα data for the Hyades M dwarfs are vitally important in this respect. For instance, it has been suggested that the low-velocity peaks in the rotation distributions of the G and K dwarfs in the young clusters might be caused by a switchover from an anomalous "rapid braking" phase to more conventional braking of the kind described above. A rapid braking mechanism operating at a rate independent of Ω on a timescale comparable with the spread in ages of the cluster stars could give a low-velocity peak and a high-velocity tail of the kind observed in α Per and the Pleiades. The presence of a low-activity peak and a high-activity tail in the Hyades M dwarf data cannot, however, be explained in this way. If as is usually the case the Hα EW is correlated with rotation rate, the presence of both rapid (active) and slow (inactive) rotators among the Hyades M dwarfs argument requires that the braking timescale for M dwarfs be comparable with both the age and the age *spread* in the cluster. This would give an age spread for the Hyades similar to the total age of the cluster, which is inconsistent with the width of the main sequence and turnoff in the cluster H-R diagram. The low-velocity peak and high-velocity tail must therefore be intrinsic features of the zero-age main sequence angular momentum distribution inherited from the T Tauri phase.

A further complication is introduced by dynamo saturation. Its implications for the existence (or otherwise) of an anomalous "rapid braking" mechanism can be tested by evolving the Sun backwards in time using an empirically-calibrated power-law braking relation as decribed above, but reducing the value of the power law index p to the value $p-2$ at some critical rotation rate $\tilde{\Omega}$ above which the dynamo saturates. For example, a saturated dynamo with mainly centrifugal driving is expected to give a very weak rotational dependence of the braking rate, of the form $\dot{\Omega} \propto -\Omega^{1/3}$ (Mestel 1988); in general a $p-2$ power-law dependence should give a reasonable representation for values of p between 2 and 3.

If we assume that the rapid rotators have forgotten their initial angular momenta by the age of the Hyades, we should be able to evolve such a star backwards until its equatorial rotation speed reaches a value of 40 km s^{-1} typical of the most rapidly rotating G stars in the Pleiades. To maintain consistency with the observations, the rotation speed of such a star should have risen to a value in excess of 140 km s^{-1} after a further 20 Myr of backward evolution, corresponding to the age difference between the Pleiades and α Per. The time required for a G star to spin down from $140\,\mathrm{km\,s^{-1}}$ to $40\,\mathrm{km\,s^{-1}}$ is very sensitive to the value chosen for $\tilde{\Omega}$. The observational evidence presented by Vilhu (1984) suggests that saturation occurs in G stars at some ten times the solar rotation rate. If so, a Pleiades G star rotating at $40\,\mathrm{km\,s^{-1}}$ will have been rotating at $44\,\mathrm{km\,s^{-1}}$ at the age of α Per, for $p = 2.7$. With $\tilde{\Omega} = 20\Omega_{\odot}$, the velocity at the age of α Per is $59\,\mathrm{km\,s^{-1}}$, which is still inadequate.

In fact, the best agreement for G stars is obtained by ignoring dynamo saturation altogether. If we consider the G stars only, crude estimates of $\dot{\Omega}$ can be derived by considering the change in the *maximum* rotation rate among stars of a given spectral type, between the ages of α Per and the Pleiades, and between these clusters and the Hyades. An additional point on the resulting $\dot{\Omega}(\Omega)$ relation is provided by the spacecraft measurements of the angular momentum flux of the solar wind published by Pizzo *et al.* (1983). A single power law with an index of 2.7 ± 0.3 gives a surprisingly good fit to the observed rotational evolution of both the rapid and slow rotators, from the age of the α Per cluster through the ages of the Pleiades, the Hyades and the Sun. It may be significant that an index close to 2.7 follows from the discussion in Mestel & Spruit (1987) when the relations $\rho_0 \propto B \propto \Omega$ are adopted.

The same is not, however, true of the K stars: a lower power-law index is required at high rotation rates to match the evolution of the high-velocity tail between the ages of α Per and the Pleiades, than can be matched to the $\dot{\Omega}(\Omega)$ relation for the slow rotators in the clusters and the low rotational velocities of old K dwarfs in the field. This might imply that dynamo saturation sets in at lower rotation rates in K dwarfs than it does in G dwarfs. If indeed saturation sets in at some critical value of the Rossby number, as Vilhu (1984) suggested, we would expect the value of $\tilde{\Omega}$ to be lower in K dwarfs than in G dwarfs.

In conclusion, it seems possible that the "rapid braking" problem may have arisen from our present ignorance of the saturation behaviour of stellar dynamos at high rotation rates. If the two are to be reconciled it would appear that the average surface magnetic field must continue to increase almost linearly with rotation rate to a value of $\tilde{\Omega}$ considerably higher than the $10\Omega_{\odot}$ suggested by the chromospheric and transition region line flux data.

5. Models with Weak Core-Envelope Coupling

If dynamo saturation does occur in G stars at $\tilde{\Omega} \simeq 10\Omega_{\odot}$ and at even lower rates in K stars, we are left with a serious inconsistency between observation and wind theory: the surface rotation rates of the rapid rotators in the young clusters are decreasing at rates an order of magnitude greater than can be accounted for by empirically-calibrated models which work well at lower rotation rates. All of these braking models assume, however, that the stars rotate as solid bodies.

Stauffer and Hartmann (1987) suggested that one possible way of resolving this apparent inconsistency is to assume that the coupling between the radiative core and convective envelope is weak. Let us suppose that angular momentum is transferred from the core to the envelope at a rate dependent on the local magnetic field strength and the rate of shear at the core-envelope interface, according to

$$I_c\dot{\Omega}_c = -K_1(\Omega_c - \Omega_e)^\beta \tag{33}$$

and the net angular momentum loss rate from the envelope is

$$I_e\dot{\Omega}_e = K_1(\Omega_c - \Omega_e)^\beta - \dot{J}_{wind}. \tag{34}$$

If at high rotation rates angular momentum is lost from the envelope at a rate faster than it can be transferred from the core, the envelope alone will be braked initially. Since the envelope of a G dwarf has less than 10% of the total moment of inertia of the star, the surface braking rate will be an order of magnitude faster than would be expected for solid-body rotation. The deeper convective envelope of a K dwarf contains a higher proportion of the stellar moment of inertia, and so gives a longer surface braking timescale.

In this picture, the "rapid braking" phase ends and the transition to slower envelope braking occurs when the envelope has slowed sufficiently that angular momentum transfer from the core balances the wind losses. The subsequent evolution of the surface rotation rate is characterised by a "plateau" of duration 10^8 to 10^9 y, while the remaining angular momentum of the core is transferred through the envelope into the wind (MacGregor 1990; Li 1991).

This picture of magnetic braking involves some theoretical difficulties. Mestel & Weiss (1987) found that the upper limits on the field strength at the core-envelope boundary are severe if asynchronous rotation of the core and envelope is to be maintained on timescales of order 10^8 yr or more. Moreover, the evolution of the surface rotation rate in models of this kind bears little resemblance to the Skumanich relation. This in itself need not be a serious problem: we should bear in mind that the rapid rotators to which these models apply are not as numerous as the intrinsically slow rotators in the young clusters. However, we need to check that the predictions of these models are consistent with the small range in rotation rates observed among the Hyades G and K dwarfs. It should be noted that if the angular momentum loss rate depends on the rotation rate of the envelope only, the early reduction in Ω_e necessarily implies that the total angular momentum loss over the star's lifetime is lower than if solid body rotation is maintained, and in fact may not be sufficient to yield the present day solar rotation. This would imply that the Sun reached the zero-age main sequence as a comparatively slow rotator.

6. Conclusions and Future Directions

At rotation rates up to a few times solar, where the wind is primarily thermally driven and a linear dynamo relation is applicable, the inclusion of a closed-field corona produces only a minor modification to the Weber-Davis power-law dependence of the braking rate on the rotation rate. To this extent, magnetic braking theory has enjoyed a great deal of success in explaining the gross features of the rotation-age relation for the slower rotators in young open clusters. At these low rotation rates, the braking behaviour is nearly independent of the mass loss rate in the wind.

The detailed physics of the wind driving mechanism and the stellar dynamo do, however, play a much more significant role at rotation rates greater than roughly ten times solar.

Even when centrifugal driving of the wind is significant, the physical dependence of the thermal driving terms on the field strength and stellar rotation rate can be important in determining the effective Alfvén radius, and hence the specific angular momentum of the escaping wind material. Although a significant portion of the stellar surface may be occupied by closed field structures, it now appears that physical processes within this dead zone may allow a substantial rate of leakage into the wind zone, allowing a modest increase in braking efficiency at high rotation rates. If the observational evidence for dynamo saturation at high rotation rates has been correctly interpreted, however, the efficiency of magnetic braking in the very rapid rotators in the young clusters will be too low to brake the rotation of the entire star at the rate inferred from observations. This may imply that although surface activity indicators saturate at roughly ten times the solar rotation rate, the mean strength of the large-scale dipole field continues to increase linearly with Ω to much higher rotation rates. Alternatively, it may be that the coupling between the radiative core and convective envelope is weak enough to allow the envelope alone to be spun down initially.

Either way, the solution to the braking problem in the rapid rotators is likely to involve aspects of dynamo theory and the dynamics of stellar interiors about which little is known at present. Real theoretical progress must therefore be made through advances in the physics of stellar dynamos and internal rotation, coronal loop dynamics and magnetic wind driving mechanisms.

On the observational side, further detailed studies must be made of the rotation distributions among G, K and M dwarfs in open clusters with a wide range of ages. These observations will place tighter constraints on the variation in braking rate as a function of rotation rate and spectral type. Detailed X-ray, (E)UV and optical studies of individual rapid rotators are also needed to probe the detailed geometry of the closed-field corona and may eventually lead to direct measurements of the angular momentum loss rates.

References

Barker, D, 1990. D.Phil thesis, University of Sussex, Brighton, England. In preparation.

Belcher, J.W. & MacGregor, K.B., 1976. *Astrophys. J.*, **210**, 498.

Benz, W., Mayor, M., & Mermilliod, J.C., 1984. *Astr. Astrophys.*, **138**, 93.

Chandrasekhar, S., 1956. *Astrophys. J.*, **124**, 232.

Collier Cameron, A., Bedford, D.K., Rucinski, S.M., Vilhu, O., & White, N.E., 1988. *Mon. Not. R. astr. Soc.*, **231**, 131–147.

Collier Cameron, A. & Robinson, R.D., 1989. *Mon. Not. R. astr. Soc.*, **236**, 57.

Collier Cameron, A. & Robinson, R.D., 1989. *Mon. Not. R. astr. Soc.*, **238**, 657.

Collier Cameron, A., Duncan, D.K., Ehrenfreund, P., Foing, B.H., Kuntz, K.D., Penston, M.V., Robinson, R.D., & Soderblom, D.R., 1990. *Mon. Not. R. astr. Soc.* In press.

Fitzpatrick, R. 1988. Private communication.

Freeman, K.C. & Mestel, L., 1966. *Mon. Not. R. astr. Soc.*, **134**, 37.

Goldreich, P. & Julian, W.H., 1970. *Astrophys. J.*, **261**, 279.

Kraft, R., 1967. In G. Herbig, editor, *Spectroscopic Astrophysics*, page 385, University of California Press, Berkeley.

Li, J., 1991. D.Phil thesis, University of Sussex, Brighton, England. In preparation.
Linsky, J.L. & Saar, S.H., 1987. In Linsky, J.L. & Stencel, R.E., editors, *Lecture Notes in Physics, Vol. 291: Cool Stars, Stellar Systems and the Sun*, page 44, Springer-Verlag.
Lüst, R. & Schlüter, A., 1955. *Z. Astrophys.*, **38**, 190.
MacGregor, K.B. 1990. These proceedings.
Mestel, L., 1961. *Mon. Not. R. astr. Soc.*, **122**, 473.
Mestel, L., 1967. *Mem. Soc. R. Sci. Liège*, **5**, 15, 351.
Mestel, L., 1968. *Mon. Not. R. astr. Soc.*, **138**, 359.
Mestel, L. & Spruit, H.C., 1987. *Mon. Not. R. astr. Soc.*, **226**, 57.
Mestel, L. & Weiss, N.O., 1987. *Mon. Not. R. astr. Soc.*, **226**, 123.
Mestel, L., 1988. In Havnes, O. *et al.* , editor, *Activity in Cool Star Envelopes*, page 1, Kluwer Academic Publishers, Dordrecht.
Noyes, R.W., Hartmann, L., Baliunas, S.L., Duncan, D.K., & Vaughan, A.H., 1984. *Astrophys. J.*, **279**, 763.
Parker, E.N., 1963. *Interplanetary Dynamical Processes.* Interscience, New York.
Pneuman, G.W. & Kopp, R.A., 1971. *Solar Phys.*, **18**, 258.
Radick, R.R., Thompson, D.T., Lockwood, G.W., Duncan, D.K., & Baggett, W.E., 1987. *Astrophys. J.*, **321**, 459.
Robinson, R.D. & Collier Cameron, A., 1986. *Proc. Astr. Soc. Australia*, **6**, 308.
Saar, S.H., 1987. In Linsky, J.L. & Stencel, R.E., editors, *Lecture Notes in Physics, Vol. 291: Cool Stars, Stellar Systems and the Sun*, page 10, Springer-Verlag.
Sakurai, T., 1985. *Astr. Astrophys.*, **152**, 121.
Skumanich, A., 1972. *Astrophys. J.*, **171**, 565.
Soderblom, D., 1983. *Astrophys. J. Suppl.*, **53**, 1.
Stauffer, J.R. 1990. These proceedings.
Stauffer, J.R. & Hartmann, L.W., 1987. *Astrophys. J.*, **318**, 337.
Vilhu, O., 1984. *Astr. Astrophys.*, **133**, 117.
Weber, E.J. & Davis, L., 1967. *Astrophys. J.*, **148**, 217.

DISCUSSION

van't Veer: I agree with you that there is a very high angular momentum loss rate for these rapidly rotating binaries, which you cannot explain easily by changing the magnetic structure properties. I think we should not forget that there are at least two other ways to increase the angular momentum loss rate: (1) increase the mass loss rate or (2) increase the degree of ionization of the wind. So, we should look for observations (or theories) regarding this point.

Cameron: I think we can answer that in two sections. First, by the time you have a wind temperature of around 100,000 K or 1,000,000 K, which we have in the Sun, increasing the degree of ionization is not going to increase the degree of linkage to the field. Most of the angular momentum transfer occurs within a couple of Alfven radii from the star, so a higher degree of ionization might conceivably increase the degree of linkage at very large distances, but the amount of angular momentum that is being transferred to the wind particles at large radii is very small. The second point is that although the decrease in the Alfven radius doesn't, in more realistic field geometries, compensate completely for an increase in the mass loss rate, the effect is in that direction and the two nearly compensate. You don't gain much of an increase in angular momentum loss rate by increasing the mass loss rate a lot.

Gough: It was my impression from reading the Pizzo et al. paper that if you calculate the angular momentum loss rate with their data, that you were not able to explain the Sun very well. Is that correct?

Cameron: No, I don't think so. Their estimate of the angular momentum loss rate is small by about a factor of two, and I would say that is not too bad. Basically, if you have the time derivative of ω (the angular rotation rate) varying as $-\omega^3$, then $\omega/(d\omega/dt)$ is proportional to twice the age of the object minus whatever zero point you have. The braking time-scale derived by Pizzo et al. (1983) was about 20 Gyr, or about four times the age of the Sun, whereas you would expect the rate for a conventional Weber-Davis ω^3 law to be twice the age of the Sun. Now, given the other uncertainties in the Pizzo et al. data, or just in the braking theories, I don't believe a factor of two is significant. Much more serious is the factor of 5 to 15 that comes out of the Pleiades.

Roxburgh: Have you thought about what happens when you put in wave driven winds?

Cameron: There are a number of problems there. How do you actually get a sufficiently large pressure term out of transverse Alfven waves to do anything effective, given that your field lines are only tied at one end? I never understood this about wave driven wind mechanisms where you only have one end of the field lines anchored in the star.

Hartmann: Maybe it is just like increasing the mass loss rate?

Cameron: But that doesn't necessarily increase the angular momentum loss rate.

Pre-Main Sequence and Main Sequence Rotational Evolution: Constraints on Models Derived from Observations

K. B. MacGregor
High Altitude Observatory
*National Center for Atmospheric Research**
Boulder, Colorado, USA

ABSTRACT. We describe a simple, parameterized model for the redistribution of angular momentum within the interiors of solar-type stars. Among other things, it enables us to treat: *(i)* the outward flow of angular momentum from the radiative interior in response to rotational deceleration of the overlying convection zone; and *(ii)* the reapportionment of angular momentum between the core and convective envelope during pre-main sequence evolution. By combining the model with a computation of the rate of angular momentum removal by a magnetically-coupled wind, we can trace the rotational histories of low-mass dwarf stars for a variety of initial conditions and parameter specifications. We present the results of calculations of the rotational evolution of a $1M_\odot$ star, from an age $\approx$ few $\times\ 10^6$ years up to the age of the present-day sun. We compare these results with what is known from observations about the internal rotation of the sun and the evolution of the surface rotation of sun-like stars. Within the context of the present model, we find best agreement with the constraints imposed by observations of main sequence stars for: *(i)* a surface magnetic field strength which is only weakly dependent on surface angular velocity Ω for rapid rotation, becoming linearly-dependent on Ω for $\Omega \approx \Omega_\odot$; and *(ii)* a time scale for angular momentum transfer from the core to the convection zone which is relatively constant throughout the evolution, with magnitude $\sim 10^7$ years. For the adopted description of rotational braking by a magnetized wind, only a small fraction ($\sim$ 10-15%) of the initial angular momentum is lost during pre-main sequence evolution, with most of this loss occurring just prior to arrival on the zero-age main sequence. As a result, Ω increases during most of the process of contraction to the main sequence. The amount of spin-up and some of the details of the subsequent rotational evolution are dependent upon the prescribed core-envelope coupling time.

1. Introduction

That the sun rotates has been known since at least the early part of the

---

*The National Center for Atmospheric Research is sponsored by the National Science Foundation.

S. Catalano and J. R. Stauffer (eds.), Angular Momentum Evolution of Young Stars, 315–331.

seventeenth century. Between 1610 and 1612, no fewer than four observers reported seeing dark spots on the disk of the sun, and all of them noticed the apparent east-to-west motion of these features (Berry 1961). It was Galileo who explained the foreshortening of sunspots at the limb as a consequence of their proximity to the solar surface, ascribing their motion to the rotation of the sun. Not until late in the nineteenth century was the solar rotation measured by alternative (spectroscopic) means.

In recent years, much progress has been made in determining the rotation properties of not just the sun's surface, but its interior as well. We now know that the solar photosphere rotates differentially in latitude, the angular velocity Ω being greatest at the equator ($\Omega_\odot \approx 3 \times 10^{-6}\,\mathrm{s}^{-1}$) and decreasing by about 20% toward the poles (see, e.g., Snodgrass 1983). Measurements of rotationally-induced splitting of p-mode oscillation frequencies reveal that this surface distribution of Ω with latitude persists throughout the convection zone (Brown *et al.* 1989, and references therein). Analysis of the oscillation data furthermore suggests that Ω is independent of both depth and latitude within the radiative interior of the sun. This inner region appears to rotate at a rate which is intermediate between the surface rotation rates of the equator and poles.

The observational determination of stellar rotational velocities is of a far more recent vintage than the earliest studies of the solar rotation. Although the possibility of deriving rates of rotation from broadened absorption profiles had been discussed in the 1870's, convincing evidence of stellar rotation was not obtained until early in the twentieth century. In a study of the eclipsing binary δ Librae, Schlesinger (1910) attributed discrepant radial velocities measured just before and after minimum light to the rotation of the brighter component. Among single stars, definitive measurements of rotation velocities by analysis of rotationally-broadened spectral lines were first obtained in the 1930's by Struve and collaborators (see Struve 1945). The first such determinations were made for a number of rapidly rotating stars of early spectral type by Elvey (1930).

More recently, it has become possible to reliably measure the surface rotation rates of low-mass stars with ages considerably less than the age of the sun. Among T Tauri stars (ages $< 10^7$ years) with masses $\lesssim 1.25\mathrm{M}_\odot$, the average observed rotational velocity is about 15 km s^{-1} (Bouvier 1990). By comparison, the distribution of rotational velocities for stars less massive than 1 $\mathrm{M}_\odot$ in the α Persei cluster contains a significant number of stars with $v \sin i$ values in the range 50-100 km s^{-1} (Stauffer 1987; Hartmann and Noyes 1987). Since the age of this cluster ($\approx 5 \times 10^7$ years) is slightly greater than the time interval between conception and arrival on the zero-age main sequence (ZAMS) for a 1 $\mathrm{M}_\odot$ star, it would appear that the rotational velocities of at least some low-mass stars increase during pre-main sequence (PMS) evolution. Observations of solar mass stars in the slightly older

Pleiades cluster (age $\approx 7 \times 10^7$ years) reveal that virtually all such stars are slow rotators, with $v \sin i$ typically < 20 km s^{-1} (cf. Stauffer 1987). Consequently, we infer that once on the main sequence, solar-type stars undergo rapid deceleration of their surface rotation, in a time $\sim$ few $\times 10^7$ years. This spin-down is presumably the result of angular momentum loss via a magnetized stellar wind (Schatzman 1962), a process which continues (albeit at a reduced rate) throughout the main sequence lifetime of the star. Indeed, in the still older Hyades cluster (age $\approx 6 \times 10^8$ years), the measured rotational velocities of G dwarfs are < 10 km s^{-1}. In addition, the spread in rotational velocity among G stars in the Hyades is remarkably small, amounting to only a few km s^{-1} (cf. Hartmann and Noyes 1987, and references therein).

From the observational synopsis given above, there emerges the following qualitative picture of the rotational evolution of solar-type stars. The presence of rapid rotators among the membership of the α Persei cluster implies that at least some solar mass stars spin-up during PMS evolution, from rotational velocities $V_{rot} \approx 10 - 25$ km s^{-1} at an age of a few $\times 10^6$ years to $V_{rot} > 50$ km s^{-1} near the time of arrival on the ZAMS. Over the next few $\times 10^7$ years, the rate of rotation of the surface layers is diminished considerably, to velocities $V_{rot} \approx$ 10-20 km s^{-1}. Subsequent evolution on the main sequence is characterized by continuous, gradual loss of angular momentum such that $\Omega \approx \Omega_\odot$ with approximately uniform internal rotation at an age $t \approx t_\odot (= 4.7 \times 10^9$ years).

In the remainder of this paper, we describe a physically sound yet computationally simple model for the rotational evolution of solar-type stars. The model includes (among other things) a parametrized treatment of the internal redistribution of angular momentum during PMS and main sequence evolution, and a calculation of the angular momentum lost as a result of the torque applied by a magnetically-coupled stellar wind. In its simplest form, implementation of the model requires specification of both the time scale τ_c which characterizes the transfer of angular momentum from the core to the convection zone, and the dependence of the surface magnetic field strength B on rotation rate Ω. Our program in ensuing sections is to trace the rotational history of a 1 $M_\odot$ star, for a variety of initial conditions and prescriptions for input quantities (i.e., for τ_c and $B(\Omega)$). Our intent is to delimit that portion of the parameter space yielding models in accord with the observational summary given above. In so doing, we provide constraints and insights for subsequent, more detailed calculations.

2. Model: Basic Properties

Because our model is discussed in some detail elsewhere (MacGregor and Brenner 1990; MacGregor and Cohen 1990), we provide only an epitome of it herein. We defer description of our approach to calculating the angular momentum evolution of stars prior to their arrival on the main sequence until section IV.

We assume that for rotational velocities like those encountered during the PMS and main sequence evolution of solar-type stars (cf. Sec. I), the physical properties of the stellar interior can be approximated by those of a non-rotating, spherical model of the same mass. By this assumption, we separate determination of the rotational evolution from calculation of the structural evolution. Furthermore, guided by the inferred internal rotation of the present-day sun, we assume that the radiative core and convective envelope each rotate rigidly, although not necessarily at the same rate. With this assumption, the angular momenta of these regions are $J_{core} = I_{core}\Omega_{core}$ and $J_{conv} = I_{conv}\Omega_{conv}$, where I_{core}, I_{conv} and Ω_{core}, Ω_{conv} are the moments of inertia and angular velocities of the core and convection zone, respectively.

To compute the evolution of J_{core} and J_{conv} in time, we first assume that the angular momentum removed from the star by the magnetized wind which emanates from its corona (see below) derives solely from the angular momentum of the convection zone. Hence, in the absence of any coupling between the rotation of the core and that of the envelope, evolution from an initial state of uniform rotation will lead to an internal rotational velocity distribution for which the specific angular momentum increases outward everywhere except at the core-envelope interface. Since this condition represents a violation of the Solberg-Hoiland (or Rayleigh) stability criterion at the interface (Tassoul 1978; Fricke and Kippenhahn 1972; Goldreich and Schubert 1967), angular momentum redistribution must take place. To treat such an occurrence, we assume that an amount of angular momentum

$$\Delta J = \frac{(I_{conv}J_{core} - I_{core}J_{conv})}{I_{core} + I_{conv}} \tag{1}$$

is transferred from the core to the envelope in a specified time τ_c. Note that the instantaneous transfer of ΔJ would equilibrate Ω_{core} and Ω_{conv}. The equations governing the time evolution of J_{core} and J_{conv} now follow directly from angular momentum conservation, and are given by

$$\frac{dJ_{core}}{dt} = -\frac{\Delta J}{\tau_c}, \tag{2}$$

and

$$\frac{dJ_{conv}}{dt} = \frac{\Delta J}{\tau_c} - \frac{J_{conv}}{\tau_J}, \tag{3}$$

where τ_J is the e-folding time for loss of angular momentum by a magnetically-coupled wind. Equations (2) and (3) do not include terms describing changes in J_{core} and J_{conv} arising from evolutionary changes in the core and envelope masses (see sec. IV).

The model is completed by incorporating a description of rotational braking by a magnetic stellar wind. For this purpose we adopt the MHD wind theory of Weber and Davis (1967, hereafter WD), according to which the spin-down time τ_J is given by

$$\tau_J = \frac{3}{2}\frac{I_{conv}}{r_A^2 \dot{M}}, \tag{4}$$

where $\dot{M}$ is the mass loss rate and r_A the Alfvén radius. As a number of authors have noted (e.g., Mestel and Spruit 1987), the WD theory probably overestimates the actual rate of angular momentum loss. Because the wind properties (i.e., r_A and $\dot{M}$) are themselves dependent upon the instantaneous stellar rotational state, integration of equations (2) and (3) with τ_J as given by (4) requires a complete WD wind solution at each time step. In the present application, this is accomplished using the methods described by Belcher and MacGregor (1976).

3. Main Sequence Rotational Evolution

In this section, we use the model described above to study the main sequence rotational evolution of a 1 $M_\odot$ star. The requisite properties of the stellar interior are taken from an evolutionary track provided by R. Gilliland. The calculation is started at the time of arrival on the ZAMS ($t_{ZAMS} = 3\times 10^7$ years), and is stopped upon reaching the age of the sun. Boundary conditions for the wind calculation are stipulated at a reference radius $r_0 (= 1.25 R_*)$ in the stellar corona. The thermal properties of the gas at r_0 are held constant throughout the evolution, at values appropriate to the WD solution which best describes the present-day solar wind. A sample of the results is given in figures 1-4; for a more complete discussion, the reader is referred to the paper by MacGregor and Brenner (1990).

In figure 1, we show the evolution of the core and surface rotation rates for a model having $B = B_\odot\,(\Omega_{conv}/\Omega_\odot)$, where $B_\odot (= 1.47$ G) is the coronal magnetic field strength of the WD solar solution. The results depicted were obtained for $\Omega_{core} = \Omega_{conv} = 25\Omega_\odot$ at $t = t_{ZAMS}$ and $\tau_c = \tau_{JO}$ at all times, where $\tau_{JO} (= 9\times 10^6$ years in the present example) is the ZAMS spin-down time. The vertical bars in the figure span the range of measured rotational velocities for G dwarfs in the α Persei, Pleiades, and Hyades clusters.

Initially, the rate of angular momentum removal from the convective envelope

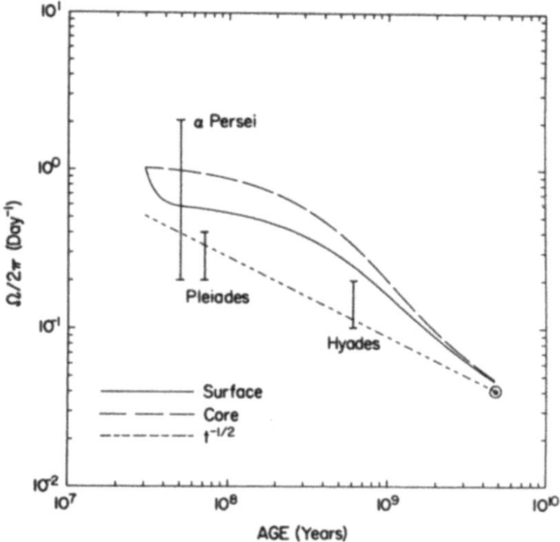

Figure 1. Main sequence rotational evolution of a model with $B \propto \Omega_{conv}$, $\tau_c = \tau_{JO}$, and initial angular velocities $\Omega_{core} = \Omega_{conv} = 25\Omega_{\odot}$.

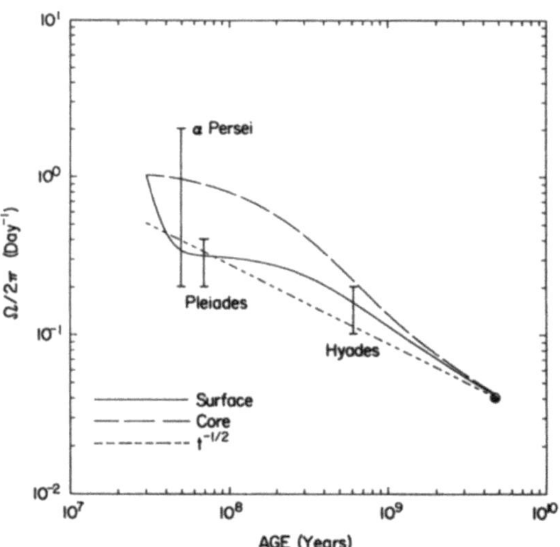

Figure 2. Main sequence rotational evolution of a model with $B \propto \Omega_{core}$, $\tau_c = \tau_{JO}$, and initial angular velocities $\Omega_{core} = \Omega_{conv} = 25\Omega_{\odot}$.

by the magnetically-coupled wind is much greater than the rate of angular momentum replenishment by transfer from the underlying core. Thus, as is also the case in more detailed models (Endal and Sofia 1981; Pinsonneault *et al.* 1989), for a period of time $\sim \tau_c$ only the rotation of the convection zone is braked. In the present example, however, despite the early onset of such rapid surface spin-down, the amount of angular momentum lost over the first few $\times 10^7$ years of the evolution is insufficient to account for the observations. The reason for this can be traced to the way in which τ_J depends on rotation during the early stages of the evolution. Up to an age of about 10^9 years, the primary source of the wind energy flux is stellar rotational kinetic energy. In this "fast magnetic rotator" (FMR) limit, it can be shown that if $B \propto \Omega_{conv}$, then $\tau_J \propto \Omega_{conv}^{-2/3}$. Hence, the rate of magnetic braking decreases with decreasing Ω_{conv}, and deficient slowing of the surface rotation results.

The tendency of models having $B \propto \Omega_{conv}$ to lose an inadequate amount of angular momentum at early times is not reversed by alternative specifications for the coupling time τ_c. If τ_c is decreased relative to τ_{JO}, the failure of the models to reproduce the observed initial rapid spin-down is aggravated since enhanced core-envelope coupling leads to increasingly rigid rotation of the entire star. Shorter ZAMS spin-down times can be obtained by increasing τ_c relative to τ_{JO}; at later times ($t \approx t_\odot$), however, such models have $\Omega_{core} \gg \Omega_{conv} > \Omega_\odot$.

The results thus far suggest that one way to moderate the decline in the rate of angular momentum loss during early main sequence evolution might be to weaken the dependence of B on surface rotation at that time. In figure 2, we show the core and surface angular velocities for a model having this property. The results depicted there were obtained by assuming $B = B_\odot\,(\Omega_{core}/\Omega_\odot)$; otherwise the initial conditions and the prescription for τ_c are identical to those of figure 1. As is apparent from figure 2, the present model undergoes a rapid reduction in the surface rotation rate when $t \approx t_{ZAMS}$, and tends toward a state in which $\Omega_{core} \approx \Omega_{conv} \approx \Omega_\odot$ when $t \approx t_\odot$. All of these attributes are in accord with the observational inferences summarized in section 1. Note that because the rotation of the core is decoupled from that of the envelope for a period $\sim \tau_c$ at the start of the evolution, B is essentially constant over this interval. At later times, as the core and envelope re-couple, B assumes a linear dependence on the surface angular velocity. By application of the FMR analysis, it can be shown that if $B \approx$ constant, then $\tau_J \propto \Omega_{conv}^{2/3}$ so that the initial rate of angular momentum loss is only weakly-dependent on rotation, $|J_{conv}/\tau_J| \propto \Omega_{conv}^{1/3}$. Stauffer and Hartmann (1987) have suggested on observational grounds that the rate of rotational braking during early main sequence evolution might be independent of rotation. For ages $\gtrsim 10^9$ years, the star is no longer a FMR. In this case, it can be shown that $\tau_J \propto \Omega_{conv}^{-2}$ and $|J_{conv}/\tau_J| \propto \Omega_{conv}^3$, so that the model asymptotically obeys the Skumanich (1972) $t^{-1/2}$ relation.

The results of model calculations which include treatment of rotational

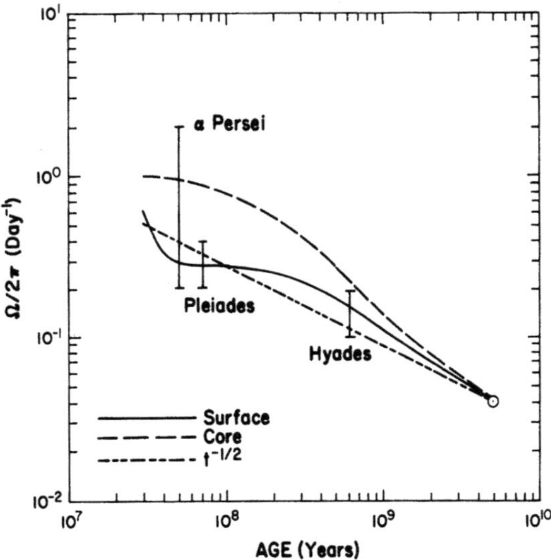

Figure 3. Main sequence rotational evolution of a model with $B \propto \Omega_{core}$, $\tau_c = 1.5\tau_{JO}$, and initial angular velocities $\Omega_{core} = 25\Omega_\odot$, $\Omega_{conv} = 15\Omega_\odot$.

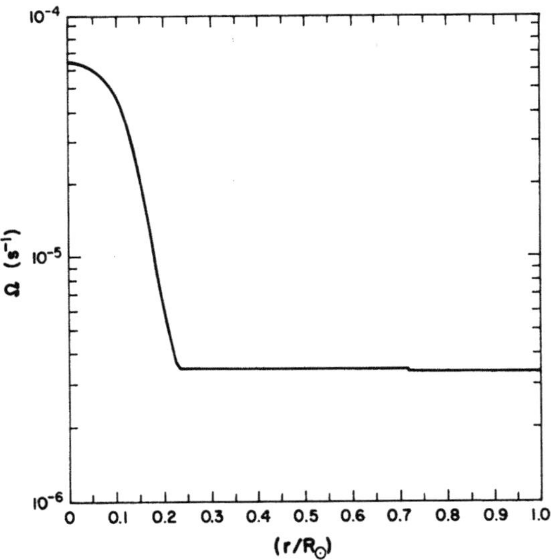

Figure 4. Angular velocity as a function of radius at $t = t_\odot$ for a model with $B \propto \Omega_{conv}$, $\tau_c = \tau_{JO}$, and $\Delta\mu/\mu_0 = 10^{-2}$.

evolution during PMS contraction (cf. Endal and Sofia 1981; Pinsonneault *et al.* 1989) indicate that significant differences between Ω_{core} and Ω_{conv} can exist by the time main sequence evolution begins. In these models, τ_J is $\lesssim$ both the contraction and coupling times during the last few $\times 10^7$ years of PMS evolution, so that $\Omega_{core} > \Omega_{conv}$ when $t = t_{ZAMS}$. We illustrate the effect of choosing differential rather than uniform rotation as the initial condition in figure 3, which shows the core and surface angular velocities as functions of time for $B \propto \Omega_{core}$, starting from $\Omega_{core} = 25\Omega_\odot$ and $\Omega_{conv} = 15\Omega_\odot$ at t_{ZAMS}. As was true for evolution from an initial state of uniform internal rotation, models with $B \propto \Omega_{conv}$ do not spin down as rapidly as observations suggest. But unlike the models considered previously, in the present example the core and envelope are strongly coupled at the start of the evolution. The source of this coupling is the angular momentum transfer described by the term $\Delta J/\tau_c$ in equations (2) and (3), since now $\Delta J \neq 0$ at t_{ZAMS}. One consequence of this transfer is a faster reduction of Ω_{core} at early times, making compliance with the young cluster observations difficult even for models with $B \propto \Omega_{core}$. This problem is easily remedied, however, by slightly increasing τ_c; the results depicted in figure 3 were obtained for $\tau_c = 1.5\tau_{JO}$ ($\approx 10^7$ years).

Finally, we note that in models that combine calculation of the internal and rotational evolution, angular momentum redistribution within the core on other than dynamical time scales is treated diffusively (cf. Endal and Sofia 1981; Pinsonneault *et al.* 1989). The diffusion coefficients of important non-magnetic transport mechanisms in the deep interior are composition dependent, and diminish in magnitude with increasing mean molecular weight μ (Endal and Sofia 1978). Hence, the development of μ gradients during main sequence evolution restricts the flow of angular momentum and leads to the formation of a small, central core having an angular velocity significantly greater than that of the surface (cf. Pinsonneault *et al.* 1989). Within the framework of the present model, we can assess the effect of excluding the angular momentum of the innermost portion of the core from participation in the overall redistribution as follows. From our 1 $M_\odot$ evolutionary track, we determine, as functions of time, the radius and moment of inertia of the spherical volume containing material whose mean molecular weight μ exceeds its initial value μ_0 by a specified amount $\Delta\mu$. As the evolution proceeds and this volume increases in size, we continuously subtract the angular momentum contained in the thin shell formed by the present and former positions of the volume's leading edge from that of the overlying radiative interior. The angular momentum contained in each such shell is subsequently conserved. As a result of this process of isolation based on chemical composition, a rapidly rotating, central core is formed, an example of which can be seen in figure 4. In the figure, we show the angular velocity at $t = t_\odot$ throughout the stellar interior for a model with $B \propto \Omega_{conv}$, $\tau_c = \tau_{JO}$, and $\Delta\mu/\mu_0 = 10^{-2}$. Models calculated with smaller values of $\Delta\mu/\mu_0$ have correspondingly larger cores. In the present example, since the inner core contains only a small fraction ($\lesssim 10\%$) of the total angular momentum, the surface rotational evolution is nearly the same as that depicted in figure 1.

4. Pre-Main Sequence Rotational Evolution

The basic model presented in section 2 can be extended to incorporate some of the processes affecting rotational evolution during contraction to the main sequence. For a 1 $M_\odot$ star, a radiative core and convective envelope are identifiable components of the internal structure for ages $\gtrsim$ few $\times 10^6$ years. Hence, we treat the PMS evolution of these regions by making the same assumptions regarding their braking, coupling, and the rigidity of their respective rotations as were made in sections 2 and 3. Unlike evolution for $t > t_{ZAMS}$, however, in this case the internal angular momentum distribution is also affected by the increasing (decreasing) mass of the core (envelope) as the base of the convection zone ($r = R_{conv}$) moves from its initial position near the center of the star toward the surface. In the present section, we include the continual reapportioning of angular momentum between the core and the envelope which results from this internal evolution; further details are available in the paper by MacGregor and Cohen (1990).

We again suppose that rotational modifications to the internal structure are small, and use spherical, non-rotating models to describe the PMS evolution of the star. During the approach to the main sequence, most of the stellar interior is converted from a convective to a radiative state (see, e.g., Iben 1964). As a result of this transformation, in an incremental time Δt, the core (envelope) mass increases (decreases) by an amount ΔM. Prior to its assimilation by the core, this material was contained in a thin shell about the radius R_{conv}, and had an average specific angular momentum $j = \frac{2}{3}\Omega_{conv}R_{conv}^2$ (approximately). Thus, over the time interval in question, the angular momentum of the core (envelope) increases (decreases) by an amount $\Delta J = j\Delta M$. By considering the limit $\Delta t \rightarrow 0$, we conclude that a term of the form $j \cdot (dM_{core}/dt)$, where (dM_{core}/dt) is the time rate of change of the core mass, should be added to the right-hand-side of equation (2). Conservation of angular momentum requires that the negative of this term appear on the right-hand-side of equation (3).

Evaluation of the quantities upon which this additional term depends is performed through the use of a 1 $M_\odot$ PMS evolutionary track calculated by D. VandenBerg. Our calculations begin at age $t_0 = 1.77 \times 10^6$ years, at which time $R_* = 2.04 R_\odot$, $R_{conv} = 0.37 R_\odot$, and $M_{core} = 0.04 M_\odot$. Arrival on the ZAMS, defined as the time after which gravitational contraction accounts for less than 1% of the luminosity, occurs for $t_{ZAMS} = 4.03 \times 10^7$ years. The evolution is started from a state of uniform rotation; for the results described below, the initial angular velocity is chosen to yield a surface rotational velocity $v_{rot} = 10$ km s^{-1} at t_0.

In figure 5, we show Ω_{core} and Ω_{conv} as functions of time for a model with $B \propto \Omega_{conv}$ and τ_c held constant at the value $\tau_c = 2 \times 10^7$ years. At the time of arrival on the ZAMS (denoted by the heavy dots on each curve), the angular velocities of the core and envelope are $\Omega_{core}/\Omega_\odot = 26.5$ and $\Omega_{conv}/\Omega_\odot = 17.3$, both having increased from $\Omega_{core}/\Omega_\odot = \Omega_{conv}/\Omega_\odot = 2.35$ at t_0. This spin-up is achieved despite the loss of about 8% of the initial angular momentum of the configuration during the contraction phase. Throughout much of the PMS and early main sequence evolution, the stellar wind properties are rather accurately represented by the FMR approximation, according to which

$$\tau_J = \frac{(I_{conv}\Omega_{conv}^{2/3})}{r_0^{8/3} B^{4/3} \dot{M}^{1/3}}. \tag{5}$$

Between t_0 and t_{ZAMS}, the magnitude of τ_J is most affected by changes in I_{conv}, a quantity which decreases by a factor $\approx$150 over this period. Primarily as a result of this variation, τ_J decreases during the approach to the main sequence, from $\tau_J = 9.5 \times 10^8$ years at t_0 to $\tau_J = 2.4 \times 10^7$ years at t_{ZAMS}. Because τ_J does not become less than the time scale for structural evolution τ_{evol} until an age $\sim 2 \times 10^7$ years, most of the angular momentum loss occurs late in the PMS contraction process. Moreover, since τ_c is greater than τ_{evol} until about the same time, the details of the early evolution are not strongly dependent upon the choice of τ_c. Shorter coupling times tend to increase Ω_{conv} somewhat while decreasing Ω_{core}; models with $\tau_c < 10^6$ years exhibit complete uniform rotation throughout PMS evolution.

In the present example, the modest amount of angular momentum lost for $t \leq t_{ZAMS}$ impedes but does not cancel the spin-up which results from the overall contraction of the star. The results shown in figure 5 also indicate that the amount of rotational braking experienced by the star for $t > t_{ZAMS}$ is insufficient to account for the Pleiades and Hyades observations. This behavior is a further manifestation of the inadequately rapid rate of angular momentum loss which characterizes models with $B \propto \Omega_{conv}$. As in section 3, we attempt to enhance the rate of magnetic braking during late PMS and early main sequence evolution by making B essentially independent of Ω_{conv} for rapid rotation.

In figure 6, this goal is accomplished using (as in section 3) the device of prescribing $B \propto \Omega_{core}$. Note that the early rotational evolution ($t \lesssim 2 \times 10^7$ years) is nearly identical to that seen in figure 5, implying little difference between the spin-down times τ_J of the respective models before this age. For a period $\sim \tau_c$ after τ_J first becomes shorter than τ_{evol} (age $\approx 2 - 3 \times 10^7$ years), angular momentum loss from the convection zone takes place at a nearly constant rate (cf. the discussion of section 3). By the time of arrival on the ZAMS, slightly

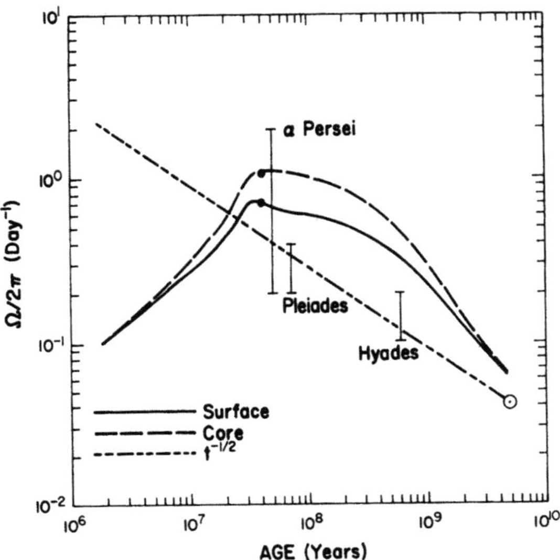

Figure 5. Rotational evolution of a model with $B \propto \Omega_{conv}$, $\tau_c = 2 \times 10^7$ years, and initial rotational velocity $v_{rot} = 10$ km s^{-1}.

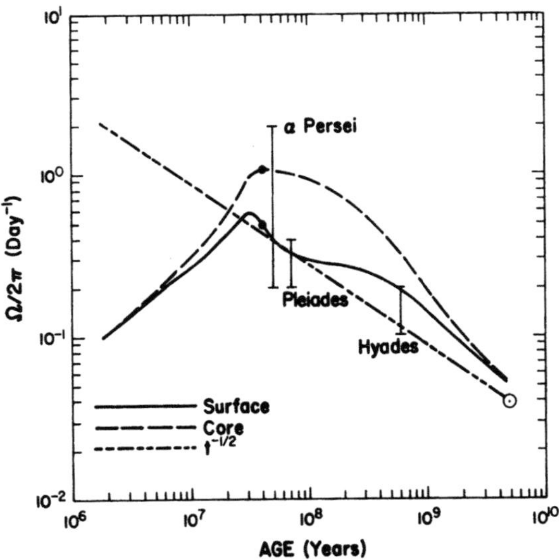

Figure 6. Rotational evolution of a model with $B \propto \Omega_{core}$, $\tau_c = 2 \times 10^7$ years, and initial rotational velocity $v_{rot} = 10$ km s^{-1}.

more than 10% of the initial angular momentum has been lost and the spin-down time is $\tau_J = 1.1 \times 10^7$ years, about half the ZAMS τ_J value for $B \propto \Omega_{conv}$. The model depicted in figure 6 attains a maximum surface angular velocity $\Omega_{conv}/\Omega_\odot = 14.3$ ($v_{rot} = 29.6$ km s^{-1}) at $t = 3.2 \times 10^7$ years, in contrast to a maximum value of $\Omega_{conv}/\Omega_\odot = 17.5$ ($v_{rot} = 34.2$ km s^{-1}) at $t = 3.6 \times 10^7$ years for the model with $B \propto \Omega_{conv}$. At the beginning of its main sequence lifetime, the model of figure 6 has $\Omega_{core}/\Omega_\odot = 26.1$, $\Omega_{conv}/\Omega_\odot = 12.5$, and $\tau_c/\tau_{JO} = 1.8$ (as opposed to $\tau_c/\tau_{JO} = 0.8$ for $B \propto \Omega_{conv}$). The details of its subsequent rotational evolution are as previously described for the case $B \propto \Omega_{core}$ in section 3.

To conclude this section, we note that models with $B \propto \Omega_{core}$ and coupling times shorter than that considered above have difficulty in matching the main sequence rotational evolution implied by observations. For example, in a model with $\tau_c = 10^7$ years the ratio of τ_c to τ_J at $t = t_{ZAMS}$ is about 0.7, a value sufficiently small that rapid early spin-down does not occur. Alternatively, models with somewhat longer coupling times have no difficulty in conforming to observations. Indeed, in a model with $\tau_c = 3 \times 10^7$ years, the surface rotational velocity becomes $\lesssim 10$ km s^{-1} at an age like that of the α Persei cluster. The same conclusions apply to models calculated with $v_{rot} = 15$ km s^{-1} at t_0.

5. Conclusions

The model calculations described in the preceding sections and the comparison of the results derived from them with observational inferences provide the following preliminary constraints on input quantities. First, the core-envelope coupling time τ_c should have a magnitude $\sim 10^7$ years. Over the course of the evolution, little variation in τ_c is required to accommodate the limits of the observations. And second, the magnetic field strength should be independent of the surface rotation rate during late PMS-early main sequence evolution, becoming linearly-dependent on Ω_{conv} for ages approaching $t_\odot$. We emphasize that within the context of the present model, the physical conclusion is that B remain approximately constant during the initial spin-down, and not necessarily that $B \propto \Omega_{core}$. In particular, we could have derived rotational histories much like those depicted in figures 2, 3, and 6 by specifying B as the appropriate function of Ω_{conv} directly. Furthermore, we note that neither of the field strength-rotation rate relations considered herein precludes the occurrence of PMS spin-up since in each case only a small fraction of the initial angular momentum of the star is lost in this phase of the evolution.

We stress again the tentative character of the conclusions enumerated above. Determination of their ultimate validity will require that a number of refinements be made to that simple model presented here. A partial list of improvements might include such things as: (i) consideration of specific, possibly magnetic (see, e.g., Mestel and Weiss 1987) mechanisms for internal angular momentum transport; (ii) treatment of time-varying coronal thermal conditions; and (iii) a more realistic calculation of the rate of angular momentum loss by a magnetically-coupled wind. In this regard, we note that the WD estimate of the rate of magnetic braking probably exceeds that appropriate to either the two-dimensional extension of the WD model or the flow from a star having a more complicated poloidal field geometry (cf. Mestel and Spruit 1987). On this basis, we would not expect the use of a more detailed wind model to change the qualitative character of our conclusions.

Acknowledgments

The author is grateful to John Stauffer for useful information concerning rotation and evolution, Ed Shoub for comments on the manuscript, Michael Brenner and Paul Cohen for computational assistance, and the workshop organizers/sponsors for financial support.

References

Belcher, J. W. and MacGregor, K. B. 1976, *Ap. J.*, **210**, 498.

Berry, A. 1961, *A Short History of Astronomy* (New York: Dover), p. 154.

Brown, T. M., Christensen-Dalsgaard, J., Dziembowski, W., Goode, P., Gough, D. O., and Morrow, C. A. 1989, *Ap. J.*, **343**, 526.

Bouvier, J. 1990, to appear in *Lecture Notes in Phys.*

Elvey, C. T. 1930, *Ap. J.*, **71**, 221.

Endal, A. S. and Sofia, S. 1978, *Ap. J.*, **220**, 279.

Endal, A. S. and Sofia, S. 1981, *Ap. J.*, **243**, 625.

Fricke, K. J. and Kippenhahn, R. 1972, *Ann. Rev. Astr. Ap.*, **10**, 45.

Goldreich, P. and Schubert, G. 1967, *Ap. J.*, **148**, 217.

Hartmann, L. W. and Noyes, R. W. 1987, *Ann. Rev. Astr. Ap.*, **25**, 271.

Iben, I. 1965, *Ap. J.*, **141**, 993.

MacGreogr, K. B. and Brenner, M. 1990, submitted to *Ap. J.*

MacGregor, K. B. and Cohen, P. 1990, in preparation.

Mestel, L. and Spruit, H. C. 1987, *M. N. R. A. S.*, **226**, 57.

Mestel, L. and Weiss, N. O. 1987, *M. N. R. A. S.*, **226**, 123.

Pinsonneault, M. H., Kawaler, S. D., Sofia, S., and Demarque, P. 1989, *Ap. J.*, **318**, 337.

Schatzman, 1962, *Ann. Astrophys.*, **25**, 18.

Schlesinger, F. 1910, *Pub. Allegheny Obs.*, **1**, 134.

Skumanich, A. 1972, *Ap. J.*, **171**, 565.

Snodgrass, H. B. 1983, *Ap. J.*, **270**, 288.

Stauffer, J. 1987, *Lecture Notes in Phys.*, **291**, 182.

Stauffer, J. R. and Hartmann, L. W. 1987, *Ap. J.*, **318**, 337.

Struve, O. 1945, *Pop. Astronomy*, **53**, 201.

Tassoul, J. L. 1978, *Theory of Rotating Stars* (Princeton: Princeton University Press), p. 167.

Weber, E. J. and Davis, L. 1967, *Ap. J.*, **148**, 217.

DISCUSSION

Stauffer: I had difficulty in interpreting one of your figures. It looked like you might actually predict too much spin down by the main sequence for rapid rotators. Would a rapidly rotating T Tauri star arrive on the main sequence with a rotation rate comparable to that observed in young clusters?

MacGregor: Yes it would. A T Tauri star with a rotational velocity of about 40 km/sec would arrive on the main sequence with a rotational velocity well in excess of 100 km/sec, and still be a quite rapid rotator by the age of the Alpha Persei cluster.

Roxburgh: If I take your favored combination of parameters, is it difficult to get as good a fit with other parameters? For instance, if I insisted that the stars rotate rigidly, but you changed the angular momentum loss rate, couldn't you get as good a fit?

MacGregor: I have tried to take a minimalist approach, and see where that leads me. However, what you say is true, I could modify the coupling and the angular momentum loss rate correspondingly, and get the desired result.

Hartmann: To rephrase your original answer, what you were saying was that you did not want to have an angular momentum loss rate that varied too violently with time, just to match the observed rotation rate.

MacGregor: I did not want to vary severely things which I did not understand physically.

Bouvier: If you start with a rotational velocity of 30-40 km/sec as a T Tauri star, so that it is a very rapid rotator upon arrival on the ZAMS, would it be a slow rotator by the age of the Pleiades?

MacGregor: I have to admit that I haven't done initial rotational velocities as large as 40 km/sec, but I have done as high as 25 km/sec. And, in those models, ZAMS rotations as high as 100 km/sec are attained, and it is possible to get them to spin down rapidly enough so that by the age of the Pleiades they are fairly slow rotators.

Gough: I don't know what Occham would say about whether one large parameter is better than two small ones, but in any case what you have convincingly shown observationally is that there must be some change in the functional form of the angular momentum loss rate of the star. Have you given any thought as to what the physics of that might be?

MacGregor: I have, but without any real result. There are both theoretical and observational results which suggest that at sufficiently high rates of rotation, magnetic field strengths may be largely independent of rotation. The apparent saturation of magnetically sensitive radiative indicators of chromospheric and coronal activity is one such observational result. Results like the Durney and Robinson study where buoyancy limited the operation of magnetic dynamos or anticipated non-linearities that might occur at rapid rotation where the Lorentz force begins to impede dynamo activity or some people suggest increased filling factors - all might cause a field strength to be independent of rotation at large rotation rates. But, these are only conjectures. I have not quantitatively examined any of them. This merits a detailed study, and I have not done that yet.

Gray: You are talking about different kinds of fields though aren't you? For braking, you are talking about open field lines, whereas for most of the observational indicators you are measuring closed field line regions. Isn't that true?

MacGregor: But presumably they have a common origin.

Cameron: One of the important questions concerning saturation, if it exists, is the rotation period at which it occurs. In a model like this, that can have a crucial effect. Once you go into the saturated regime, the linear dependence of the field strength on rotation is actually set by the value of the Weber-Davis braking rate at the rotation period where saturation sets in. So, in fact, with quite a small difference in age - and you can replace age by rotation rate - at which the dynamo saturates, you can get quite a large difference. We need to investigate this question and obtain observational constraints on the rotational velocity for saturation, perhaps as a function of effective temperature and/or gravity.

MacGregor: In these particular models, the saturation sets in at about 8 to 10 times the current solar rotation rate. If it was made to occur earlier, there wouldn't be adequate early spindown; if it were made to occur at higher rotation rates, there would be too much spindown. But what you are saying is true, and we do have to worry about this.

Cameron: Can you modify when saturation occurs but keep the core and envelope rotationally well coupled and still match the observations?

MacGregor: I haven't tried that.

EVOLUTIONARY MODELS OF ROTATING STARS

SABATINO SOFIA, MARC PINSONNEAULT, AND CONSTANTINE P. DELIYANNIS
Center for Solar and Space Research
Yale University
P.O. Box 6666
New Haven, CT, USA

ABSTRACT. In this paper we review the principal elements of the Yale code for evolving rotating stars (YREC). The various mechanisms to distribute angular momentum in the solar interior used in the code are discussed, both in regard to their properties and their consequences in the stellar models. In particular, two general types of mechanisms exist, one which depends on the velocity of rotation, and the other which depends on the radial gradient of this quantity. The regions where these mechanisms operate effectively, and the times when this occurs, and possibly the efficiency of these transfer mechanisms to cause material mixing, depend on which of the two types they belong.

The main effects of rotation in stellar evolution are on the surface velocity and the surface abundance of trace elements and CNO as a function of time. The effects on the internal structure, and therefore the evolution, as shown in an H-R diagram, are much smaller. Because of the structure differences in stars of different composition, the consequences also differ with population type.

We have computed stellar models including the effects of rotation for the Sun, open cluster stars, and halo stars. Solar-calibrated models with rotationally-induced mixing can reproduce the main features of the surface lithium depletion pattern and surface velocity both as a function of stellar mass within an open cluster and as a function of time from cluster to cluster, and also in halo dwarfs. Furthermore, our halo star models have substantial differential rotation with depth. This differential rotation provides them with enough internal angular momentum to explain the high observed rotation velocities of evolved metal-poor horizontal branch stars, while at the same time they rotate slowly enough at the surface to be consistent with the low upper limits on the rotation of metal-poor main sequence stars. By contrast, models constructed assuming rigid rotation in the main sequence cannot reconcile the high rotation velocities of evolved stars with the low rotation velocities of their main sequence progenitors.

The solar models constructed in this fashion show that the present Sun rotates with increasing angular velocity in the deep interior.

S. Catalano and J. R. Stauffer (eds.), Angular Momentum Evolution of Young Stars, 333–356.

1. Introduction.

The current version of the Yale code to study the structure and evolution of rotating stars (YREC) is a direct descendent of the code developed over a decade ago by Endal and Sofia (1976, 1978, 1981) for that same purpose. Besides the numerical and detailed physical improvements that are usually incorporated in an "evolving" code, YREC was totally rewritten (Pinsonneault 1988) starting out not from the Endal and Sofia code, or its predecessor, the Paczynski code, but rather from an early Yale (also known as Prather [1976]) code. The salient feature of YREC that differ from all non-rotating codes are the following:

1. Geometry. Because of rotation, spherical geometry is no longer valid. In order to avoid a fully two-dimensional treatment, prohibitively expensive on numerical resources, it uses the suggestion of Kippenhahn and Thomas (1970) to use equipotential surfaces as the independent variable. The formulation of this suggestion developed by Endal and Sofia allows a treatment of rotationally distorted geometry valid up to near critical velocities by means of numerical resources closer to the one-dimensional than the two-dimensional regime.

2. Differential rotation. YREC requires that shells bounded by different equipotential surfaces rotate as a solid body, but it allows different shells to rotate with different velocities. This choice is justified in terms of the effects of processes that distribute angular momentum within the stellar interior.

3. Initial angular momentum, and its subsequent variations. Our models must start out with a prescribed angular momentum, and they can deal with angular momentum loss, e.g. through a magnetized stellar wind.

4. Transport of angular momentum and the associated mixing. YREC allows the transport of angular momentum and the mixing induced by this process, and computes it on the basis of a number of circulation processes, and fluid dynamical instabilities.

In the formulation of the Yale Rotating Evolution Code, the equations of stellar structure and evolution constitute an initial value problem. Like all stellar evolution codes, they must be started before any nuclear processing has taken place within the star. However, unlike the non-rotating methods, it is not sufficient to start the evolution at the ZAMS stage, since by then a low mass star already possesses a radiative core and a convective envelope whose rotational state cannot be easily inferred. Instead, the calculations must be started at an early pre-MS stage when the star is fully (or nearly fully) convective. Although the specific rotational configuration of a convective region remains controversial, that configuration must be established very quickly because of the very short turnover timescale for convective eddies. In YREC, it is assumed that convection zones rotate rigidly.

The objective of YREC is to follow the effects of rotation into the detailed structure and evolution of the stars, in particular the evolution of the surface velocity, and the effects of rotational mixing on the surface composition of trace elements. Besides the intrinsic importance of the last two features, they turn out to provide the most sensitive diagnostics of the accuracy of our starting models, as well as of our treatment of the angular momentum losses and of the redistribution of angular momentum within the star.

Section 2 gives a brief description of the models for angular momentum redistribution and loss, as well as our choices for the initial angular momentum. Section 3 contains a summary of results for the evolution of the internal rotation curve for the solar case, Population I stars, and for halo stars. Section 4 discusses the role of light trace elements as a probe of our evolution scheme, and considers implications for other areas in astronomy, such as cosmology. Finally, Section 5 gives a summary and conclusions.

2. Mechanisms for Mixing and Transport of Angular Momentum in the Stellar Interior.

It is broadly understood that a stellar wind forced by a magnetic field to co-rotate up to the Alfven surface, transfers angular momentum from the stellar surface to the interstellar medium, thus producing a slowing down of the stellar envelope. The observational verification of this idea is provided by the Skumanich relation (1972). What is not simply explainable, or testable by direct observations, is how this slowing down of the surface transfers to the deeper layers, or how effective this process is at mixing the material in the stellar interior.

Under the above circumstances, our approach has been to calculate the transfer of angular momentum in our models by assembling all known processes for which sensible estimates for the efficiency of the process exist.

We have recognized three types of processes. One type operates in the presence of rotation, independently of the existence of angular velocity gradients. An example of this process is the Eddington-Sweet circulation. The magnitude of this effect is proportional to ω^2. The second type depends on $d\omega/dr$, and it often manifests itself as an instability. The third type involves the effects of non radial oscillations.

The details of the formulation of the angular momentum transfer in our stellar models are contained in Endal and Sofia (1978) and updated in Pinsonneault *et al.* (1989, hereafter PKSD). In all our treatment we assume that the ultimate effect of all transport mechanisms is to locally decrease the differential rotation. We differentiate the transport processes in two classes depending on their timescales. For processes which operate on dynamical timescales, we assume instantaneous redistribution of angular momentum. The new distribution is that which is marginally stable in terms of the process being considered. For processes which operate on longer timescales, we describe the angular momentum transport as a diffusion process whose coefficient depends on the velocity and scale length of the instability. Our formulation neglects magnetic fields and oscillations, as well as the abcd and doubly diffusive instabilities, besides any other process that remains to be identified. The reasons for the omissions vary from uncertainties regarding the efficiency of a given mechanism, to the lack of reliable diffusion coefficients in the literature.

Besides the obvious incompleteness of the transport mechanisms, there remains a question regarding what fraction of the angular momentum transport results in actual material mixing in the stellar interior. Whereas a circulation or turbulent transfer may fully mix the material, some magnetic field configurations can transfer angular momentum without any mixing. Does it make any sense to proceed in the presence of all these uncertainties?

The answer is positive on account that the problem is observationally overdetermined. We try to identify at least a process of each possible type, and its effects are computed by means of an adjustable coefficient. In practice, the diffusion coefficients for each of the processes we include are left as found in the literature, because we have found that varying them over an order of magnitude in either direction produces negligible changes in our results. However, in order to obtain self-consistent results on the abundances of several trace elements on different stellar populations of different ages, it is necessary to adjust the diffusion coefficient for material mixing to be only approximately 5 percent of the diffusion coefficient for angular momentum transport (designated as $f_c \sim 0.05$). This means that about 95 percent of the angular momentum transfer in the stellar interiors is due to magnetic fields or waves which occur without material mixing.

We can trust the results of our calculations despite the existence of all these adjustable parameters, because we self-consistently use the identical parameters, first calibrated for the solar case, for all our other stellar studies. For example, different processes are efficient at mixing material at different depths, during different stages of evolution. Since different trace elements are destroyed at different depths in the stellar interior, their surface abundances as a function of age constitute a powerful test for our transport scheme. Moreover, as already mentioned above, the results are robust to parameter variations.

Our formulation computes stellar structure and evolution as an initial value problem. We start the stellar evolution on the Hayashi track with an angular momentum given by the Kraft (1970) curve, or its extrapolation to lower masses. We subject the convective envelope (assumed to rotate as a solid body) to a slowing down torque as given by Kawaler's (1988) parameterization of Mestel's (1984) general angular momentum prescription. The mixing and angular momentum transfer is done as described in PKSD. The results presented in the next two sections have all been obtained with the identical formulation. As this scheme continues to be able to explain a broad set of observations, we gain confidence on the soundness of the approach. We actually seek any discrepancy, because these allow us to refine the scheme. Of course, we also seek unreconcileable discrepancies, since those would prove that the approach is incorrect. To date, we have not found any observation which are patently at odds with this scheme.

3. Observational Constraints on the Angular Momentum Evolution of Stars.

Rotational stellar models must satisfy the observational constraints imposed by stellar surface rotation rates. In fact, some aspects of the models (such as the angular momentum loss rates) are empirically calibrated. In this section we review the observational constraints on angular momentum loss and internal angular momentum transport in main sequence stars. The surface rotation rates of evolved stars, however, are even more powerful : *The rotation of evolved stars can teach us about the internal rotation of their main sequence predecessors.* As we will see, there is evidence for differential rotation with depth from observations in a variety of evolved stars.

3.1. ANGULAR MOMENTUM LOSS.

The observed surface rotation rates of low mass stars are a function of both the degree of angular momentum loss from the surface by a magnetic stellar wind and the amount of angular momentum transported into the surface convection zone from the interior. Both of these properties are difficult to constrain from first principles. The angular momentum loss from the magnetic wind depends on such uncertain properties as the external magnetic field geometry and the relationship between the field strength and the stellar rotation rate. Angular momentum transport depends on the characteristic time scales and stability conditions of hydrodynamic rotational instabilities, which are not precisely known. Angular momentum could also be transported by other mechanisms, such as waves or internal magnetic fields.

However, there is now a substantial database of observed rotation rates (both surface rotation velocities and rotation periods) for low mass stars. This data is available in clusters that span a wide range in age, and this is crucial for determining the angular momentum loss rate, and applying it to the models. When the models are forced to lose angular momentum from the surface in accordance with the observations, an angular velocity gradient develops between the core and envelope as the surface convection zone is spun down. Instabilities can be triggered which result in angular momentum transport from the interior to the surface. This internal transport can be an important influence on the surface rotation rate.

The relative importance of angular momentum loss and transport is different in stars with different mass. For stars of decreasing mass, the depth of the surface convection zone and the fraction of the moment of inertia in the surface convection zone both increase. The angular momentum content of the radiative core relative to the convective envelope therefore decreases, and the importance of angular momentum transport from the interior declines with decreasing mass. For sufficiently deep surface convection zones, the angular momentum content of the core is so small that it has little impact on the surface rotation rate. The surface rotation rates of relatively low mass (i.e. 0.8 $M_\odot$ and below) stars are therefore primarily a function of the angular momentum loss law (Kawaler 1988; Pinsonneault *et al.* 1990, hereafter PKD). The amount of angular momentum transport can be inferred in higher mass stars by using the loss law determined from the properties of the lowest mass stars (PKD).

Young clusters provide the best test for the angular momentum loss rate of rapid rotators. The angular momentum loss rate, dJ/dt, is assumed to be proportional to the surface rotation rate ω to some power, consistent with the observations as interpreted by a simple magnetic model (Kawaler 1988). It can be shown analytically that the time dependence of the surface rotation velocity determines the exponent of ω in the loss rate (Kawaler 1988). For example, the observed spin-down in stars of clusters of different ages (Skumanich relation $v \sim t^{-1/2}$) implies $dJ/dt \sim \omega^3$.

T Tauri stars also have a range of observed rotation velocities (Bouvier *et al.* 1986; Hartmann and Stauffer 1989). The simplest explanation for these observations is that stars of the same mass, age, and composition are born with a range in initial angular momentum. To intepret the observations, it is therefore necessary to consider models with a range of initial angular momentum. Because the steep dependence of the angular momentum loss rate on the surface rotation rate implies severe angular momentum loss for rapid rotators, the angular momentum loss law acts to erase a range in surface rotation rates caused by different initial conditions by an age of order a few times 100 Myr (PKSD). In fact, there is only a small range in observed rotation period at fixed T_{eff} in the Hyades (Radick *et al.* 1987), consistent within the errors with a unique relationship between period and mass. Furthermore, the predicted surface velocities from our models are in good general agreement with the observations both as a function of mass within each cluster and from cluster to cluster.

However, it is not clear that this law applies to all stars. In particular, some K stars in the Pleiades are rapid rotators (Stauffer and Hartmann 1987) contrary to the expectations of our models (see also §4.1.2). This implies that angular momentum loss in at least some rapidly rotating stars is not as efficient as prescribed by the law described above. This might be an indications that magnetic field configurations in such rapidly rotating stars are more complex (Collier Cameron and Mestel 1991; MacGregor 1991).

The old halo stars provide a test of the validity of the angular momentum loss law in a lower metallicity regime. Based on an extrapolation of the $t^{-1/2}$ law, one would expect rotation of order 1 km/s in the hottest halo dwarfs, with slower rotation for the cooler dwarfs. For a sample of 30 main sequence halo dwarfs there are only upper limits on the surface rotation velocity of 8 km s^{-1} (Peterson, Tarbell, and Carney 1983). (Work in progress by S. Ryan and C. Deliyannis is expected to bring this down to at least below the 3 km s^{-1} level.) The only metal poor star with a published rotation rate, Groombridge 1830, has a v sin i of 1 ± 1 km s^{-1} (Smith 1978). The slow rotation of these stars is therefore consistent with the angular momentum loss law calibrated on open clusters and the Sun.

It is nevertheless possible that the halo stars rotate more rapidly than one might expect based on applying the Population I angular momentum loss rate for low mass stars. The rotation of these stars may therefore be an interesting test of the Durney and Latour (1978) hypothesis for explaining the absence of angular momentum loss in massive stars. In order to maintain a coherent dynamo, the surface rotation period must be shorter than the convective turnover time scale multiplied by some arbitrary scaling factor. In the presence of such a criterion, angular momentum loss would shut down at progressively higher rotation rates for progressively thinner convection zones with shorter turnover time scales. Because the hot halo stars have thin surface convection zones and are old, they might have dropped below this threshold; such stars would therefore rotate more rapidly than they would with continued angular momentum loss.

3.2. INTERNAL ANGULAR MOMENTUM TRANSPORT.

The degree of internal angular momentum transport is difficult to quantify precisely because the inferred amount of transport depends on the assumed amount of angular momentum loss, which is itself uncertain. Nonetheless, we have several potential independent tests in main sequence stars. The observations outlined below indicate that angular momentum transport is ineffective in young stars but becomes effective by an age of a few hundred million years. In principle, then, one could determine the internal rotation of stars from main sequence rotation velocities; in practice, the observational constraints are too weak and the theoretical uncertainties are too large. Nevertheless, main sequence rotation velocities do provide a qualitative picture of the rotation of the outer layers of stars; as we will see in §3.3, the surface rotation of evolved stars is the best means of determining the rotation of the deep interior of stars.

The young G stars have thin surface convection zones relative to lower mass stars. This provides a potential means of determining the degree of angular momentum transport in young stars. Initally, angular momentum is removed only from the surface convection zone of stars. Angular momentum transport from the core to the envelope can then occur once an angular velocity gradient has developed between the two regions. If the characteristic time scale for this process is shorter than the lifetime of a cluster, such as the Pleiades, then the entire star must have spun down by angular momentum loss. By contrast, if the characteristic time scale is much longer than the model's lifetime, then only the convection zone is spun down by angular momentum loss. In the limit of short time scale angular momentum transport relative to the age of the system, the higher mass stars with larger total moments of inertia will rotate more rapidly than lower mass stars. In the limit of long time scale angular momentum transport relative to the age of the system, the higher mass stars with smaller surface convection zone moments of inertia will rotate more slowly than lower mass stars.

Stauffer and Hartmann (1986, 1987) found that G stars rotate more rapidly than K stars in the young Pleiades (age 50-70 Myr) and that the situation is reversed in the older Hyades cluster (age 350-700 Myr). They noted that these observations imply that the characteristic time scale for angular momentum transport is between 50 and 500 Myr (see also PKD). This constraint is especially strong because the rapid rotation in some Pleiades K stars implies that angular momentum loss is weaker for rapid rotators than predicted by an extrapolated angular momentum loss law (see §3.1); angular momentum transport must therefore be even less efficient to match the observations. Our basic models presented in PKD consistently predict more rapid rotation for higher mass stars than for lower mass ones; this implies that the time scale for angular momentum transport in our rapidly rotating models is too short.

The slow rotation of K stars relative to G stars in the Hyades does imply that there is strong coupling between the core and envelope by this age. It does not, however, require rigid rotation in these stars. First of all, the moment of inertia of the core is small, so the degree of rotation in the deep core has little impact on global angular momentum transport properties. Second, a differentially rotating model can spin down while still maintaing differential rotation.

The rotation of F stars relative to G stars in clusters such as the Hyades also provides evidence for transport of angular momentum on an intermediate time scale, and may in practice prove the best means to constrain angular momentum transport in open cluster stars.

3.3. THE INTERNAL ROTATION OF EVOLVED STARS : EVIDENCE FOR DIFFERENTIAL ROTATION WITH DEPTH.

It has often been claimed that because of the action of magnetic fields, stars rotate rigidly. However, besides our models, there is much evidence against rigid rotation even for main sequence stars. The most convincing evidence comes from evolved stars, as descibed below.

3.3.1. Subgiants. After the main sequence phase of evolution, stars undergo a transition on the subgiant branch from a central hydrogen burning energy generation source to a hydrogen burning shell with an inert helium core. During this subgiant phase of evolution, the core contracts while the outer layers expand. The surface convection zone steadily deepens until (on the giant branch) it extends down to near the hydrogen burning shell. These properties make subgiant branch stars an excellent test of the rotation as a function of depth in the outer layers of low mass stars (PKSD).

Angular momentum loss should be negligible in subgiants whose main sequence progenitors rotated slowly. The time scale for internal angular momentum transport should also be relatively long. As a result, the rotation periods of subgiants with convection zones of different depths (and therefore different effective temperatures) depend primarily on the internal rotation as a function of depth in their main sequence precursors. As the convection zone deepens, it will incorporate material which was within the radiative interior on the main sequence. If this material was rotating more rapidly than the outer layers on the main sequence, then adding it will tend to increase the rotation rate of the convection zone. If rigid rotation was enforced throughout the whole star, on the other hand, the depth of the convection zone has no impact on the surface rotation velocity (which is simply related to the moment of inertia of the star as a whole).

The observed subgiant rotation rates are few in number, but they are consistent with a slow envelope and a smooth transition to rapid core rotation. If the entire radiative interior rotated much more rapidly than the surface convection zone, then there would be a sharp drop in the rotation period as high angular momentum material was introduced into the surface convection zone. Instead, for cooler subgiants, the rotation period does not increase as rapidly as one would expect for rigid rotation (PKSD). The observational sample is small, however, and consists of periods estimated from chromospheric activity levels rather than measured from periodic variations in Ca emission line strength. A larger sample is needed; cluster stars or field stars with known parallax would be particularly helpful. The large sample of Coravel subgiant data, which consists of measured surface rotation velocities, should be analyzed in detail as well. Preliminary indications are that the observations are consistent with differential rotation with depth in the interior (de Medeiros 1990).

3.3.2. Horizontal Branch Stars. The rapid rotation observed in horizontal branch stars provides compelling evidence for differential rotation with depth in low mass stars.

Because of the importance of this result, it is the subject of a detailed journal article (Pinsonneault, Deliyannis, and Demarque 1991a, hereafter PDD1).

Horizontal branch stars have experienced substantial post-main sequence mass loss, of order 0.15 $M_{\odot}$. This has two important consequences. First, the material at the surface of the stars on the horizontal branch was well below the surface on the main sequence. Second, mass loss on the giant branch would have drained most of the angular momentum from the convective envelope, even neglecting angular momentum loss from a magnetic wind.

If rigid rotation is enforced on a time scale which is shorter than the giant branch lifetime, then the angular momentum content of the small radiative core is minimal. Slow main sequence rotation therefore implies slow horizontal branch rotation. If differential rotation with depth is permitted, on the other hand, then a rapidly rotating core can provide an internal reservoir of angular momentum which can survive until the horizontal branch phase. Slow main sequence rotation on the surface with a rapidly rotating core can therefore produce rapid surface rotation on the horizontal branch.

Peterson (1983, 1985a,b) found evidence for rapidly rotating horizontal branch stars in globular clusters. At the same time, only low upper limits have been obtained for the rotation velocities of main sequence metal-poor stars. This combination is inconsistent with rigid rotation enforced on a time scale shorter than the lifetime of the giant branch; in fact, it is inconsistent with rigid rotation enforced over the main sequence lifetime of the metal-poor stars. The rotation in the evolved stars is much too rapid. In fact, if the upper limits on the rotation of main sequence stars can be reduced to 3 km/s or less, then rigid rotation can be ruled out even in the unlikely case that stars can lose mass but somehow still retain all the angular momentum.

The models of PDD1 develop naturally differential rotation with depth, and thus a reservoir of internal angular momentum. This reservoir provides two to three times the amount required to explain the surface rotation velocity of horizontal branch stars, assuming they rotate rigidly. This excess may in fact be required if the horizontal branch stars themselves have some differential rotation.

Even with such a reservoir of internal angular momentum, *the effects of rotation on the structure (and therefore ages) of globuler cluster stars is negligible* (Deliyannis, Demarque and Pinsonneault 1989).

3.3.3. Planetary Nebula Nuclei (PNN). The properties of evolved PNN provide intriguing evidence for differential rotation with depth in asymptotic giant branch stars, although the results are difficult to quantify at present. Cool PNN near the asymptotic giant branch are spherically symmetric; this is consistent with a slowly rotating envelope. As one looks at progressively hotter PNN, there is a transition from spherical symmetry to axisymmetry. This is exactly the pattern one would expect if a rapidly rotating core was embedded in a slowly rotating envelope, but is a puzzle if the asymptotic giant branch precursor rotated rigidly (Willson and Bowen 1988).

3.3.4. White Dwarfs. Pilachowski and Milkey (1987) measured rotation velocities of order 50 km/s in some white dwarfs. Given the many potential ways for stars to lose angular momentum on their way to the white dwarf phase, this detection provides only a lower bound to the core rotation in earlier phases of evolution. However, the detection of any measurable rotation in these stars requires differential rotation with depth in their asymptotic giant branch precursors. Rigid rotation during the asymptotic giant branch phase would have left a white dwarf remnant with virtually no angular momentum, with a rotation period typical for giants of many years and a tiny radius. Once again, the observations stand in strong contradiction to the hypothesis of rigid rotation enforced on a short time scale.

3.4. IMPLICATIONS FOR THE INTERNAL ROTATION OF THE SUN.

We have seen that stellar data both requires some differential rotation with depth, even in very old stars, and requires that angular momentum transport occur over a typical time scale measured in the hundreds of millions of years. When YREC is used to evolve solar models to the present age, we find that *models of the present Sun show differential rotation with depth* (see Figs. 4, 7, 9, 10, and 12 in PKSD). The outer envelope of our solar model rotates slowly and nearly as a solid body in agreement with current helioseismological data (Hill 1987). The deep interior, on the other hand, rotates substantially faster. We predict that such differential rotation will be observed if and when g modes (or very low-l p modes) will be detected.

4. The Light Element Tracers : A Probe of Stellar Evolution.

The elements Li, Be, and B are important tracers of physical processes occuring in the outer layers of stars because these elements are very fragile : they are destroyed by (p,α) reactions at temperatures of only a few million degrees. Furthermore, because Li, Be, and B (in that order) survive to pregressively greater depths in stellar interiors, when used together they become powerful probes of stellar structure and evolution. As theoretical models proliferate (as they have in recent years) that attempt to include physics not usuallly included in the standard model, knowledge of light element abundances on the surface of the Sun and stars can be used effectively to discriminate observationally between realistic and less realistic scenarios.

4.1 LITHIUM.

Of the light element tracers, by far the largest body of work has concentrated on lithium. This is so partially because lithium is easier to observe than are the other elements, and partially because its importance was recognized early (see e.g. the classic paper by Greenstein and Richardson [1951] in which they conclude that mixing must take place in the Sun because the solar lithium abundance is depleted). Successful modern stellar evolutionary models must reproduce the observations as a function of mass, as a function of time, as well as a function of metallicity. We review hereafter some of the key properties of lithium abundances in Population I stars and compare them to those of

Population II stars, and then compare the predictions of the rotational models to the observations.

4.1.1 Observed Lithium Abundances in Stars of Different Mass, Metallicity, and Age. Observations of the surface lithium abundance in Population I stars are far more numerous than those in Population II stars. Observations exist for stars with T_{eff} ranging from 10,000 K down to about 3000 K, which yields information about surface Li abundance as a function of mass. Furthermore, observations have now been made in many clusters of different ages, allowing us to ascertain the evolution of the surface Li abundance with time.

It is convenient to summarize a few of the important features of the lithium evolution observed in Population I stars in broad terms as a function of age : Stars form with an abundance of at least 3 – 4. (Li, Be, and B abundances are given on a logarithmic scale by number relative to hydrogen, where hydrogen is 12, and symbolized by brackets. Thus, for element X, $[X] \equiv 12 + \log (N_X / N_H)$.) Lower mass stars are the first to deplete (during the pre-MS), and they continue to do so during the MS. A sharp drop in the Li abundance (the Li "dip") develops in F stars during the MS, leaving a Li peak between them and the lower mass stars. The peak decreases in abundance during the early MS, approaching asymptotically a value of 2.4 – 2.6 for ages greater than 3 Gyr. The few hot dwarfs (> 7000K) observed so far seem to have an abundance near or slightly higher than 3, and it is unclear whether this is an evolved abundance. However, subgiants inferred as originating from hot (> 7000K) turnoff stars show large dispersion in abundance (Balachandran 1988).

We now expound in slightly more detail, making reference to specifc clusters.

1) T Tauri stars exhibit remarkably high Li abundances, ranging from 3 to 5 or even 6. However, the latter values are generally not believed to be realistic because of the many uncertainties in determining the abundance (Strom *et al.* 1989; Magazzu and Rebolo 1989; discussion in §7 of Deliyannis 1990).

2) A few observations in IC 2391 (age ~ 20-30 Myr) show a maximum abundance at ~3.4 for G stars (6000 K – 5500 K) and declining abundance for K stars (≤ 5000 K, Stauffer *et al.* 1989), providing direct evidence for pre-MS depletion as predicted long ago by Bodenheimer (1965) using standard models.

3) The Pleiades (age ~ 50-70 Myr) show a reasonably uniform plateau at ~ 3.0 – 3.3 for F and G stars (7000 K – 5500 K; Pilachowski *et al.* 1987; Boegaard, Budge, and Ramsay 1988); cooler stars have declining abundance and an increasing dispersion at fixed T_{eff} (Duncan and Jones 1983; Butler *et al.* 1987).

4) Ursa Major (age ~ 300 Myr) shows the development of a dip for F stars with substantial (at least two orders of magnitude) Li depletion (Boegaard, Budge, and Burck 1988). Thus, the dip develops during the main sequence, not before.

5) The Hyades (age ~ 500 Myr) is perhaps the cluster with the most precise observations (Cayrel *et al.* 1984; Boesgaard and Tripicco 1986; Boesgaard and Budge 1987; Rebolo *et al.* 1988; Soderblom *et al.* 1990). Only a few hot normal stars (≥ 7000 K) have been observed, and they have [Li] ~ 3.2 with a possible dispersion of ~ 0.5 dex. The F stars show a deep dip with maximum depletion of at least two orders of

magnitude, with considerable dispersion at fixed T_{eff}. There is a Li peak for G stars at ~ 3. A short period tidally-locked binary has an abundance as much as 0.7 dex above this peak (depending on its precise T_{eff}), and another star has an abundance about 0.7 dex below this peak. Progressively cooler stars exhibit progressively lower abundance and more dispersion. A short-period tidally-locked binary at ~ 5400 K lies 0.3 – 1 dex above the mean (again, depending on its T_{eff}). The lowest detection lies at 5000 K with [Li] = 0.0 (Hobbs and Pilachowski 1988b).

6) NGC 752 (age ~1.7 Gyr) shows similar qualitative features as the Hyades. However, the G star peak is lower (~ 2.6) and the cool stars are more depleted (Hobbs and Pilachowski 1986).

7) Cool dwarfs in M67 (age ~ 4 Gyr) appear to be still more depleted; the peak in M67 and in NGC 188 (age ~ 5 – 7 Gyr) appears at about 2.4 – 2.6 (Hobbs and Pilachowski 1988b).

The Sun is the only star for which we have a confident initial abundance. This is the meteoric abundance of 3.3; the solar abundance near 1.0 implies that the Sun has been depleted approximately by a factor of 200. The solar Li abundance is at least an order of magnitude lower than that of stars of comparable masses and ages (M67 and NGC 188).

Unfortunately, because Population II stars are generally much older than Population I stars, it is not possible to map out the Li evolution in as much detail for Population II stars. In fact, we are restricted to the properties observed in very old stars, and it is not possible to compare directly stars of the same age from the two populations. Nevertheless, it is possible to ascertain that the evolution of the Li abundance must have been quite different for the two populations.

We first restrict attention to the properties of the lithium abundances in extreme halo dwarfs (the Group A stars as defined and compiled in Deliyannis, Demarque, and Kawaler [1990], hereafter DDK, and in Deliyannis and Demarque 1991a, hereafter DD; for sources see Spite and Spite 1982, 1986; Spite *et al.* 1984, 1987; Boesgaard 1985; Hobbs and Duncan 1987; Hobbs and Pilachowski 1988a; Rebolo *et al.* 1987, 1988). These properties are both very striking and very different from those of Population I stars. Instead of a peak for disk G stars, after many Gyr of evolution halo stars exhibit an extended plateau. Instead of considerable dispersion for disk G stars, halo stars exhibit a remarkable uniformity in the plateau, possibly with a small slope (DD) and with at most only a small dispersion (§II.c in DDK; §III.E in Pinsonneault, Deliyannis, and Demarque 1991b, hereafter PDD2). The cool halo stars are also depleted, and the (few) subgiants are in good agreement with dilution predicted from the deepening of the convection zone in standard subgiant models (DDK). The halo star properties are even more striking when one considers that each of the three major observational projects exhibits them independently, with each study having used different instrumentation and mostly different model atmospheres.

Currently there are few observations in intermediate metallicity stars. Nevertheless, it is clear that as progressively less extreme stars are added to the Group A sample, progressively more Li dispersion is seen (§II in DDK). Furthermore, a short period tidally-locked binary has a detected abundance of ~ 0.7 at a T_{eff} of ~ 4750 K, where other dwarfs show only substantially lower upper limits.

4.1.2. Comparison of Observation with Theory. For rotationally-induced Li depletion in the Sun see PKSD, for open cluster stars see PKD, and for halo stars see PDD2 and Deliyannis 1990. See also Pinsonneault (1991) in this volume.

By the time a model reaches the ZAMS, lithium survives only in the outermost few percent (by mass fraction) of the model. The surface abundance is thus affected by how the base of the convection zone evolves relative to the Li preservation boundary, and by how additional mechanisms (such as rotationally-induced mixing) transport Li-poor/rich material to the base of the convection zone. Thus, the surface Li abundance is affected at least as much as described by standard models; the action of rotational models is in addition to that of standard models.

The young open cluster models assume an initial abundance of 3.6, derived by fitting to the data. Li depletion in the models occurs first during the early pre-main sequence at the base of the convection zone itself. The burning is negligible in higher mass models, but is an increasing function of decreasing mass (following the increasing convection zone depth with decreasing mass), in good agreement with the IC 2391 data.

Rotational breaking begins later during the pre-main sequence and results in rotationally-induced mixing and Li depletion. This effect continues to be important through the early main sequence. The amount of Li destruction depends on both, the amount of angular momentum transported through the interior, and its timing. Because stars of a given mass are known to form with a range in initial angular momenta, J_o, and because models with different J_o spin down to the about same *surface* rotation value after only a few hundred Myr, *our models naturally produce a dispersion in Li abundance at fixed T_{eff}.* Low mass stars have a range in J_o of about an order of magnitude; we label the boundaries in this range as J0 and J2, and an in-between value (by a factor $\sqrt{10}$) as J1. The observed distribution in J_o seems to be highly peaked near J1, with a tail toward higher values that implies more severe Li depletion for these stars. The fact that the Sun is more Li-depleted than its analogues suggests that its J_o was high; the calibration of f_c assumes $J_o \sim$ J2 for the Sun.

The total lithium preservation region is only a weak function of stellar mass. Therefore, the "buffer" zone in which lithium is preserved between the base of the convection zone and the Li preservation boundary is an increasing function of mass. This, coupled with the shorter pre-main sequence evolutionary time scale for higher mass models (and thus for the recession of their convection zones) implies that by the ZAMS, rotationally-induced Li depletion is not yet effective in higher mass models; their Li abundances form a plateau with little dispersion. By contrast, progressively lower mass models show progressively more rotationally-induced Li depletion and the associated dispersion. These features are in good agreement with the Pleiades data. The low mass rapid Pleiades rotators (§3.1) have high Li abundances (Butler *et al.* 1987); this, at least, is consistent with our expectation that stars that have not yet spun down will not have suffered rotationally-induced Li depletion.

The rotational models provide a natural explanation for the Li dip.

Rotationally-induced Li depletion becomes effective during the main sequence for higher mass models. Furthermore, for stars more massive than about mid- to late F, J_0 becomes an increasing function of mass. These trends account for the sharp Li decline (and dispersion) on the cool side of the dip. The middle of the dip coincides with the break in the Kraft (1970) curve, where progressively hotter stars lose a progressively smaller fraction of their initial angular momentum and thus suffer less rotationally-induced Li depletion. This accounts for the Li rise on the hot side of the dip. Stars hotter than early F probably do not spin down at all; however, this does not imply that they are not lithium depleted : in future work we will investigate whether meridional circulation might be important for Li depletion in the corresponding models (it is not in lower mass models).

The dip becomes deeper and wider with age in the models. At the same time, cool stars continue to deplete Li during the main sequence, in agreement with the Hyades, NGC 752, and M67 data. Thus the peak gets narrower, and its maximum decreases with time. However, because most of the rotationally-induced mixing occur early during the main sequence, the rate of the peak abundance depletion decreases with time, in agreement with the open cluster data. The Li dispersion at fixed T_{eff} observed in clusters is naturally explained by the models, particularly the severely depleted stars below the mean trend. The higher abundances seen in the short period tidally-locked binaries is also understood in terms of their having circularized during the early pre-main sequence (Zahn and Bouchet 1989), thus preventing the rotationally-induced depletion from spin down that the single star models experience.

Rotational models can reproduce the very different Li abundance pattern observed in halo dwarfs. The differences in rotationally-induced Li depletion between halo stars and Population I stars can be traced to the dependence of the evolution of the internal structure on metal abundance. Most of the Li preservation region in halo stars is a buffer zone; Population I stars of comparable mass have no (or almost no) buffer zone. This and other factors combine to produce a uniform depletion of about an order of magnitude for halo dwarfs, which is much smaller than that for Population I stars. Because the initial conditions are erased during the pre-main sequence when mixing is inefficient, the dispersion is also much smaller than in Population I stars. Both are consistent with the observations (Fig. 5 in PDD2).

Whereas standard Population I models have difficulty accounting for the Li observations, standard halo models agree well with the observations. We must therefore be alert to the possibility that some mechanism that is not effective in Population I stars might at least partially inhibit rotationally-induced mixing in halo stars. (Helium diffusion, for example, could conceivably create a sufficient mean molecular weight gradent to inhibit some of the rotationally-induced mixing; conversely, rotational turbulence could render diffusive motions inefficient. The effects of helium diffusion can also be traced and constrained by observations of Li, since it diffuses very similarly to helium [Deliyannis and Demarque 1991b].) Future work will address this issue, especially because of the critical implications for cosmology.

The inferred primordial Li abundance from standard stellar models renders the standard model of big bang nucleosynthesis remarkably self-consistent. However, the higher primordial abundance inferred from rotational stellar models hints at an inconsistency, which can perhaps be resolved if (among other possibilities) neutrinos are sufficiently massive (Deliyannis, Pinsonneault, and Demarque 1991). Knowledge of the primordial Li abundance also constrains alternate cosmological scenarios, such as inhomogeneous nucleosynthesis (e.g. Malaney 1991).

Preliminary work on intermediate metallicity rotational models shows a larger dispersion than the halo models, apparently consistent with the observations.

4.1.3. Lithium-6. At least two classes of cosmological models (Dimopoulos et al. 1988; Carlson *et al.* 1990) predict a high primordial ^{6}Li abundance. These models can be constrained by searching for ^{6}Li in halo stars at the hot end of the Spite plateau where ^{6}Li can survive in observable quantities. Although standard models preserve so much ^{6}Li that much of the parameter space in such cosmologies can be constrained (Deliyannis *et al.* 1989), rotational models deplete lithium more severely and do so preferentially for ^{6}Li relative to ^{7}Li (Deliyannis and Demarque 1991c). According to the rotation scenario, to constrain such cosmologies, it is necessary to observe a large number of hot plateau stars to ensure inclusion of stars with low J_0.

4.2 BERYLLIUM.

4.2.1 Observed Beryllium Abundances. Beryllium observations offer an independent diagnostic of stellar interior processes. Furthermore, because ^{9}Be survives to a greater depth than ^{7}Li, these two elements provide severe contraints for stellar evolution when they are used together; most scenarios proposed to date fail to meet these constraints.

There are far fewer Be observations than there are Li observations, and the patterns they reveal are more subtle. Once again, the meteoric abundance of 1.42 ± 0.04 (2σ limits; see discussion in Anders and Grevesse 1989) provides the only confident initial stellar abundance, and suggests that the solar abundance of 1.15 ± 0.2 (e.g. Chmielewski *et al.* 1975 and references therein) might be depleted . If this is indeed so, successful stellar evolution scenarios for the Sun must be able to deplete both Li and Be, even though these elements survive to different depths. In field stars, most of the observed abundances cluster around the solar value, with a few as high as the meteoric value. The abundance uncertainty is of order 0.3 - 0.5 dex, so it is unclear whether this spread is intrinsic. Upper limits in some F stars indicate that these are depleted in Be by at least an order of magnitude relative to the meteoric abundance, and that they are also very Li-deficient as well (Boesgaard 1976; Boesgaard and Lavery 1986). Li and Be detections in one key F star, 110 Her, show a Be depletion of an order of magnitude and a Li depletion of two orders of magnitude (Boesgaard and Lavery 1986). In the Hyades, Be abundances range from slightly below solar to perhaps as little as 1/5 solar in F stars that are roughly an order of magnitude depleted in Li (Boesgaard and Budge 1989); note, however, that the most Li-depleted Hyades dip stars have not yet been observed for Be. In halo stars, upper limits exist which are substantially below the meteoric abundance (Ryan *et al.* 1990).

4.2.2 Comparison of Observation with Theory. Contrary to standard models (and other scenarios) that deplete Li catastrophically before depleting Be, rotationally-induced mixing depletes Be while observable amounts of Li still remain in the atmosphere. This results from the timescale for mixing in the outer layers being comparable to the evolutionary timescale during which mixing is important (late pre-MS and early MS). In fact, the Be depletion relative to Li depletion is an increasing function of mass and of J_0.

The solar calibration of f_c using the depleted solar Li abundance also gives a Be depletion of order 0.5 dex, which is not inconsistent with the observations. F star models exhibit a wide range of Be depletion, depending on J_0. For the age of the Hyades, the depletion ranges from only 0.1 dex to over an order of magnitude; models with a value of J_0 (between J1 and J2) *that reproduces the Li depletions are consistent with the Be observations in the same stars*, assuming Be is depleted from an initial abundance at or slightly higher than the meteoric level. The higher G star Li abundances (and possibly higher Be abundances) are also consistent with the rotational models. Field stars are probably older than the Hyades on average; older F star models can account for the severely Be (and Li) depleted F stars. The (essentially) undepleted F stars agree with either low J_0 old models or young models with a larger possible range in J_0. Particularly striking is the agreement between the models and the Li, Be depletions observed in 110 Her. For more details see Deliyannis and Pinsonneault (1991). Future Be observations in Li-depleted stars will provide important information for further testing of our models.

Although halo star models deplete some Be, a conservative upper limit of 0.5 dex in the Be depletion can be ascertained. Together with the upper limits of Ryan *et al.* (1990), this implies that the Be abundance has evolved during the history of the Galaxy (Deliyannis and Pinsonneault 1990). Improved observations together with maximum upward corrections from our models will be able to constrain inhomogeneous big bang nucleosynthesis models that predict an observably high primordial Be abundance. Cosmic rays are generally regarded as the source of Be that enriched the primordial abundance to the present levels. Future observations (combined with our models) will be able to map out the Galactic Be chronology, and to provide tests for cosmic ray theory.

4.3 BORON

Very few boron observations exist and, with the possible exception of HD 140283 (Molaro 1987), they generally cluster around the solar value. Boron survives to a still greater depth than Be, so it is potentially another very powerful probe of stellar evolution. It can also be used to test cosmology and cosmic ray theory. High quality observations are now possible using HST. With stellar evolution theory pressing ahead, the time is ripe to make boron observations, particularly in Li-depleted and Be-depleted cluster and field stars, and in progressively more metal-poor stars.

5. Summary and Conclusions.

The YREC formulation of stellar evolution with rotation is an initial value problem which is both mathematically well-posed and observationally well-constrained. The current detailed formulation of the variety of mechanisms that redistribute angular momentum is neither final nor necessarily complete. In the coming years, we are sure that great efforts will be needed to remedy this. The strength of the method, however, lies in its use of self-consistent modelling of stars with a variety of masses, ages, compositions, and evolutionary states. Using such models, we have been able to explain a large variety of observations, and we have produced a series of predictions for new observations which will further test and refine the theory. The broad success of the models gives us confidence that they have some real physical basis, and gives credence to their predictive power.

As described above, surface rotation rates and light element abundances of main sequence stars can simultaneously be explained for a variety of masses, as a function of time, and for different metallicities. Evolved stars (Population I subgiants, white dwarfs, planetary nebulae, Population II horizontal branch stars) provide strong and consistent evidence for differential rotation with depth. Our models naturally possess differential rotation, in good quantitative agreement with the observations. This formulation produces a model for the current Sun with significant differential rotation in the deep interior. We therefore predict that g-mode oscillations will show evidence for differential rotation in the solar core (if and when they can be detected).

In the future, we plan to both improve the physical model and to apply it to more cases. In particular, we are expanding our work to include post-main sequence modeis and the effects of mixing on CNO elements. We also plan to include additional tracers, such as boron. In terms of refining the method, we plan to study the following : 1) inclusion of microscopic diffusion; 2) inclusion of magnetic fields; 3) the impact of improvements in the model physics, such as new opacities, equation of state, and convection theory; 4) the effects of an improved treatment of rotational instabilities and angular momentum loss.

References

Anders, E., and Grevesse, N. (1989), 'Abundances of the Elements : Meteoritic and Solar', *Geochim. Cosmochim. Acta*, **53**, 197-214.

Balachandran, S. (1988), *Lithium Depletion and Rotation in Main Sequence Stars*, Ph. D. Dissertation, University of Texas.

Bodenheimer 1965, 'Studies in Stellar Evolution II. Lithium Depletion During the Pre-Main Sequence Contraction', *Ap.J.*, **142**, 451-461.

Boesgaard, A. M. (1976), 'Beryllium in Main-Sequence Stars' *Ap. J.*, **210**, 466-474.

Boesgaard, A. M. (1985), 'Lithium in Halo Dwarfs', *Pub. A. S. P.*, **97**, 784-789.

Boesgaard, A. M., and Budge, K. G. (1987) 'Lithium in the Hyades, the Hyades Moving Group, and Praesepe', *Ap. J.*, **332**, 410-420.

Boesgaard, A. M., and Budge, K. G. (1989), 'Beryllium Abundances in Lithium-Deficient Hyades F Stars', *Ap. J.*, **338**, 875-887.

Boesgaard, A. M., Budge, K. G., and Burck, E. E. (1988), 'Lithium and Metallicity in the Ursa Major Group', *Ap. J.*, **325**, 749-758.

Boesgaard, A. M., Budge, K. G., and Ramsay, M. E. (1988), 'Lithium in the Pleiades and α Persei Clusters', *Ap. J.*, **327**, 389-398.

Boesgaard, A. M., and Lavery, R. J. (1986), 'Abundances in Beryllium-Deficient F Stars', *Ap. J.*, **309**, 762-770.

Boesgaard, A. M., and Tripicco, M. J. (1986), 'Lithium in the Hyades Cluster', *Ap. J.*

(Letters), **302**, L49-L53.
Bouvier, J., Bertout, C., Benz, W., and Mayor, M. (1986), 'Rotation in T Tauri Stars', *Astr. Ap.*, **165**, 110.
Butler, R.P., Cohen, R.D., Duncan, D.K., and Marcy, G.W. (1987), 'The Pleiades Rapid Rotators : Evidence for an Evolutionary Sequence', *Ap. J. (Letters)*, **319**, L19-L22.
Carlson, E. D., Ezmailzadeh, R., Hall, L. J., and Hsu, S. D. (1990), 'Black Hole Nucleosynthesis and $\Omega_b = 1$', *Phys. Rev. Letters*, **65**, 2225-2228.
Cayrel, R., Cayrel de Strobel, G., Campbell, B., and Däppen, W. (1984), 'The Lithium Abundance of Hyades Main Sequence Stars', *Ap. J.*, **283**, 205-208.
Chmielewski, Y., and Müller, E. A., and Brault, J. W. (1975), 'The Solar Beryllium Abundance', *Astr. Ap.*, **42**, 37-46.
Collier Cameron, A. and Mestel, L. (1991), 'Theory of Magnetic Braking of Late-type Stars', in S. Catalano and J.R. Stauffer (eds.), *Angular Momentum Evolution of Young Stars*, Kluwer Academic Publishers, Dordrecht, in press (this volume).
de Medeiros (1990), private communication.
Deliyannis, C. P. (1990), *Primordial Lithium and Stellar Evolution : Lithium in Halo Stars from Standard, Diffusive, and Rotational Stellar Evolution and Implications for Cosmology,* Ph. D. Dissertation, Yale University.
Deliyannis, C. P., and Demarque, P. (1991a), *Ap. J. (Letters)*, in press.
Deliyannis, C. P., and Demarque, P. (1991b), *Ap. J.*, submitted.
Deliyannis, C. P., and Demarque, P. (1991c), *Ap. J.*, submitted.
Deliyannis, C. P., Demarque, P., Kawaler, S. D. (1990), 'Lithium in Halo Stars from Standard Stellar Evolution', *Ap. J. Suppl.*, **73**, 21-65.
Deliyannis, C. P., Demarque, P., Kawaler, S. D., Krauss, L. M., and Romanelli, P. (1989), 'Primordial Lithium and the Standard Model(s)', *Phys. Rev. Letters*, **62**, 1583-1586.
Deliyannis, C. P., Demarque, P., and Pinsonneault, M. H. (1989), 'The Ages of Globular Cluster Stars : Effects of Rotation on Pre-Main Sequence, Main Sequence, and Turnoff Evolution', *Ap. J. (Letters)*, **347**, L73-L76.
Deliyannis, C. P., and Pinsonneault, M. H. (1990) 'Beryllium in the Galactic Halo : Surface Abundances from Standard, Diffusive, and Rotational Stellar Evolution, and Implications', *Ap. J. (Letters)*, **365**, L67-L71.
Deliyannis, C. P., and Pinsonneault, M. H. (1991), 'Beryllium in Population I Lithium Dip F Stars : Constraints on Stellar Evolution', *Ap. J.*, submitted.
Deliyannis, C. P., Pinsonneault, M. H., and Demarque, P. (1991a), 'Primordial Lithium, Stellar Evolution, Big Bang Cosmology, and Dark Matter', *Ap.J.*, submitted.
Dimopoulos, S., Esmailzadeh, R., Hall, L. J., and Starkman, G. D. (1988), 'Is the Universe Closed by Baryons? Nucleosynthesis with a Late-Decaying Massive Particle', *Ap. J.*, **330**, 545-568.
Duncan, D. K. and Jones, B. F. (1983), 'Lithium Abundance and Age Spread in the Pleiades', *Ap. J.*, **271**, 663-671.
Durney, B. R., and Latour, J. (1978), *Geophys. Ap. Fluid. Dynamics*, **9**, 241.
Endal, A. S., and Sofia, S. (1976), 'The Evolution of Rotating Stars. I. Method and Exploratory Calculations for a 7 $M_\odot$ Star', *Ap. J.*, **210**, 184-198.
Endal, A. S., and Sofia, S. (1978), 'The Evolution of Rotating Stars. II. Calculations with Time-Dependent Redistribution of Angular Momentum for 7 and 10 $M_\odot$ Stars', *Ap. J.*, **220**, 279-290.
Endal, A. S., and Sofia, S. (1981), 'Rotation in Solar-Type Stars. I. Evolutionary Models for the Spin-Down of the Sun', *Ap. J.*, **243**, 625-640.
Greenstein, J. L., and Richardson, R. S. (1951), 'Lithium and the Internal Circulation of the Sun', *Ap. J.*, **113**, 536-546.

Hartmann, L. W., and Stauffer, J. R. (1989), 'Additional Measurements of Pre-Main Sequence Stellar Rotation', *A. J.*, **97**, 873-880.

Hill, F. (1987), 'The Equatorial Rotation Rate in the Solar Convection Zone', in B. R. Durney and S. Sofia (eds.), *The Internal Solar Angular Velocity*, Reidel, 45-50.

Hobbs, L. M., and Duncan, D. K. (1987), 'The Lithium Abundance in Halo Stars', *Ap. J.*, **317**, 796-809.

Hobbs, L. M., and Pilachowski, C. (1986), 'Lithium in NGC 752', *Ap. J.(Letters)*, **309**, L17-L21.

Hobbs, L. M., and Pilachowski, C. (1988a), 'Lithium in an Extreme Halo Star', *Ap. J.*, **326**, L23-L26.

Hobbs, L. M., and Pilachowski, C. (1988b), 'Lithium in Old Open Clusters : NGC 188', *Ap. J.*, **334**, 734-745.

Kawaler, S. D. (1988), 'Angular Momentum Loss in Low-Mass Stars', *Ap. J.*, **333**, 236-247.

Kippenhahn, R. and Thomas, H.-C. (1970), 'A Simple Method for the Solution of the Stellar Structure Equations Including Rotation and Tidal Forces', in A. Slettebak (ed.), *Stellar Rotation*, Reidel, Dordrecht, 20-29.

Kraft, R. P. (1970), 'Stellar Rotation', in G. H. Herbig (ed.), *Spectroscopic Astrophysics*, University of california Press, Berkeley, pp. 385-422.

MacGregor, K. B. (1991), 'Pre-main Sequence and Main Sequence Rotational Evolution : Constraints on Models Derived from Observations', in S. Catalano and J.R. Stauffer (eds.), *Angular Momentum Evolution of Young Stars*, Kluwer Academic Publishers, Dordrecht, in press (this volume).

Magazzu, A., and Rebolo, R. (1989), 'Lithium in T Tauri Stars', *Mem. S. A. It.*, **60**, 105-109.

Malaney, R. (1991), 'Probing the Early Universe : A Review of Non-Standard Primordial Nucleosynthesis', submitted to *Physics Reports*.

Mestel, L. (1984), 'Angular Momentum Loss During Pre-Main Sequence Contraction', S. L. Baliunas and L. Hartmann (eds.), in *3rd Cambridge Workshop on Cool Stars, Stellar Systems, and the Sun*, Springer-Verlag, Berlin, 49-59.

Molaro, P. (1987), 'Upper Limit to the Boron Abundance in the Population II Star HD 140283', *Astr. Ap.*, **183**, 241-246.

Peterson, R. C. (1983), 'The Rotation of Horizontal Branch Stars. II. Members of the Globular Clusters M3,M5, and M13', *Ap. J.*, **275**, 737-751.

Peterson, R. C. (1985a), 'The Rotation of Horizontal Branch Stars. III. Members of the Globular Cluster M4', *Ap. J.*, **289**, 320-325.

Peterson, R. C. (1985b), 'The Rotation of Horizontal Branch Stars. IV. Members of the Globular Cluster NGC 288' *Ap. J. (Letters)*, **294**, L35-L37.

Peterson, R. C., Tarbell, T. D., and Carney, B. W. (1983), 'The Rotation of Horizontal Branch Stars. I. Members of the Field', *Ap. J.*, **265**, 972-981.

Pilachowski, C. A., Booth, J., and Hobbs, L. M. (1987), 'The Abundance of Lithium in Pleiades F Stars', *Pub.A.S.P.*, **99**, 1288-1291.

Pilachowski, C. A., and Milkey, R. W. (1987), 'The Rotational Velocities of White Dwarfs', *Pub. A. S. P.*, **99**, 836-838.

Pinsonneault, M. H. (1988), *Evolutionary Models of the Rotating Sun and Implications for Other Low Mass Stars*, Ph.D. Dissertation, Yale University.

Pinsonneault, M. H. (1991), 'Rotational Mixing and Lithium in Young Stars', in S. Catalano and J. R. Stauffer (eds.), *Angular Momentum Evolution of Young Stars*, Kluwer Academic Publishers, Dordrecht, in press (this volume).

Pinsonneault, M. H., Deliyannis, C. P., and Demarque, P. (1991a), 'Evolutionary Models of Halo Stars with Rotation : I. Evidence for Differential Rotation with Depth in

Stars', *Ap. J.*, in press (Jan 20 issue). (PDD1)
Pinsonneault, M. H., Deliyannis, C. P., and Demarque, P. (1991b), 'Evolutionary Models of Halo Stars with Rotation : II. Effects of Metallicity on Lithium Depletion, and Possible Implications for the Primordial Lithium Abundance', *Ap. J. Suppl.*, submitted. (PDD2)
Pinsonneault, M. H., Kawaler, S. D., and Demarque, P. (1990), 'Rotation of Low Mass Stars : A New Probe of Stellar Evolution', *Ap. J. Suppl.*, **74**, 501-550. (PKD)
Pinsonneault, M. H., Kawaler, S. D., Sofia, S., and Demarque, P. (1989), 'Evolutionary Models of the Rotating Sun', *Ap. J.*, **338**, 424-452. (PKSD)
Prather, M. (1976), *The Effect of a Brans-Dicke Cosmology Upon Stellar Evolution and the Evolution of Galaxies*, Ph. D. Dissertation, Yale University.
Radick, R. R., Thompson, D. T., Lockwood, G. W., Duncan, D. K., and Baggett, W. E. (1987), 'The Activity, Variability, and Rotation of Lower Main Sequence Hyades Stars', *Ap. J.*, **321**, 459-472.
Rebolo, R., and Beckman, J. E. (1988), 'Lithium and Rotation in the Hyades Late F and G Stars', *Astr. Ap.*, **201**, 267-272.
Rebolo, R., Beckman, J. E., and Molaro, P. (1987), 'The Lithium Abundance in the Extremely Metal-Deficient Dwarf G 64-12', *Astr. Ap.*, **172**, L17-L19.
Rebolo, R., Molaro., P., and Beckman, J. E. (1988), 'Lithium Abundances in Metal-Deficient Dwarfs', *Astr. Ap.*, **192**, 192-205.
Ryan, S. G., Bessell, M. S., Sutherland, R. S., and Norris, J. E. (1990), '^{9}Be Abundances in Population II Stars : Implications for Light Element Nucleosynthesis', *Ap. J. (Letters)*, **348**, L57-L60.
Skumanich, A. (1972), 'Timescales for CaII Emission Decay, Rotational Braking and Lithium Depletion', *Ap. J.*, **171**, 565-567.
Smith, M. A. (1978), 'An Anticorrelation Between Macroturbulence and Age in G stars Near the main Sequence', *Ap. J.*, **224**, 584-594.
Soderblom, D. R., Oey, M. S., Johnson, D. R. H., and Stone, R. P. S. (1990), 'The Evolution of the Lithium Abundances of Solar-Type Stars. I. The Hyades and Coma Berenices Clusters', *A. J.*, **99**, 595-607.
Spite, M., Maillard, J. P., and Spite,. F. (1984), 'Abundance of Lithium in Another Sample of Halo Dwarfs, and in the Spectroscopic Binary BD - 0° 4234', *Astr. Ap.*, **141**, 56-60.
Spite, F., and Spite,. M. (1982), 'Abundance of Lithium in Unevolved Halo Stars and Old Disk Stars : Interpretation and Consequences', *Astr. Ap.*, **115**, 357-366.
Spite, F., and Spite,. M. (1986), 'Lithium Abundance in Nitrogen-Rich Halo Dwarfs', *Astr. Ap.*, **163**, 140-144.
Spite, M., Spite,. F., Peterson, R. C., and Chaffee, F. H. Jr. (1987), 'Lithium Abundance in Two Extreme High-Velocity Metal-Poor Dwarfs', *Astr. Ap.(Letters)*, **172**, L9-L10.
Stauffer, J. R., and Hartmann, L. W. (1986), 'The Rotational Velocities of Low Mass Stars', *Pub. A. S. P.*, **98**, 1233-1251.
Stauffer, J. R., and Hartmann, L. W. (1987), 'The Distribution of Rotational Velocities for Low Mass Stars in the Pleiades', *Ap. J.*, **318**, 337-355.
Strom, K. M., Wilkin, F. P., Strom, S. E., and Seaman, R. L. (1989), 'Lithium Abundances Among Solar-Type Pre-Main Sequence Stars', *A. J.*, **98**, 1444-1450.
Willson, L. A. and Bowen, G. H. (1988), 'Atmospheric Structure and Stellar Winds from Luminous Late-type Variables', in G.W. Coyne *et al.* (eds.), *Polarized Radiation of Circumstellar Origin*, University of Arizona Press, Tuscon, p. 485.
Zahn, J.–P., and Bouchet, L. (1989), 'Tidal Evolution of Close Binary Stars II. Orbital Circularization of Late-Type Binaries', *Astr. Ap.*, **223**, 112-118.

DISCUSSION

Vauclair: I have a comment about beryllium. You said that you could explain beryllium and lithium both in the Sun, and that other processes cannot do that. I think that any turbulent diffusion coefficient which decreases with depth rapidly enough can have lithium abundances decrease by a factor of 100 and beryllium abundances decrease by less than a factor of 3. And, with a Zahn coefficient and overshooting, my models provide this result.

Sofia: I agree - any turbulent diffusion coefficient will do. But I don't agree with overshooting - I think you need a different amount of over-shooting for each element if you are going to match the observations.

Vauclair: Well, it depends on how you treat over-shooting.

Sofia: If you have over-shooting by one amount, you can fix lithium, but not beryllium then.

Vauclair: You mean if the over-shooting is just an increase in just the convection zone completely mixed?

Sofia: Yes, that has been one of the ideas. Otherwise, I agree with you.

Vauclair: Your process for depleting lithium consists of transport of matter which is 20 times less than the transport of angular momentum. If the angular momentum is transported by magnetic fields, there is no reason for the transport of matter to be related to the transport of angular momentum. What physical processes do you think would provide the kind of mixing you want?

Sofia: Probably it is a combination of them. My guess would be that there is one mechanism which is responsible for 95% of the angular momentum transport and does not mix at all, and the other mechanisms which transport both angular momentum and matter equally account for the last 5%.

Vauclair: But, if it is different mechanisms, there is no reason they should be correlated, and so the factor of 20 should vary from star to star.

Pinsonneault: What one requires from the observations is that the time-scales for the angular momentum loss and the lithium depletion are correlated, so some mechanism which responds to the gradient in angular velocity is responsible for both. So, in that sense therefore you can say that is why the constant factor produces reasonable agreement with the observations of lithium.

Schatzman: What is the depth of the convection zone which you have with your parameters?

Sofia: We compute the depth of the convection zone in our models as a function of time from the pre-main sequence tracks onward. There is no assumption involved.

Gough: He is asking what the answer is.

Schatzman: What is the depth of the convection zone which you obtain?

Sofia: I believe it is about 70%.

Gough: Saying that you match standard models is ok, but standard models vary. The point is that if you are diffusing lithium, and it starts to burn not very deep, at least compared to the location of the bottom of the convection zone, then the amount by which you have to change the diffusion coefficient depends on the difference between the depth of the convection zone and the depth of the lithium burning region, which is quite a large variation compared to the relative variation with respect to the differences between standard models.

Schatzman: No, my question was whether your convection zone depth fits the helioseismology results.

Sofia: Yes, and it is not affected by rotation.

Gough: But 70% is not the right answer!

Pinsonneault: It is close enough.

Schatzman: What do you use for the mixing length to scale height parameter?

Sofia: We used something like 1.37.

Schatzman: I am surprised, because most models to obtain the observed depth of the convection zone use something around 2.0.

Gough: Different people use different mixing length theories, so the absolute value of alpha is meaningless.

Sofia: You use what you need to fit the solar radius at solar age.

Dziembowski: I agree that it is more elegant to just rely on one constant for your alpha and one constant in your diffusion equation, but it doesn't mean that it is the correct approach. However, these parameters are not like the gravitational constant, and it is quite feasible that they could vary with time.

Sofia: I agree with you. This is obviously an over-simplification. However, we need to limit how many things we can vary. If we just change constants to fit the observations, the whole modelling approach becomes meaningless. Until we understand physically why we would be changing these parameters, we should not vary them.

Rodonó: Do you have any results about angular rotation evolution for very late spectral types? I saw in a review paper on rotation that your results did not seem to work when you go to late type stars. The evolution with age of the angular rotation rate did not agree with the observations. Do your new models agree with the observations?

Pinsonneault: That depends on what you mean. I think you are talking about the rapid rotators in the Pleiades. Is that correct?

Rodonó: The figure I am talking about is in the review by Hartmann and Noyes.

Sofia: We knew we had a problem (in the original Endal and Sofia paper) and we identified the problem. It was that we did not treat the shear instability correctly. The new models fit the observations much better.

Roxburgh: What would you expect the quadrupole moment of the Sun to be, and what would you expect its angular momentum to be? These parameters might be amenable to determination by space missions in the not too distant future.

Sofia: I don't have the exact numbers, but they are compatible with current upper limits. We can also get the neutrino rate.

Kraft: You have talked about lithium and beryllium for the young stars. For old stars, more serious elements get to be interesting, and conventional models do not do well explaining them, particularly with regard to crossing chemical composition barriers where you have to wait a long time during giant branch evolution before you see surface abundance anomalies - in contradiction to the observations. What do your models say about this?

Sofia: The new models do address this, and we compute CNO abundances in giants and subgiants. What happens is that a given distribution of an element is widened, and so a convection zone that would not have reached down far enough to get to processed elements would do so now. We are in the process of looking at this in detail.

Kraft: How early can you get such a thing to happen?

Sofia: The widening of the abundances happens as the material is produced. You must wait to see that until the convection zone widens to get to that level.

Kraft: You are talking about the thermally driven convection. Is there any mixing driven by something else?

Sofia: No. It is just a combination of the wider distribution of processed matter and the normal growth of the convection zone during post main sequence evolution.

Gough: A comment to Roxburgh. If the gravity mode oscillations are like they might be, then the variations in the quadrupole moment might well be of order the expected quadrupole moment, so any observational measurement of that quantity might have to be treated delicately.

Roxburgh: While there may not be a British school in solar physics, there appears to be

one in experimental gravitation. I refer you to my article on the time dependent quadrupole moment of the Sun for project Vulcan.
When one includes mixing in stellar evolution, it is easy to forget that it requires energy to do it, rather than to just impose some condition and assume it takes place. Have you ever gone back to check those cases where you have assumed there is mixing, to see if there is enough energy available from the mechanism to which you are appealing for the mixing to take place?

Sofia: I have not done it explicitly, but when you have a large gradient in angular velocity, that should provide sufficient energy. We can test that. The differential rotation itself will provide the energy.

Pinsonneault: Two comments. The bulk of the mixing and the angular momentum transport take place very early in the models - when they are less than one billion years old - and mean molecular weight gradients in that time period are very small.

Second, we do take the mean molecular weight gradient instabilities into account in our stability conditions, and there is a claim in the papers from which we have taken these criteria that the energetics have been considered.

Gough: You should still look into it, because the kind of energetics that produce some instabilities is not necessarily the same which produce mixing.

THE INTERNAL ROTATION OF THE SUN: IMPLICATIONS ON THE HISTORY OF ITS ANGULAR MOMENTUM

L. PATERNO'
Istituto di Astronomia, Universita' di Catania, Italy
CNR - Gruppo Nazionale di Astronomia, UdR Catania, Italy

ABSTRACT. Recent helioseismological data have revealed that the internal rotation of the Sun may not differ very much from that observed on its surface. This implies that the angular momentum loss must be compatible with a slowing-down of the entire Sun. If we look at the rotation of solar type stars we cannot avoid considering the difficulty of reconciling the initial sharp spin-down in a very short time scale with the subsequent soft one leading to the present angular velocity of the Sun, unless some very effective internal instability mechanism takes place. This may involve the presence of magnetic fields in the Sun's core and is related to the location of the activity cycle dynamo mechanism.

1. Past Surface Rotation

The possibility of tracing back the surface rotation of the Sun is offered by the observations of stars with about the same mass of the Sun at earlier evolutionary stages.

Four main stellar evolutionary stages can be used to follow the past rotational behaviour of the Sun. Each of them is represented by a differently aged class of stars: T Tauri stars, and α Persei, Pleiades and Hyades clusters.

T Tauri of about one solar mass are believed to be young still fully convective stars, just at the end of the Hayashi track with ages of about 10^6 years and radii of the order of 3 $R_\odot$. Their rotation is essentially low, ranging from twice to five times the present surface rotation of the Sun $\Omega_\odot$ (Hartmann and Noyes 1987; Bouvier 1991).

The stars in α Persei cluster are aged about 5 10^7 years, and they burn hydrogen steadily, having developed a radiative core which has grown from post T Tauri phase to cover about 98% of the mass. In the standard models of the Sun, their position in the H-R diagram should correspond to that of maximum luminosity, before the star undergoes a further small contraction to settle onto the ZAMS.

These stars are characterized by a large spread in their rotational velocities, which, for one solar mass stars, range from few to fifty times $\Omega_\odot$ (Stauffer 1988).

S. Catalano and J. R. Stauffer (eds.), Angular Momentum Evolution of Young Stars, 357–364.

Pleiades are stars aged about 7 10^7 years in the evolutionary stage immediately prior to entering the ZAMS, which, for the Sun, is nominally located at about 10^8 years, assuming the Iben's definition based on thermal-gravitational energy reduction to 1% of the total energy output.

The rotational velocities of Pleiades with masses very close to the Sun's mass are concentrated in a narrow strip, whose average rotational velocity is about 5 $\Omega_\odot$ (Hartmann and Noyes 1987; Stauffer 1988).

Hyades are main sequence stars aged about 8 10^8 years. Those with masses close to the Sun's mass are slow rotators with typical velocities of the order of 2-3 $\Omega_\odot$, with a small spread in their values (Hartmann and Noyes 1987). Clusters of about the same age, such as Coma Ber and Praesepe, also show a similar rotational behaviour (Benz et al. 1984).

If the stellar samples we have considered really trace the history of the past surface rotation of the Sun, we can distinguish three main evolutionary phases: i) spin-up from T Tauri to α Persei in about 5 10^7 years; ii) rapid spin-down from α Persei to Pleiades in about 2 10^7 years; iii) slow spin-down from Pleiades through Hyades to the present Sun's rotation.

The typical rotational velocities, in terms of the present average solar surface rotation $\Omega_\odot \simeq 2.6$ µrad s^{-1}, are represented by open circles in Figure 1.

2. Present Internal Rotation

While there is no way to measure the internal rotation of stars other than the Sun, Helioseismology allows us to know, with a good degree of accuracy, the behaviour of solar rotation through the layers underneath the surface up to a distance of 0.3 $R_\odot$ from the center. The main result of the inversion of the helioseismological data is that the surface differential rotation is maintained throughout the convection zone, at the base of which there is a tendency towards a rigid rotation, the abrupt change occurring in a thin layer 0.05 $R_\odot$ thick (Morrow 1988).

For a discussion on the angular momentum of the Sun, we consider the rotation of the layers below the convection zone, which contain almost the whole Sun's mass. Figure 2 (squares) reports the equatorial velocity, from 0.75 to 0.38 $R_\odot$, as an average of two series of data obtained in 1986 and 1988 (Libbrecht and Woodard 1990). There is no clear indication, from p-mode rotational splittings, of what happens below 0.4 $R_\odot$, due to the very large errors in determining the rotational splittings of low degree modes, which are the most penetrative ones (Duvall and Harvey 1984). However, a large angular velocity increase in the central regions of the Sun might be excluded. The inversion of data may be consistent with a core rotating at most few times faster than the surface (Duvall et al. 1984).

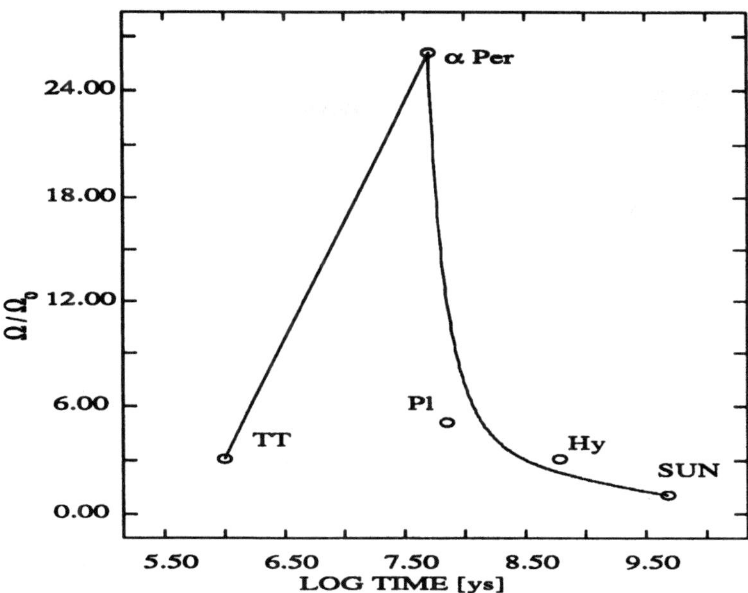

Fig. 1. Observed typical rotation rates of a sample of one solar mass stars in earlier evolutionary stages (open circles) and the theoretical behaviour of the surface angular velocity (solid line).

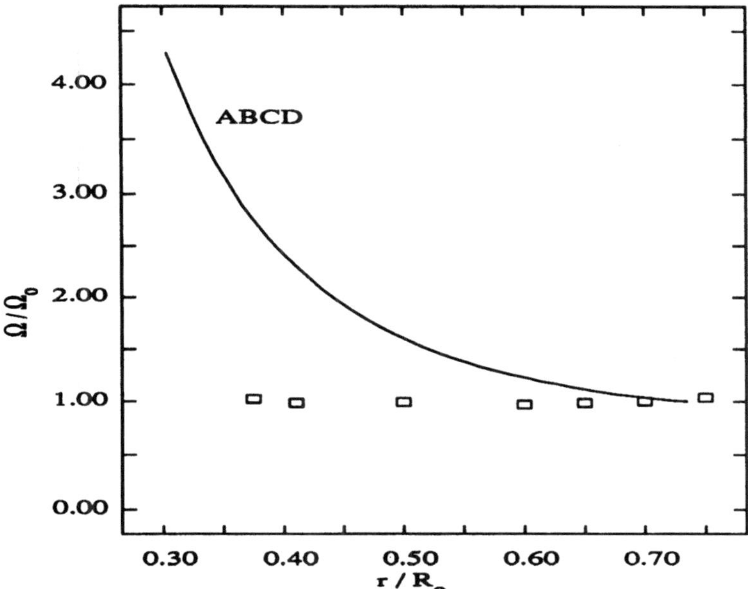

Fig. 2. Mean equatorial angular velocity (squares) from 1986 and 1988 helioseismological data (Libbrecht and Woodard 1990) and its marginal gradient for ABCD instability (solid line).

3 Angular Momentum Loss

If we assume a rigidly rotating polytropic structure with index n = 1.5, the angular momentum of one solar mass T Tauri stars is $J = h^2 MR^2 \Omega \simeq 1.3\ 10^{50}$ g cm^{2}s^{-1}, where h^2 = 0.2 is the gyration radius in terms of stellar radius R^2. The present angular momentum of the Sun, assuming a rigid rotation $\Omega_\odot$, is $J_\odot \simeq 1.8\ 10^{48}$ g cm^{2}s^{-1}. If we suppose that the inner part of the Sun ($r < 0.3\ R_\odot$) rotates at a different rate than the surface, the angular momentum can be expressed as $J_\odot = I_i\Omega_i + I_o\Omega_o$, where $I_i = h_i^2 M_i R_\odot^2$ and $I_o = h_o^2 M_o R_\odot^2$ are the moments of inertia , with h_i^2 = 0.028, h_o^2 = 0.17, M_i = 0.75 $M_\odot$ and M_o = 0.25 $M_\odot$ as deduced from a standard model of the Sun, and Ω_i and Ω_o the angular velocities of rotation of the inner and outer regions respectively. If the core of the Sun rotates 3 times faster than its surface, the total angular momentum of the Sun does not differ very much from that given above ($J_\odot \simeq 2.7\ 10^{48}$ g cm^2 s^{-1}).

If the Sun had to conserve its angular momentum, the core should now rotate at a rate two orders of magnitude faster than the surface, which is absolutely not admittable.

Considering $\Omega_i = 3\Omega_\odot$, one can conclude that about 98% of the initial angular momentum of the Sun went lost during its evolution.

The spin-up phase from T Tauri to α Persei may be consistent with angular momentum conservation and rigid rotation. We have in fact:

$$\Omega_2/\Omega_1 = (h_1^2/h_2^2)\cdot(R_1/R_2)^2 \simeq 12 \qquad (1)$$

where suffices 1 and 2 refer respectively to the initial and final conditions and h_1^2 = 0.2, h_2^2 = 0.15 are relative to two polytropic structures with n_1 = 1.5 (T Tauri) and n_2 = 2 (α Persei), while $R_1/R_2 \simeq$ 3. The assumption of rigid rotation is consistent with a rapid redistribution of the angular momentum in largely convective stars, owing to the high turbulent diffusion.

The structure of the star does not change very sensibly from α Persei phase to the present one, except for a slightly larger mass concentration in the main sequence, which makes h^2 decrease by a factor of two and consequently the moment of inertia is reduced by the same factor. Therefore the angular momentum loss is almost entirely caused by the rotational braking.

During the evolution subsequent to the α Persei phase, the angular momentum should have been removed at an average rate of about 10^{-8} $J_\odot$/year to meet the present $J_\odot$. However, during the fast spin-down phase, in order to meet the observed rotation of Pleiades, the removal of the angular momentum should have been of the order of 2 10^{-6} $J_\odot$/year, if the internal regions had to spin-down at about the same rate as the surface.

4 Magnetic Braking

The theory of magnetic braking for explaining the angular momentum loss in active stars was first proposed by Schatzman (1962) and then refined by Mestel (1968). In this framework, the angular momentum variation with time can be expressed as:

$$\frac{dJ}{dt} = \frac{2}{3}\frac{dM}{dt}\,\Omega\, R^2 \left[\frac{r_A}{R}\right]^k \tag{2}$$

where:

$$\frac{dM}{dt} = -\,4\pi\rho_A u_A r_A^2 \tag{3}$$

is the mass loss, r_A the Alfvèn radius of a surface within which the magnetic field is strong enough to force the gas to co-rotate with it, while outside the gas flow drags the field to follow itself, ρ_A and u_A are respectively the wind density and velocity at r_A and k determines the magnetic field geometry inside the surface (close to open field line ratio). The consequent angular velocity variation, neglecting the small change of the moment of inertia caused by the mass loss, is given by:

$$\frac{d\Omega}{dt} = \frac{N}{I(\Delta M)} \tag{4}$$

where $N = dJ/dt$ is the torque, $I(\Delta M) = h^2(\Delta M)R^2\Delta M$ the moment of inertia of the mass braked ΔM and $h^2(\Delta M)$ the relative gyration radius in terms of R^2. Equation (4) is likely to be correct, because there is no evidence of strong mass losses during the evolutionary phases we are concerned with.

Castellani and Paterno' (1984) calculated the angular velocity variation for stars of about the same mass of the Sun evolving from ZAMS to the red giant branch, in the cases of dipole and quadrupole magnetic configurations. They calculated the strength of the surface magnetic field produced by an α-ω dynamo working in the deepest layers of the convection zone, by comparing the flux tube rise time with the dynamo amplification time (Durney and Robinson 1982), and solved the relevant equations of a coronal model in order to determine dM/dt and ρ_A, u_A, r_A during the evolution. For a dipole field, the results were consistent with mass losses of the order of the present Sun's mass loss ($\simeq 3\ 10^{-14}\ M_\odot$/year) and $r_A/R_\odot$ of the order of the present value ($\simeq$ 25-30). Having assumed the slowing-down of the entire star, they found a drop in the angular velocity by a factor of 4 from ZAMS to the age of $5\ 10^9$ years. This means that, even with modest mass losses, it is possible to obtain substantial angular momentum losses if the Alfvèn radius is sufficiently far from the surface.

The results are consistent with the observed rotational velocity drop from Pleiades to the present Sun and with an angular momentum loss of the order of 10^{-8} $J_\odot$/year.

It is not easy to explain the rapid spin-down phase if the whole Sun had to be braked by the external magnetic torque. This latter should have been some 200 times stronger than that in action during the soft spin-down phase to be consistent with a rate of loss of 2 10^{-6} $J_\odot$/year. Assuming co-rotation inside the Alfvèn surface, with a dipole configuration, and magnetic flux conservation, it turns out from equations (2) and (3) that the torque depends only on the square of the surface magnetic field strength. This would imply that, during the rapid spin-down phase, the dynamo should have been so active to produce mean surface toroidal fields more than one order of magnitude stronger than those produced in all the subsequent phases. It is not easy to reconcile this fact with the small change in the structural characteristics of the stars evolving from α Persei to Pleiades phases. On the other hand, it cannot be stated that faster rotators produce stronger magnetic fields, since fast rotation tends to inhibit convection, reducing both the α-effect and differential rotation (Gilman 1983).

5. Angular Momentum Flux in the Interior

The most natural explanation of the rapid spin-down problem is that Pleiades retain almost all the angular momentum of α Persei and therefore of T Tauri, possessing a rapidly spinning core almost decoupled from the surface. If only the convection zone, which contained some 10 % of the moment of inertia, had to be braked, it would not be difficult to account for the rapid decrease of Ω without invoking a much stronger torque.

Considerations of Stauffer et al. (1985), based on the rotational characteristics of K dwarfs in α Persei and Pleiades clusters, strongly support the present point of view.

This scenario implicitly requires the existence of a mechanism for coupling the interior with the surface layers in the course of the subsequent evolution until the present situation of almost rigid rotation was reached, as the helioseismological data seem to indicate. To be consistent with the observations and calculations of the slowing-down in the main sequence (Endal and Sofia 1981; Castellani and Paterno' 1984), it is necessary to assume that an effective mechanism for transferring the angular momentum towards the exterior acted immediately after the Pleiades phase and reached stability in a short time scale with respect to the time spent on the main sequence.

The solid line in Figure 1 represents a possible theoretical evolution of the surface rotation of the Sun as a function of time. The line connecting T Tauri to α Persei has simply been drawn, under the constraint of angular momentum conservation, slightly changing the value of the ratio of the radii in equation (1) to fit the points. The descending branch from α Persei to the present Sun has been constructed by integrating equation (4), using a constant torque consistent with the values found by Castellani and Paterno' (1984). A continuous increase of I(ΔM) with time has been numerically simulated to account for the progressive linking of the surface to the core. The simulating function has been constructed in such a way as to give a rapid initial increase of the linking and then adjusted in order to fit the present surface rotation of the Sun.

As far as the physics of the angular momentum transport in the interior is concerned, an assortment of hydrodynamic and hydromagnetic instabilities in the radiative regions below the convection zone has been proposed (Spruit 1987; Zahn 1987). In Figure 2 (solid line) the marginal gradient of the angular velocity is reported for the Axisymmetric Baroclinic Diffusivity instability (ABCD), which is one of the most severe hydrodynamic ones, as calculated by Dziembowski et al.(1986) for a standard model of the Sun. It is evident, from a comparison with the observations shown in the same Figure, that hydrodynamic instabilities alone may not transfer angular momentum at a sufficient rate for explaining the present rotation.

A magnetic field in the core may offer a wide source of instabilities. It has been shown (Spruit 1987) that even a field of a few Gauss can be so efficient in transporting momentum to produce an angular velocity gradient compatible with the observations. However the major problem lies in the long term stability of such a field.

6. References

Benz,W., Mayor,M., Mermilliod,J.C.: 1984, Astron.Astrophys. **138**, 93

Bouvier,J.: 1991, *These Proceedings*

Castellani,V., Paterno',L.: 1984, Astron. Nachr. **305**, 251

Diembowski,W.A., Paterno',L., Ventura,R.: 1986, in *Seismology of the Sun and Distant Stars*, D.O. Gough ed., Reidel, Dordrecht, p. 257

Durney,B.R., Robinson,R.D.: 1982, Astrophys. J. **253**, 290

Duvall Jr,T.L., Harvey,J.W.: 1984, Nature **310**, 19

Duvall Jr,T.L., Dziembowski,W.A., Goode,P.R., Gough,D.O., Harvey,J.W., Leibacher,J.W.: 1984, Nature **310**, 22

Endal,A.S., Sofia,S.: 1981, Astrophys.J. **243**, 625

Gilman,P.A.: 1983, in *Solar and Stellar Magnetic Fields: Origins and Coronal Effects*, J.O. Stenflo ed., Reidel, Dordrecht, p. 247

Hartmann,L.W., Noyes,R.W.: 1987, Ann.Rev.Astron.Astrophys. **25**, 271

Libbrecht,K.G., Woodard,M.F.: 1990, Nature **345**, 779

Mestel,L.: 1968, Monthly Notices Roy.Astron.Soc. **138**, 359

Morrow,C.A.: 1988, ESA-SP 286, p. 91

Schatzman,E.: 1962, Ann.Astrophys. **25**, 1

Spruit,H.C.: 1987, in *The Internal Solar Angular Velocity*, B.R. Durney, S. Sofia eds., Reidel, Dordrecht, p. 185

Stauffer,J.R.: 1988, in *Cool Stars, Stellar Systems, and the Sun*, J.L. Linsky, R.E. Stencel eds., Springer-Verlag, Heidelberg, p. 182

Stauffer,J.R., Hartmann,L.W., Burnham,J., Jones,B.F.: 1985, Astrophys. J. **289**, 201

Zahn,J.-P.: 1987, in *The Internal Solar Angular Velocity*, B.R. Durney, S. Sofia eds., Reidel, Dordrecht, p. 201

DISCUSSION

Duncan: Just a comment. I wish to remind people that the only cluster data between the T Tauri stage and the main sequence are the stars in Orion which I have observed, which have ages between about 1 Myr and 10 Myr, and those stars do not show the strong increase in surface rotation which you might have expected. So, things may be even more complicated than they seemed.

Paternó: I agree. As a theoretician, I have just tried to use the observations which I was aware of. Things may be more complicated.

Angular Momentum Transport, Rotational Instabilities, Magnetic Fields and Mixing

IAN W ROXBURGH
Astronomy Unit,
Queen Mary and Westfield College,
University of London
Mile End Road,
London E1 4NS., UK.

ABSTRACT. The energetics of mixing material through a chemical composition gradient is considered. In regions where nuclear reactions are significant the rate of energy input required to maintain mixing against the growing composition gradient is of the order of the internal energy per nuclear time scale, this is greater than the rate of energy available from most sources except nuclear reactions themselves. Thus angular momentum transport by material motion is only likely to be effective in convective zones (and any associated overshoot regions), and possibly by instabilities themselves driven by nuclear reactions. Even though the angular velocity gradient may be unstable, transport of angular momentum by rotationally driven instabilities will be inhibited. Transport by magnetic fields does not require material motion through a composition gradient and is likely to be very effective in establishing almost uniform rotation. If the rotation is not constant then the differential rotation produces a toroidal field which reacts back on the differential rotation leading to torsional oscillations with a period of the local Alfven travel time. Neighbouring field lines rapidly get out of phase with each other creating a large gradient in magnetic field and substantial dissipation. This is likely to establish almost uniform rotation as this is the lowest energy state for a given angular momentum.

1. Results

The principle points made in this article are:

a) Mixing of material requires a substantial energy input to overcome the stabilising effect of the chemical composition gradient.

b) For a given angular momentum uniform rotation is the lowest energy state.

c) Differential rotation may drive instabilities but these do not mix material or transport angular momentum in the radial direction, since the energy input required to overcome the chemical composition gradient exceeds that available from the differential rotation.

d) Nuclear reactions produce more than sufficient energy to sustain vertical mixing and thereby redistribute angular momentum. Matter and angular momentum can be redistributed in convectively unstable layers since the energy source is ultimately the nuclear reactions.

e) In solar type stars the ^{3}He instability has enough energy, through the burning of excess ^{3}He to ^{4}He, to cause mixing against the stabilising effect of the chemical composition gradient. Whether or not mixing or diffusion occurs requires further investigation.

f) Magnetic fields can transport angular momentum without mixing matter and therefore provide a mechanism for redistributing angular momentum inside stars. Differential rotation produces a toroidal field, this field reacts back on the differential rotation leading to torsional oscillations with a period of the local Alfven travel time. Neighbouring field lines get out of phase with each other creating a large gradient in magnetic field and substantial dissipation. This is likely to establish almost uniform rotation as this is the lowest energy state for a given angular momentum. In the authors opinion this is the most likely scenario.

S. Catalano and J. R. Stauffer (eds.), Angular Momentum Evolution of Young Stars, 365–372.

2. Some Energy Considerations.

I first emphasise a point made several times before (Roxburgh 1984a, 1984c, 1985a, 1985b) that the mixing of material inside stars requires a substantial energy input to overcome the stabilising effect of the chemical composition gradient.

The radiative regions of a star are stable, the work required to raise matter adiabatically through the gravitational field being of the same magnitude as the internal (or gravitational) energy. However displacements that are sufficiently slow, or on a sufficiently long time scale (eg fingers), are able to exchange heat with their surroundings and overcome the stabilising effect of the thermal stratification. But in regions of chemical inhomogeneity, energy still has to be provided to overcome the composition gradient. This can be quantified as follows.

Consider a "blob" of fluid of mass $\delta M = \rho\ \delta V$ raised "slowly" through a distance z in the vertical direction, and let the molecular weight be a function of radial distance r, $\mu = \mu(r)$. If the displacement is slow enough the pressure and temperature inside the blob are the same as the surroundings; the density excess is therefore

$$\frac{\delta\rho}{\rho} = \frac{\delta\mu}{\mu} = -\frac{\underline{z}.\nabla\mu}{\mu} \tag{1}$$

The excess force on the blob is g $\delta\rho$, the work per unit mass done in raising the blob through a distance k is therefore

$$w = \int_0^k g_z \frac{\delta\rho}{\rho}\, dz = -\int_0^k g_z \frac{z.\nabla\mu}{\mu}\, dz = -\frac{1}{2}(\underline{k}.\underline{g})\frac{\underline{k}.\nabla\mu}{\mu} \tag{2}$$

Since such a displacement also requires an equivalent mass to be moved in the opposite direction the total energy (per unit mass) required for such an interchange is 2w. Any process that mixes regions of a star has to provide this energy.

If the major part of the star is to be mixed then the distance k is of order of magnitude the radius of the star and the total energy required is of the order

$$W = \frac{GM^2}{R}\frac{\Delta\mu}{\mu} \tag{3}$$

where $\Delta\mu$ is the mean change in molecular weight over the whole star of mass M and radius R. For stars which have a substantial degree of chemical inhomogeneity $\Delta\mu/\mu \approx 1$ and the energy required for mixing is comparable to the gravitational energy of the star and is considerably greater than the kinetic energy of rotation. Since $\Delta\mu \approx (d\mu/dt)\ t_o$, where t_o is the age of the star, mixing would require a continuous rate of input of energy per unit mass per unit time of approximately $\Phi\ (1/\mu)\ d\mu/dt$, where Φ is the absolute value of the gravitational potential. This is considerable greater than the rate available from kinetic energy in differential rotation which is $< \Omega^2\ \varpi^2/2t_o$. However the rate of production of energy per unit mass from nuclear sources is considerably in excess of that required to overcome the composition gradient.

3. Energy available in differential rotation.

If a star is rotating differentially it has a potential source of energy that could in principle cause mixing of material and angular momentum. To quantify this I first show that for a given angular momentum the lowest energy state of a rotating star is uniform rotation. This is straight forward, we wish to find

$$\mathrm{Min}\left(\frac{1}{2}\int \Omega^2 dI\right) \quad \text{subject to} \quad J = \int \Omega\, dI = \text{constant} \tag{4}$$

The Euler-Lagrange equations yield $\Omega = J/I$ = constant. A simple alternative derivation is to consider two bodies M_1, M_2, with moments of inertia I_1, I_2, and angular velocities Ω_1, Ω_2. The total angular momentum and kinetic energy are $J = I_1\Omega_1 + I_2\Omega_2$, $E = (I_1\Omega_1^2 + I_2\Omega_2^2)/2$. Eliminating Ω_2 and minimising E gives $\Omega_1 = J/(I_1 + I_2)$. The energy released by this transition is $\Delta E = I_1 I_2 (\Omega_1 - \Omega_2)^2 /[2(I_1 + I_2)]$.

Consider now the interchange of 2 elements of fluid in the equatorial plane over a distance $\underline{k}$ inside a star where $\Omega = \Omega(\underline{r})$; since $I_1 \approx I_2 \approx \rho\,\delta V\,\varpi^2$, where ϖ is the distance from the rotation axis, and $(\Omega_1 - \Omega_2)^2 \approx (\underline{k}.\nabla\Omega)^2$, the energy released per unit mass from local mixing of matter over a distance $\underline{k}$ is $\approx (1/4)\,\varpi^2\,(\underline{k}.\nabla\Omega)^2$. Combining this result with condition (2) for the energy required to overcome the stabilising effect of the molecular weight we deduce that if differential rotation is to provide the energy for mixing then

$$\varpi^2(\underline{k}.\nabla\Omega)^2 > -4\,(\underline{k}.g)\left(\frac{\underline{k}.\nabla\mu}{\mu}\right) \quad \text{or} \quad \left(r\frac{d\Omega}{dr}\right)^2 > -4\,g\,\frac{1}{\mu}\frac{d\mu}{dr} \quad \text{for } \Omega = \Omega(r) \tag{5}$$

where the last condition follows from considering a displacement in the equatorial plane. In typical central regions of a star this requires such steep gradients of angular velocity that the core would be rotating faster than the limit of dynamical stability, $(\Omega^2/2\pi G\rho_c \approx 0.1)$. In practice the chemical composition gradient prevents mixing and therefore angular momentum transport by material motion.

4. Rotational Instabilities

The condition just derived is essentially the instability condition derived by Goldreich and Schubert (1967) and Fricke (1968) but in the presence of chemical composition gradients (Roxburgh 1975). Other instabilities may occur which are are not inhibited by the composition gradient (Shibahashi 1980, Knobloch and Spruit 1983, Roxburgh 1984b). A stability analysis allowing for the fact that surfaces of constant entropy, effective gravity and composition do not necessarily coincide, and including viscosity (ν), heat conduction (κ) and diffusion (η), gives 4 modes which are stable if

$$\{(X_i + Y_i)R_1 \sin\Gamma + Z_i R_2 \sin(\lambda-\phi)\}^2 > 4 R_1 Y_i \sin(\lambda-\Gamma)\{X_i \sin\theta + Z_i R_2 \sin\phi\} \tag{6}$$

where the values of (X_i, Y_i, Z_i) for the 4 modes, and the other variables are

$$(X_i, Y_i, Z_i) = (\sigma_1\sigma_2, \sigma_2, \sigma_1,),\quad (\sigma_1+\sigma_2, 1+\sigma_2, 1),\quad (1, 1, 0),\quad (1+\sigma_1, 2\sigma_1, -1) \tag{7}$$

$$\sigma_1 = \frac{\nu}{\kappa}, \quad \sigma_2 = \frac{\eta}{\kappa}, \quad R_1 = \frac{N_\Omega^2}{N_T^2}, \quad R_2 = \frac{N_\mu^2}{N_T^2}, \tag{8}$$

$$\mathbf{g} - \Omega^2 \varpi = g\,(\cos\lambda, 0, \sin\lambda), \qquad \nabla S = \frac{N_t^2}{g}(\cos\theta, 0, \sin\theta) \tag{9}$$

$$\frac{\nabla(\Omega^2 \varpi^4)}{\varpi^3} = N_\Omega^2(\cos\Gamma, 0, \sin\Gamma), \qquad \frac{\nabla\mu}{\mu} = -\frac{N_\mu^g}{g}(\cos\phi, 0, \sin\phi) \tag{10}$$

where the vectors are referred to cylindrical polar coordinates (ϖ, Ψ, z). With $\sigma_1 \approx \sigma_2 \approx 10^{-6}$, and taking $\lambda = \phi$, and $\Omega = \Omega(r)$, then for $R_2 > \sigma$ the first unstable mode occurs when

$$\left(r\frac{d\Omega}{dr}\right)^2 > 8\,\sigma N_T^2 \tag{11}$$

This has been called the ABCD (Axisymmetric-BaroClynic-Diffusive) instability by Spruit, Knoblock and Roxburgh (1983), who calculated the critical angular velocity distribution in a model of the present Sun. This instability is not suppressed by a chemical composition gradient. However the resulting motion is almost horizontal, and from the previous arguments it cannot have enough energy to produce vertical mixing and redistribution og angular momentum.

In fact this instability can be readily suppressed by a slight (almost horizontal) readjustment of chemical composition. Returning to condition (6) but with $\lambda \neq \phi$ we have stability if

$$\{(1+3\,\sigma)\,R_1 \sin\Gamma - R_2 \sin(\lambda - \phi)\,\}^2 < 8\,\sigma \sin(\lambda - \Gamma)\,R_1\,\{(1+\sigma)\sin\theta - R_2 \sin\phi\} \tag{12}$$

and this is satisfied if

$$\sin(\lambda - \phi) = \frac{R_1}{R_2}(\,1 + 3\sigma)\sin\Gamma \tag{13}$$

That is there provided the composition gradient is large enough, so that $R_1/R_2 < 1$, there exists an angle ϕ such that the differential rotation is stable. Thus a slight readjustment in the distribution of μ can stabilise the differential rotation.

5. Instabilities driven by nuclear reactions

For effective mixing the energy input must be comparable with the gravitational energy of the star over the evolution time. Nuclear energy sources are more than sufficient since they effectively generate this amount of energy in a thermal time scale, but there has to be a mechanism that can draw on this energy source. One effective mechanism is convection which converts the radiative energy into kinetic energy which is therefore more than energetic enough to mix the star. In this case one would expect angular momentum to be effectively redistributed by the convective motions; the actual distribution of angular momentum being determined by the interaction of rotation and convection (see Roxburgh 1991, in this volume).

Another possibility is that the nuclear reactions excite some secularly unstable, or overstable mode. The known example of this is the ^{3}He instability discovered by Gough and co-workers (Dilke and Gough, 1972, Christensen-Dalsgaard et al 1974), but I am not convinced that the stability analysis has been done with sufficient rigour to rule out other possibilities. The ^{3}He instability is an overstable g-mode in which the energy input by nuclear reactions over an oscillation period exceeds the energy losses by radiation; the source of energy being the very temperature sensitive ^{3}He burning. The instability sets in early in the evolution of a star (3 10^8 years for the a solar model) when the molecular weight gradient is still small, but the ^{3}He abundance is large enough to excite the mode. An estimate of the energy available is obtained from calculating the rate of production of excess ^{3}He at the time instability sets in and assuming this to be converted into kinetic energy. This is more than sufficient to overcome the stabilising effect of the composition gradient.

The consequences of this instability have not been fully explored. One possibility, explored by myself (Roxburgh 1985b, 1987), is that it leads to finite amplitude oscillations where the excess rate of burning of ^{3}He during an oscillation maintains the distribution near the marginally stable state; this requires quite large amplitudes. The centre of the star is a node of the oscillation so the contribution to the energy generation from regions away from the centre, where the amplitude of the temperature oscillation is large, is substantially enhanced; such models therefore have a lower central temperature and a higher central hydrogen abundance. Once the oscillation ceases, (at about 3 10^9 years for a solar model) the star evolves in a standard inhomogeneous manner, but compared to a standard solar model the central hydrogen abundance is higher and the central temperature is lower. A simple model by Roxburgh (1985b, 1987) incorporated these effects by taking the amplitude of the temperature perturbation as

$$\left(\frac{\delta T}{T}\right)^2 = A\left(\frac{t}{t_o} - 0.065\right)\sin^2\!\left(2\pi\frac{r}{R}\right), \quad 0.065 < \frac{t}{t_o} < 0.75 \tag{14}$$

Resulting models of the present Sun had a higher central hydrogen abundance, lower central temperature and could give a Solar Neutrino Flux as low as 3 SNU.

Another possibility, perhaps more relevant to the present discussion on mixing, is that the oscillation breaks down into mild turbulence which diffuses the excess ^{3}He into the centre where it is burnt (Roxburgh 1984a). The diffusion coefficient can be determined from the need to transport excess ^{3}He into the central regions on the evolution time scale, where it is burnt to ^{4}He . Since the gradient of ^{3}He is large, the diffusion coefficient is small, ($\nu_{turb} \approx 60$ cm^2/sec) and insufficient to diffuse angular momentum on the evolutionary time scale unless the angular velocity gradient is as steep as the gradient of ^{3}He. Evolutionary models of the Sun including this diffusion do not show any significant changes over standard solar models. At the present time one cannot be sure whether this instability, or others driven by nuclear reactions are effective in mixing and transporting angular momentum. More work is needed on this problem.

6. Transport of angular momentum by magnetic fields

There is no reason known to the author for ignoring the effect of magnetic fields on the rotation of stars. Magnetic fields are present in the material from which stars are formed, strong fields are observed on the surface of some stars and weak fields may exist at the surface of all other stars. Moreover most stars go through some period when they have substantial convective regions and we may expect a dynamo to operate in this phase and build up a strong field. If a star had no magnetic field then the build up of differential rotation, due to the slowing down of the surface layers, or just to the differential contraction of the central parts of the star with evolution, will of itself generate a magnetic field due to the "battery effect" (Biermann 1950, Roxburgh 1966). This could provide the seed field for enhancement by a dynamo that could takes place in the convective core that almost all stars have during the early phase of their evolution (from burning ^{12}C to ^{14}N).

The interaction of differential rotation and magnetic fields is not well understood, in particular there is the possibility that instabilities limit the growth of the field. Ignoring this possibility for the moment, differential rotation would generate a large toroidal field from a small poloidal field in a relatively short time scale. For example if the poloidal field were of 1 Gauss, and the rotation period of the order of days, and the rotation varied by a factor of two, the the toroidal field would grow linearly on a time scale of the rotation field and would reach 10^7 Gauss in 3 10^5 years. At this stage it would be strong enough to react back on the differential rotation and set up torsional oscillations. If $S(r,\theta)$ is the field line function of the poloidal field then neglecting diffusion the oscillation is governed by the equations

$$\underline{B}_p = \frac{1}{\varpi} \underline{k} \times \nabla S(r,\theta) \tag{15}$$

$$\frac{\partial \Omega}{\partial t} = \frac{1}{4\pi\rho r^3 \left(1-\mu^2\right)} \left[\frac{\partial S}{\partial r}\frac{\partial C}{\partial \mu} - \frac{\partial S}{\partial \mu}\frac{\partial C}{\partial r} \right] \tag{16}$$

$$\frac{\partial C}{\partial t} = \left(1-\mu^2\right) \left[\frac{\partial S}{\partial r}\frac{\partial \Omega}{\partial \mu} - \frac{\partial S}{\partial \mu}\frac{\partial \Omega}{\partial r} \right] \tag{17}$$

where $C = B_\phi\, r \sin\theta$ and $\mu = \cos\theta$. These equations describe toroidal oscillations with a period given by the local Alfven speed and the build up from an initial state with $\Omega = \Omega(r)$, $B_\phi = 0$ at $t = 0$, and a given poloidal magnetic field.

But diffusion cannot be neglected; different regions of a star will have different oscillation periods and would therefore get out of phase, this would create large gradients in the field and enhanced dissipation of energy. Even if the growth of the field were limited by instabilities, these would also produce enhanced dissipation. The energy that is dissipated comes from the differential rotation, so the magnetic field is a means by which energy can be extracted from the rotation. But as we have seen above, the lowest energy state of a rotating star is one with uniform rotation, so the dissipation would have the effect of equalising the angular velocity, that is of transporting angular momentum. Of course the less the differential rotation the longer the time required for the toroidal field to be produced, so that some (small) differential rotation would remain. Detailed calculations to demonstrate that effectiveness of this "phase mixing" are currently being undertaken

7. Conclusion

The most likely scenario (in the authors opinion) is that in radiative regions magnetic fields are effective in redistributing angular momentum, establishing (almost) uniform rotation. In convective zones mixing takes place and redistributes angular momentum. The structure of the boundary region between convective and radiative zones is not understood; there will be some convective overshooting, possibly gravity waves and so probably some effective means of coupling the convective region to the magnetically controlled radiative region. There may be some mild diffusive mixing of material in solar type stars due to the ^3He instability.

References

Biermann L 1950, *Zeit fur Naturforschung A*, 5,65.

Christensen-Dalsgaard J, Dilke F W W and Gough D O, 1974, Monthly Notices Roy Astron Soc, **169**, 429.

Dilke F W W and Gough D O 1972, *Nature*, **240**, 262.

Fricke K 1968, *Zeitschrift fur Astrophysik*, **68**, 317.

Goldreich P and Schubert G 1967. *Astrophysical Journal*, **150**, 571.

Knoblock E and Spruit H 1983 *Astronomy and Astrophysics*, **125**, 59.

Roxburgh I W 1966, "On Stellar Rotation III" *Mon Not Roy Ast Soc* , **132**, 201.

Roxburgh I W, 1975, 'The Bubble Derivation of the Thermal Stability Criteria for a Rotating Star'", *Memoires de la Societe Royale des Sciences de Liege*, 6e serie, **8**, 69.

Roxburgh I W 1984a, On Turbulent Mixing", *Observational Tests of the Stellar Evolution Theory*, ed A Maeder and A Renzini, 519, D Reidel Pub Co., Dordrecht.

Roxburgh I W 1984b, Rotational Instabilities in the Solar Interior, Turbulent Diffusion and the Solar Neutrino Problem, *Mem Soc Astro Ital*. 55, 273.

Roxburgh I W 1984c, "The Stability of Differentially Rotating Stars". paper dedicated to Paul Ledoux and presented at the 25th Liege Symposium, 1984.

Roxburgh I W 1985a, "Instabilities, Mixing and Solar Neutrinos", *Solar Neutrinos and Neutrino Astronomy*, ed M L Cherry, W A Fowler and K Lande, Am Institute Physics, 88.

Roxburgh I W 1985b, "Present Problems of the Solar Interior", *Solar Physics*, **100**, 21.

Roxburgh I W 1987, "Problems of the Solar Interior", in *The Internal Solar Angular Velocity*, ed B Durney and S Sofia. D Reidel Publishing Company, p 1.

Spruit H , Knoblock E and Roxburgh I W 1984, *Nature*, **304**, 320.

Shibahashi H 1980, Publ Asrton Soc Japan, 22, 341.

DISCUSSION

Pinsonneault: Just a comment. The energetics you used was for the present day Sun. Is that correct?

Roxburgh: Yes.

Pinsonneault: Ok. Then for a much younger, more rapidly rotating star, you might get different answers. Second, the hypothesis that uniform rotation is enforced on a relatively short time-scale is a very testable thing. It makes definite predictions for evolved stars. In particular, it predicts that white dwarfs should have very long rotation periods and that horizontal branch stars should rotate very slowly. Subgiants should also be slow rotators. In all of these phases, we see some evidence for substantial rotation. Is there any way to reconcile these observations with your model?

Roxburgh: The only one of these observations which I have thought about, because it is the one which the Yale school has pushed the most, are the horizontal branch stars. I was told that it is only about 1% of the horizontal branch stars which are rapid rotators. This is numerically compatible with the evolution of close binaries.

Kraft: Without grinding any axes, I think the correct statement is that only about 1% of the stars have been looked at for rotation. For those that have been examined, the average rotational velocity is about 15 km/sec, which is unexpectedly high.

Dziembowski: One comment about rotation in evolved stars. The only good evidence of the rotation of the cores comes from white dwarfs. In all white dwarfs for which rotation has been measured, it is always very low. Now, a question on torsional oscillations. According to my estimates, the decay time for the fundamental mode of torsional oscillations is about 1 Gyr. I would like to hear your argument why you think it is much faster than that.

Roxburgh: That is presumably because you are just thinking of one global torsional oscillation. The point I am emphasizing is that different parts of the star have their own torsional oscillation and their local Alfven period, and they soon get out of phase and develop sharp gradients and enhanced diffusion.

Dziembowski: Ok. I understand.

THE SPIN-DOWN OF MAIN SEQUENCE STARS BASED ON OBSERVED MAGNETIC FIELD STRENGTH

C. TRIGILIO, G. UMANA
Istituto di Radiostronomia CNR Bologna,
Stazione VLBI Noto, ITALY

S. CATALANO
Istituto di Astronomia
Città Universitaria, 95125 Catania, ITALY

E. MARILLI
Osservatorio Astrofisico di Catania
Città Universitaria, 95125 Catania, ITALY

ABSTRACT: In the present paper we try to reproduce the observed dependence of rotation decay of main sequence stars in the mass range 1.1–0.5 $M_\odot$. We adopt a radial magnetic field structure which yields the $t^{-1/2}$ relation and the observed magnetic field values as given by Saar(1990). We assume that the effective magnetic field at photospheric level is given by the product of the equipartition field B_{eq} by the filling factor f, and that the latter is proportional to the angular rotation ω. The results are consistent with the observed rotation rate decay as a function of the mass and do suggest that the whole star is slowed down.

1 Introduction

The decay in the rotation rate which a star apparently experiences during its main sequence lifetime, is a fundamental parameter to understand the evolutionary scenario of outputs such as the stellar magnetic fields, stellar winds and energetic particles. The early studies of stellar rotation clearly showed that stars of solar and lower masses suffer a dramatic angular momentum loss with age (Kraft 1967, Skumanich 1972). Soderblom (1983) confirmed the Skumanich's $t^{-1/2}$ rotation decay relation from a broad range of ages in solar-mass stars. Moreover, we have shown (Trigilio et al. 1986, Catalano et al. 1988) that the $t^{-1/2}$ dependence holds for low-mass stars in the range 1.1–0.5 $M_\odot$, with a decay rate larger for lower mass stars.
As first proposed by Schatzman(1962) the rotation decay of low mass stars with outer convective zones is due to a magnetic wind braking. The $t^{-1/2}$ relation is explained by a simple wind model and a dynamo generated surface magnetic field (Weber and Davis 1967). However, more general and refined analyses predict that other decay dependences are possible (Belcher and MacGregor 1976, Mestel 1984).

S. Catalano and J. R. Stauffer (eds.), Angular Momentum Evolution of Young Stars, 373–378.

Critical aspects of these models are the magnetic field dependence on rotation and field topology, as derived from dynamo models, and the stellar wind model.
The physical quantities entering the braking model equations, like the magnetic field strength, the covering factor and the convective turn-over time, are now available from observations. Therefore we are in the postion of constraining models of rotation decay and checking for consistency the different empirical dependences deduced from the observations.
In this work, starting from simple physical assumption, we will try to reproduce the rotation decay observed for main sequence stars in the mass range of 1.1–1.2 $M_\odot$, using the magnetic field observed at the stellar surface.

2 Observational data

The data we use here refer to:

i) *observed rotation period dependence on mass and age*
Studies of the rotation decay with time for different stellar masses done with accurate values of rotation periods and ages (Trigilio et al. 1986, Catalano et al. 1988) have led us to the following results, also displayed in Figure 1 and Figure 2:
- the decay of rotation with $t^{-1/2}$ holds for low-mass stars in the range 1.1–0.5 $M_\odot$;
- the rate of spin-down (i.e. the increase in the rotation period) is larger for lower mass stars;
- the rotation period for a given low-mass star is tightly determined by its mass and age.

It appears from Figure 1 that for some group of masses the linear relation is fairly well established, while for lower mass star groups the statistical linear correlation coefficients are rather poorly determined, due to the lack of data. Consequently the error bars in Figure 2, where the slopes of the linear fit of Figure 1 for the different masses are plotted as a function of square root of the age, are much larger.

ii) *observed magnetic field dependence on mass and angular rotation*
Magnetic field intensity has recently been determined for late type stars from the analysis of line profiles broadened by the Zeeman effect (Saar and Linsky 1986, Saar 1990). It has been found that the effective magnetic flux density comes from magnetic regions covering a fraction f of the stellar surface with a mean magnetic field strength B_{eq} consistent with the equipartition condition between magnetic and kinetic energy, i. e.:

$$B \propto f B_{eq} \propto f P_{gas}^{0.5} \tag{1}$$

Besides, the magnetic field filling factor is linearly correlated with the Rossby number $R_o = \tau_c \omega$, i.e.:

$$f \propto \tau_c \omega \tag{2}$$

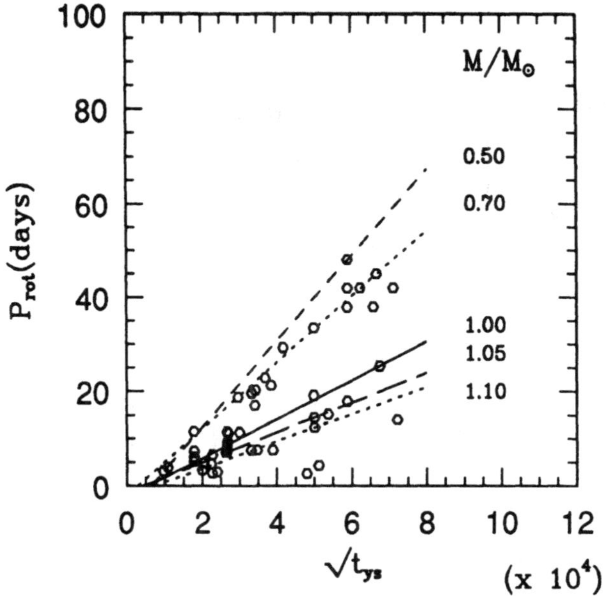

Figure 1: Observed rotation period, in days, versus square root of the age. Lines refer to constant mass stars (Catalano et al. 1988).

where τ_c is the turn-over time for the convection eddies.

iii) *empirically derived turn-over time*

The chromospheric emission flux of main sequence stars has been found to be well correlated with the Rossby number, (e.g. Noyes et al. 1984,). However, there is still some discrepancy in the α value, that is the convection length to pressure scale-height ratio, adopted to compute the turn-over time yielding the best correlation. We have deduced empirical convective turn-over time values as a function of the mass from chromospheric emission luminosity-rotation period correlations (Marilli et al. 1986, Catalano 1988). Our values agree very well with the values computed by Gilman(1980) for $\alpha = 2$. Since our τ_c values are obtained also for spectral types cooler than those given by Gilman, we will adopt them in the present analysis.

3 The theoretical model

We will calculate the evolution of rotation rate in main sequence low mass stars with the assumption of angular momentum loss via magnetic braking (Schatzman 1962, Weber and Davis 1967). Since the observed t $^{-1/2}$ dependence, displayed in Figure 1, is consistent with a model of radial magnetic field and a radial wind flow, as shown by Weber and Davis, we will take this assumption. The angular momentum loss due to the stellar wind is:

$$\frac{dJ}{dt} = -\dot{M}R_a^2\Omega \tag{3}$$

where $\dot{M}$ is the mass loss rate and R_a is the Alfven radius.
Under the assumptions of a radial stellar wind flow and radial magnetic field, we have $B_o R_o^2 = B_a R_a^2$, $B_a^2 = 4\pi\rho v_a^2$ and $\dot{M} = 4\pi\rho v_a^2$ (where the subscripts refer to values at the stellar surface and at the Alfven radius, and v_a is the wind speed at R_a).
Combining these conditions one gets, in terms of the stellar radius R and of the mean magnetic field at the stellar surface:

$$\frac{d\Omega}{dt} = -\frac{B_o^2}{I}\frac{R_o^4}{v_a}\Omega \tag{4}$$

It is reasonable to assume that the radius and the momentum of inertia remain nearly constant during the main sequence phase. On the other hand, if the effect of ω on v_a through the modulation of magnetic heating of the corona or the efficiency of the acoustic energy transport (Mestel 1984) is negligible, v_a also can be considered constant.
Let us now make use of the empirical relations between the observed magnetic fields and the stellar parameters (Saar 1987, 1990) for main sequence stars, as given in relation (1) and (2). We also assume that the surface magnetic field strength is defined by an effective field $B_{eff} = B_{eq}f$. Then the rotation decay expression can be written as:

$$\frac{d\Omega}{dt} = -k\frac{B_{eq}^2}{I}\tau_c^2 R^4 \Omega^3 \tag{5}$$

with k being a constant including the coefficients in the previous relation.
Integrating, we obtain in terms of the rotation period:

$$P = \left[k\frac{B_{eq}}{\sqrt{I}}\tau_c R^2\right]\sqrt{t+t_0} \tag{6}$$

with t_o being a constant defining the initial value of the rotation period. For large enough ages t_o is negligible, i. e. larger than the Hyades age, and the relation can be simply written as:

$$P \approx f(M)\sqrt{t} \tag{7}$$

where the coefficient f(M) has to be compared with the slope of regression lines of Figure 1, as plotted in Figure 2.

4 Results and discussion

We have computed the coefficient f(M) using B_{eq} and the empirical values of τ_c as given by Trigilio et al. (1986) in two extreme hypotheses: i) only the convection zone

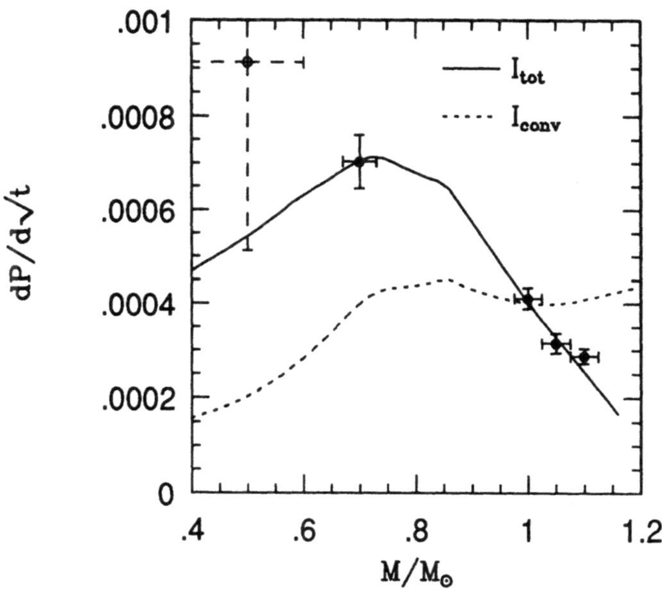

Figure 2: Empirical coefficients of the rotation decay at different mass groups (points and error bars), and computed coefficients for a stellar spin-down confined to the convection zone (dotted line) and a spin-down of the whole star (solid line).

is spun-down, ii) the whole star is spun-down. In order to compare the calculated coefficients with the observed ones we normalized the former to the solar mass value. Coefficients f(M), computed in the assumpion that only the convection zone (i.e. inserting the momentum of inertia of the convection zone in the relation 6) is braked during the main sequence lifetime, are completely discordant with respect to the observed coefficients (dotted line in Figure 2). The values computed in the hipothesis that the whole star is slowing down, are in very good agreement with the observed ones up to stellar masses of 0.7 $M/M_\odot$ (solid line in Figure 2). These results do show that a simply radial field model is adequate to describe the angular momentum loss in the interval 1.1–0.7 $M/M_\odot$ and that the whole star or nearly the whole star is spun-down, in agreement with the results on the solar internal rotation (Libbrecht and Woodard 1990). Moreover the observed rotation rate decays are consistent with the other observed physical parameters entering the model, i.e.:

- the effective field strength given by the equipartition value and weigthed by the filling factor;
- the magnetic field filling factor linearly determined by the Rossby number.

The behavior of the observed coefficient for mass smaller than 0.6 $M_\odot$ is not well defined so that the discrepancy with the computed values may not be so large as the present results seem to show. Any detailed discussion appears to be premature, obviously more data are needed to better define observed rotation decay rate for this range of mass.

Acknowledgements: This work has been supported by the Ministero dell' Università, della Ricerca Scientifica e Tecnologica through the University of Catania, the Catania Astrophysical Observatory and the CNR (Istituto di Radiostronomia di Bologna and Gruppo Nazionale di Astronomia).

References

Belcher J. W., MacGregor K. B.: 1984, Astrophys. J. **210**, 498
Catalano S.: 1988, Irish Astron. J. **18**, 265
Catalano S., Marilli E., Trigilio C.: 1988, in *NATO ASI Conf. Proc., Formation and Evolution of Low-mass Stars*, eds. A. K. Dupree and M. T. V. T. Lago, Kluwer, Dordrecht, p. 377
Gilman P. A.: 1980, in *Stellar Turbulence*, eds. D. F. Gray and J. L. Linsky, Springer, Heidelberg, p. 19
Kraft R. P.: 1967, Astrophys. J. **150**, 551
Libbrecht K. G., Woodard M. F.: 1990, in *Progress of Seismology of the Sun and the Stars*, ed. Shibahashi, Springer–Verlag, in press.
Marilli E., Catalano S., Trigilio C.: 1986, Astron. Astrophys. **167**, 297
Mestel L.: 1984, in *Cool Stars, Stellar Systems and the Sun*, eds. S. L. Baliunas and L. Hartmann, Springer–Verlag, Berlin, p. 49
Noyes R. W., Hartmann L. W., Baliunas S. L., Duncan D. K., Vaughan A. H.: 1984, Astrophys. J. **279**, 763
Saar S. H.: 1987, in *Cool Stars, Stellar Systems and the Sun*, eds. J. L. Linsky and R. E. Stencel, Springer–Verlag, New York, p. 10
Saar S. H.: 1990, *IAU Coll. 138, Solar Photosphere: Structure, Convection and Magnetic Field*, ed. J. O. Stenflo, Kluwer, Dordrecht, p. 427
Saar S. H., Linsky J. L.: 1986, in *Cool Stars, Stellar Systems and the Sun*, eds. M. Zeilik and D. M. Gibson, Springer, New York, p. 278
Schatzman E.: 1962, Ann. d'Aastrophys. **25**, 18
Skumanich A.: 1972, Astrophys. J. **171**, 265
Soderblom D. R.: 1983, Astrophys. J. Suppl. **53**, 1
Trigilio C., Catalano E., Marilli E.: 1986, Adv. Space Res. **6**, 203
Weber E. J., Davis L.: 1967, Astrophys. J. **148**, 217

OBSERVATIONAL PERSPECTIVES

L. HARTMANN
Harvard-Smithsonian Center for Astrophysics
60 Garden St., Cambridge, MA, USA

ABSTRACT. The slow rotation of T Tauri stars was an unexpected observational result not predicted by theory. The observations pose particular problems for models in which T Tauri stars are actively accreting substantial amounts of mass from circumstellar disks. Stellar winds have been invoked to carry off the required angular momentum, but the rapid rotation of α Persei and Pleiades stars implies that stellar wind angular momentum loss is relatively inefficient. The newest model suggests that interaction of a T Tauri magnetosphere with the disk limits the spinup of the star, in analogy with pulsars. Direct observational tests of this model are suggested. If magnetosphere-disk interactions last somewhat longer than a few million years, it may be possible to prevent some stars from spinning up during contraction to the main sequence, which would help explain the large dispersion in rotational velocities of low-mass stars seen in clusters like the Pleiades and α Per. It seems quite possible that low-mass stars may have differing rotational histories, and in particular that the Sun may never have been rapidly rotating.

1. Introduction

The "angular momentum problem" of star formation has been "solved" several times. The first solution was magnetic braking of the parent gas cloud (e.g., Mestel and Spitzer 1956; Mouschovias 1978), which, however important in star formation, is unlikely to eliminate enough angular momentum to enable collapse to stellar dimensions. A large amount of angular momentum is present in binary systems, but since most multiple stellar systems are widely separated, binary formation cannot account for the effective elimination of spin angular momentum (see Hartmann *et al.* 1986).

From the observational point of view, it has become obvious over the last few years that circumstellar disks are a common by-product of star formation (Adams, Lada, and Shu 1987; Kenyon and Hartmann 1987; Bertout, Basri, and Bouvier 1988). The frequent presence of disks tells us that the angular momentum of the parent cloud usually *is* a barrier preventing direct collapse to a star. *Accretion* disks are engines which can accomplish the necessary outward redistribution of angular momentum (Lynden-Bell and Pringle 1974). There is mounting observational evidence for substantial protostellar disk accretion (Hartmann and Kenyon 1985, 1987a,b; Bertout, Basri, and Bouvier 1988) and so

S. Catalano and J. R. Stauffer (eds.), Angular Momentum Evolution of Young Stars, 379–395.

it is plausible that disk accretion is the mechanism which in a basic sense permits stars to form (Mercer-Smith, Cameron, and Epstein 1984; Lin and Pringle 1990; Adams, Ruden, and Shu 1989).

However, disk accretion cannot be the whole story. The outward transport of angular momentum in a standard viscous disk requires that the angular velocity of material decrease outward (cf. Pringle 1981). Pringle (1989) has argued that stars can form while rotating at breakup, so that the outward viscous transport of angular momentum can be effective. In other words, the outer radius of the star is considered as the inner radius of an accretion disk, for which it is known that near-Keplerian motion does not prevent accretion. More detailed calculations supporting this possiblity have been presented by Popham and Narayan (1990). Thus, the argument runs, angular momentum loss beyond what is accomplished by a disk is not needed to form a star. While this is reasonable as far as it goes, the unfortunate observational fact is that the youngest stars we can see are rotating an order of magnitude or more below breakup. There must be yet another mechanism of angular momentum loss, which may not be "needed", but operates nontheless and may have important effects on stellar evolution.

A magnetically-coupled stellar wind is usually invoked to explain T Tauri angular momentum loss, in analogy with the solar wind and with the observed spin-down of main sequence stars (e.g. Schatzman 1962; Mestel 1968; Weber and Davis 1967; Belcher and MacGregor 1976). But the observations of rapid rotation in some Pleiades and α Per dwarfs implies that some stars spin *up* as they contract to the main sequence on timescales $\gtrsim 10^7$yr (Stauffer and Hartmann 1987; Stauffer, this volume). It is difficult to understand why stellar wind angular momentum loss is *efficient* on timescales $\lesssim 10^6$yr in the T Tauri phase but *inefficient* on timescales $\gtrsim 10^7$yr during post-T Tauri contraction.

Of course, some T Tauri stars show evidence for very massive winds (e.g., Kuhi 1964; Kuan 1975; Hartmann, Edwards, and Avrett 1982; Natta, Giovanardi, and Palla 1988). But the observations seem to show that the signatures of strong winds occur only in those systems accreting rapidly from disks (Croswell *et al.* 1987; Cabrit *et al.* 1990; Hartigan *et al.* 1990), suggesting that T Tauri outflow may be from the disk rather than from the star. Indeed, there is an independent reason to believe that the stellar winds of T Tauri stars and low-mass, "zero-age" main sequence stars are similar. The levels of X-ray emission are comparable in T Tauri stars (with and without disks) and stars of Pleiades age (Walter *et al.* 1988; Strom *et al.* 1990). If the levels of X-ray emission can be used to indicate the level of stellar magnetic and coronal activity, which in turn is related to the flux of a solar-type wind, one would expect little difference in outflow properties.

The addition of angular momentum from accreted material could eventually cause the central star to spin up close to breakup velocities, which would produce a very strong, cold (i.e., non-coronal) wind (Shu *et al.* 1988). However, we are fairly complete in surveys of T Tauri rotation, and no stars are observed to rotate this fast.

The most recent proposal to solve the slow rotator problem posits an interaction of the stellar magnetosphere with the disk, permiting the outward transfer of angular momentum even when the star is slowly rotating (Camenzind 1990; Königl 1991). This idea is an extension of magnetospheric models for accreting pulsars developed by Ghosh and Lamb (1979a,b; see also Arons *et al.* 1984). Because this model does not require a strong stellar wind to slow T Tauri rotation, it is easy to understand why the (diskless) star spins up during its relatively slow contraction to the zero-age main sequence. The physics of the disk-magnetosphere interaction are extremely complicated, and it is difficult to make detailed predictions from these calculations. However, a number of general observational effects are predicted by this class of models, and it appears possible to place quantitative constraints on the efficiency of this mechanism, as discussed below.

2. Disks, Magnetospheres, and Spindown

2.1 DISKS AND ACCRETION

The standard model for the T Tauri stars is the accretion disk model (Lynden-Bell and Pringle (1974), in which the infrared excess is interpreted as arising from a disk, while the optical-ultraviolet excess emission is interpreted as radiation from the boundary layer between the rapidly-rotating disk and the equator of the slowly-rotating star.

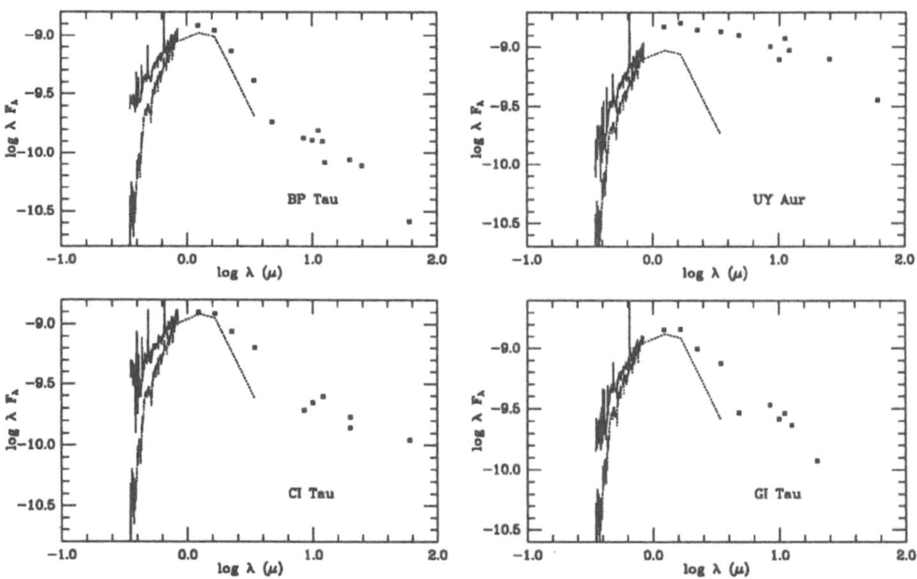

Figure 1. Optical and near-infrared spectral energy distributions of four T Tauri stars. The dotted lines indicate the estimated spectral energy distribution of the underlying stellar photosphere, showing that substantial optical and infrared excess emission is present. From Hartigan *et al.* (1991).

The accretion rate can be estimated from the luminosity,

$$L_{acc} = \frac{GM\dot{M}}{R}. \tag{1}$$

In the simplest disk models for slowly-rotating stars, this luminosity is split equally between disk emission in the infrared and boundary layer emission in the optical-uv. The determination of the accretion rate from the excess luminosity is not straightforward, because much of the infrared excess could arise from reprocessing of stellar radiation in the disk, i.e. the absorption of short-wavelength light from the central star and reradiation at longer wavelengths (Adams, Lada, and Shu 1987). The boundary layer luminosity is similarly difficult to interpret because of the need to make extinction corrections and subtract the stellar continuum luminosity. The best estimates sugges $\dot{M} \sim 10^{-7} M_{\odot}\, yr^{-1}$ (Bertout, Basri, and Bouvier 1988; Basri and Bertout 1989; Hartmann and Kenyon 1990; Hartigan *et al.* 1991). With this kind of accretion occuring over $\sim 10^6$ yr, we would expect the star to spin up to about 1/2 breakup velocity, or $\sim 130\,\mathrm{km\,s^{-1}}$ (Hartmann and Stauffer 1989), far above what is observed.

A disk wind could potentially remove the angular momentum that would otherwise be added to the star (e.g., Pudritz and Norman 1983; Pringle 1989). The principal difficulty for this mechanism is the relatively slow wind velocities $\sim 150-200\,\mathrm{km\,s^{-1}}$ observed in most T Tauri stars. Since the Keplerian velocity of the inner disk is thought to be $\sim 250\,\mathrm{km\,s^{-1}}$, this means that the Alfvén radius of the wind cannot be much larger than the inner disk radius; otherwise, the wind would tend to acquire a velocity $\sim (R_A/R_{inner}) \times 250\,\mathrm{km\,s^{-1}}$. With a relatively small Alfvén radius, one must eject a large fraction of the infalling material to carry away most of the angular momentum, and this is unlikely on energetic grounds.

2.2 STARSPOTS AND MAGNETIC FIELDS

Photometric investigations have shown that several pre-main sequence stars exhibit periodic light variations (Rydgren and Vrba 1983; Rydgren *et al.* 1984; Bouvier *et al.* 1986; Bouvier and Bertout 1989; Vrba *et al.* 1989). Most of these photometric periods are consistent with the rotational periods expected from the observed projected rotational velocities (Bouvier *et al.* 1986; Hartmann *et al.* 1986; Hartmann and Stauffer 1989). The best explanation of this behavior so far invokes the presence of magnetic structures or "spots" on the stellar surface, possibly analogous to sunspots but covering $\gtrsim 10\%$ of the stellar photosphere.

The weak-emission line T Tauri stars (WTTS) tend to have light curves consistent with the presence of a cool spot on the surface, suggesting a giant magnetic starspot(s) at least qualitatively similar to the cool spots on the Sun. The strong emission stars (CTTS), in contrast, have light curves seemingly explainable only in terms of "hot spots" (e.g., Bertout, Basri, and Bouvier 1988; Bouvier and Bertout 1989; Vrba *et al.* 1989). Bertout, Basri, and Bouvier (1988) and Bouvier and Bertout (1989) argued that the "hot spots" of CTTS were due to "magnetically- controlled accretion", due to the influence of the magnetic field of a large spot or spot group on the stellar surface. In this case, the ultraviolet and optical excess emission comes not from an equatorial "boundary layer" surrounding the star at the edge of the disk, but rather arises in some sort of accretion column as disk material slowed from Keplerian rotation strikes the star.

Bertout, Basri, and Bouvier (1988) and Bouvier and Bertout (1989) suggested that the magnetic pressure of the spot field might be sufficient to hold off the disk to large radial distances, as much as $\sim 5R_*$. Such a large Alfvén radius could have very interesting effects on the angular momentum transfer between star and disk. For typical T Tauri parameters, the Keplerian velocity at the stellar surface is $\sim 250\,\mathrm{km\,s^{-1}}$. However, the angular velocity of the disk at $5R_*$ is $5^{3/2} \sim 11$ times smaller than the angular velocity at the stellar surface. Thus, if the star were rigidly connected to the disk at this radius by magnetic fields, angular momentum would be transferred outward from star to disk if the star had an equatorial rotational velocity $\gtrsim 22\,\mathrm{km\,s^{-1}}$, and over time would try to spin down the star to this velocity, which is not far above the median rotational velocity observed (Bouvier, this volume).

The situation is more complicated when one considers accretion. The material at the inner edge of the disk must lose angular momentum to accrete onto the star. In the Ghosh and Lamb (1979a,b) model, the star must rotate more slowly than the inner disk edge, so that angular momentum can be transferred from inner disk to the star, allowing the inner disk gas to fall in. This angular momentum is then transmitted to the *outer*, more slowly-rotating disk by other magnetic field lines from the star. The range of relative rotation through which such effects occur depends upon the radial range over which the stellar magnetic field lines penetrate the disk (e.g., Königl 1991). Recent treatments of the disk-magnetic field structure suggest that the field penetrates a smaller range of disk radii than in the original Ghosh and Lamb model (Camenzind 1990; Spruit and Taam 1990), but the physics are quite uncertain and complicated. Whatever the case, it seems likely that magnetospheric spindown will be effective only if the magnetospheric radius, that is, the radius to which the magnetic field holds off the disk, is close to the radius where the disk corotates with the star. For typical T Tauri parameters, we expect $R(corotation) \simeq 5R_*$.

2.3 DISK HOLES AND SPECTRAL ENERGY DISTRIBUTIONS

If the inner disk is held off to $5R_*$ because of the stellar magnetic field, one might hope to see some effect of this structure on the spectral energy distribution. In Figure 2 I show some calculations by Scott Kenyon and myself on the simplest possible model of a disk with an inner hole. The disk is presumed to be flat, radiate like a blackbody, and has a negligible accretion rate. Thus the disk radiates only because it absorbs light from the central star. For a typical M0 T Tauri star, with an effective temperature of 4000 K, the inner disk temperatures are in the range of 3000 - 1000 K. Removing this material thus reduces the excess emission at near-infrared wavelengths. The models show that the effect on spectral energy distribution is modest unless the hole is quite large ($R \gtrsim 5R_*$). Furthermore, since the inclination is usually not known, one might interpret the reduced infrared excess simply in terms of a larger inclination angle (the flat disk flux is proportional to $\cos i$).

A comparison of Figures 1 and 2 indicates that it will be difficult to demonstrate the existence of inner disk holes $< 5R_*$ from the spectral energy distribution, even adopting the simplest possible model for disk emission. Moreover, there are other effects not included in this simple model which could increase the amount of near-infrared excess emission. To begin with, accretion will cause the disk temperatures to be larger than in these purely reprocessing models, and this will have the effect of shifting the "dip" in the energy distribution to shorter wavelengths, making it harder to distinguish from the stellar photospheric emission. In addition, the stellar magnetic field threading the inner disk may dissipate energy in excess of that produced in a simple accretion disk model. The gas near the magnetospheric radius must not only lose potential energy but kinetic energy as well to fall in nearly radially, and some of this energy might be deposited locally, heating the gas (Camenzind 1990). Finally, if inner disk material piles up along magnetic field lines near the disk boundary, it will be more directly illuminated by the star than flat disk material obliquely heated by light from the stellar photosphere (e.g., Königl 1991). All of these processes could produce more emission at short wavelengths than assumed in the

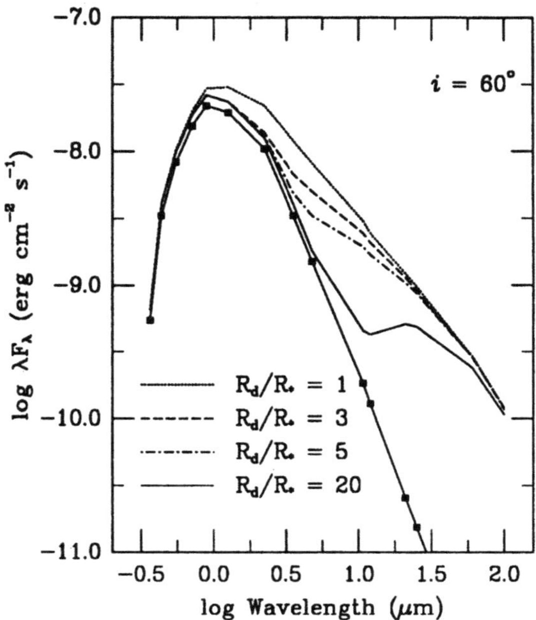

Figure 2. Optical and near-infrared spectral energy distributions for a series of disk models with inner holes, all viewed at an inclination of 60°. The disks are assumed to radiate purely by reprocessing radiation absorbed from the central M0 star. The heavy solid line denotes the stellar photospheric emission, and the other curves are labeled by the inner disk radius in terms of of the stellar radius. From Kenyon and Hartmann (in preparation).

models shown in Figure 2, but the magnitude of these effects is uncertain. Thus, there is no firm prediction yet for the effects of a small hole in the inner disk on the spectral energy distributions of T Tauri stars. In fact, some T Tauri stars actually show a larger near-infrared excess than can be easily explained with simple disk models *without* holes; the effects of magnetic energy dissipation or pile up of material could potentially account for such extreme near-infrared excess emission.

2.4 MODELS FOR INVERSE P CYGNI PROFILES

If magnetic accretion is initiated at a distance much larger than the stellar radius, large infall velocities must develop. In this connection the so-called "YY Orionis" stars are of particular interest. These objects often show "inverse P Cygni" behavior (red-shifted absorption) in the profiles of many lines. Originally, it was thought that these inverse P Cygni profiles might represent the final buildup of the star from its parent gas cloud (e.g., Walker 1972; Wolf *et al.* 1977). The original models developed to account for the inverse P Cygni line profiles observed in YY Ori stars either assumed spherically symmetric infall (Bertout 1977, 1978; Bastian 1982) or infall from a rotating cloud, which, at large distances, was essentially spherically symmetric (Ulrich 1976; Bertout 1979a,b). However, Herbig (1977) pointed out that almost all T Tauri stars, including the YY Orionis class, generally showed *blueshifted* absorption in Hα, constituting evidence for the outflowing T Tauri wind. It now appears that mass loss is occurring from even the YY Orionis objects (cf. Appenzeller *et al.* 1984; Edwards *et al.* 1987). The spectral features from which mass loss is inferred tend to be formed over much larger radii than those lines in which infall is observed (Hartmann 1986).

Hence, the general picture is one in which the outer envelope of a T Tauri star is outflowing at large distances, but the velocity field is more complicated close to the star. One model which could explain the observations supposes that magnetically-controlled accretion from the inner disk causes the inverse P Cygni profiles seen in lines formed close to the star (Krautter *et al.* 1990; Camenzind 1990; Königl 1991), while a more general outflow (of uncertain geometry) exists on large scales.

An essential aspect of a magnetospheric model is that the infall must occur over a fairly restricted area for two reasons. (1) With a strong magnetosphere, no boundary layer is present; so one must obtain the observed blue continuum excess emission in T Tauri stars in the region where the infalling material comes to rest at the stellar surface. Observational estimates of emitting area responsible for the blue continuum are $\sim$ 5 - 10% of the surface area of the star (Bertout *et al.* 1988; Basri and Bertout 1989; Hartigan *et al.* 1991). (2) If the stellar magnetic field is to be strong enough to hold off the disk to several stellar radii, spinning down the star to values well below breakup, the field is likely to be strong enough to channel the material into fairly narrow accretion cones.

Nuria Calvet and I (Calvet and Hartmann 1991) have computed the line profiles for infalling material in a "cone" geometry, i.e. infall is essentially radial within restricted latitudes. This is highly simplified compared to the real magnetospheric case, but probably contains the essential features. Figure 3 shows calculations for radial infall starting from rest at r = $3R_*$. The material is assumed to have a temperature of $10^4 K$. We have considered two density distributions; one corresponding to a spherically-symmetric mass infall rate of $10^{-8} M_\odot\, yr^{-1}$, the other for $10^{-7} M_\odot\, yr^{-1}$. The final infall rate depends upon the range of latitudes assumed.

The models show that, as expected, red-shifted P Cygni absorption is observed when the accretion column is along the line of sight. Conversely, a polar accretion column observed at large inclination results in modest or undetectable redshifted absorption. Figure 3 shows that the redshifted absorption in Hα in the $10^{-8} M_\odot\, yr^{-1}$ model is difficult to detect because of thermalization effects. This is in agreement with observations that Hα rarely shows redshifted absorption, even when the higher Balmer lines exhibit such absorption. (This effect was also found in a particular model by Bastien [1982]). But what is most surprising is that in the $10^{-7} M_\odot\, yr^{-1}$ model, *none* of the first three members of the Balmer series show redshifted absorption. Instead, they show profiles which are suggestive of "outflow", i.e., more emission on the red side of the line profile than on the blue side. The reason for this effect is thermalization at the high densities of this model, which produces a line source function that is greater than the photospheric level; and thus the red-shifted material is seen in *emission* rather than *absorption*.

Our results suggest that infall may be more common than supposed by direct observation of Balmer lines for two reasons. First, if infall is confined to magnetic poles, we may see the redshifted absorption only when the magnetic pole is along the line of sight. Many objects may be seen in other inclinations where the infall is not so apparent. And if the magnetic field is oriented obliquely to the disk, then the infall would be seen only at certain rotational phases. Second, if the infall gas is dense and hot enough, no absorption

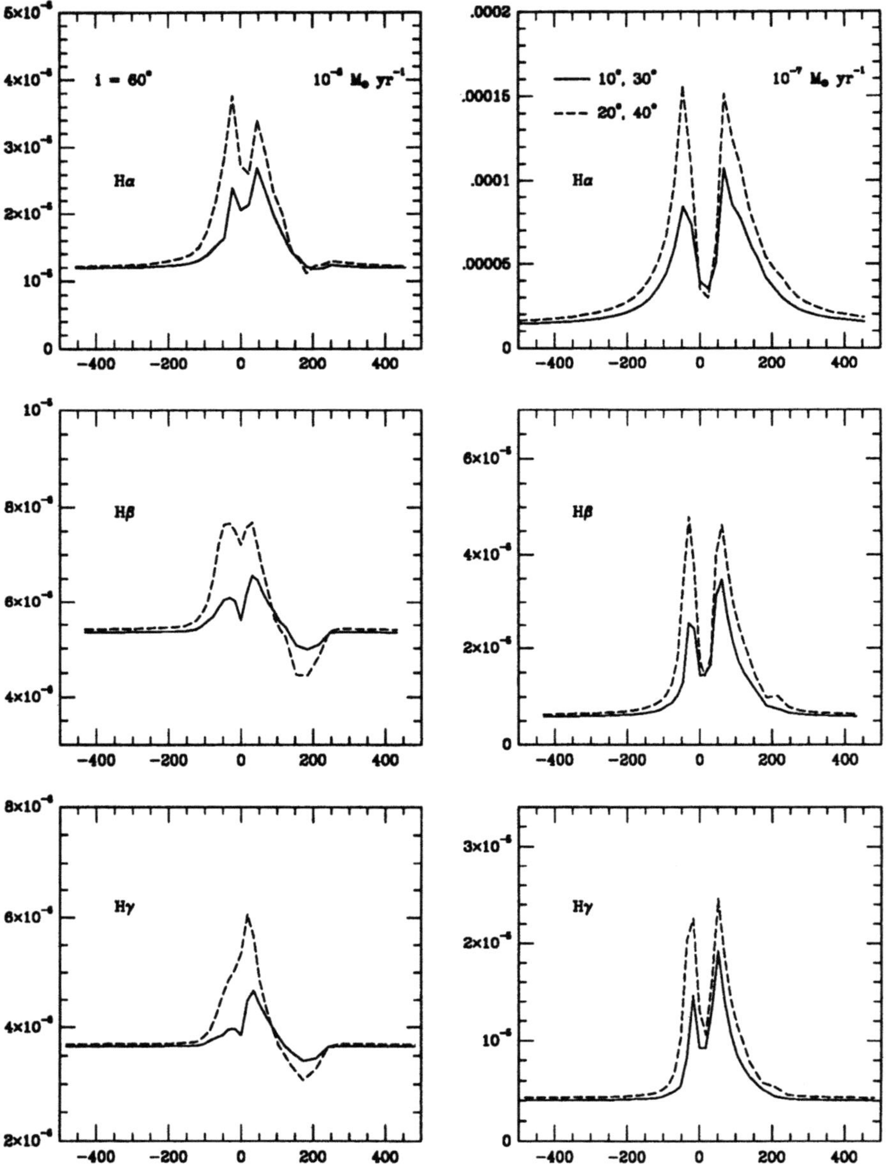

Figure 3. Balmer line profile calculations for infalling gas with a temperature of 10^4K onto a T Tauri star. The model assumes purely radial infall within restricted latitudes, i.e., in a "cone"; the radial velocity field assumes ballistic infall starting from rest at r = $3R_*$. The latitude limits are indicated in the figures; in all cases the system is observed at an inclination of 60°. The labels indicate the mass infall rates corresponding to the density distributions if the geometry were spherically-symmetric. The left panels are for $\dot{M} = 10^{-8} M_\odot\, yr^{-1}$; the right panels correspond to $\dot{M} = 10^{-7} M_\odot\, yr^{-1}$. From Calvet and Hartmann (1991).

would be seen at *any* rotational phase.

Ulrich and Knapp (1979) pointed out that several T Tauri stars exhibit red-shifted absorption in the Na I resonance lines. In Figure 5 I show model Na I profiles calculated in the spherically-symmetric limit. For the same case ($\dot{M} = 10^{-7} M_{\odot}\, yr^{-1}$) where the Balmer lines showed no evidence of redshifted absorption, such absorption is seen clearly in Na I, suggesting that the latter line may be a more reliable test of infalling material in high mass accretion rate objects. The simple model calculations compare reasonably well with some observed line profiles John Stauffer and I have obtained over the last couple of years (Figure 5).

Note in particular the similarity of the two observations of BP Tau. There are night-to-night variations in the presence of the redshifted absorption in this object; yet spectra taken one year apart, denoted by the solid and dotted lines, match up perfectly, suggesting that the infall is a persistent phenomenon. Of course, if the spot were not at the pole of the star, the rotation of the star could carry the infalling column around, easily producing night-to-night variations (the expected rotational period is ~ one week), but we do not have enough spectral coverage to demonstrate any possible periodicity.

It is amusing to note that, if this general picture is correct, efforts to obtain mass loss rates based on the strength of the hydrogen emission lines (e.g., Kuhi 1964; Kuan 1975; Hartmann *et al.* 1982; Hartmann *et al.* 1990; Natta, Giovanardi, and Palla 1988) are misleading; one may actually be measuring the *infall* rate instead. The fact that these estimates seem to agree roughly with estimates based on the blue-shifed forbidden lines (e.g., Edwards *et al.* 1987) may simply mean that the mass loss rate is of the same order of magnitude as the accretion rates, and our models for these processes are not accurate enough to distinguish smaller differences.

2.5 PROBLEMS WITH MAGNETOSPHERIC MODELS

The magnetospheric model is attractive in that it can explain some of the photometric and spectroscopic behavior of T Tauri stars in a natural way. However, a number of issues remain to be resolved:

(1) The most convincing evidence for a magnetosphere would be to demonstrate correlated, if not periodic, photometric and spectroscopic behavior with a timescale comparable to the stellar rotational period. If there truly is an accretion column, one would expect to see red-shifted absorption at the times when a "hot spot" comes into view. The evidence for this appears to be conflicting. Walker (1972) found no evidence for a correlation between the brightness of YY Ori stars and the presence of inverse P Cygni absorption. Mundt (1979) found that the red-shifted absorption strengths in the Hδ and Hϵ lines in S CrA increased when the star was bluest, i.e., had the most evidence for "boundary layer" or accretion continuum emission. However, Bertout *et al.* (1982) pointed out that S CrA and other YY Ori stars are variable on very short (hour) timescales, making it difficult to test the hypothesis of rotational modulation. Clearly, further monitoring is needed.

(2) From the data that John Stauffer and I have obtained, the fraction of classical T Tauri stars exhibiting redshifted Na I absorption appears to be ~ 25% of the total. This is *consistent* with assuming that most or all TTS have large magnetospheres, but that we tend not to observe along accretion columns because such columns cover a modest range of solid angle. Obviously, consistency is not the same as proof.

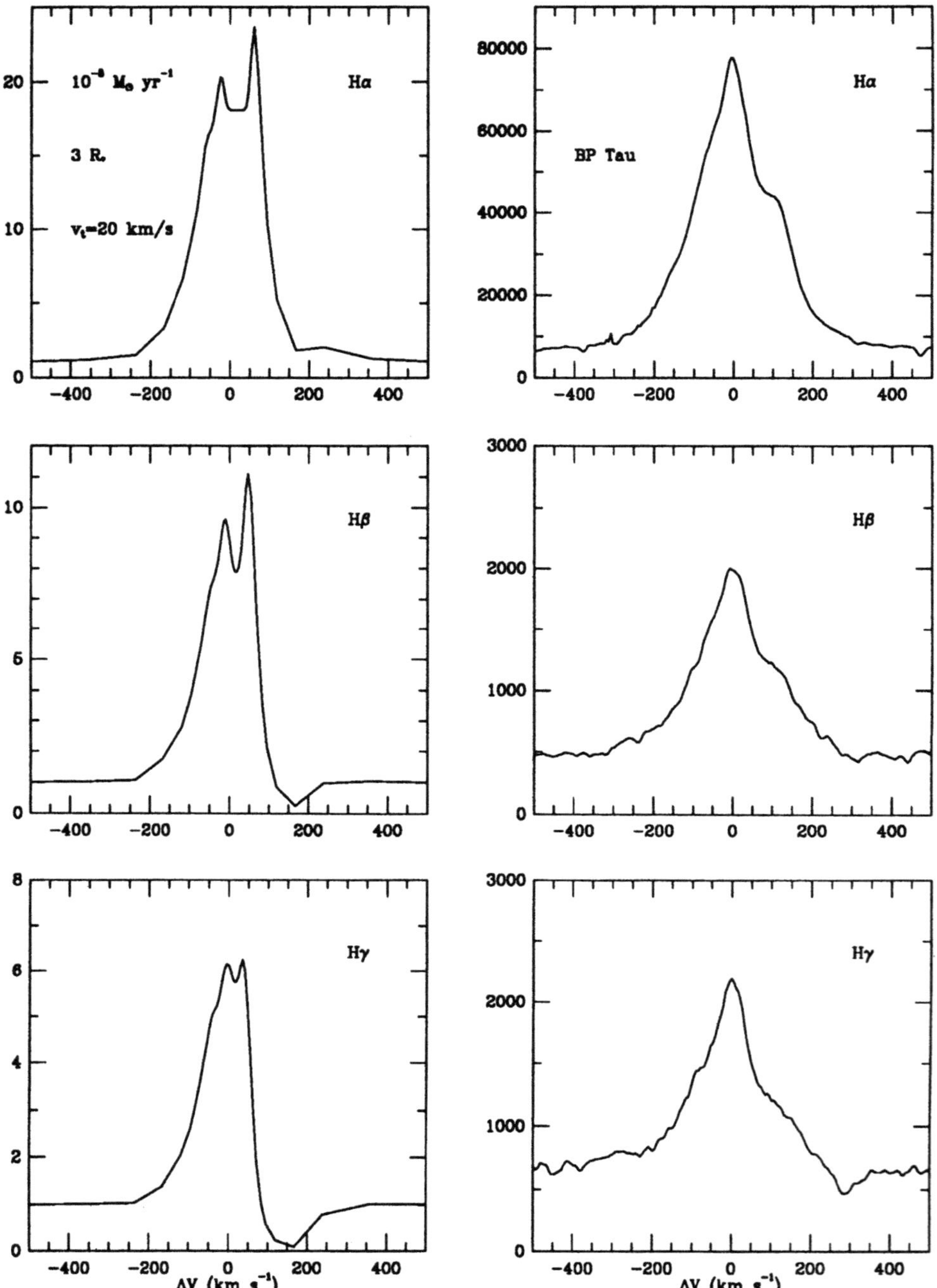

Figure 4. Comparison of predicted Balmer line profiles for a spherically-symmetric infall model ($\dot{M} = 10^{-8} M_\odot\, yr^{-1}$) with observations of BP Tau (Edwards, private communication).

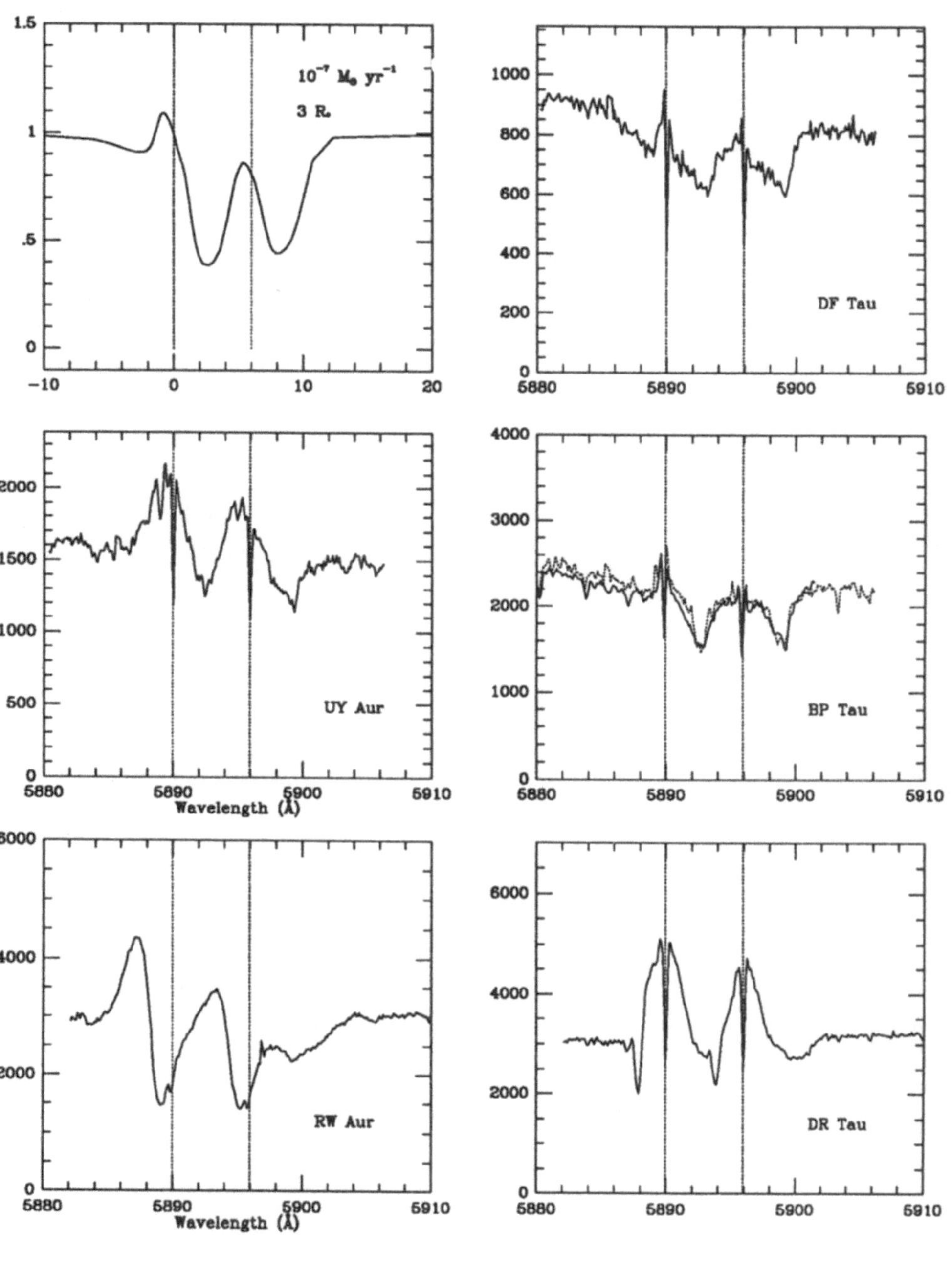

Figure 5. Observed Na I line profiles in selected T Tauri stars compared with profiles for a simple spherically-symmetric infall model.

(3) It is not completely obvious that the stellar magnetic field is sufficiently strong to hold off the disk to the corotation radius. In the original Ghosh and Lamb models, the magnetospheric radius corresponds to the region where the azimuthal motion of the disk departs substantially from Keplerian rotation. Typical estimates in this theory result in a predicted magnetospheric radius of approximately

$$R_{in} \sim 2R_* \, {B_{*,3}}^{4/7} \, R_{2.5}^{5/7} \, M_{0.8}^{1/7} \, \dot{M}_{-7}^{-2/7} ,$$

where the stellar mass and radius are measured in units of $0.8\ M_\odot$ and $2.5\ R_\odot$, respectively, and $B_{*,3}$ is a typical estimated surface stellar dipole magnetic field (Basri and Marcy 1990) and $10^{-7} M_\odot\, yr^{-1}$ is a typical mass accretion rate. In the Ghosh and Lamb models, the corotation radius is estimated to be $r_{co} \sim 2 \times R_{in}$ (cf. Königl 1991), so that for the above parameters one would predict a steady-state rotational velocity of $\sim$ 30 km s^{-1}. While this result seems reasonably encouraging, it depends crucially on having enough magnetic field lines thread the outer disk to carry off the required angular momentum. If, instead, $r_{co} \sim R_{in}$, one would require large-scale surface magnetic fields of $\gtrsim$ 3 kG, which may be too large to be consistent with present observations (Basri and Marcy 1990). In addition, the above estimates assume that the stellar magnetic field is dipolar, but it might easily be of higher order if it is distributed among more than one starspot. The stellar field might be concentrated in the disk to much higher values than would be present in a simple dipole field (Camenzind 1990), but such concentration would mean that few field lines thread the disk at large distances, reducing the coupling to the outer, slowly-rotating disk.

In this context the observed infall velocities provide an extremely important clue to the size of the magnetosphere. If the infall is nearly ballistic, as seems likely, the velocities require infall from radii $\gtrsim 3R_*$, suggesting that the magnetosphere is large enough to play a role in establishing the stellar spin angular momentum. The physics is complicated, messy, and magnetospheric radii are unlikely to be determined by *a priori* calculations.

2.6 OBSERVATIONAL TESTS

The case for magnetospheres must be made observationally. We need further photometric and spectroscopic studies to see how common accretion columns really are. I think it would also be important to see if we can find evidence for magnetic concentrations or spots on stars of intermediate mass, since they also rotate below breakup velocities (although not by as much as the low-mass T Tauri stars; see Bouvier's review). It would be nice not to invent another mechanism for rotational spindown in 2 - 3 $M_\odot$ stars. Praderie *et al.* (1986) found evidence for a periodic wind variation in AB Aur, such as might be produced by a magnetically-channelled high speed stream; further observational tests would be desirable. Obviously, more observations to determine the frequency of inverse P Cygni profiles are desirable.

3. THE SLOW ROTATORS IN YOUNG CLUSTERS AND THE EVOLUTION OF THE SUN

As a final comment, I would like to remind theoreticians that the rotational velocity distributions for G stars in the α Persei cluster are not to be taken as "error bars" for models of rotational evolution to pass through, but represent real ranges of rotational velocities. The range of v sin i for main sequence G stars in the α Per cluster and for main sequence K stars in the Pleiades (Stauffer, this volume) appears to be somewhat greater than the range of rotation observed in T Tauri stars (Bouvier, this volume). The feedback of increasing angular momentum loss with increasing rotation rate apparently causes all stars to have the same equatorial velocity by the age of the Hyades. Why then should this feedback not suppress the rotational velocity range in clusters like the Pleiades? Perhaps there is some other mechanism operating between the T Tauri phase and the zero-age main sequence which plays a role in determining the stellar spin angular momentum. It is tempting to speculate that disks might last longer in some stars, and that a disk-magnetosphere interaction might enhance the spread in angular momentum as T Tauri stars contract toward the main sequence. Given the relative numbers of slow and rapid rotators in α Per, it seems quite possible that the Sun was never rapidly rotating at any phase in its evolution.

I am grateful to Nuria Calvet for comments on an early version of this manuscript, and to Suzan Edwards for general discussions and for permission to show the $H\beta$ and $H\gamma$ profiles of BP Tau in Figure 4. This research was supported by NASA Grant NAGW-511 and by the Scholarly Studies Program of the Smithsonian Institution.

4. REFERENCES

Adams, F.C., Lada, C., and Shu, F.H. 1987. *Astrophys. J.* **312**, 788.
Adams, F.C., Ruden, S.P., and Shu, F.H. 1989, *Astrophys. J.* **347**, 959.
Appenzeller, I., Jankovics, I., and Ostreicher, R. 1984, *Astron. Astrophys.* **141**, 108.
Arons, J., Burnard, D.J., Klein, R.I., McKee, C.F., Pudritz, R.E., and Lea, S.M. 1984, in *High Energy Transients in Astrophysics*, AIP Proceedings No. 115, ed. S.E. Woosley, p. 215.
Basri, G., and Bertout, C. 1989, *Astrophys. J.* **341**, 340.
Basri, G., and Marcy, G. 1990, in *I.A.U. Symposium 130*, in press.
Bastian, U. 1982, *Astron. Astrophys.* **109**, 245.
Belcher, J.W., and MacGregor, K.B. 1976, *Astrophys. J.* **210**, 498.
Bertout, C. 1977, *Astron. Astrophys.* **58**, 153.
Bertout, C. 1979a, *Astron. Astrophys.* **80**, 138.
Bertout, C. 1979b, *Mitt. Astron. Ges.*, **43**, 176
Bertout, C., Basri, G., and Bouvier, J. 1988, *Astrophys. J.* **330**, 350.
Bertout, C., Carrasco, L., Mundt, R., and Wolf, B. 1982, *Astron. Astrophys. Suppl. Ser.* **47**, 419.
Bouvier, J., and Bertout, C. 1989, *Astron. Astrophys.* **211**, 99.
Bouvier, J., Bertout, C., Benz, W., and Mayor, M. 1986, *Astron. Astrophys.* **165**, 110.
Cabrit, S., Edwards, S., Strom, S.E., and Strom, K.M. 1990. *Astrophys. J.* **354**, 687.
Calvet, N., and Hartmann, L. 1991, in preparation.
Camenzind, M. 1990, *Reviews in Modern Astronomy*, Springer-Verlag, Berlin, **3**, in press.
Clarke, C.J., Lin, D.N.C., and Pringle, J.E 1990, *Mon. Not. Roy. Astron. Soc.* **242**, 439.
Edwards, S., Cabrit, S., Strom, S.E., Heyer, I., and Strom, K.M. 1987, *Astrophys. J.* **321**, 473.
Ghosh, P., and Lamb, F.K. 1979a, *Astrophys. J.* **232**, 259.

Ghosh, P., and Lamb, F.K. 1979b, *Astrophys. J.* **234**, 296.
Hartigan, P., Hartmann, L., Kenyon, S., Strom, S.E., and Skrutskie, M. 1990, *Astrophys. J. Lett.* **354**, L25.
Hartigan, P., Kenyon, S., Hartmann, L., Strom, S.E., Edwards, S., and Stauffer, J.R. 1991, in preparation.
Hartmann, L. 1986, in *Fundamentals of Cosmic Physics,* **11**, 279.
Hartmann, L., Calvet, N., Avrett, E.H., and Loeser, R. 1990, *Astrophys. J.* **349**, 168.
Hartmann, L., Edwards, S., and Avrett, E.H. 1982, *Astrophys. J.* **259**, 180.
Hartmann, L., Hewett, R., Stahler, S., and Mathieu, R.D. 1986, *Astrophys. J.* **309**, 275.
Hartmann, L., and Kenyon, S. 1985, *Astrophys. J.* **299**, 462.
Hartmann, L., and Kenyon, S. 1987a, *Astrophys. J.* **312**, 243.
Hartmann, L., and Kenyon, S. 1987b, *Astrophys. J.* **322**, 293.
Hartmann, L. and Kenyon, S.J. 1990, *Astrophys. J.* **349**, 190.
Hartmann, L., and Stauffer, J.R. 1989, *Astron. J.* **97**, 873.
Herbig, G.H. 1977, *Astrophys. J.* **214**, 747.
Kenyon, S.J., and Hartmann, L. 1987, *Astrophys. J.* **323**, 714.
Königl, A. 1990, preprint.
Krautter, J., Appenzeller, I., and Jankovics, I. 1990, *Astron. Astrophys.*, in press.
Kuan, P. 1975, *Astrophys. J.* **202**, 425.
Kuhi, L.V. 1964, *Astrophys. J.* **140**, 1409.
Lin, D.N.C., and Pringle, J.E. 1990, *Astrophys. J.* **358**, 515.
Lynden-Bell, D., and Pringle, J.E. 1974. *Mon. Not. Roy. Astron. Soc.* **168**, 603.
Mercer-Smith, J.A., Cameron, A.G.W., and Epstein, R.I. 1984, *Astrophys. J.* **279**, 363.
Mestel, L. 1968, *Mon. Not. Roy. Astron. Soc.* **138**, 359.
Mestel, L., and Spitzer, L. Jr., *Mon. Not. Roy. Astron. Soc.* **116**, 503.
Mouschovias, T. Ch. 1978, *Astrophys. J.* **224**, 283.
Mundt. R. 1979, *Astron. Astrophys.* **74**, 21.
Natta, A., Giovanardi, C., and Palla, F. 1988, *Astrophys. J.* **332**, 921.
Popham, R., and Narayan, R. 1990, preprint.
Praderie, F., Simon, T., Catala, C., and Boesgaard, A.M. 1986, *Astrophys. J.* **303**, 311.
Pringle, J.E. 1981, *Ann. Rev. Astron. Ap.* **19**, 137.
Pringle, J.E. 1989, *Mon. Not. Roy. Astron. Soc.* **236**, 107.
Pudritz, R.E., and Norman, C.A. 1983, *Astrophys. J.* **274**, 677.
Rydgren, A.E., and Vrba, F.J. 1983, *Astrophys. J.* **267**, 191.
Rydgren, A.E., Zak, D.S., Vrba, F.J., Chugainov, P.F., and Zajtsevea, C.V. 1984, *Astron. J.* **89**, 1015.
Schatzman, E. 1962, *Ann. d'ap*, **25**, 18.
Shu, F.H., Lizano, S., Ruden, S.P., and Najita, J. 1988, *Astrophys. J. Lett.* **328**, L19.
Spruit, H.C., and Taam, R.E. 1990, *Astron. Astrophys.* **229**, 475.
Stauffer, J.R., and Hartmann, L. 1987, *Astrophys. J.* **318**, 337.
Strom, K.M., Strom, S.E., Wilkin, F.P., Carrasco, L., Cruz-Gonzales, I., Recillas, E., Serrano, A., Seaman, R.L., Stauffer, J.R., Dai, D., and Sottile, J. 1990, *Astrophys. J.* **362**, 168.
Ulrich, R.K. 1976, *Astrophys. J.* **210**, 377.
Ulrich, R.K., and Knapp, G.R 1979, *Astrophys. J. Lett.* **230**, L99.
Vrba, F.J., Rydgren, A.E., Chugainov, P.F., Shakovskaya, N.I., and Weaver, W.B. 1989, *Astron. J.* **97**, 483.

Walker, M.F. 1972, *Astrophys. J.* **175**, 89.
Walter, F.M., Brown, A., Mathieu, R.D., Myers, P.C., and Vrba, F.J. 1988, *Astron. J.* **96**, 297.
Weber, E.J., and Davis, L. Jr. 1967, *Astrophys. J.* **148**, 217.
Wolf, B., Appenzeller, I., and Bertout, C. 1977, *Astron. Astrophys.* **58**, 163.

DISCUSSION

Soderblom: Two comments and a question. First, you can argue that the Sun is relatively a slow rotator, but it is not a slow rotator compared to other old G dwarfs - it has an average rotation for its age and mass. Second, it almost looks as if there is a problem with the Pleiades slow rotators - there are too many of them and they are rotating too slowly compared to the Hyades stars. This may be a problem for our simple model of things. Finally, I wonder if the strength of the frozen-in field may be an important initial condition for stars. Is it possible that the spread in rotation in the open clusters could be due to a spread in braking efficiency rather than a spread in initial angular momentum? Do you have any thoughts on this?

Hartmann: The best we have for a proxy for anything related to magnetic fields for the very young stars is probably the X-ray emission. Even then, there is a large theoretical gap in going from X-ray emission to angular momentum loss - so it is difficult to say. It is possible that the magnetic field topology might give very different angular momentum loss rates. What I worry about is that we want something that spins down the T Tauri stars in about 1 million years, and then we want to turn off that large angular momentum loss rate for the next 10 Myr as the star contracts to the main sequence. Perhaps one could arrange for the relict field to decay away on short time-scales.

Van't Veer: If I remember, you studied an even younger cluster - IC 2391. How does this cluster fit into your diagram?

Hartmann: We have measured so few stars in that cluster that we really can't say anything definitive. There are some slow rotators even at that age, however (the age is about 30 Myr).

Bouvier: I would like to comment about the hot spots. If your interpretation is correct, we can learn something about the process by measuring the time dependence of the spot behavior. For DF Tau, but also for a couple of other cases, it is true that the hot spot is there only intermittently. Therefore, whatever process that is producing the hot spot has fairly rapid variability.

Hartmann: Could it also be that perhaps some disk instability is causing the variability by modifying the mass inflow, rather than a varying magnetic field structure?

Bouvier: Yes,that is possible.

Vaiana: With regard to the magnetospheric modelling, am I correct that you are saying that we should not consider H alpha as a wind indicator at all? Where does the H alpha come from? Is it just from infall?

Hartmann: I think that is possible.

Vaiana: Then the infall is totally dominant? Can you fit the X-ray emission then? For instance, I note that there is an anti-correlation between X-ray luminosity and H alpha emission. Can you explain that? Could there still be a wind present? Could you have a model somewhat like that for coronal loops on the Sun, where there is both infall and outflow and the mass outflow from the tops of the loops could be providing the angular momentum loss that is needed.

Hartmann: I don't have any clear idea at this point of what the geometry of the magnetic field structures are, other than a very simplified model. Perhaps the observations can give some clue. I would guess that the bulk of the emission measure is in the infall and not in the wind.

Vaiana: That has to be demonstrated; you are just inferring it now.

Dziembowski: We can perhaps learn a lesson from X-ray pulsars, which show that accretion may well be intermittent. Now a question. It does look like young main sequence stars are divided into two groups - a set of slow rotators and a set of rapid rotators. Do you see any trace of this dichotomy among T Tauri stars?

Hartmann: I think the answer to that is yes, we do see some trace of such a division, though it depends to some extent on how you treat the upper limits. Bouvier showed that the lower mass stars rotate at a substantially smaller fraction of break-up than the higher mass stars. Among the low mass stars alone, there is a hint of a peak at low rotation though it is not nearly as evident as for open clusters.

Palla: Does the redshifted absorption in the sodium line also appear in the weak-lined T Tauri stars (i.e. those which do not show strong IR excesses or other evidence of disks)?

Hartmann: No.

PRISMA: A SPACE FACILITY FOR STUDYING ROTATION AND ACTIVITY

P. LEMAIRE [1], T. APPOURCHAUX [2], C. CATALA [3], S. CATALANO [4], S. FRANDSEN [5], A. JONES [6], W. WEISS [7]
[1] *Institut d' Astrophysique Spatiale, PB 10, F-91371 Verrieres le Buisson Cedex, France*
[2] *ESA ESTEC/SSD, P.O. Box 299,2200 AG Noordwijk, The Netherlands*
[3] *DESPA Observatoire de Paris-Meudon, F-92100 Meudon, France*
[4] *Instituto di Astronomia, Cittá Universitaria, I-95125 Catania, Italy*
[5] *Astronomisk Institut, DK 8000 Aarhus, Denmark*
[6] *Instituto de Astrofisica de Canarias, E-39200, La Laguna Tenerife, Spain*
[7] *Institut für Astronomie, Turkenschnstrasse 17, A-1180 Wien, Austria*

ABSTRACT. Here we report on PRISMA, a space mission under assessment study at ESA, devoted to the study of the internal structure, rotation and activity of stars. Simultaneous observations of oscillation modes and activity for several stars well distributed over the HR diagram are expected to allow the mapping of the stellar structure from core to corona, as a function of mass and evolution.
Instrument performancies, observation strategies and mission concepts will be described.

1 Introduction

We already know and it was widely stressed at this meeting, that dynamical effects can be important in the evolution, even in a slow rotator like the Sun. Mixing process due to rotation in the solar interior have been shown to play an important role in redistributing chemicals and angular momentum. Much stronger effects are expected for stars rotating up to 10 times faster than the Sun. Moreover, theories of dynamo generation of magnetic fields must be constrained by observations of surface magnetic fields and internal hydrodynamics(convection, rotation) Therefore models of stellar structure and evolution definitely need confrontation to observational data other than effective temperature and luminosity.

The PRISMA (Probing Interior of Stars: Microvariability and Activity) mission has been proposed (Lemaire et al. 1989) in response to the ESA call for Medium Size Mission, with the aim of providing for the first time an integrated facility suitable for filling this gap. The proposed mission has been selected for an Assessment Study. The objective of the mission is to sound the stellar interiors and to constrain stellar structure models by surface microvariability and surface magnetic field mapping.

This paper describes the scientific objectives of the mission, the measurements to be done, the technique to be used and gives the main characteristics and performances of the proposed model payload.

S. Catalano and J. R. Stauffer (eds.), Angular Momentum Evolution of Young Stars, 397–403.

2 Scientific Objectives

The primary scientific objectives of PRISMA mission will be the study of the structure, evolution and dynamics of stars from the core to corona. Let us emphasize here the impact of PRISMA on the main topics of this workshop, i.e. stellar rotation and magnetic field generation by dynamo process. This scientific goal will be acheived by using two complementary tools:

- *Asteroseismology*

High precision photometry will be the base for the observation of frequencies, amplitudes and lifetimes of eigenmodes of oscillation,. which allow to meet the following goals:

- Test stellar structure and evolution through the determination of oscillation frequencies for many stars with different mass and age. This will lead to a considerable improvement of the present stellar models, and will allow a better parametrization of physical processes (equation of state, opacities, mixing length)

- Derive radial and latitudinal internal rotation from oscillation frequency splitting and broadening. This will give an observational basis to models on angular momentum transport, such as those presented at this meeting (MacGregor 1991, Sofia et al. 1991).

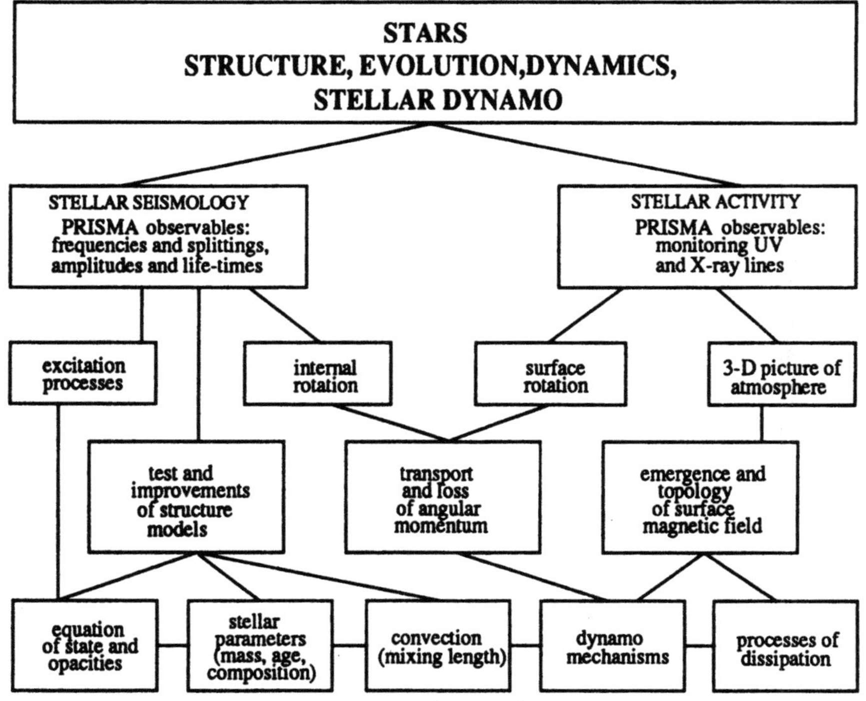

Figure 1. Summary of scientific objectives

- Study of mode excitation processes. It is generally agreed that, solar p modes are excited by convection. Amplitudes and lifetime of modes should allow to monitor the convection parameters along the evolutionary tracks.

- *Activity Monitoring*

Stellar activity monitoring will be performed through observations in several UV and X-ray emission lines, probing the various levels in the stellar atmosphere (chromosphere, transition region and corona). The resulting time-resolved 3D picture of the atmosphere will provide the topology of the surface magnetic field, the vertical structure of the atmosphere and the the time-scale magnetohydrodynamic processes. The major observational aspects are:

- The precise determination of surface rotation rates from the rotational modulation of activity diagnostics, which, when combined with results on internal rotation derived from asteroseismology, will yield estimate of radial differential rotation. This will pose constraint on the depth of dynamo action.

- The location of active regions from light curve deconvolution and Doppler imaging. To the extent that structures are distributed in latitude, the differential rotation can be measured. Moreover activity belts at different rotation rate and age can be defined.

- Mapping of the large scale magnetic structures, will allow to estimate the relative weight of open and closed magnetic field structures. This will help to understand the angular momentum loss process and efficiency.

3 Measurements and Observation Strategy

- *Asteroseismology*

Stellar oscillation studies require accurate determinations of frequencies, amplitudes and width of eigen-modes. In principle either Doppler shifts in photospheric lines or flux variation measurements are able to provide the needed quantities. This is has been proven to be the case for the Sun, even if a small discrepancy has been found between the two techniques (Libbrecht, 1990). However, in the stellar case Doppler shift measurements are made very difficult by the lack of photons, and the rotational broadening of spectral lines. All the efforts to observe oscillations in solar-type stars by Doppler measurements from the ground resulted in the detection of power above the noise for Procyon (Brown and Gilliland 1990). Velocity variations on Arcturus that could correspond to solar-type p modes have been also detected (Belmonte et al. 1990)

The observations of brightness fluctuations from space has now proven to be a a very powerful tool to sound the solar interior (Woodard and Hudson, 1983, Fröhlich and Toutain, 1990). This method is the most promising one to perform an extensive survey of stars in the H-R diagram, because the use of wide wavelength band provides a large photon flux and because there are no limitation due to rotational velocity. Mode detection requires a photometric stability of the order of 0.1 t 1 ppm, line-width and line separation require 1 to 0.4 μHz frequency resolution.The observational strategy to meet this goal assume 10-30 days uninterrupted sequence and sampling of 60 sec. The accuracy and associated science reachable are listed in TABLE 1a.

Stellar Activity

Stellar activity studies have been pursued with reference to all the empirical evidences we have of magnetic phenomena seen on the Sun, namely magnetic field measurements, local inhomogeneities (spot and plages), time-evolving manifestations (flares) and global structure (Rodonó 1986). PRISMA ability in doing stellar activity studies take advantage of the following main features:

- High sensitivity and stability in photometry
- Multiwavelength coverage of activity tracers formed at different atmospheric layers
- Sufficient time resolution (short-lived events detection).

While stellar spots will be detected as dips in the photometric curves (at a level of 0,1% for the Sun seen as a star), plages, during their transit on the visible part of the star, will enhance emission flux in UV lines. Activity will be monitored simultaneously in the Mg II, He II, C IV and Ly-α lines and the XUV at 170Å (several coronal lines of Fe VII to Fe X). The Doppler imaging technique applied to Mg II lines will provide the location and the size of individual plages on the faster rotating stars(Neff et al. 1989). Using a combination of photometric and spectroscopic data, it is possible to estimate the size and location of active regions at the different atmospheric levels, and therefore to determine the topology and structure of the surface magnetic field. Moreover, activity is a source of noise for solar photometry. Stars may be even more active, at level several orders of magnitude higher than in the Sun, therefore a simultaneous registration of asteroseismology and activity measurements will be necessary to interpret the photometric data. Observables and accuracy of the activity segments are listed in TABLE Ib.

Since the understanding of the Sun is one of our prime goals, the target coverage will be especially refined around the Sun, i.e. at spectral types F, G, K, and luminosity classes III, IV, V. In order to understand the dynamics in the interior of the Sun and stars, and in order to provide tests of dynamo theories, stars with different rotation rates will be observed. A reasonable test of stellar structure models can be achieved if one or several of the basic stellar parameters are known independently. For this reason, high priority targets for PRISMA are stars for which independent information is available :

- *members of open clusters, for which a good estimate of the age is available.*
- *members of some binary systems, for which the mass, and sometimes the radius, are well-known.*
- *stars observed by HIPPARCOS, for which we have a good determination of the absolute luminosity.*

The observation of a significant sample (more than 100 targets) in a two-year mission, with exposure times of 1 month per target imposes the choice of a strategy in which several stars are observed simultaneously. In the course of the preliminary target selection, we discovered that the simultaneous use of two photometers was necessary to reach the required number of targets.

TABLE 1a. Photometric segment

Observables	Range	Accuracy	Scientific keywords
Frequencies	0.02mHz-0.1Hz	0.1μHz	Stellar structure,g modes, He content, neutrino physics
Amplitudes	10^{-6}-10^{-1}	10^{-7}-10^{-5}	Excitation, stellar spots, and plages, non linear pulsators?, convection
Line-widths	1-10μHz	0.05μHz	Excitation, convection, surface structure
Frequency separations	>0.3μHz	0.1μHz	Basic stellar parameters, stellar structure, rotation, magnetic fields

TABLE 1b. Spectroscopic and simultaneity segment

Observables	Range	Accuracy	Scientific keywords
UV flux	10^{-10}-10^{-14}erg/sec/cm ^{2}	5%	3-D pictures
UV line profiles	1200-2800 Å	0.2Å$_{resolution}$	Doppler imaging
170Å X-ray flux	10^{-11}-10^{-14}erg/sec/cm ^{2}	20%	Corona, flares
Rotation	0.4-12μHz	0.4μHz	Angular momentum loss

4 The Model Payload

The model payload proposed to achieve these scientific objectives, comprises 4 instruments:

- A Large Photometer (**LP**), with a 40 cm diameter effective aperture, has a 1.5 x 1.5 ^{o} field of view. It collects simultaneously the broad band visible flux of stars brighter than 8 magnitude to detect micro-magnitude flux variations. This LP drives the pointing axis of the UV Spectrometer and XUV Telescope imager near the anti-Sun direction.

- A Small Photometer (**SP**) with a 15 cm diameter effective aperture has a 3 x 3 ^{o} field of view. It has the same purpose and design as the LP except that micro-magnitude flux variations will be detected on stars brighter than 6 magnitude. A internal mirror allows a wide selection of fields offset from the line of sight of the other instruments.

- An Ultra Violet Spectrometer (**UVS**), aligned with the Large Photometer, has a 40 x 30 cm aperture and covers the 1200-2800Å wavelength range and that includes selected chromospheric and transition region lines. A high-resolution (0.1-0.2 Å) cross-dispersed echelle system allows observation of one star at a time in a 1.5 x 1.5 ^{o} field of view corresponding to the LP field.

- An eXtreme Ultra Violet Telescope (**XUVT**) which images the field of the LP at the high coronal temperatures sampled by the Fe VIII-XI lines included within the bandpass of the filter near 170 Å. Multilayer mirrors allow imaging of an 1.5 x 1.5 ^{o} field of view at

170 Å with a 30 Å bandwidth.

The photometers will use CCD, while the UVS and XUVT will use detectors working in photon counting mode for imaging purposes.

5 Conclusion

PRISMA is a space experiment well suited for probing the interior of stars by measuring microvariability and activity. Long time series of continuous observations (typically one month) will provide the resolution needed to detect eigen modes of stellar oscillations, and also to cover completely the phase of stellar rotation for the activity measurements. The main scientific impact expected for PRISMA includes: test of stellar structure and evolution models; convection and oscillation mode excitation, magnetic dynamo, angular momentum distribution and angular momentum transport.

6 References

- Belmonte, J.A., Jones, A.R., Pallé, and Roca Cortès, T., 1990, *Astrophys. J.*, **358**, 595
- Brown, T.M., Gilliland, R.L., 1990, *Astrophys. J.*, **359**,839
- Frőhlich, C., Toutain, T., 1990, in *Progress of Seismology of the Sun and the Stars*, ed. H. Shibahashi, Springer Verlag, in press
- Lemaire,P., C. Catala, C., Christiansen-Dalsgaard,J., Hammer, R., Isaak, G., Mangeney, A., Praderie, F., Rodonó, M., Roxburg, I.W., 1989 Proposal to ESA, November 1989
- Libbrecht, K.G., 1990, *Astrophys. J.*, **359**,232
- MacGregor, K., 1991, this volume
- Neff, J.E., Walter, F.M., Rodonó, M., Linsky, J.L., 1989, *Astron. Astrophys.*, **215**, 30
- Rodonó, M., 1986, in *Cool Stars, Stellar Systems and the Sun* eds., M. Zeilik and D.M. Gibson, Springer-Verlag, Berlin p. 475
- Sofia, S., Pinsonneault, M., Deliyannis, C., 1991, this volume
- Woodard, M.F., Hudson, H.S., 1983, *Solar Phys.*, **82**, 67

DISCUSSION

Soderblom: How many stars can you get over the two year life-time?

Catalano: For photometry, about 100 stars.

Vaiana: What sets that limit? What's the distance limit for a G stars or a B star?

Catalano: We are limited just by the total time available. It takes about one month for each star in order to have a long enough time base-line to sample the frequency space we wish to cover, so only a few stars can be done well in the two year mission life. Rather than distance, we should speak about limiting magnitude.

Vaiana: So, the selection of stars is crucial. Can you start with a survey of some sort, then pick the list based on the initial observations? How many stars are bright enough for measurement? What is your limiting sensitivity?

Catalano: The limiting signal-to-noise depends on what you are looking for as well as how bright the star is. The theoretical limit is just set by the photon noise, but also for stars with close companions by the pointing jitter.

Stauffer: What is the earliest possible launch date?

Catalano: About 1999. A preliminary selection will be made near the start of 1991, when approximately 3 out of 6 proposals will be selected for the phase A. Then, in 1992 or 1993, one of the proposals will be accepted by ESA for the realization.

Soderblom: What limits your mission life-time to two years?

Catalano: Money. The cost of ground operations. The spacecraft will work for at least 5 years. The UV spectrometer on board could be used in an IUE mode, though there may be a telemetry problem. For the first year of the mission, we will observe objects selected by our team. After that, there will be a Call for Proposals and some of the observing time will be allocated to guest observers.

THE SPECTRUM–UV PROJECT

M. RODONÓ[1], E.G. TANZI[2]
on behalf of the Spectrum–UV Project Team, Italy

A.A. BOYARCHUK[3], N.V. STESHENKO[4]
on behalf of the Spectrum–UV Project Team, USSR

1) *Osservatorio Astrofisico and*
Istituto di Astronomia Università di Catania, Italy
2) *Istituto di Fisica Cosmica, CNR, Milano, Italy*
3) *Institute of Astronomy, Moscow, USSR*
4) *Crimean Astrophysical Observatory, Nauchny, Crimea, USSR*

ABSTRACT. A feasibility study of the Ultra–Violet Space Telescope SPECTRUM–UV, an international UV Observatory to be launched in a highly excentric (4 to 7 days) orbit in the mid–1990s, is being carried out by an International Team which includes scientists from Canada, Germany, Italy and the USSR. The main telescope features a 170–cm diffraction limited aperture for spectroscopy and imaging from the Lyman limit up to the visible domain. Two additional 50–cm telescopes for imaging and spectroscopy in the 400 to 1200 Å range and four 20–cm, multilayer coated telescopes for narrow band imaging in the 100 to 400 Å range are coaligned with the main telescope.

1 The Spectrum–UV Mission

An Ultraviolet Observatory, dubbed SPECTRUM–UV, is to be launched in 1995 by a PROTON booster in a 4 to 7 days orbit aboard a spacecraft of the SPECTRUM series. The scientific payload will consist of a main telescope (T–170) of 170–cm aperture for imaging and spectroscopy in the 912 to 3500 Å range, two iridium coated 50–cm telescopes (T–50) for imaging and low resolution spectroscopy in the 400 to 1200 Å range, and four multilayer coated 20–cm telescopes (T–20) for imaging in narrow bands to be selected in the $\sim$100 to $\sim$ 400 Å interval.

A $\simeq$ 2.5 arcsec stability of the spacecraft will be provided on three axis. For the main T–170 telescope, a $\simeq$ 0.1 arcsec pointing and tracking accuracy will be obtained by tilting the secondary mirror around the neutral point. A Fine Guidance System (FGS) will control the actuators of the secondary mirror.

A real time operation concept is adopted to ensure maximum flexibility and to take full advantage of the long, uninterrupted exposures allowed by the deep orbit. A backup automated operating mode is however envisaged in case of occasional unavailability of the Deep Space Communication Network. A Mission Operation Center will bear the overall responsibility of all spacecraft related tasks, while a Science Operation Center, under direct control of the USSR Academy of Sciences, will be responsible for the observations.

S. Catalano and J. R. Stauffer (eds.), Angular Momentum Evolution of Young Stars, 405–408.

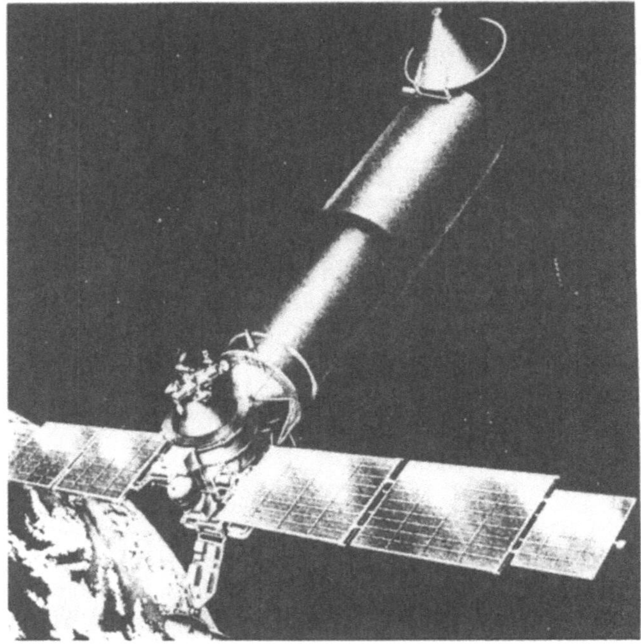

Figure 1: SPECTRUM-UV concept

A feasibility study of the Project is being carried out by an international Team which includes scientists from Canada, Germany, Italy and the USSR.

2 The T–170 telescope

The T–170 telescope is in the Ritchey–Chrétien configuration with primary mirror diameter of 170 cm and equivalent focal length of 17.0 m. The primary mirror consists of a thin (10 cm) meniscus of equal thickness, with a 58 cm central hole. The hyperbolic secondary is at 3.5 m from the primary apex and can be rotated on two axis around the neutral point as to achieve the fine pointing and tracking required, while keeping to a minimum the aberrations due to decentering.

The on axis overall image quality will be kept close to the diffraction limit over the whole wavelength range while at the edge of the ± 20 arcmin field of view a 70 % EE in 1 arcsec will be achieved. The main parameters of the T–170 telescope are summarized in Table 1.

The focal instrument complement includes:

A dual echelle spectrograph (similar to that aboard of IUE) to achieve high resolving power ($\Re \simeq 4 \times 10^4$) for the 120–190 nm and 190–350 nm intervals. A low resolution ($\Re \simeq 10^3$) mode is obtained by inserting a flat mirror in front of the echelle.

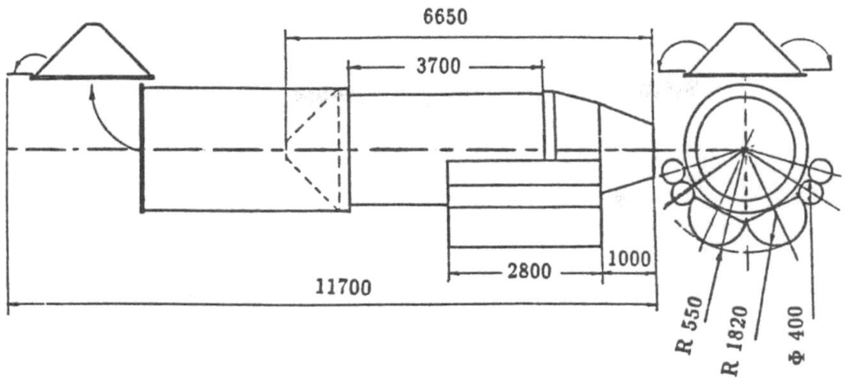

Figure 2: Telescope assembly layout

A spectrograph in the Rowland mounting to provide a resolving power $\Re \simeq 10^4$ in the Lyman region (912 to 1150 Å) where, in spite of the low normal-incidence reflectivity, an accurate control of the thickness of the Al+MgF_2 coating and the large collecting surface can provide a sizable effective area. An option to add a $\Re \simeq 3 \times 10^3$ and/or $\Re \simeq 100$ mode over the 120–320 nm interval is being considered.

A nebular spectrograph with a 200 × 2 arcsec^2 slit, to provide $\Re \simeq 10^2$ in the 400 to 900 nm range.

A direct imaging camera with broad band filters in the region 912 to 3500 Å and f.o.v. of $\sim$ 6 arcmin will be mounted on-axis. A $\sim$ 16 arcmin f.o.v., off-axis focal reducer in the 190–350 nm range to feed a low resolution imaging camera is also being considered.

3 The T–50 and T–20 Telescopes

The complementary telescopes T–50 and T–20 will be co–aligned with the T–170 telescope.

The T–50 telescopes will consist of two off-axis paraboloids (each made of a half 80–cm mirror) iridium coated to enhance reflectivity in the 400 ÷ 1200 Å range. One of the two will be equipped with a direct imaging camera to achieve a few arcsec angular resolution over the whole 4 arcmin f.o.v.. while a very low resolution ($\Re \simeq 20$) grating objective will provide dispersed images on the other.

The four 20–cm, multilayer coated, f/12.5 spherical mirror telescopes will provide low resolution imaging capabilities in four different narrow bands ($\Delta\lambda/\lambda = 0.05$), to be selected within the 100 ÷ 400 Å region.

4 The Spacecraft

The design of the spacecraft for the SPECTRUM–UV mission is under the joint responsibility of the Babakin Research Center and the Lavochkin Science and Industry Corporation. It is a common design platform to be used also for the RADIO–ASTRON and the SPECTRUM–X–Gamma Missions.

TABLE 1

T-170 Telescope Parameters	
Primary Mirror	
Diameter	1.7 m
Focal ratio	f/2.8
Curvature	$0.10589{\cdot}10^{-3} mm^{-1}$
Asphericity	1.0539
Secondary Mirror	
Diameter	0.5 m
Curvature	$-0.29545{\cdot}10^{-3} mm^{-1}$
Asphericity	3.6829
Overall Parameters	
Equivalent focal length	17 m
Plate scale	12.13 $arcsec \cdot mm^{-1}$
Field of view (2 β)	40 $arcmin$
PSF on axis	diffraction limited
PSF 20 $arcmin$ off-axis	1.0 $arcsec$ (70%EE)
Mirrors separation	3500 mm
Focal extraction	900 mm
F.P. Curvature	1270 mm

The platform hosts scientific payloads up to 2.5 tons, providing an operational life in excess of 3 years. The telemetry bit rate is of 64 kbit sec^{-1} up to a distance of the satellite of 5×10^5 km.

The basic attitude determination system of the spacecraft consists of high precision gyros. In order to derive absolute pointing direction and to correct for drifts, the spacecraft relies on sun and star sensors. The actuators are reaction wheels periodically unloaded by gas jets. The spacecraft will be 3-axis stabilized in an inertial coordinate system. Its stability under the control of the gyros will be 1 to 3 arcsec on 1 minute time scale and 30 to 40 arcsec on a time scale of 24 hours, due to temperature instabilities.

The spacecraft will be launched by a PROTON booster on a high apogee orbit with initial inclination of 51°. Both solar and chemical batteries will supply 1 kW for the scientific payload for any attitude of the spacecraft.

SUMMARY OF THE WORKSHOP

ROBERT P. KRAFT
University of California Observatories/Lick Observatory
Board of Studies in Astronomy and Astrophysics
University of California, Santa Cruz, CA 95064

1. A Stroll Down Memory Lane

The poor devil selected as "summary speaker" is saddled with a daunting task: how to mention every speaker's work (so nobody is angry with him!), and at the same time how to cram it all into 45 minutes or less (so that people leaving immediately by plane are able to get to the airport in time!). Obviously, there's no solution: my selection of topics has to be personal and no doubt will be regarded as cranky.

At the outset I want to say that I learned a lot at this workshop and found the proceedings highly stimulating. What impressed me is how vigorous the subject now is, if "vigor" is to be measured by the plethora of new observational facts and the degree to which these remain unexplained. In the old days, we could work only in the optical domain. Nowadays, significant new facts bearing on the way in which stars solve the angular momentum problem as they evolve have been uncovered in the X-ray, IR, and mm-wave regimes. The "facts" run way ahead of the theory and much remains to be understood.

A short stroll down memory lane seems in order, if only to emphasize the extraordinary progress in observational technique that has taken place during the past 25 years. Once upon a time, we could work profitably only with blue sensitive ($\lambda\lambda$ 3500–5000 Å) photographic emulsions, baked in ovens to speed them up! (Yellow and red-sensitive plates were comparatively slow.) At conventional Cassegrain and coudé spectrographs, a resolution R $\simeq$ 20,000, corresponding to $\Delta\lambda \simeq 0.2$Å $\simeq$ 12 km/sec, was the best we could do, meaning that the V_{rot} sin i resolution limit was also 12 km/sec. Higher resolution required using a narrower spectrograph entrance slit, sometimes approaching 0.1 arcsec, to maintain spectral purity. This meant throwing away a lot of light at the slit, even at places with good seeing, such as Lick or Mount Wilson. Nevertheless, I still recall the thrill of inspecting a plate of a Pleiades G0 dwarf, fished fresh out of the hypo in the Mount Wilson 100–inch telescope dark room: the lines of the closely-spaced Fe I doublet at λ4272 were smeared together, indicating a rotational velocity in excess of 25 km/sec! Who could have imagined we would find a solar-type dwarf rotating at 10 times the solar rotational velocity? But what a price we had to pay. This will amuse David Gray and Lee Hartmann today: it took 8 hours of 100–inch coudé time to get that plate and the seeing was good, too!

S. Catalano and J. R. Stauffer (eds.), Angular Momentum Evolution of Young Stars, 409–420.

What else did we know (or *think* we knew) in the mid-1960s? 1.) It seemed pretty clear that the mean specific angular momentum $J/\mathcal{M}$ on the upper main sequence went with mass about as $\mathcal{M}^{2/3}$, but that below $\mathcal{M} \sim 1.25\mathcal{M}_\odot$, $J/\mathcal{M}$ took a sharp drop. 2.) Oke and Greenstein (1954) had shown that giants rotated slowly and had V_{rot} sin i–values more-or-less compatible with conservation of J for stars in post-main sequence evolution. 3.) George Herbig (1957) had followed a handful of pre-main sequence T Tauris along the Henyey (mostly radiative) tracks and found J again essentially conserved.

In the light of Brandt's (1966) discovery that the solar wind carried an angular momentum flux having an interestingly short time-scale, Schatzman's (1962) bold hypothesis that rotation could be "braked" by a magnetized wind, and the development of the Weber/Davis (1967) [and slightly later, Mestel (1968)] solar wind model, little imagination was required to propose (Kraft 1967b) that solar-type stars were in fact "braked" by the action of stellar winds on a time-scale of order 10^8 yr (to judge by the then-accepted ages of the Hyades and Pleiades clusters). Popular then was a contrary idea espoused principally by Su-Shu Huang, as I recall, *viz.*, that the break in main sequence rotation near $\mathcal{M} \sim 1.25\mathcal{M}_\odot$ signaled the onset of stars having planetary systems. This seemed natural enough, since the addition of Jupiter's orbital angular momentum to the sun would have bought its J-value up to the $J/\mathcal{M} \sim \mathcal{M}^{2/3}$ line defined by upper main-sequence stars. However, the observational demonstration that rotation in F/G dwarfs in young clusters gradually declined with advancing age (Kraft 1967b) appeared to fit well with the idea of a magnetized "wind-brake" model and Skumanich (1972) showed that the decline followed a $t^{-1/2}$ dependence. The existence of planetary systems could thus be viewed as independent of the break in $J/\mathcal{M}$, so that the "Jupiter effect" became an interesting coincidence.

2. Early Stages

What we learned at this workshop is that much is still to be done if we are to solve the fundamental problem of which the "Jupiter effect" is only a small piece: how can we start with a $1000\mathcal{M}_\odot$ diffuse molecular cloud having $J/\mathcal{M} \sim 10^{24}$ cm^2 s^{-1} and wind up with the present sun in which $J/\mathcal{M} \sim 10^{15}$ cm^2 s^{-1}? Peter Bodenheimer began our proceedings with a remarkably complete treatment of the earliest stages of the chain by addressing four fundamental questions:

1.) What determines the J-distribution among newly formed stars?

2.) How does one get from molecular cloud cores with dimensions of order 0.1 pc and $J/\mathcal{M} \sim 10^{21}$ cm^2 s^{-1} to T Tauri stars with $J/\mathcal{M} \lesssim 10^{17}$ cm^2 s^{-1}?

3.) What happens as contracting stars pass through the "valley of no information," i.e., why is it that when T Tauris emerge from their cocoons, their $J/\mathcal{M}$–values are only 10% of the breakup value?

4.) Under what conditions do single stars form as opposed to binaries?

Evidently, fragmentation of diffuse molecular clouds occurs when the ambient interstellar magnetic flux becomes too weak, as a result of ambipolar diffusion, to support the gas against gravity. At that point, angular momentum in each fragment (cloud core) is conserved, so the cores have only (!) to solve the problem of disposing of four orders of magnitude of $J/\mathcal{M}$, i.e. from 10^{21} cm^2 s^{-1} to 10^{17} cm^2 s^{-1}. What processes are possible?

Bodenheimer outlined the problems associated with taking up the excess angular momentum into orbital motion (binary and multiple stars) or into a rotating circumstellar disk. He reminded us that forming a wide binary was possible under the expected isothermal conditions, but that forming rather ordinary binaries with P∼ days, weeks, or even a few years was inhibited by the adiabatic conditions encountered in such cases. On the other hand, scenarios for disk formation and evolution encounter severe problems, too. Bodenheimer showed us a calculation in which one wound up with a disk of radius ∼ 120 AU having a mass much larger than that of the central star. This set the stage for what may have been *the most serious conundrum of the workshop*. There is plenty of evidence from mm-wave and IR studies (e.g., Strom, Margulis, and Strom 1989; Beckwith, *et al.* 1990) that such disks are ubiquitous in the early stages of star formation; their properties were discussed in detail by Steve Strom later in the workshop. Steve also outlined observational evidence suggesting that the disks were unusually massive. But there is also plenty of evidence — from radial velocities, peculiarities of the emerging flux distribution, etc. — that T Tauris accrete substantial material from such disks. The leading problem is: how do T Tauris get rid of the accreted angular momentum, so as to wind up with $J/\mathcal{M}$ at only 10% of the breakup value? We know that T Tauris eject matter into well-collimated jets and antijets, which presumably provide a mechanism for angular momentum losses. A magnetic braking wind mechanism is often invoked. But what is the physics that determined the relationship between

$$-\left(\frac{dJ}{dt}\right)_{wind} \text{and} + \left(\frac{dJ}{dt}\right)_{accretion} ?$$

A little later Jerome Bouvier gave us a very complete discussion of the rotation of pre-main sequence stars of all masses, both T Tauris and the Herbig Ae and Be stars, and traced the evolution of $J/\mathcal{M}$ to the main sequence. Providing a further constraint on the dJ/dt-problem just mentioned, he reminded us that pre-main sequence stars which appear to accrete from a disk do not rotate faster than stars having no evidence for the existence of an accretion disk. He also showed that the more massive pre-main sequence stars (Ae's and Be's) had rotational velocities consistent with conservation of $J/\mathcal{M}$ as they contracted to the main sequence, and that the average initial angular momentum of low-mass T Tauris was consistent with extrapolation of the $J/\mathcal{M} \sim \mathcal{M}^{2/3}$—relationship to stars of the lower main sequence. If then low-mass T Tauris are the progenitors of late-type stars in young clusters such as the Pleiades and α Persei, T Tauris must retain most of the angular momentum and rotate like solid bodies up to the ZAMS. He also remarked that the rotational properties of pre-main sequence stars do not change from one star-forming region to another, but Steve Strom advised caution, noting that the mean rotational velocities of B-type stars in certain associations vary from one subregion of the association to another.

3. The Main Sequence

In a session on angular momentum evolution starting in the pre-main sequence era, continuing onto main sequence, and ending in the immediate post-main sequence giant-star stage, John Stauffer discussed the main sequence spin-down problem in greater detail. Conservation of $J/\mathcal{M}$ for low mass stars should lead to rotational velocities of 100 to 150 km/sec in the late main sequence. What do we see? Stauffer reviewed (e.g., Stauffer, Hartmann, and Jones 1989;

Stauffer *et al.* Stauffer and Hartmann 1987) work largely by himself, Lee Hartmann and their associates in such young clusters as α Persei and the Pleiades, along with the somewhat older Hyades. As is by now well-known, the A and F-type star velocity distribution is essentially the same in all three clusters (more-or-less Maxwellian), but in the youngest cluster (α Per), rapid rotational velocities are found at all spectral types, even among the G, K and M-type main-sequence stars. In the slightly older Pleiades, early G-types rotate slowly, but many K's and M's rotate rapidly; in the still older Hyades, almost all stars less massive than $\sim 1.25\mathcal{M}_\odot$ are slow rotators, except possibly for some faint M dwarfs. Moreover, the V_{rot} distributions among G, K and M-type stars are no longer Maxwellian — there are a large number of very slowly rotating stars along with the significant, but smaller number of rapid rotators. A picture emerges in which stars arrive quickly on the ZAMS with $J/\mathcal{M}$ conserved, but then spin down, i.e., the outer convection zone spins down, on a time-scale which grows longer in proportion to the deepening of the convection zone with advancing spectral type. If age is the only difference between the clusters, then the spindown time-scales can be estimated as $\sim 10^7$, a few $\times 10^7$, and $\sim 10^8$ yr for the α Per, Pleiades, and Hyades clusters, respectively.

In looking around for possible explanations of the peculiar rotational velocity distributions in the Pleiades and α Per clusters, one finds the invocation of non-coeval star formation. It has also been invoked to explain the spread in Li abundances in a given cluster and its apparent relationship to rotation, i.e., high Li goes with large V_{rot} (e.g., Butler *et al.* 1987). Since Li destruction at the base of the convection zone takes a while, the implication is that the high rotation stars in a given cluster are younger than the stars with low rotation. However, recent observations (Soderblom and Jones, this volume) showing that there are many Li-rich slow rotators among faint Pleiades stars, suggest that rotation is not the sole mechanism driving Li abundances.

A related problem has to do with the actual interpretation of the Li line strengths themselves. Dave Soderblom, in a paper that was a high point of the meeting for me, reviewed the present state of Li abundance determinations in the Pleiades. It is the case that there is a wide spread in Li resonance line strengths at a given T_{eff} in the Pleiades, and the stars with the strongest Li lines actually yield up abundances considerably higher than the canonically-accepted galactic disk Li abundance. Soderblom explored the ground state resonance KI line at λ7699Å in these stars — KI being the same sort of chemistry and atomic term-scheme as Li — and found that the analysis of λ7699 leads to the same kind of spread in K-abundances as was found in Li. Furthermore, the strength of the Li resonance line in absorption seems to be correlated with the strength of Ca II emission. This in turn suggests that Li abundances must not be taken literally, and that some atmospheric effect (lowered boundary-layer temperature? large spotted areas?) drives at least a portion of Li resonance line strengths, not solely burning of Li at the base of the convection zone.

As long as we are dwelling on this point, let us examine a little more closely the hypothesis of non-coeval star formation. Naturally, at some level, corresponding to the star-crossing time ($\sim 10^6$ to10^7 years) in a cluster, non-coeval star formation must exist; the question is whether it can be invoked on Δt-levels exceeding 10^7 yr. In a recent study now in press, Stauffer, Klemola, Prosser and Probst (1991) conclude that the observed spread in the main sequence of K and M type stars in the Pleiades limits Δt to something surely not more than

2×10^7 yr, and possibly something considerably less. (Comparison here is made with van den Berg's most recent isochrones for low-mass stars contracting to the ZAMS.) But a better cluster for addressing the problem is α Per. Why?

To assess the non-coeval star formation question, the observer examines the spread in V-magnitude at a given color [say (V-I)] in the color-magnitude diagram of the cluster. This is compared with the spread in the theoretical isochrones for contracting stars. One has to allow for visual binaries, which naturally confuse the picture. In an earlier paper, Stauffer (1984) called any star a visual binary if it lay more than 0.3 mag above the "observational" main sequence of the cluster. Using this criterion, one finds that about 30% of faint Pleiades stars are binaries, but of course it is possible that stars lying only slightly above the ZAMS are also binaries; one cannot tell without an extensive and possibly extremely time-consuming photometric and/or spectroscopic investigation. However, it has been known (or suspected) for some time that α Per has a sparse binary population — investigations conducted in the 1960s and 1970s, both spectroscopic [Heard and Petrie (1967); Kraft (1967a)] and photometric (Crawford and Barnes 1974), gave evidence for a relative paucity of binaries. Recently, Lick graduate student Charles Prosser has derived a new V versus V-I diagram for α Per, down to M5 dwarfs, based on a proper motion-selected sample. There is little scatter and relatively few binaries, based on the Stauffer criterion. Indeed, I summarize in the Table the binary frequency situation for α Per vis-a-vis the Pleiades, based on the Prosser c-m array and the Stauffer criterion. Inspection of the Table suggests that α Per is the cluster to study if one wishes to explore the non-coeval star formation question. In any case, Prosser concludes that in α Per, $\Delta t \simeq 2 \times 10^7$ years and possibly is much smaller.

TABLE 1. Binary stars in Pleiades and α Per clusters
(Stauffer Criterion)

$(V\text{-}I)^{o}$-range	Percent			
	Pleiades		α Per	
		No.		No.
early-type stars ($\leq$ F5V)	22-30	(101)	5-15	(84)
1.0-2.0 (K5-M2)	25	(83)	14-23	(57)
> 2.0 (>M2)	36	(85)	16	(31)

What has kept the idea of non-coeval star formation alive for more than 25 years? Aside from convenience as a hypothesis, not much that is very solid, I fear. Earlier the discordance between the nuclear and contraction time-scales for clusters was often cited, but that seems to be disappearing as theoreticians invoke rotation and convective overshooting to lengthen the nuclear time-scale (cf. e.g., the recent paper of Mazzei and Pigatto 1989). Otherwise, one fell back on the enormous spread in the V vs B-V diagram for faint Pleiades stars. But as early as 1966, Kraft and Greenstein (1969) showed that this spread was not reflected in a spectral-type vs V diagram, and it has long been known that the early photographic photometry of the Pleiades was seriously flawed because of inadequate allowance for the reflection nebulosity. So as far as I can see, there is nothing to keep the idea of non-coeval star formation going at this time (at the > 10^7 yr level) beyond wishful thinking.

Turning to other matters, we come back again and again to the observationally difficult problem of measuring rotational velocities. For faint late-type stars in clusters, one needs rather high spectral resolution; V_{rot} sin i is hard to measure directly. We learned that there are other techniques besides high resolution spectroscopy that lead directly or indirectly to estimates (or measurements) of V_{rot}. Obviously, V_{rot} could be directly measured without the sin i encumbrance when van Leeuwen and Alphenar made their seminal discovery (1982) of periodic variability of Pleiades late main sequence stars; this discovery may be thought of as the one ushering in the whole modern era of stellar angular momentum studies. But there are indirect surrogates for V_{rot} as well. Of great importance are the Einstein observations of X-rays emitted by pre-main sequence stars in the Taurus dark clouds. Damiani, Micela, and Vaiana (cf., e.g., Pallavicini, *et al.* 1981; Bouvier 1990) described their work which established a definitive relationship between X-ray flux and rotation, i.e., $L_X \propto P^{-2}$, where P is the rotational period. They found an inverse relationship between Hα emission-line strengths expressed in units of the 25 μm IR flux and the X-ray luminosity L_X. If the Hα emission in some sense measures the wind and the 25 μm flux the disk emission, then this correlation indicates that the wind is inhibited in the most X-ray active T Tauris. X-rays are produced, following the solar example, in closed magnetic loops; the anti-correlation therefore suggests the existence of an open field magnetic topology that is associated with the wind outflow.

Stauffer explored the Hα–emission strengths in the Hyades and showed that for stars bluer than M0 V for which *photometric* rotation periods are known, the equivalent width of Hα emission is uniquely correlated with period. This suggests that EW (Hα) is a good surrogate for V_{rot} and can be used to measure V_{rot} if appropriately calibrated. For Hyades stars later than M0 V, there is a wide, but more-or-less continuous, spread in EW (Hα) which, by the Stauffer hypothesis, would indicate a wide spread in Hyades V_{rot}. At the same time, Prosser, Stauffer, and Kraft (1991) recently showed that EW (Hα) in the Pleiades M-dwarfs is, on the average, about twice as large as EW (Hα) in Hyades M dwarfs, but the *spread* in EW (Hα) is large in both clusters so that there is considerably overlap in Hα emission strength. The Stauffer hypothesis then implies that a large fraction of Pleiades M-type dwarfs should have velocities similar to those of Hyades M-type dwarfs.

On the other hand, these stars also show Ca II (H and K) in emission, and one might suppose, in the light of studies of dwarf star spectra by Herbig (1985), that Ca II and Hα emission line strengths would be positively correlated. If so, there emerges a curious anomaly when we plot EW (K_2) vs spectral type for Pleiades and Hyades dwarfs of type K7V and later. These are taken from a 22–year old paper by Kraft and Greenstein and are illustrated in Figure 1. We see that the two clusters have quite a *discrete* separation of EW (K2) at a given spectral type, whereas a similar plot of EW (Hα) would show a good deal of overlap between the two distributions, as already noted. The sample is small and needs to be enlarged, but raises the question of whether chromospheric emission strength is driven solely by V_{rot} (via, e.g., the dynamo mechanism) or something more indirect such as age.

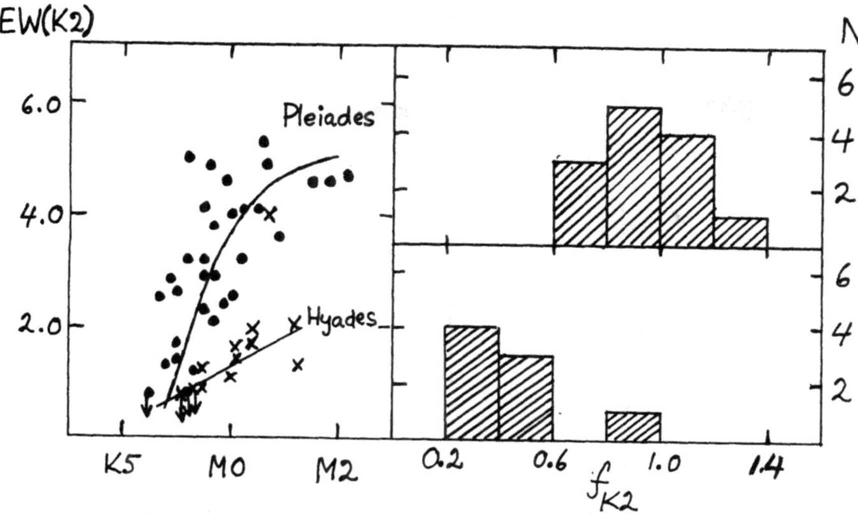

Figure 1. EW (K2) emission as a function of spectral type in the Pleiades and Hyades. The quantity f_{K2} is the emission strength in units of the ordinate of the smooth curve drawn through the Pleiades points in the left-hand portion of the figure, and N is the number of stars in the sample, per 0.2 in f_{K2}.

4. Solar Rotation and the Solar Wind

A lot of workshop participants were already convinced that magnetic solar-wind models were applicable to the problem of main sequence stellar spin-down; the problem remaining, they supposed, was the question of just what, in fact, was spun down — the whole star or simply the outer convection zone? We therefore heard quite a bit about models of rotating stars and whether $\Omega(r)$-gradients could exist. In a delightful review, Douglas Gough, self-styled surrogate observer, examined the results of the solar oscillation experiments and explained why they tell us little (at least at the moment) about the $\Omega(r)$ distribution in the deep interior of the sun. Indeed, there seems now to be no evidence that $\Omega(r)$ is anything except constant in the sun. Gough showed that a very small magnetic field in the deep interior would be effective in producing rigid rotation on a time-scale less than 5×10^9 yr. Earlier on, Ian Roxburgh pretty much convinced us that, no matter what you do, stars in convective equilibrium will rotate very nearly uniformly. Professor Belvedere told us a good deal about his group's work on solar activity belts and how their behavior (latitude distribution, migration, etc.) in the sun could be modeled as a natural consequence of dynamo action at the base of the solar convection zone. Presumably observations of latitude distribution and migration of active regions [the kind of work pursued by Professor Rodonò (e.g., Rodonò 1987) and his associates in Italy and Steve Vogt in California, for example] might reasonably allow us to infer the $\Omega(r)$-distribution in stars as well. Although I cannot speak with authority about these matters, I came away with the impression that, much as the observers might like a non-uniform $\Omega(r)$ in stars for a variety of reasons, the theoreticians don't find much support for the idea when they look into the physics of rotating stellar models.

We were treated to a brilliant review of the (by now "classical") theory of magnetic wind braking in late-type stars by Andrew Cameron, which included the closed magnetic-loop "dead zone" effects introduced some years ago by Leon Mestel. The observations suggest, as we've already noted, that a solar-type star reaches the main sequence with $J/\mathcal{M}$ conserved (rapid spin-up) on a time-scale of a few $\times$ 10^7 years, and this is followed by a longer epoch of wind-induced spindowns. But this must be divided into two parts. If the spindown is to follow the observational data on V_{rot} in the α Per, Pleiades and Hyades clusters, and if we assume in the usual way that $|B| \propto \Omega_{conv}$ where Ω_{conv} is the angular velocity of the external convection zone, then $|B|$ will decline rapidly in the early stages. The decline is rapid and a very long epoch of subsequent spindown is induced — indeed, so long as to spoil the fit of V_{rot} to the observations of Hyades stars (indeed, one may in certain circumstances wind up with excessive solar rotation as well).

This point was emphasized very neatly in the series of parameterized model calculations described by Keith MacGregor. He found that, for a wide range of values of the ratio of the time-scale τ_c for transport of angular momentum from radiative core to convective envelope to the time-scale $\tau_J(0)$ for the removal of angular momentum in the magnetized wind, and values of $\Omega_{initial}/\Omega_\odot$ ranging as high as 25, it was difficult to reproduce simultaneously the sharp drop in mean rotation between α Per and the Pleiades and the small decline from the Pleiades to the Hyades, as long as $|B| \propto \Omega_{conv}$. A satisfactory fit could be obtained, however, on the assumption that $|B| \propto \Omega_{core}$.

On the other hand, part of the problem may reside in the assignment of cluster ages. If we adopt the longer nuclear ages recently advocated by Mazzei and Pigatto, we can modify the fit of one of the MacGregor models, as shown in Figure 2. We see that α Per, the Pleiades, the Hyades, and the sun can all be accommodated with $\tau_c = \tau_J(0), \Omega_{initial}/\Omega_\odot$ ~15 to 20, and still retain $|B| \propto \Omega_{conv}$ rather than $|B| \propto \Omega_{core}$. Obviously, life would be simpler if the real cluster ages would please stand up. And it would help too if we could find and identify the faint stars in a rich cluster younger than α Per. [IC 2391 (Stauffer, *et al.* 1989) is poorly populated and thus not very helpful.]

5. Evolutionary Models; Post-Main Sequence Stages

Evolutionary models for rotating stars were calculated by Pinsonneault, Sofia and their collaborators and reported by Sofia. These models attempt to reproduce the present sun, from which unknown astrophysical parameters are scaled (e.g., ratio of mixing length to scale height, etc., etc.). This may or may not yield parameters applicable to non-solar type stars, and the procedure was criticized during the informal discussions that followed the presentation. The models do appear to develop Ω(r) gradients in the interior; such gradients have often been invoked to explain a number of other more poorly understood observational facts. Examples are the anomalously high rotation of horizontal branch stars (Peterson 1983) and the evidence for deep mixing and dredgeup of C $\rightarrow$ N and O $\rightarrow$ N processed material in globular cluster giants (Carbon, *et al.* 1982; Pilachowski 1988), far in excess of that predicted by stellar models having ordinary thermally-driven convection zones. Suntzeff (1988) has called attention to evidence for C $\rightarrow$ N processing and dredgeup into the atmospheres even of main sequence stars (!) in certain globular clusters, and Sweigart and Mengel (1981) have shown how such effects might come about as a result of Ω(r) gradients in low-mass, low metallicity post-main sequence

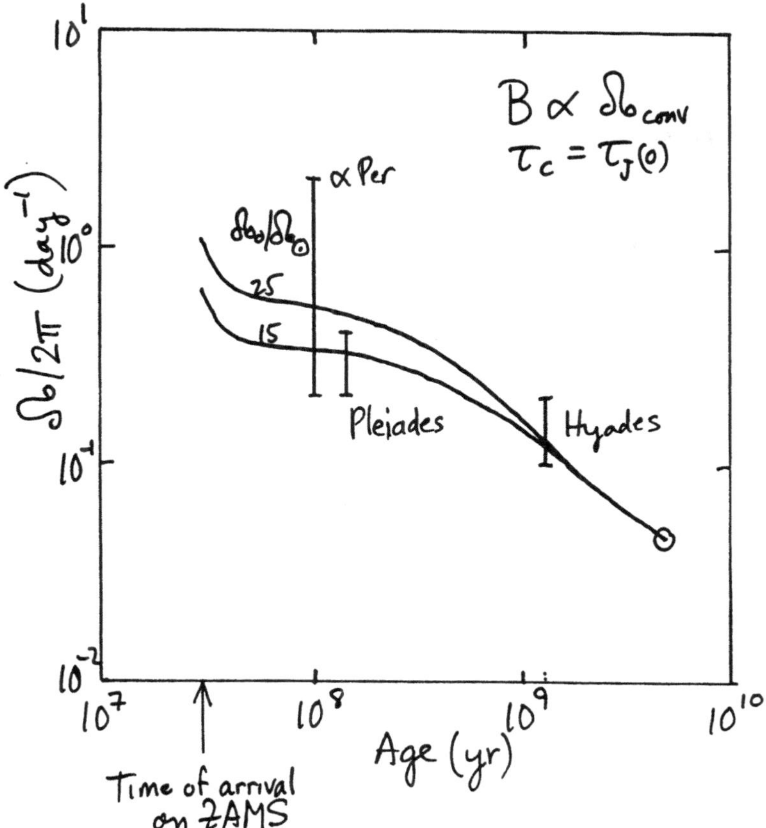

Figure 2. One of MacGregor's models having $|B| \propto \Omega_{conv}$ with cluster ages increased by a factor of 2, following Mazzei and Pigatto (1989).

stars. Of course, one cannot be sure the excess mixing is a result of such Ω(r)–gradients, but the anomalously high rotation of HB stars makes one suspicious, despite the evidence in favor of a flat Ω(r)-distribution in the sun. Observers certainly await the guidance that the Pinsonneault/Sofia models promise to provide.

Finally, I come to one of the more interesting confrontations of the workshop. In a brilliantly persuasive lecture, David Gray traced the post-main sequence evolution of class III and IV giants, showing that stars conserved their angular momentum as they evolved from the A and F dwarf star domain roughly to G0, but later than G0, their V_{rot} sin i values dropped sharply and dramatically. This virtual discontinuity, stretching over only 0.2 of a spectral class, was identified with the turn-on of magnetic wind braking.

But then de Medeiros and Mayor discussed their interpretation of Coravel-based material covering more-or-less the same ground. Sample sizes were not the same, but the topology of the V_{rot} sin i–distributions was not materially different. Based on a study of the shape and spread of velocity ellipsoids, de Medeiros and Mayor argued that the stars on the two sides of the discontinuity at G0 belong to different stellar populations, i.e., age groups, the slow rotators evolving from essentially solar-type stars.

A possible test of scenarios might come from spotted-star models and analysis of the kind pioneered by Rodonò or Vogt and their collaborators. If David Gray is right, a G0 III-star should increase its rotation period by as much as 1 sec/yr, on the average, as deduced from Iben's evolutionary track for a star with $\mathcal{M} = 2.5\mathcal{M}_{\odot}$. One should look for changes either in spot migration through the line profiles or spot-induced photometric variations. It's true that, even if a favorable case were found, one might have troubles with latitude migration and phase changes. Nevertheless, an effort made to detect the predicted period change might prove worthwhile and could plausibly be conducted with a relatively small, but dedicated, telescope.

6. Final Remarks

It's perhaps not inappropriate to mention in the end a few disappointments. For example, we didn't hear much about binary stars. What about the discovery of such objects in the pre-main sequence stage? Are the statistics comparable with the number of binaries found on the main sequence? Is wind-driven J-loss from a star in a binary system modified by the existence of a tidal-couple exerted between the components? The last issue was addressed by Drs. Maceroni and Van't Veer, who found that angular momentum losses increased above the Skumanich relation in tidally-locked binaries, i.e., those with $P \leq 4^d$. Otherwise little was said about possible interactions between rotational and orbital angular momenta.

Finally, Ap and Am stars seem to have dropped out of sight. The former have large, ordered dipolar magnetic fields and many of the latter are binaries. What should their pre-main sequence progenitors look like and how would we recognize such objects? They should not be particularly rare.

But if a few untouched topics remain, so much the better. This will give us an excuse to gather once again in the not too distant future!

References

Beckwith, A., Sargent, A., Chini, R., and Güsten, R. 1990, *A. J.*, **99**, 924.

Bouvier, J. 1990, *A. J.*, **99**, 946.

Brandt, J. 1966, *Ap. J.*, **144**, 1221.

Butler, R., Cohen, R. D., Duncan, D. K., Marcy, G. 1987, *Ap. J. Letters*, **319**, L19.

Carbon, D., Langer, G., Butler, D., Kraft, R., Trefzger, C., Suntzeff, N., Kemper, E., Romanishin, W. 1982, *Ap. J. Suppl.*, **49**, 207.

Crawford, D. and Barnes, J. 1974, *A. J.*, **79**. 687.

Heard, J. and Petrie, R. 1967, IAU Symposium No. 30, p. 179.

Herbig, G. H. 1957, *Ap. J.*, **157**, 612.

Herbig, G. H. 1985, *Ap. J.*, **289**, 269.

Kraft, R. P. 1967a, *Ap. J.*, **148**, 129.

Kraft, R. P. 1967b, *Ap. J.*, **150**, 551.

Kraft, R. P. and Greenstein, J. 1969, in "Low Luminosity Stars," ed. S. Kumar (New York: Gordon & Breach), p. 65.

Mazzei, P. and Pigatto, L. 1989, *A. A.*, **213**, L1.

Mestel, L. 1968, *MNRAS*, **138**, 359.

Oke, J. B. and Greenstein, J. 1954, *Ap. J.*, **120**, 384.

Pallavicini, R., Golub, L., Rosner, R., Vaiana, G., Ayres, T., Linsky, J. 1981, *Ap. J.*, **248**, 279.

Peterson, R. 1983, *Ap. J.*, **275**, 737.

Pilachowski, C. 1988, in "The Abundance Spread Within Globular Clusters," eds. C. De Strobel, M. Spite and T. L. Evans, IAU Gen. Assembly — JCM5 and CM 37/3, p. 1.

Prosser, C., Stauffer, J. and Kraft, R. 1991, *A. J.*, (in press).

Rodonò, M. 1987, Invited Review at the "Fifth European Solar Physics Meeting" on "Solar and Stellar Activity" (Titisee, FRG).

Schatzman, E. 1962, *Ann. d'Ap*, **25**, 18.

Skumanich, A. 1972, *Ap. J.*, **171**, 265.

Stauffer, J. 1984, *Ap. J.*, **280**, 189.

Stauffer, J. and Hartmann, L. 1987, *Ap. J.*, **318**, 337.

Stauffer, J., Hartmann, L. and Jones, B. 1989 *Ap. J.*, **346**, 160.

Stauffer, J., Hartmann, L., Jones, B. and McNamara, B. 1989, *Ap. J.*, **342**, 285.

Stauffer, J., Klemola, A., Prosser, C. and Probst, R. 1991, *A. J.*, (in press).

Strom, K., Margulis, M. and Strom, S. 1989, *Ap. J. Letters*, **346**, L33.

Suntzeff, N. 1988, in "The Abundance Spread Within Globular Clusters," eds. C. De Strobel, M. Spite and T. L. Evans, IAU Gen. Assembly — JCM5 and CM 37/3, p. 71.

Sweigart, A. and Mengel, J. 1979, *Ap. J.*, **229**, 624.

Van Leeuven, F. and Alphenar, P. 1982, ESO Messenger No. 28, p. 15.

Weber, E. and Davis, L. 1967, *Ap. J.*, **148**, 217.

DISCUSSION

Bouvier: H-alpha seems to be a good indicator of rotation for stars on the main sequence. However, it is not at all correlated with rotation during the T Tauri stage. So, one should be careful when attempting to use it as a rotation index.

Kraft: I agree.

Vaiana: X-rays are a good indicator of stellar activity, particularly for open clusters. Admittedly, this is an expensive type of observing, but it can also be a very powerful technique. This was illustrated in fact by the CCD color magnitude diagram you showed for the Pleiades field, which showed the very large number of field stars projected on to the very small number of Pleiades stars when one images the Pleiades regions to faint limits in the optical. The same fields imaged to high sensitivity in the X-rays would show almost entirely just the Pleiades members. And in the X-rays the M dwarfs are nearly as bright as the G and K dwarf Pleiades members. So, X-rays allows you to search for low mass Pleiades members very well.

Kraft: I completely agree. My only comment was the question of using X-rays as surrogates for rotation. There are of course enormously important other types of information to be gained from the X-ray data.

Schatzman: Can you give the reference for the NGC 6752 main sequence abundance anomalies?

Kraft: Yes, it is from a paper by Suntzeff from the Baltimore IAU Symposium on Abundances in Globular Clusters, edited by Mme. Cayrel.

PANEL DISCUSSION: INITIAL ANGULAR MOMENTUM

Moderator: *P. Bodenheimer*

Panel: *J. Bouvier, L. Hartmann, S.E. Strom.*

Bodenheimer: We know that T Tauri stars are all quite slowly rotating compared to the break-up velocity. We also think that all stars, somewhere in their evolution, probably went through a disk evolution stage, even those stars that do not show evidence for disks now. The questions are:

- *What is the role of disks in establishing the initial angular momentum of stars?*

- *Is there any relation between these two facts?*

- *Is there a cause/effect relation?*

- *Do we need the disks to spin down stars?*

It is not obvious that we do. There certainly are ways to spin down stars without disks, as we have heard earlier this week.

Strom: I want to re-emphasize the observational points I made before. At all masses, not just for the T Tauri stars, observations drive you to the conclusion that about half the stars - when you are first able to see them - show evidence of disks. I would conclude that it is likely that all single stars had disks. Secondly, when you look at stars just emerging from cores (e.g. IRS5, HL Tau) and look at objects still in cores (e.g. HH 34, SVS13), the disk masses are 0.1 to 1.0 solar masses, with uncertainties to be sure from the conversion of the 1.3 mm continuum observations to mass. Nevertheless, the disk masses are high. So, combining the fact that most, if not all single stars, are surrounded by disks and that the disk mass is comparable to the mass of the star, I would take the leap of saying that stars are assembled from disks, and if they are assembled from disks, then disks have to be the agent in which you solve the angular momentum problem for stars. Another corollary is that the stars that do and do not show disks (CTT's and WTT's) have the same *vsini*'s, and it would seem to me that they must have solved their angular momentum problem in the same way, and I would presume that is through the disk.

Bodenheimer: One could still say that the disk is a hindrance, and the star spins itself down in spite of the disk. That the disks actually add angular momentum to the star, at least that there is some evidence that might happen.

Hartmann: One thing that is worrisome if you wanted to appeal to a stellar wind to brake the star and you wanted a braking time of 10 Myr or so, is that such a braking mechanism might then prevent the spin–up that we think we need to get the rapid rotators in young clusters. One could appeal to arguments that perhaps it is the interaction of the disk with the star which enhances a stellar wind, but there is not any evidence that

S. Catalano and J. R. Stauffer (eds.), Angular Momentum Evolution of Young Stars, 421–433.

the magnetic activity of classical T Tauri stars, as judged from their X-ray emission for instance, is really much different for the weak-lined T Tauri stars (which generally do not have appreciable disks). It seems to me that it is a simple hypothesis to say that the angular momentum loss through the stellar wind is a process that operates on a time-scale of 100 Myr, enough to allow the stars to spin up as they evolve to the main sequence and then spin down subsequent to that. And, then, try to solve the T Tauri angular momentum problem, which is on a much shorter time-scale, with a completely different mechanism.

Gough: Whether the presence of the disk exacerbates or helps, the solution to the problem presumably depends on the relative rate of contraction of the disk and the star. What is that rate?

Hartmann: Well, we cannot measure that directly, but we see excess luminosity, which we interpret as accretion luminosity, and the accretion rates that Bertout, Bouvier, Basri, Kenyon and my collaborators all get are about 10^{-7} solar masses per year. Basically the accretion luminosity is comparable to the stellar luminosity. So, this presumably indicates an angular momentum deposition onto the star, and if you had that go on for a million years, that would start to spin-up the star.

Gough: I thought that the accretion luminosity came from the redistribution of matter within the disk, and that the disk falling onto the star was not a necessary part of producing that luminosity, and therefore I do not understand how you can make that deduction.

Strom: There is clear evidence of some matter falling on the star (for CTT's) by unknown paths, because you see boundary layer emission. In fact, the accretion rates to which Hartmann refers are not only accretion rates estimated by looking at the infrared excess luminosity produced by the disk itself, but also from boundary layer emission produced as material lands by undetermined paths onto the stellar surface.

Gough: Is this X-ray emission?

Strom: No, this is from the ultraviolet and optical continuum excesses.

Hartmann (showing the model fit to the spectral energy distribution for a classical T Tauri star, see Figure on p. 384): The red excess is presumed to be from the disk, and there is also a blue excess which is presumed to be from stuff falling onto the star - which has been interpreted as boundary layer emission, though perhaps it is not that at all.

Gough: We know that when you have magnetic fields on stars, you get activity. If you are going to have the star connected to the disk by magnetic fields, could you not be getting activity in the disk that produces the ultraviolet excess but it does not actually come from a boundary layer.

Hartmann: People are now beginning to think of magnetospheric models, more like pulsar models, in which this excess comes from stuff crashing onto the star from some distance, but it is coming along a magnetic column and so the angular momentum transfer you can have is obviously much different in some way which I do not really understand than the simple boundary layer model where all the angular momentum goes onto the star from the inner Keplerian disk.

Bodenheimer: Can I ask a question? The angular momentum problem is not during the CTT stage, where these accretion rates of 10^{-7} are derived. It must really come at some stage before the star emerges from the birthline, when it is in the embedded core stage, and the accretion rate then may be even larger.

Hartmann: That is true. There is at least the problem we have with the CTT's, but there could be an even worse problem earlier on.

Bouvier: Most of the spin-down must come before the TT stage because they all rotate slowly. There must also be removal of angular momentum during the TT stage to prevent the stars from spinning up. I think one clue to this mechanism is that since the accretion rates vary from star to star, in some way this mechanism must be self-regulating. As the accretion rate increases, the angular momentum loss rate also increases. Otherwise, the CTTs would be more rapidly rotating than the WTTs.

van't Veer: Do the life-times of the disks depend on the mass of the star, and are the life-times not longer than 10^7 years?

Strom: If by a disk you mean an optically thick accretion disk, then the answer is yes. You can show two things: (1) you can diagnose optically thick disks from the observed infrared excesses, and (2) you can look at measures of accretion. For example, the optical continuum excess, or boundary layer emission, that Hartmann talked about. When the disks go optically thin, the signatures of accretion onto the stellar surface appear to stop, and the time-scales associated with the survival of disks as optically thick structures are 10^7 for T Tauri stars and probably less than 10^6 for more massive stars.

van't Veer: When you want to make planets out of these disk, then you would be in a hurry since planet formation is thought to be rapid. Does any of you have a comment on this?

Bodenheimer: It seems like Jupiter has to be formed very quickly - it is probably the first major planet to be formed. Various pieces of evidence point to that time-scale being about a million years, maybe a bit longer. That means you have to accrete a core of solid particles of up to 20 earth masses in that time. The disk life-time estimates indicate that you have an upper limit for planet formation of about 10 Myr. It seems like runaway is required. If you just take normal accretion rates, you would derive time-scales that are much too long to build up the cores of the giant planets. It probably gets worse as you go from Jupiter to Saturn and the outer planets. To get runaway accretion, you need to have a massive disk. That is one way to solve the problem of rapid planetary accretion times. The difficulty is that if the disk is massive, it also evolves real quick and so the high surface density you start with is going to decrease on a short time-scale, and that will slow down the runaway. This is a conflict which has not been resolved very well yet by the theorists.

Strom: I have to concentrate on the existence of disks and how they affected the angular momentum problem. But one of the implications of my remarks on the B stars was (1) the survival times were very short, less than a million years for the B stars, and (2) that there are extant structures, analogous to the β Pic disk, that are apparent in stars 7-10 Myr old in Sco-Cen and if these IR excesses come from disks, then these disks must be constantly replenished, either from a ring of cold material yet to be detected because we have not yet looked with sufficient sensitivity at sub-mm and mm wavelengths, or from collisions among larger bodies in the inner part of the disk. So, the implication is because you have to replenish the grains in an optically thin disk, you may have to have larger bodies. So, you may be driven by the existence of optically thin disks around stars of ages of order 10 Myr to the existence of at least larger grains that collide and produce the small grains which account for the infrared excesses. In this context, it is useful to look at the B and A stars in young open clusters to try to constrain the evolutionary time-scale of optically thin disks. Backman, Stauffer and Witteborn have begun that task using IRAS data. Perhaps Stauffer would like to comment on their program.

Stauffer: As well as you can do at 12 microns with IRAS, and being optimistic as to what is a real detection and what is not, there is something like 20 percent of the A stars in Alpha Per and the Pleiades which are detected by IRAS at 12 microns above what you would expect from a photosphere. So, there appears to be that much in the way of remnant disks around A stars at 50-70 Myr. The F stars are too faint for IRAS.

Strom: Those are optically thin disks with the kinds of excesses you are looking at. While Stauffer's description made the excesses sound modest - and they are in some sense, at 30-40 percent above the photospheric level at 12 microns - those excesses are well greater than the excess shown by β Pic at 12 microns.

Lamzin: I see two problems. As I know, the typical mass loss rate for T Tauri stars is of the order of 10^{-7} solar masses per year. From other data, we have estimates of typical disk masses of less than 0.1 solar mass for T Tauri. If we divide the second mass by the first, we obtain that all disk mass must disappear in a time that is as short as the T Tauri stage.

Hartmann: Yes. But the masses people derive depend critically on the opacity they use in the sub-millimeter range,and different groups have gotten factors of five larger masses than what you are quoting by using a more normal opacity law for the dust in the disk.

Lamzin: Yesterday, everyone who spoke said that the disk masses for T Tauri stars were of the order of 0.1 solar masses, so I still think there is a problem.

Strom: I think the way I would argue it, given the current precision in the determination of masses from submillimeter continuum measurements, is that just your argument rules out disk masses as small as 0.01 solar masses and drives you in the direction of disk masses of 0.1 to 1.0 solar masses. And the point of my opening remarks was that we should not be thinking about little disks around massive stars, but we should be thinking about stars that are built up from massive disks.

Lamzin: When you speak about a boundary layer, do you just mean a layer where the velocity drops from Keplerian to something characteristic of the rotational velocity of the surface of T Tauri stars?

Hartmann: I think now, all we are saying is that we are just referring to this blue excess, whatever region makes it, which may not be a boundary layer.

Lamzin: Yes, but when you use this term, what do you mean now? Only a hot region? Or, a region where the velocity decreases from Keplerian to something much slower?

Hartmann: Observationally, the first thing we thought of was the boundary layer because that was the Lynden-Bell and Pringle prediction, so that was what we called it. Now, all we can say for sure is that we see this excess, we cannot tell what its geometric structure is. It is consistent with it just being a hot region covering a small fraction of the surface of the star.

Strom: Hot means 8-10 thousand degrees for a T Tauri star, covering just a few percent of the star.

Sofia: There is an interesting consequence of all this, and it is that material is in the accretion disk when it is supercritical and it is not when it is subcritical, so the transition is when it becomes just the critical value. Below that, it just falls onto the star. That means that when we discover the magic mechanism that slows down, we will not have any problem with initial angular momentum because by definition if the star is formed by accretion it starts out with material which is just critical. So, the minute we understand that mechanism, we will not have to worry about what differences to begin with.

Strom: That is right.

Bodenheimer: Let us go on with the next question:

- *Why do we have the Kraft Law, and when is it established?*

Bouvier: (referring to a viewgraph showing J/M versus M for PMS stars) This is again the total angular momentum as a function of mass for pre-main sequence stars, and this shows that at least when a star first becomes visible as a T Tauri or as a Herbig Ae/Be star for high mass stars, they already obey the Kraft law. Now, about the spread about the relation. Is it intrinsic, or is it due to uncertainties? First, there is an uncertainty due to *vsini*, but I do not think it contributes much to the spread because when you calculate the angular momentum from the rotation periods instead of *vsini*, you get the same spread. The other source of uncertainty is of course the stellar masses, which appear in both axes of my figure. Consider the HR diagram for these stars. The masses are estimated by comparing the position of the T Tauri stars to evolutionary tracks computed by theoretical models. Our evolutionary tracks are taken from the 1965 paper by Iben. There are other evolutionary tracks that are somewhat different from these, for example those by Vandenberg which tend to be shifted from Iben's. So, depending on which set

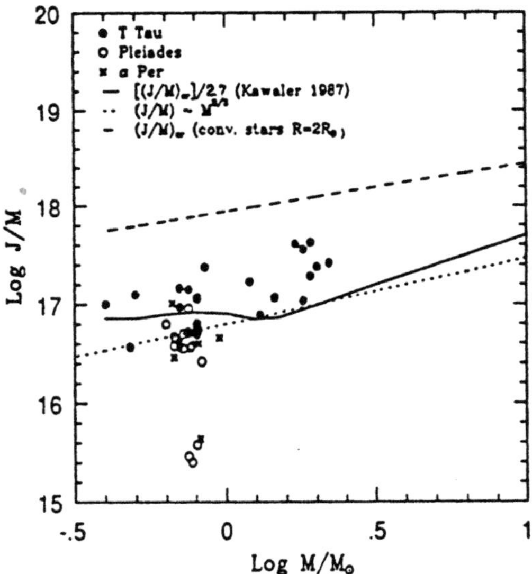

Figure A: Specific angular momentum as a function of mass for T Tauri stars and young K stars in the Pleiades and α Persei clusters, based on photometric rotational periods and rigid rotation assumption. Different angular momentum dependences are shown for comparison.

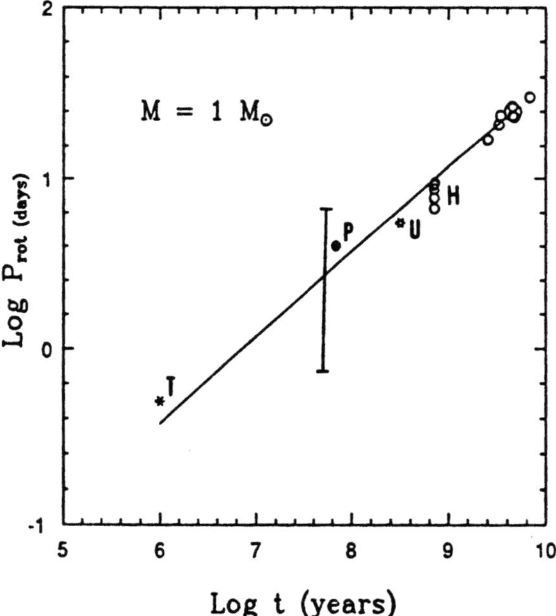

Figure B: Observed rotation periods versus age for one solar mass stars. Symbols: T - T Tauri; α - Alpha Persei; U - Ursa Major; H - Hyades; open circles - field stars. The value for T Tauri stars is an average value scaled for the momentum of inertia change to the main sequence.

of tracks you use, you will get somewhat different masses. However, in general the tracks are shifted as a group, so relative masses should not be greatly in error. I also want to point out that, for example, there is one rapid rotator on the right hand side of the HR diagram - so, presumably it is a very low mass star, yet it is a rapid rotator. There are more massive stars which are near the birth line but are very slow rotators. So, I think that much of the spread in rotation at a given inferred mass is real, and indicates a spread in the initial angular momentum of pre-main sequence stars. Let me note that the Kraft law can be reformulated in other terms, in particular into a relation with break-up velocity. This indicates that T Tauri stars are rotating at about 1/10 breakup velocity, while massive pre-main sequence stars are rotating at about 1/3 breakup. Why this is true, I do not know.

Catalano: For a different representation of the angular momentum of pre-main sequence stars, I plotted the specific angular momentum of T Tauri stars and some Pleiades and Alpha Persei stars as a function of mass(Figure A). I assumed solid body rotation and used only those stars for which period determinations have been made, so the largest uncertainty is with the radius of the star. I estimated radii in two ways - first, with the Barnes-Evans relation, and then as a check using the known *vsini* determinations to infer radii. So, the J/M values should be well defined. I plotted these J/M's versus mass inferred from the HR diagram positions of these stars and standard evolutionary tracks. I then compared the positions of these stars to the Kraft Law, to Kawaler's suggested revision to the Kraft Law, and to break-up velocity, all computed for solid body rotation. The result is that the T Tauri stars are between break-up and the Kraft law. Notice that the Pleiades and Alpha Persei rapid rotators are at the lower value for the T Tauri star J/M's, and if we computed J/M for the slow rotators they would be even lower. Therefore, we can infer that there has been some angular momentum loss between the T Tauri stage and the open cluster ages.

We can make a similar comparison by plotting the rotational periods of the low mass open cluster stars along with periods for T Tauri stars if they were evolved to the main sequence with no angular momentum loss versus the ages of these stars(Figure B). One can extrapolate from the position of the T Tauri stars to the open cluster age by use of the Skumanich relation. Such a diagram could be used to infer the amount of internal rotation that the slowly rotating low mass open cluster stars have retained in their cores.

Roxburgh: Given that just putting in break-up produces a power law similar to that for the Kraft law, is there any significance to the precise exponent used? It is not clear to me that J/M proportional to M squared would not be just as good.

Bouvier: You are right that there is considerable latitude as to the exact exponent you could use to fit the T Tauri data.

Palla: Just a comment about the difference between the rotational velocities observed for the more massive pre-main sequence stars compared to the T Tauri stars. Could that just be due to a mismatch in the relative accretion rates and the outflow rates determined for the two classes? That is, the inferred accretion rates for the T Tauris are of the same order as the mass loss rates. In other terms, there is a factor f between the accretion rate and the mass loss rate, which is of order unity for T Tauri stars. In the other case, and this is suggested by the observations alluded to by Strom yesterday, it seems that the accretion rates are much higher for high mass PMS stars, while the mass loss rates are comparable to those for T Tauri stars. So, the factor f is much larger. Perhaps the accretion rate for the more massive stars is so high that there is essentially no time to slow them down significantly.

Strom: My only answer is that I do not have a lot of confidence in the mass loss rates to factors of 10.

Palla: It becomes crucial then to improve those estimates.

Bodenheimer: And now go on with the problem of the initial angular momentum.

- *What determines the spread in the initial angular momentum of stars of a given mass?*

- *Is the initial angular momentum distribution a function of the environment?*

Bodenheimer: It is hard to believe it is a function of the environment because there is such a large difference between the J/M in the molecular cloud cores and the stars when they first become visible. I think it is likely that there is essentially a loss of memory about what happens between the molecular cloud and protostar stage and when a star arrives at the birth-line. It seems to me it must have to do with the spin-down mechanism itself, rather than with initial conditions in the molecular cloud.

Strom: Unfortunately, as far as I am aware, there is no observational data that allow you to compare the frequency with which disks appear - you could either use direct measures such as IR excesses or indirect proxies such as Hα luminosity - but so far as I know, there is no complete set of such data for the Trapezium cluster itself which you could compare to similar data for other pre-main sequence populations where the stellar densities are much lower. I think this would be extremely interesting. I made a very simple estimate of the time-scale for a 100 AU target (a disk, for instance) to collide with another 100 AU target in a cluster as dense as the Trapezium, and the number I got was 20 Myr. I assume the time-scale for a tidal interaction would be considerably shorter. Do these interactions between disks take place on a time-scale short compared to the time-scale in which the angular momentum distribution is set up in regions like the Trapezium cluster? Unfortunately, no such data is available yet.

Stauffer: Would you care to make a prediction that there will be a significant difference between the Trapezium rotational velocities and those for stars in much less dense regions?

Strom: There are already claims in the literature that B stars in some portions of Orion do differ from each other. In particular, the Orion Id region (which includes the Trapezium cluster) shows no slowly rotating B stars, whereas the Orion Ic region shows quite a few. A similar distinction is found for the Sco-Cen region, in which the upper Scorpius region contains only rapidly rotating B stars, whereas the Centaurus region contains both rapid and slow rotating B stars. The question is, is that built in or the result of angular momentum evolution among the B stars?

Bodenheimer: Is it clear from the observations that there is any correlation between mean rotational velocities and the density of the region from which the stars were formed?

Strom: That was claimed in Wolff, Edwards and Preston, but I do not think the data are sufficiently compelling to support that.

Pinsonneault: Are the cluster-to-cluster differences you mentioned statistically significant?

Strom: That was the claim in Guthrie's Monthly Notices papers.

Lamzin: You are trying to examine the angular momentum distribution function in the same way that we try to determine the stellar mass function. However, there is a big difference. For the mass function, you have a scalar quantity, whereas for the angular momentum function you have a vector quantity - but you only measure a projected rotation generally. I thus think it is hopeless to determine the true angular momentum distribution function given that you only measure projected values. Can you comment?

Bouvier: The best way to get the inclination angle is to measure both the *vsini* and the rotation period (from photometric variability). What we lack most now are more rotational periods. We are working on this problem. The present indications, however, are that for large groups of stars the axial inclination distribution is random and thus one can infer the true angular momentum distribution from the *vsini* measures.

van't Veer: Is there any evidence for non-random alignment of inclination angles?

Strom: The relevant thing to look for is actually magnetic souvenirs, because there are at least some regions within molecular clouds where there appear to be very strong alignments (in projection of course) among jets presumably emanating from the poles of circumstellar disks. So, if the jets are surrogates for the projected directions of the angular momentum axis, then at least very early on there are some regions of clouds in which the degree of alignment among the jets is very high. So, there may be some indication that at least in some molecular clouds there is a high degree of alignment initially of angular momentum vectors.

Bodenheimer: Over what length scales are you talking about and how good is this alignment?

Strom: Well, for example, a GMC like L1641, where there are now 10-12 outflows seen and you can define jet position angles fairly well, they lie within 30 degrees of the same position angle in a region which is 10's of parsecs in size.

Vauclair: What is your present estimation of the dispersion of the initial angular momentum in young stars and do you think it should have been the same for Pop II stars?

Hartmann: My guess is that there is a factor of five spread in initial angular momentum. I have no idea what the spread might be for Pop II, particularly because we have no idea as to what causes the spread for Pop I.

Bouvier: It is hard to accurately assess the possible spread because one has to account for observational errors, an incomplete knowledge of the inclination angles, and we also cannot be certain what masses to assign to the stars.

Hartmann: Another thing to consider is that there are even problems when we look at the main sequence where we have better statistics. It is hard to understand the large spread in rotation seen among the Pleiades low mass stars. Either you have to have a stochastically random variation in the angular momentum loss rates from star to star, or an age spread (which we do not believe), or say that there is some seed initial angular momentum distribution - and I think the last idea is the best hypothesis which is consistent with the observations.

Strom: Unfortunately, no one has talked about the high mass stars so far, and I am not prepared to discuss them adequately myself. However, my reading of Guthrie's papers is that he claimed to find a very small spread in the rotational velocities for the early B stars in at least several of the OB associations where he looked. And that seems consistent with what Bouvier has shown for the Ae/Be stars. So, the distribution of *vsini*'s for the main sequence early B stars in at least a number of associations seems consistent with a single value of rotational velocity, viewed at random inclination angles. (to Bob Kraft) - Have I described Guthrie's work accurately?

Kraft: I think you have - I was going to ask why the dispersion in angular momentum among the low mass stars seemed so much higher than for the high mass stars. The only thing I would have to say is that you have to be cautious in interpreting the old rotation papers, because there can be very large systematic effects when you compare the rotational velocities derived by one group with another.

Mayor: First, a comment to Vauclair. It is certainly true that we cannot say much about the rotation of Pop II stars, but perhaps we can infer something from their duplicity. If you have some relation between fragmentation and the distribution of orbital elements of stars, then you can find some constraint from this point. For many years, people believed that the duplicity rate of halo stars was completely different from that for Pop I stars. In the last few years, this belief has completely changed and apparently recent measurements no longer indicate any strong difference between the two. So, this is perhaps weak evidence that their initial rotations may also have been similar.

My second point is a question to Hartmann and Bouvier. Do you have difficulty in measuring the *vsini* of some T Tauri stars? That is, are your *vsini* distributions for T Tauri stars complete or are you possibly missing something? This is important if we have to compare the T Tauri distribution to that for older stars.

Hartmann: It is difficult to measure the rotation of very strong emission line stars, but we have now some estimates with CCD's and long exposure times for some very strong emission stars. We cannot measure it too well, but they seem to be of the order of 20 km/s for DG Tau, DR Tau is about 10 km/s, and HL Tau is also narrow lines. I think we are pretty close to being complete in Taurus, at least to 15th magnitude. There is no large incompleteness.

Kraft: How do you handle the continuous radiation in such stars?

Hartmann: When you do a cross-correlation analysis, the zero level does not really matter. You have to excise the regions with strong emission lines. You might worry about this somewhat, but it seems to work since you get the same answer as for the other stars.

Kraft: Can I come back to the Pop II business, because it raises some issues related to the environmental question that was discussed earlier. You could also ask if the rotational velocities of field halo stars differ from those for stars in globular clusters. In globular clusters, encounters between stars are frequent, and it might be possible for these stars to be spun up by the encounters. Then you might find a different rotational distribution for field halo stars compared to globular cluster members.

Bodenheimer: Let us proceed to the last set of questions.

- *What is the realistic accuracy in our assignment of masses and ages for PMS stars?*

- *What is the effect of accretion on PMS tracks?*

Hartmann: If you look at an HR diagram for Taurus where we believe we are reasonably complete and overlay some representative PMS tracks, you can see discrete clumping in the horizontal (effective temperature) axis which is not real but is due to the discretization produced by spectral typing. Spectral type errors alone of 1-2 subclasses can then lead to errors in the assigned mass of 0.1 to 0.2 solar masses. The luminosities are much less certain that the effective temperatures, but they do not tend to matter too much in terms of the mass determination if the theoretical tracks are correct because the tracks are nearly vertical (at typical T Tauri ages).

What I think is a more serious problem are the systematic errors that may occur in these theoretical models because they depend sensitively on the treatment of the outer envelope. The curvature of the PMS tracks in the log L, log T(eff) plane for Mazzitelli's models is different from that for the other tracks shown during this meeting, and so just the uncertainty in the shape of the tracks may lead to a possible error of 0.1 to 0.2 solar masses. Then there is also the question of an overall shift in the horizontal location of the tracks - which is possible according to the model makers, which could introduce an additional error of 0.1 to 0.2 solar masses.

There is another reason to consider the possibility that our mass estimates for PMS stars may be systematically wrong, and that is that there is an age problem in Taurus. The crossing time for Taurus is of order 10^7 years, but we do not seem to have very many 10^7 year old stars as judged by the canonical isochrones. This could have been an observational selection effect, so a group of us (Hartmann, Jones, Stauffer and Kenyon) decided to conduct an unbiased proper motion survey in the central region of Taurus and to look for these old stars. We found some new Taurus members, but they fell just where the previously known members had in terms of their inferred ages - we did not identify any older members, despite our survey being designed to find them. The natural question to ask then is whether the Hayashi track ages are correct. If we could make the ages of the known Taurus members closer to 10^7 years, then we would solve our program. Since on a Hayashi track, the age is proportional to a Kelvin time and so the age depends sensitively on the mass - if we could change the inferred mass by a factor of 2 or 3, we could modify the ages significantly.

There is another possible effect, which is that perhaps the Hayashi tracks are not right at all. Bodenheimer will address this question.

Bodenheimer: It all depends on what fraction of the luminosity of a star that is measured is due to accretion luminosity and what fraction is due to the intrinsic stellar core. If the accretion luminosity is significant, then one could be quite a bit off in the age determination by the conventional method. The results of a theoretical calculation done by Guy Stringfellow at Santa Cruz illustrate this. This is a model for a low mass star accreting mass at a constant rate starting at 0.1 solar masses and continuing to accrete up to 1.0 solar masses. The idea for the model is that the accretion takes place in a disk, so the effective temperature is just the photospheric temperature of the star (there is no outer envelope of optically thick material which might reprocess the photospheric light. Stringfellow considered accretion rates of 10^{-5} to 10^{-7} solar masses per year. Even at 10^{-6} solar masses per year, the accretion luminosity completely dominates over the intrinsic "core" luminosity of the star, and the ages you would infer would be completely different from standard Hayashi track ages.

If you compare observed stars in Taurus to Stringfellow's model, you see that an accretion rates of about 10^{-7} solar masses per year fits the data pretty well (which agrees with other inferences about accretion rates for T Tauri stars.

Hartmann (to Bodenheimer): At any given instant, if you had a good effective temperature, would that give you the right mass if you compared to Hayashi tracks?

Bodenheimer: The stars migrate from lower mass Hayashi tracks up to higher mass Hayashi tracks, so a good effective temperature would give you a pretty good mass estimate at that point in time.

Roxburgh: One has to worry about the initial conditions, it seems to me, because the contraction time-scale is about the same as the thermal readjustment time-scale, so where you start a star determines its evolutionary track. You started your star on a Hayashi track with a given mass. If you took some other initial condition, you would get a considerably different result.

Bodenheimer: Once we pick an initial mass, about the only initial parameter we set is the initial radius, which determines where we fall on that track.

Roxburgh: But you could start the model of the star such that it is not in thermal equilibrium - for example it might have a radiative core but still be very young, and then its evolutionary track will not be like a Hayashi track.

Bodenheimer: These stars are started at low mass - about one tenth of a solar mass, and they definitely have to be on Hayashi tracks then.

Roxburgh: That is because you are placing them on Hayashi tracks.

Bodenheimer: We are placing them on it because when a star of one tenth of a solar mass has that luminosity, it would be fully convective.

Hartmann (to Roxburgh): If you had a star which had whatever conditions you want, and just let it go would that go to a Hayashi track on a thermal time-scale?

Roxburgh: No!

Bodenheimer: Yes!

Roxburgh: I think these two statements are reconcilable. The question is where it would hit the Hayashi track.

Strom: These concerns about relative ages and masses may be addressable observationally because there exist binary pairs in star forming regions. One can place the stars in those pairs in HR diagrams and ask whether those stars fall on conventional isochrones or require some very different evolutionary path. There are about 25-30 accessible pairs (almost certainly physical pairs) and these should let us test how bad the assumptions of the conventional models are.

Pinsonneault: Speaking as one of the modelers, you can indeed shift these tracks quite considerably - I do not even think that a factor of two error in the mass would be too much of a problem. You could make a 0.5 solar mass stars really a 1.0 solar mass star. You cannot play with the luminosities much though, so at some level you have limits. There are uncertainties in the mixing length theory, in molecular opacities, and even atomic opacities. These can be large effects.

PANEL DISCUSSION: DYNAMOS AND INTERNAL ROTATION

Moderator: *E. Schatzman*

Panel: *D. Gough, K. MacGregor, S. Sofia, D. Gray, I. Roxburgh, and L. Paternò.*

Schatzman: Let us start with the first question:

- *Is there a dynamo, and if so, how does it work? Also, is there a magnetic field in the core of the Sun?*

Gough: I have not a clue if there is a dynamo. I said in my talk that there seems to be little evidence of one. In stars, the fossil fields last for typically the main sequence lifetime or even longer. However, the processes that dynamo theorists discuss, particularly in turbulent convective regions the magneto-hydrodynamic processes that twist the field, regenerate it, augment it, diffuse it and everything else - these processes, which have become known as the dynamo process even though they do not necessarily produce dynamos, are no doubt going on in the convection zones of all stars. As some of us have argued, all stars contain magnetic fields; all stars contain convection zones; and therefore, all stars contain regions in which convection is stretching the magnetic field lines and augmenting the magnetic fields. Whether or not the augmentation is greater than the decay determines whether or not there is a dynamo. I do not believe there is any compelling evidence - either for or against the hypothesis that stars actually do this. On the other hand, I do not think that it really matters because the time-scales are so long. What does matter is that these processes are taking place. And it is these processes that determine what the star looks like on the outside and determines to some extent (though perhaps not completely) what field there is on the outside - both magnitude and geometry - and therefore the extent to which braking takes place. Therefore, I think it is perhaps a red herring to ask whether there is a dynamo. What you should really ask is what the star is doing to the magnetic field that it possesses.

Sofia: There are several issues here. One is, do you believe in any of the proposed dynamo processes and what is their ultimate role? Are they the ultimate source of the magnetic energy? Undoubtedly there are regions with a high magnetic field inside, and whether they leak out or not has not been empirically established. But one thing that has been established during analysis of the last several solar cycles is that the poloidal field which existed at the previous minimum is very well linearly related to any measure of the activity at the following maximum. In a simple model view of any dynamo in which a source poloidal field is twisted into a toroidal field, such a relation is to be expected and it works quite well. So, there must be a grain of truth there, just because it keeps repeating itself. Whether or not there is a further source of magnetic energy deeper in the star or whether the amplification is done elsewhere somehow, or whether any specific dynamo mechanism is at work, I do not know.

Paternò: As far as the dynamo is concerned, one has to distinguish between the classical

S. Catalano and J. R. Stauffer (eds.), Angular Momentum Evolution of Young Stars, 435–450.

alpha-omega dynamo (which is a linear model), and the attempts to model the problem in a non-linear way. The alpha-omega dynamo works well to explain most things, but fails in a few cases probably. In particular, it cannot predict the strength of the magnetic field. The non-linear models are certainly more correct and can explain also the period in which there is no activity. The major problem with the classical alpha-omega dynamo is that the period of the cycle cannot be reproduced unless the magnetic diffusivity is set very low. The reason for that is that the field tends to be buoyant as soon as it is amplified. Therefore, it is difficult to think that the classical alpha-omega dynamo can work, even in the deepest part of the convective zone. A possible alternative to that is that the field is amplified in the overshooting layer below the convective zone. Probably the overshooting layer does exist, because many meteorological phenomena indicate that on the Earth such a penetration of convection into the stable layers can be possible. The problem is to intensify the field in order to exceed greatly the equipartition value.

Generally the most unstable mode, even considering the spherical geometry, is a toroidal mode. The only possibility is a toroidal mode which becomes unstable before the field exceeds the equipartition value. It is possible that this toroidal mode will slip towards higher latitudes; possibly the magnetic tension is sufficient to keep this torus inside, to be amplified, and to emerge as a flux tube.

In my opinion, a magnetic field in the core may be possible. An almost constant magnetic field, over-modulated by some other mechanism, not necessarily in the same core, probably is necessary to explain the fact that the total magnetic flux of the Sun does not change from maximum to minimum periods. Whichever kind of dynamo model you choose it modulates the flux.

Gray: I am a pragmatist. I do not care if there is a dynamo working or not. There is certainly a great deal of observational data connecting magnetic fields and rotation - for example, all of the correlations between Rossby number and activity, not only for main sequence stars but also for evolved stars. It seems to me from the observational point of view that it is a moot point whether these fields are manipulated by the rotation in some way or if a dynamo is involved. Certainly using the dynamo as a model seems to work. If we can take some other mechanism, other than a dynamo, and have it make as good a prediction, fine. Otherwise, it seems to me that we should go ahead and use the dynamo formalism.

Gough: That is just the point in a way. Whether you actually have a dynamo or you just have the same magneto-hydrodynamic processes going on without there being a dynamo, is the question. Something is clearly going on. But we do not know whether the magnetic field is being totally regenerated or if it is just slowly decaying away. And it largely does not matter. So, I agree; let us use those concepts. It is just that the word "dynamo" may be a misnomer, perhaps.

MacGregor: I would adopt a middle ground. The basic physics of the alpha-omega dynamo is the production of a toroidal field in the presence of shear produced by rotation and the injection of helicity through the action of Coriolis force on convecting elements. There may be difficulty getting that mechanism to work as originally planned in the solar convection zone due to the apparently short buoyant tube which would be produced. That difficulty may be alleviated by placing the seat of the dynamo at the interface between the convective zone and the radiative interior and using cyclonic convective overshoot to inject helicity. For that reason, I do not see any overt reason not to believe in the dynamo. By the same token, I see no overt reason to believe in the absence of magnetic flux in the radiative interior of the Sun. Certainly some flux was lost from the Sun during star formation due to ambipolar diffusion. And early in the Sun's pre-main sequence evolution, it may be possible that while in the fully convective state flux can be expelled through turbulent reconnection and the formation of buoyant elements that can subsequently rise to the surface. So, when I say I adopt the middle ground, I see no clear reason to not believe in the existence of fields in the radiative interior and the subsequent maintenance and perhaps even amplification of these fields by dynamo activity.

Schatzman: You mean in the convective zone or everywhere?

MacGregor: I mean at the interface between the convective zone and the radiative core. Dynamo modeling is only now in its infancy. We are only just beginning to be able to treat non-linear dynamos, and perhaps at the advent of this era of being able to do non-linear dynamo modeling we should seek to pose questions which can be illuminated by the observations. That is, what are the aspects of a fully non-linear dynamo that are amenable to some observational test?

Roxburgh: Long ago, Dennis Schommer accused me of not believing in anything. And this is a fair statement, so I certainly do not have an active faith. What I would like to say though, is that the dynamo theories, which are used, are turbulent mean field dynamos, and they are cookbooks - so let us not pretend that we are really modeling the physics here any more than we are when we use a mixing length theory for the convective zone. We are sweeping the unknowns into some parameterization and then adjusting the parameters to get the answers that we want, and even that is turning out to be more difficult than expected. So, I would treat with caution any extrapolation from models that have been adjusted on uncertain physics to try to fit the Sun to predict what would happen in other situations. I think the theoretical grounds are too weak to justify such an extrapolation. I would put much more faith in what inferences can be drawn from the observations rather than from theoretical extrapolation. Like Gough, I do not care whether one calls it a dynamo or not. As I said earlier, I do not think that you can get rid of a magnetic field in the core of the Sun, and if you did, another one would be regenerated.

Schatzman: Let me add that from the point of view of people working in magnetohydrodynamics, there still remains to understand exactly the way in which the reconnection of lines of force takes place at a speed which is sufficiently large to match the general description which is obtained in the mean field electrodynamics. So, I must say that this question of the dynamo from basic aspects is not yet solved, even if we have to accept that there is some system producing magnetic field.

Gough: Let me just make a remark to support the Yale school - this is principally to show that there is not a British school - and that is that if you can introduce enough parameters to describe an elephant and then if you can make it walk without introducing another parameter, then perhaps you are getting somewhere.

Roxburgh: Maybe the elephant has three legs, but it can still walk. It might be a bit dangerous to extrapolate that all elephants have three legs.

Gough: It might be a spherical elephant.

Schatzman: I think that, related to this discussion, we need to consider other questions:

- *How the magnetic field topology is finally responsible for the amount of angular momentum loss from the star?*

- *How is it related to the total number of open lines of force which leave the surface of the star?*

Does anyone have any comments on this?

Roxburgh: Some years ago, I looked at the consequences of using dipole and quadrupole and in principle other multipole fields. We have heard a lot of people use the Weber-Davis radial field model here. But it is not the case that all things which are true for that model remain true if you go to different field topologies. If you go to a higher multipole field, then the angular momentum loss rate depends on parameters that it does not depend on in the radial case, and in general the momentum loss rate decreases as something like the square of the multipole order. So, the angular momentum loss rate does depend critically on the field topology.

MacGregor: I am not even sure what the angular momentum loss rate would be in even just the two dimensional extension of the Weber-Davis model. The Weber-Davis model is based on a solution obtained in the equatorial plane and an extrapolation of that solution to higher latitudes. It relies on, for example, assuming no meridional dependence of the mass flux density. There are things which almost certainly would change if I did a radial field model and allowed the flow to have a meridional component as well as azimuthal and radial components. In the Weber-Davis model, if extended out of the equatorial plane, the largest component of the azimuthal field would certainly be in the rotational equatorial plane and diminish towards the pole, and the resulting gradient in the azimuthal field must produce a magnetic pressure force which would cause the flow to diverge out of the rotational equatorial plane. I do not have any clear conception of how the flow topology and rate of angular momentum loss are affected by things like that in even the simplest case.

Roxburgh: There has been quite a lot of work on this, in fact we heard a talk today where this was being done for dipole fields, and indeed the Neumann-Kopp solution in the first place was extended to include rotation at least in the domain where it is valid to add that on. You can do similar sorts of things with different structural bases . Of course, what you say are the base conditions, and the mass loss per flux tube is undetermined. But you could do it by saying, for example, that it is an isothermal case as for Neumann-Kopp. Robertson did a similar sort of thing for a conducting model of the solar wind as well.

Gray: My comment from an observer's point of view is that what few magnetic field measurements we have so far are for dwarfs, and this is because the spectral lines are narrow enough to let us just detect the Zeeman broadening. Those fields come out to be typically 1-2 kilogauss. If you assume that that is the kind of field that is involved for giants as well, then certainly the open field line structures need only be a very small fraction of that field - a few tens of gauss in order to supply the braking that is apparently needed to dissipate the angular momentum. I have not done the calculations for dwarfs, but I suspect it is also a small fraction of the surface field that is involved. So, whatever kind of topology is involved, we do not need to have it all open field line structure or anything like that, we just need a few open field lines compared to the total.

Paternò: With the premise that the magnetic field of the Sun is much more complicated than what one can imagine with a dipole or quadripole field, I constructed models of angular momentum loss rates for quadrupole field, as a functions of field topology in the same way as Roxburgh did, and my experience was that the mass and the angular momentum loss rates for a quadrupole mode are much less than for a dipole mode. This is in apparent contradiction with the supposed observation that the great majority of field lines are closed. Maybe it is true, because it is observed, but if you do the calculations, dipole modes lose a lot of mass and angular momentum, and quadrupole modes lose much less (perhaps by a factor of 10 or so).

Roxburgh: I agree, the smaller the fraction of the surface that is open, the more rapidly the field lines diverge, and that is why.

Sofia: I have always been scared of "beatnik" magnetic fields, and that seems to be to a large extent the observed topology - "hairy" magnetic fields. Somehow with my understanding of Maxwell's equations they should not be there. In particular, at the end of a cycle for a few months you can have a south pole at both ends of the Sun. The only thing that does not scare me is that in order to recover the Skumanich relationship our modeling requires $n = 1.5$, which is not quite a radial configuration but rather some sort of dipole. How this relates to the observations of magnetic fields on a star has to be taken with a bit of caution because very high polarity fields may be very strong but they die out and it is the open field lines that do the damage, and those may be weaker.

Cameron: One point which I think people often fail to take into account is that these fields are not static. We talk about dipole and quadrupole fields, and assume that closed field regions are dead. But, if you look at the Sun, you will see that any closed field region tend to evolve towards progressively larger loop structures, which then burst open, releasing what was formerly closed field lines into open field regions of the wind. So, if you take a time average of a stellar atmosphere it may well be that the time averaged flux that is emerging from the surface of the star will at some stage in its career get to be part of the open field. I think that when this effect is taken into account, it will significantly alter the effectiveness of closed field regions at reducing the efficiency of the wind.

Gray: But there are also some regions that are seen to go back in.

Cameron: On very small scales, yes. But, by and large you tend to get an evolution towards larger and larger loop structures and we are now getting to the stage where we can see this sort of thing going on in some very rapidly rotating field stars where we can do Doppler imaging above the surface - so again we tend to see this feature of loop structures evolving towards larger sizes and apparently then bursting into open field structures.

Gough: Just a small point concerning the field topology elsewhere, other than the visible regions of the one star in which we have detailed observations. And that is that, in any star where we believe that at the base of the convective region there is some shear, the so-called dynamo theories imagine that first of all field is pushed down to the base of the convection zone, it is then spread out into a toroidal field loop by the shear, and then somehow rises up buoyantly to the surface. If however the field is a remnant field from the interior core, and it is slowly emerging and getting torn off by the shear, it too will be wound up at the base of the convective zone in a toroidal loop; then, whatever magnetohydrodynamical processes the dynamo theorists talk about, buoyancy will then take over and make it rise to the surface. I say this to point out that whatever you see at the surface for the two ideas are probably very similar and this makes it very difficult to determine which is the dominant effect.

Hartmann: Can I ask Gough to explain to me simply - if it is a remnant field then how do you get the change of polarity in the two cycles?

Gough: Well, I do not know the answer to that question, but then I do not really know the answer if it takes place in the convective zone alone. As Roxburgh points out, people invent toy equations that have that property, but these were invented for the purposes of producing the result.

Gray: If it is a remnant field, why does it correlate with rotation?

Gough: I know of no observation that correlates the interior field of a star with the rotation of the photosphere.

Schatzman: Now for the 3rd question:

- *What inferences can one draw regarding stellar activity and magnetic fields from observations of the correlation of activity and rotation with age?*

Catalano: I will make a comment on the observations of main sequence stars and angular momentum loss. The observations of rotational velocities of main sequence dwarfs with outer convective zones indicate that the decrease in their rotational velocities with time is a steady process. The Skumanich relation describes this process, though we found a different rate of rotational decay for different stellar masses. The rate increases as the mass of the star decreases. This is understandable in terms of angular momentum loss from the entire star, not just from the outer convective envelope. Probably there may be a central region where rapid rotation is maintained, as suggested by helioseismology observations for the Sun, but this region contains only a very small fraction of the total angular momentum of the star and so it is a good approximation to just consider that the entire star is being spun down as a whole.

Paternò: A comment from a theoretical point of view concerning mass loss as a function of time from arrival on the main sequence to the red giant phase. What happens early in a star's life is that the mass loss rate is small, but the angular momentum loss rate is very high. As the star leaves the main sequence and evolves to the red giant branch, exactly the opposite is true. The angular momentum loss rate is small, but the mass loss rate is extremely large. These are the results of a theoretical model, and they agree with the observations both for the mass loss rate and the terminal wind velocity. The mechanism which determines this fact is essentially the position of the Alfven radius, which is far from the star to begin with and inside the star at the end of the evolution (at the helium flash).

Schatzman: I would like to get to the second series of items.

- *Let us start with the question of the magnetic field in the radiative zone, and what its relation to the magnetic field in the convective zone might be?*

Roxburgh: The naive answer to the first question which you have asked is of course that the magnetic field has the same sort of Von Zeipel problems as rotation. Any general departure from spherical symmetry has the same sort of effect. The magnetic field would need to be larger than the sort of numbers we have talked about if it is to provide a perturbing force greater than the rotational perturbing force. So, if you have a reasonably strong magnetic field then of course it has the same sort of problems.

Schatzman: Do you believe that the interior magnetic field, which is strong enough and has the right geometry, could prevent meridional circulation?

Roxburgh: Of course, that is possible. It relates to what I was going to say, which is that deep inside the star, where in principle meridional circulation might take place, the circulation, which we now think of from a long time ago work by Leon Mestel, redistributes the chemical composition such that it stops the circulation itself. You only need a marginal readjustment in chemical composition over horizontal surfaces to stop the circulation. If you started from the beginning and said: is there a distribution of angular velocity and chemical composition gradients such that there is no circulation, then the answer is yes. If you ask: is there a combination of rotation and magnetic fields such that there is no circulation? The answer to that is also yes. In fact, I think that was the first problem that Leon Mestel gave me as a research student. But it had been elaborated in considerable detail previously by Dave Moss and others. Whether or not you evolve to such a system, it is a different issue. I do not know the answer to that. In fact, my guess would be that in the layers beneath the convective zone, if there is a comparable strength magnetic field, there is some meridional circulation, and it is broken down by horizontal shear turbulence in the same way everything else is.

MacGregor: As I said previously, I see no reason why there should not be a magnetic field in the radiative interior of the Sun and other solar type stars. As Roxburgh pointed out in his talk, a pre-existing poloidal field when acted upon by rotational shear will generate a toroidal field which has associated with it a current which has a component perpendicular to the original poloidal field, so there is a Lorentz force which tends to act against the pre-existing shear in the rotation.

One question that I have, is what about dissipation in that system? We have talked about magnetic oscillations in the sense of differential rotation, but I recall some papers by Taylor that dealt with the stability of purely poloidal and toroidal fields in radiative stars, pointing out that such fields configurations were, on an energy principle analysis, apparently susceptible to instabilities of the sausage or kink type. Does that really happen in a rotating, radiative interior and does that provide the dissipation that will be required to actually transport angular momentum and dump it some place?

Roxburgh: The first remark on that is that no stable magnetic field structure is known on the inside of a star. What is known I think is that you have to have a linked toroidal and poloidal field, and that there are arguments on the conservation of total helicity that demonstrate that there has to be some non-zero lowest energy state, but no one has found one as far as I am aware.

Gray: As an honest observer, I cannot see into the inside of a star. So, I really do not know if there are magnetic fields in the radiative cores of stars. But, as a philosopher, I would say the following: because the weight of opinion now seems to indicate that there is a field, I would suspect there is none.

Roxburgh: Can I make an interjection on something completely different? A long time ago, George Gamow had a poll between 30 cosmologists as to whether the universe was in a steady state or started off with a Big Bang. The voting was 16 for a Big Bang and 14 for a steady state. He had a second question which was: Is this the right way to settle the issue? And the answer was that 30 said no.

Paternò: Just a very short comment. Since I notice that Gough likes dynamos very much, why not put one in the core of the Sun?

Gough: If a dynamo is in the core of the Sun then changes the structure of the Sun, and thus changes the neutrino flux, then that would be great.

Pinsonneault: If the recent results of the Soviet gallium experiment are confirmed, we have a deficit not only of the ^{8}B neutrinos, which one can adjust by playing with the central temperature to some extent, but also we have a deficit of the p-p neutrinos which are not a free parameter in the solar model. So, if this is confirmed, we know we have some sort of exotic neutrino physics going on.

Gough: I make this remark taking a leaf out of Dziembowski's book. Since these preliminary results have not yet been confirmed, and in the very near future will either be confirmed or refuted and we will know the answer, we must quickly take the opportunity to enjoy ourselves before the issue is closed by the observers and make new theoretical models.

Sofia: I would like to describe a very simple calculation. One cannot infer very much about the possibility of wrapping up magnetic fields due to differential rotation in the radiative zone. We ran one of those calculations without taking into account feedback, which means one cannot believe the results. But, one of the interesting things that came out was that the magnetic field amplification was not helter-skelter as one usually visualizes in order to stop rotation but rather was concentrated in very specific tori and areas, and if that is going to be the concentration one can envision differential rotation being locally destroyed but not throughout the core. Because in the calculation we do not have feedback in the Lorentz force, I do not know what is going to happen in the long term. We are going to do the full calculation soon, but we do not have any results yet.

Paternò: Do you include turbulence?

Sofia: Yes, we include turbulence, dissipation, etc. and differential rotation.

Roxburgh: I will wish you good luck, because I have tried that calculation, and it was horrendous and it did not work. You can follow through the evolution but... maybe the Yale school is better than the London school. I gave it up after a while.

Schatzman: I would like to make a comment on the idea of having a dynamo in the radiative zone. In order to have a dynamo, you need a turbulent flow. If you want to have a turbulent flow, it must satisfy certain conditions. One can possibly consider the origin of the turbulent flow from meridional circulation. It turns out that in order to have turbulence, the dynamo number is less than one, whereas to have a dynamo you need a dynamo number larger than one. So that, I personally have doubts about the possibility of regenerating the magnetic field by turbulence produced by circulation. Let us focus on how the magnetic field in the radiative zone and the magnetic field in the convective zone might interact and the way in which angular momentum might be transferred. Does anyone wish to comment further on this? There is a very great difficulty because the diffusivity of the magnetic field is extremely different in the convective zone and the radiative zone, so to connect the two systems would be very difficult.

Roxburgh: There is nothing very constructive I can say, but I will say something anyway. In 1978, I drew a picture of the Sun at the European solar physics meeting, and that had a field inside and a very fuzzy area at the bottom of the convective zone where somehow the field in the middle has to connect to the field in the convective zone. I still do not know how that takes place. If anyone else does, please tell me.

Sofia: Let me tell you how we tried to handle this question in our calculations, because of course if there is no interaction then you cannot do anything. The early idea was postulating just enough overshoot to link one to the other so there were fingers going into the radiative zone. That perhaps was a bit too arbitrary, so we did something equally arbitrary which was to postulate a boundary layer where the velocity was dropping by 8-10 orders of magnitude and we fit some sort of exponential linkage, etc., and said it was a boundary layer. Basically we do not know what happens at the interface.

Rodonò: I have been listening with great attention and reverence to all this discussion on what happens inside the Sun. Gough previously said to Gray that he could not answer some question because the observations could not tell him what the internal magnetic field was. So, let us not ask observers impossible things. What I would like to have happen during the discussion group is for the theoreticians to indicate the observable surface parameters which are of interest. When I go to buy a car, I care about how it looks and how it handles. I do not care what happens inside the engine. Tell us the most important parameters which observers should provide in order to test theories to determine whether there is a dynamo, how turbulence affects things, etc. In the end, you should tell us what comes out at the surface where we can give you some facts.

Sofia: What I always tell observers is that they should continue whatever it is they are doing, but just do it more frequently and more precisely.

Gray: That sounds very personal.

Schatzman: I quite agree with what Rodonò has just said. The problem is that what the observers can observe and what the theorists need do not always match. A very important constraint on dynamo theory, since the first observations by Wilson, has been the good correlation of the rotation period of stars with their chromospheric activity. All the observations of recent years where the stellar rotation is observed through the influence of spots on the stellar luminosity are also very important. But this is only part of the picture. We do not yet know how the magnetic field is brought from the bottom of the convection zone to the surface. We do not understand how the spot pairs evolve and then dissipate, and then something else appears at the same place one or two rotation periods later. There are a number of things which we do not understand, there is much still to be done, but right now I do not see what new specific things theoreticians can suggest to observers other than to keep working.

Gray: To go back to the coupling between the two zones, has anybody actually done the calculation that would tell if you only have magnetic coupling between the two zones, how large would that field be expected to be (for it to rotationally link the two)?

Gough: That depends on what the time-scale is and the topology of the field. The field winds up two π revolutions per sunspot cycle. So, therefore you can now estimate the torque with ease after a certain amount of time. Roughly speaking, you get a time-scale of the order of the age of the Sun with an initial field of a microgauss.

Schatzman: I would like to raise a question on the exchange of angular momentum in binary stars. Can you think that instead of tidal effects, namely the dissipation which is related to the internal waves generated by the tidal effects, the internal magnetic field can play a role too, by helping to produce the circularization and the synchronization?

MacGregor: What you are wondering about is whether the tidally-induced shear flow due to the gravitational interaction would further interact with the magnetic field in the interior.

Schatzman: I am not so precise in what I am saying. I just compare the problem of the transfer of angular momentum from the convection zone to the radiative zone through possible magnetic fields with the question of the transfer of angular momentum which is induced by the tidal effects of a companion in a short period binary system.

Does the magnetic field play a role there? Because if it plays a role inside a rotating star, perhaps it would play a role in the binary.

Roxburgh: You mean having the principle effect being dissipation in a convective region but having the magnetic field couple the whole star together? It is a logical consequence of the first premise I believe.

van't Veer: The difficulty in answering this question is that the transfer of angular momentum from the orbit to the components requires knowing if all the layers of the components are also spun up. If it is only the outer layers, then you need less angular momentum than if you are doing the whole star. So this question is difficult to settle at the moment because we do not know how much of the star is being spun up. Always the same problem. I can say that when you try to solve the question from observations, then we find that the parameters in the formulation of Zahn and Papaloizou are too small. So there are instabilities that alter the magnetic field or other factors which influence the transfer of angular momentum which are not included in the theory.

Rodonò: I think we already have evidence for magnetic interaction between the components of close binaries. This comes from very detailed VLA and VLBI radio maps which indicate that coronae extend over several stellar radii - often involving the entire system. So, some kind of magnetic interaction should be expected. And, a few observed stellar flares can be explained only with a source much larger than the area provided by a single star - possibly indicating the source is from interconnecting loops between the two stars. This may help in channelling of matter and perhaps the exchange of angular momentum.

Cameron: There is one way to get an observational lower limit on magnetic spin-orbit coupling in close binaries, and that is to go away and look at the AM Her objects - where the white dwarf is synchronised with its binary companion, presumably by the action of a magnetic field, which is extremely strong in these objects. The orbital periods for these systems are of order 100 minutes.

Kraft: A less spectacular example of what is seen in the AM Her's is the whole industry of cataclysmic variables which is built on the wind driving mechanism which we have been talking about. The problem is that you cannot get the matter out of the late type component in a cataclysmic variable by the old-fashioned processes of either stellar evolution - the swelling of the outer envelope during post-MS evolution, because the masses are too small - or by gravitational waves, because the stars are not close enough together. So, the problem is how do you get the mass transfer? The current answer which people seem to prefer is that the late type component has a magnetic wind of just the kind we have talked about for the young stars. The evolving star fills its inner Lagrangian surface and tidal forces are being exerted. This takes angular momentum out of the rotation of the star, but this is compensated for by taking angular momentum out of the orbit. So, the lobe shrinks down on the star, and the matter spills out of the inner Lagrangian point and goes into a disk around the white dwarf. So, it is just the reverse of the whole angular momentum/disk-wind problem for Tauri stars.

Sofia: For the benefit of our colleagues in high energy astronomy, there is an item which has not been discussed yet, and that is the 154-day periodicity which I believe still persists for very large γ-ray producing flares. This seems to indicate some preferred longitudes where things happen for a fairly long period of time. This is born out by the fact that even though active regions do not last long, if you look at a trend of 100 years, you find a very strong peak at 27 days. If the active regions were distributed randomly, that would be washed out - and it is not. So, in addition to the nasty things we are dealing with, there appears to be a very strong preferential longitude where for a long period of time there is coherence and perhaps every six rotation periods a big flare is produced. This was discovered with data from SMM, and if the conclusions of that work are still correct for very small flares there is no periodicity, but for quite high energies there is a tremendously strong peak at 154 days, which I presume is about 6 times 25 days.

Rodonò: Is this in any way connected to the fact that hard X-ray flares can only be observed when these are at the limb?

Sofia: I do not think so - I believe it is genuine.

Schatzman: You say this has been observed for 100 years?

Sofia: No, the γ-rays only since SMM. But, if you plot the record of sunspots for 100 years, you see a tremendous peak, whereas if they arrived and disappeared randomly on the surface of the star they would just wash out.

Schatzman: This reminds me of the question of the red spot on Jupiter, which has been there for a very long time and is probably related to some very interesting property of the turbulent flow. That is, the production of large scale structures from small scale ones, and reaching a more or less stable structure.

Duncan: One thing that is shown very clearly by some of the Mt. Wilson and other observations, is that even on single stars you can have active regions which persist for more than a decade - so long lived stable regions do exist on other stars.

Rodonò: Actually, we should distinguish between wave-like light curves and stable active regions, because you may have a stable light curve, as I have been observing since my 1963 thesis on RS CVn and this is still going on, but this does not mean there is one active Jupiter-like spot. There is a spot forming region that migrates possibly, but certainly it is not always the same. We have not enough resolution to say that.

Schatzman: That is one of the questions we can put on the list for the observers to provide to the theorists.

Gray: I would like to ask a question about flares. As I understand it, magnetic fields lines get twisted up and annihilate each other and bang - you get a flare. At the bottom of the convective zone, we are talking about magnetic fields being very strong and there is a lot of convection. Why do not we get tremendous flare activity there, and might this not affect the star's convection zone in some way?

Schatzman: This is the question of buoyancy, which you raise, but in a more non-linear aspect than usually people think of, because usually people think that if the magnetic field becomes large enough, buoyancy will cause the field lines to rise, but if there is some intersection with lines of force, which is quite possible, with a local flare at the bottom of the convection zone, it will probably rise up also.

Roxburgh: The velocities are very low though.

Schatzman: That is true. So, I do not think we will see very much from this.

Sofia: The field is ordered instead of having opposite polarity, also.

Gough: But, nevertheless, if you get a winding up of the magnetic field particularly in the vicinity of say 35 degrees latitude, you will get a reversal of the directions in which you are winding it up and you will get points where there will be rapid diffusion and the breaking of magnetic field lines, and that is what a flare is - so, you will get flares at the base of the convection zone. They will not explode of course because they are being held in by the pressure of the material around them, which is absolutely dominating the flow down there whereas up in the atmosphere of the star the magnetic field is dominant, so the final outcome looks very different but the same processes are going on presumably.

Gray: How much energy can be released that way?

Gough: Very little compared to the energy density of the material around.

Schatzman: I would like us to consider now a number of observational questions.

- *Why do the young open clusters show the peculiar distribution of rotational velocities, with more than half of the stars having very small rotational velocities, while the rest have a wide range of rotational velocities up to 200 km/s?*

- *What determines M dwarfs to come in dM and dMe varieties, and what role does rotation play?*

- *How can we explain the rotational velocities of giants?*

- *Why are white dwarfs slow rotators, when simple conservation of angular momentum would suggest they should be rapid rotators?*

Does anyone wish to comment on any of these questions?

Bouvier: I think the real challenge is to explain how PMS stars manage to spin up very fast and then spin down very fast on the main sequence. That is an observational fact-T Tauri stars spin relatively slowly, but when we first see stars on the main sequence some stars rotate at up to 200 km/sec. And, soon after that they rotate quite slowly again.

Explaining this is a real challenge for theoreticians.

Duncan: If I can just second that comment, the whole reason I started observing stars in Orion was that I could not understand theoretically why stars should get to the main sequence rotating rapidly. From all that I understand about angular momentum loss mechanisms, I should think that in the time just prior to arriving on the ZAMS, most of these mechanisms for effectively losing angular momentum on the main sequence should also be working on the PMS, so it seems difficult to understand how the rapid spin up could take place.

Sofia: I can comment on that. Obviously one of the big changes that happens is the relatively large change in the moment of inertia - and that will spin it up, and if there is a saturated loss mechanism that might decrease the slow down during PMS evolution. Remember the contraction to the main sequence is quite rapid. Also, remember that things are even worse than described by Bouvier. One of the big problems is how to get rid of the vast supply of angular momentum in a diffuse cloud while forming a star, which apparently stars manage to do so that they are slow rotators by the T Tauri stage. It would be easy to make T Tauri stars rapid rotators - accretion from a disk would do that very well. But somehow this is not happening according to the observations.

Duncan: The weak line T Tauri stars do not rotate systematically differently from the T Tauri stars with disks. Perhaps the weak line T Tauri stars once had disks, and so these are not completely different. But still you would expect them to be more slowly rotating because they had less material to accrete or because they accreted for a smaller amount of time.

Gough: Is the observed spin up inferred from the open cluster data consistent with having the T Tauri stars contract to the main sequence while simply conserving angular momentum and with no substantial redistribution of angular momentum within the star?

Sofia: Yes, it is I believe.

Roxburgh: That is without coupling between the envelope and the growing radiative core?

Pinsonneault: It is assuming solid body rotation.

Roxburgh: But that is with coupling, so what you are saying is that it is consistent with the decrease in change in moment of inertia as you approach the main sequence. You cannot then appeal to that on the way down and then dismiss it subsequently.

Sofia: Oh, yes you can, because a convective star behaves very differently from a radiative star.

Roxburgh: Oh, no, this is long after. The moment of inertia only changes after the fully convective stage.

Bouvier: A one solar mass star is already almost completely radiative at 10 MYR, so there is still 20 MYR to reach the ZAMS and another 20 MYR before the age of Alpha Per cluster. This gives 40 MYR when the star is almost radiative, and in this portion of its evolution the moment of inertia does not decrease greatly compared to the decrease prior to an age of 10 MYR. So during this 40 MYR time period, it must still spin up from 10-30 km/sec to up to 150-200 km/sec, and then perhaps in another 20 MYR it must decrease back to 30 km/sec or less. So I think, qualitatively, you may be able to explain what is going on, but the quantitative problem is really trying to fit these numbers.

Schatzman: Let me note that the structure of the star is changing during the radiative track evolution because the convective zone is becoming shallower. The density at the bottom of the convective zone is changing. The result is that the efficiency of the loss of angular momentum changes during this radiation phase. I have not made the calculation, but when the star contracts sufficiently - if nothing else happens during this contraction phase, it should rotate faster. The question is how much angular momentum is lost during this phase, and this depends very much on the way in which the angular momentum loss rate is generated and this is a long pending question.

Pinsonneault: I have a comment on the age scale that is being applied to all of these young open clusters. A lot of the results we are getting, particularly for the Alpha Per cluster versus the Pleiades, are very dependent on this difference of say between 50 Myr and 70 Myr in ages. You can get very different properties, and very different interpretations by assigning different, but defensible, ages to these cluster stars. In particular, some people - myself among them - would tend to argue that Alphe Per is much younger and that the difference in properties between these two clusters can therefore be more easily explained. At any rate, it is an uncertainty that is important and must be taken into account.

MacGregor: Attempting to forestall the spin up that results from the diminishing moment of inertia during PMS contraction requires a spin down time-scale less than 10 Myr or so, and within the context of any of the thermal-centrifugal magnetic wind theories, this is very difficult to attain. The amount of angular momentum loss in any treatment is rather modest and does not occur until the latter part of the PMS evolution when the moment of inertia of the convective zone is reduced to make the braking time actually be the dominant time scale of the problem. What I think is a tremendous problem, and one to which I have no answer, is the degree of coupling between the convective zone and the radiative interior, which must affect the subsequent zero-age and main sequence rotational evolution. That is, if the convective zone and radiative interior were initially magnetically coupled, can they be uncoupled and how would that take place?

Stauffer: A comment with regard to the age of the Alpha Perseus cluster. To the extent you can trust PMS evolutionary models, and I will use Don Vandenberg's models, if you compare theoretical model isochrones to the observed color magnitude diagram for loss mass stars in Alpha Persei (using data from Charles Prosser's thesis), you get a contraction age for Alpha Persei of order 70 Myr, probably plus or minus 20 Myr or so. Very few low mass Alpha Per stars could be as young as 20 or 30 Myr if Vandeberg's tracks are approximately correct.

Pinsonneault: That depends to some extent on the mixing length you choose.

Stauffer: Yes, but pushing it to 20 or 30 Myr would still be quite difficult.

Gough: Let me give a partial answer to MacGregor's question. This is no longer for a star but for 2 solid, coaxial cylinders, both rotating and both electrically conductive and initially having a uniform magnetic field pervading them. Then what happens if you now spin the inner cylinder, then it twists the interior magnetic field (I am doing this at high magnetic Reynold's number), and produces very sharp shear around the edge and so the torque goes up tremendously. You then develop a very thin boundary layer and the field starts to cut off and as it cuts off there is less field cutting through.

MacGregor: You mean there is actual reconnection?

Gough: Yes. And then the torque goes down and eventually reaches some asymptotic level. And so what happens is that you get an initial rise with very strong coupling and then it decays off to much weaker coupling and goes to some constant value depending upon the parameters of the problem.

Strom: I have a rather innocent question. We have been talking about fairly complex objects which undergo fairly complex evolutionary histories; namely, solar type stars as they evolve from the pre-main sequence to main sequence phases. I wonder what can be learned from looking at earlier type stars, in particular, the A and B stars, where you do not have this complex change of structure and in which there are at least hints of significant evolution of angular momentum as well. In particular, there are hints, though I am not sure how well established they are observationally, that among the B and A stars there are significant decreases in rotational velocity on the main sequence - there is a small amount of evidence for this among the older B stars and better evidence among the A stars. I think the evidence for this is fairly conclusive. I wonder what these results imply for spin down mechanisms, particularly since these stars lack outer convective zones and lack the same sorts of structures that we find for solar type stars.

Schatzman: If we believe the work of Praderie and Mangeney, they extend the role of the convective zone up to B stars if I remember correctly, which implies a very deep convective zone which is not close to the surface. So, I think this remains to be investigated.

AUTHOR INDEX

OBJECT INDEX

SUBJECT INDEX

MIX
Papier aus verantwortungsvollen Quellen
Paper from responsible sources
FSC® C105338

If you have any concerns about our products,
you can contact us on
ProductSafety@springernature.com

In case Publisher is established outside the EU,
the EU authorized representative is:
Springer Nature Customer Service Center GmbH
Europaplatz 3, 69115 Heidelberg, Germany

Printed by Libri Plureos GmbH
in Hamburg, Germany